FUZZY AND
NEURAL APPROACHES
IN ENGINEERING

Adaptive and Learning Systems for Signal Processing, Communications, and Control

Editor: Simon Haykin

Werbos / THE ROOTS OF BACKPROPAGATION: From Ordered Derivatives to Neural Networks and Political Forecasting

Krstić, Kanellakopoulos, and Kokotović / NONLINEAR AND ADAPTIVE CONTROL DESIGN

Nikias and Shao / SIGNAL PROCESSING WITH ALPHA-STABLE DISTRIBUTIONS AND APPLICATIONS

Diamantaras and Kung / PRINCIPAL COMPONENT NEURAL NETWORKS: Theory and Applications

Tao and Kokotović / ADAPTIVE CONTROL OF SYSTEMS WITH ACTUATOR AND SENSOR NONLINEARITIES

Tsoukalas and Uhrig / FUZZY AND NEURAL APPROACHES IN ENGINEERING

Hrycej / NEUROCONTROL: Toward an Industrial Control Methodology

Beckerman / ADAPTIVE COOPERATIVE SYSTEMS

FUZZY AND NEURAL APPROACHES IN ENGINEERING

Lefteri H. Tsoukalas

Purdue University

Robert E. Uhrig

The University of Tennessee

A Wiley-Interscience Publication

JOHN WILEY & SONS, INC.

New York / Chichester / Weinheim / Brisbane / Singapore / Toronto

Library of Congress Cataloging in Publication Data:
Tsoukalas, Lefteri H.
    Fuzzy and neural approaches in engineering / Lefteri H. Tsoukalas,
  Robert E. Uhrig
        p.   cm.
    "A Wiley-Interscience publication."
    Includes index
    ISBN 0-471-16003-2 (cloth : alk. paper)
    1. Neural networks (Computer science)   2. Fuzzy systems
  3. Engineering—Data processing.   I. Uhrig, Robert E., 1928-
  II. Title.
  QA76.87.T76   1996
  620'.0028563—dc20   96-14102

Printed in the United States of America

10 9 8 7 6 5 4

To
Demetra
and
Paula

CONTENTS

FOREWORD   xiii

PREFACE   xvii

1 Introduction to Hybrid Artificial Intelligence Systems   1

    1.1   Introduction / 1
    1.2   Neural Networks and Fuzzy Logic Systems / 2
    1.3   The Progress in Soft Computing / 3
    1.4   Intelligent Management of Large Complex Systems / 5
    1.5   Structure of this Book / 7
    1.6   Problems and Programs Available on the Internet / 8
    References / 9

I FUZZY SYSTEMS: CONCEPTS AND FUNDAMENTALS   11

2 Foundations of Fuzzy Approaches   13

    2.1   From Crisp to Fuzzy Sets / 13
    2.2   Fuzzy Sets / 15
    2.3   Basic Terms and Operations / 17
    2.4   Properties of Fuzzy Sets / 28
    2.5   The Extension Principle / 30
    2.6   Alpha-Cuts / 34
    2.7   The Resolution Principle / 37
    2.8   Possibility Theory and Fuzzy Probabilities / 38
    References / 45
    Problems / 46

3 Fuzzy Relations   49

3.1 Introduction / 49
3.2 Fuzzy Relations / 52
3.3 Properties of Relations / 57
3.4 Basic Operations with Fuzzy Relations / 60
3.5 Composition of Fuzzy Relations / 65
References / 74
Problems / 75

4 Fuzzy Numbers   77

4.1 Introduction / 77
4.2 Representing Fuzzy Numbers / 79
4.3 Addition / 84
4.4 Subtraction / 90
4.5 Multiplication / 95
4.6 Division / 99
4.7 Minimum and Maximum / 101
References / 102
Problems / 103

5 Linguistic Descriptions and Their Analytical Forms   105

5.1 Fuzzy Linguistic Descriptions / 105
5.2 Linguistic Variables and Values / 113
5.3 Implication Relations / 120
5.4 Fuzzy Inference and Composition / 125
5.5 Fuzzy Algorithms / 136
References / 141
Problems / 142

6 Fuzzy Control   145

6.1 Introduction / 145
6.2 Fuzzy Linguistic Controllers / 151
6.3 Defuzzification Methods / 163
6.4 Issues Involved in Designing Fuzzy Controllers / 176
References / 185
Problems / 187

II NEURAL NETWORKS: CONCEPTS AND FUNDAMENTALS   189

7 Fundamentals of Neural Networks   191

7.1 Introduction / 191
7.2 Biological Basis of Neural Networks / 192

7.3   Artificial Neurons / 193
7.4   Artificial Neural Networks / 196
7.5   Learning and Recall / 203
7.6   Features of Artificial Neural Networks / 211
7.7   Historical Development of Neural Networks / 213
7.8   Separation of Nonlinearly Separable Variables / 221
References / 227
Problems / 227

8 Backpropagation and Related Training Algorithms   **229**

8.1   Backpropagation Training / 229
8.2   Widrow–Hoff Delta Learning Rule / 234
8.3   Backpropagation Training for a Multilayer
        Neural Network / 238
8.4   Factors That Influence Backpropagation
        Training / 248
8.5   Sensitivity Analysis in a Backpropagation Neural
        Network / 255
8.6   Autoassociative Neural Networks / 257
8.7   An Alternate Approach to Neural Network
        Training / 266
8.8   Modular Neural Networks / 270
8.9   Recirculation Neural Networks / 274
8.10 Functional Links / 279
8.11 Cascade-Correlation Neural Networks / 280
8.12 Recurrent Neural Networks / 281
References / 285
Problems / 287

**9 Competitive, Associative, and Other Special Neural
    Networks**   **289**

9.1   Hebbian Learning / 289
9.2   Cohen–Grossberg Learning / 290
9.3   Associative Memories / 296
9.4   Competitive Learning: Kohonen Self-Organizing
        Systems / 306
9.5   Counterpropagation Networks / 315
9.6   Probabilistic Neural Networks / 319
9.7   Radial Basis Function Network / 325
9.8   Generalized Regression Neural Network / 326
9.9   Adaptive Resonance Theory (ART-1) Neural
        Networks / 328
References / 331
Problems / 332

10   Dynamic Systems and Neural Control   **333**

10.1   Introduction / 333
10.2   Linear Systems Theory / 333
10.3   Adaptive Signal Processing / 341
10.4   Adaptive Processors and Neural Networks / 345
10.5   Neural Network Control / 353
10.6   System Identification / 363
10.7   Implementation of Neural Control Systems / 368
10.8   Applications of Neural Networks in Noise Analysis / 374
10.9   Time-Series Prediction / 380
References / 382
Problems / 383

11   Practical Aspects of Using Neural Networks   **385**

11.1   Selection of Neural Networks for Solution
       to a Problem / 385
11.2   Design of the Neural Network / 386
11.3   Data Sources and Processing for Neural Networks / 395
11.4   Data Representation / 391
11.5   Scaling, Normalization, and the Absolute Magnitude
       of Data / 395
11.6   Data Selection for Training and Testing / 399
11.7   Training Neural Networks / 401
References / 405

III   INTEGRATED NEURAL–FUZZY TECHNOLOGY   407

12   Fuzzy Methods in Neural Networks   **409**

12.1   Introduction / 409
12.2   From Crisp to Fuzzy Neurons / 410
12.3   Generalized Fuzzy Neuron and Networks / 414
12.4   Aggregation and Activation Functions in Fuzzy
       Neurons / 416
12.5   *AND* and *OR* Fuzzy Neurons / 418
12.6   Multilayer Fuzzy Neural Networks / 421
12.7   Learning and Adaptation in Fuzzy Neural Networks / 423
12.8   Fuzzy ARTMAP / 431
12.9   Fuzzy-Neural Hybrid Data Representation / 434
12.10 Survey of Engineering Applications / 437
References / 440
Problems / 442

13 Neural Methods in Fuzzy Systems **445**

13.1 Introducing the Synergism / 445
13.2 Fuzzy-Neural Hybrids / 447
13.3 Neural Networks for Determining Membership Functions / 450
13.4 Neural-Network-Driven Fuzzy Reasoning / 455
13.5 Learning and Adaptation in Fuzzy Systems via Neural Methods / 461
13.6 Adaptive Network-Based Fuzzy Inference Systems / 466
References / 468
Problems / 470

14 Selected Hybrid Neurofuzzy Applications **471**

14.1 Introduction / 471
14.2 Neurofuzzy Interpolation / 472
14.3 General Neurofuzzy Methodological Developments / 474
14.4 Engineering Applications / 476
14.5 Diagnostics in Complex Systems / 477
14.6 Neurofuzzy Control Systems / 478
14.7 Neurofuzzy Control in Robotics / 481
14.8 Pattern Recognition and Image Enhancement / 482
14.9 Medical and Environmental Imaging Using Neurofuzzy Methodologies / 483
14.10 Transportation Control / 484
14.11 Adaptive Fuzzy Systems / 485
14.12 Inspection Using Neurofuzzy Methods / 486
14.13 Neurofuzzy Methods in Financial Engineering / 486
14.14 Commercial Neurofuzzy System Software / 487
References / 488

15 Dynamic Hybrid Neurofuzzy Systems **493**

15.1 Introduction / 493
15.2 Fuzzy-Neural Diagnosis for Vibration Monitoring / 495
15.3 Decision Fusion by Fuzzy Set Operations / 500
15.4 Hybrid Neurofuzzy Methodology for Virtual Measurements / 504
15.5 Neurofuzzy Approaches to Anticipatory Control / 510
References / 516

IV OTHER ARTIFICIAL INTELLIGENCE SYSTEMS **521**

16 Expert Systems in Neurofuzzy Systems **523**

16.1 Introduction / 523
16.2 Characteristics of Expert Systems / 524

16.3   Components of an Expert System / 525
16.4   Knowledge Representation and Inference / 527
16.5   Uncertainty Management / 529
16.6   State of the Art of Expert Systems / 531
16.7   Use of Expert Systems / 532
16.8   Expert Systems Used with Neural Networks and Fuzzy Systems / 534
16.9   Potential Implementation Issues for Expert Systems / 535
References / 537
Problems / 538

17 Genetic Algorithms   **539**

17.1   Introduction / 539
17.2   Basic Concepts of Genetic Algorithms / 540
17.3   Binary and Real-Value Representations of Chromosomes / 542
17.4   Implementation of Genetic Algorithm Optimization / 544
17.5   Fitness Functions / 546
17.6   Application of Genetic Algorithms to Neural Networks / 552
17.7   Fuzzy Genetic Modeling / 554
17.8   Use of Genetic Algorithms in the Design of Neural Networks / 556
References / 557
Problems / 559

18 Epilogue   **561**

18.1   Introduction / 561
18.2   Is Artificial Intelligence Really Intelligent? / 562
18.3   The Role of Neurofuzzy Technology / 563
18.4   Last Thoughts / 564
References / 565

APPENDIX: *T* NORMS AND *S* NORMS   567

INDEX   575

FOREWORD

To say that *Fuzzy and Neural Approaches in Engineering* is an important work is an understatement. With skill, authority, and insight, Professors Tsoukalas and Uhrig share with us their expertise in a new field that holds much promise and offers a fertile ground for the development of unorthodox techniques and novel applications.

Basically, the book reflects the proliferation and wide-ranging impact of systems that achieve a high level of performance through the employment of the methodologies of fuzzy logic and neurocomputing, singly or in combination. Systems in which fuzzy logic and neurocomputing are used in combination have come to be known as neurofuzzy systems. Takagi and Hayashi in Japan were among the first to describe such systems in 1988. Today, neurofuzzy systems are growing rapidly in number, visibility, and importance.

Viewed in a broader perspective, neurofuzzy systems constitute a subclass of systems based on "soft computing." The essence of soft computing (SC) is that unlike the traditional, hard computing, it is aimed at an accommodation with the pervasive imprecision of the real world. Thus, the guiding principle of soft computing is: Exploit the tolerance for imprecision, uncertainty, and partial truth to achieve tractability, robustness, low-solution cost, and better rapport with reality. In the final analysis, the role model for soft computing is the human mind.

Soft computing is not a single methodology. Rather, it is a consortium. The principal members of the consortium at this juncture are fuzzy logic (FL), neurocomputing (NC), genetic computing (GC), and probabilistic reasoning (PR), with the latter subsuming evidential reasoning, belief networks, chaotic systems, management of uncertainty, and parts of machine-learning theory. Within SC, the main contribution of FL is a methodology for dealing with imprecision, approximate reasoning, rule-based systems, and computing with words; that of NC is system identification, learning, and adaptation; that of GC is systematized random research and optimization; and that of PR is decision analysis and management of uncertainty.

In the main, FL, NC, GC, and PR are synergistic and complementary rather than competitive. For this reason, it is frequently advantageous to use FL, NC, GC, and PR in combination rather than exclusively, leading to so-called "hybrid systems." Today, the most visible systems of this type are neurofuzzy systems. We are also beginning to see fuzzy-genetic, neuro-genetic, and neurofuzzy-genetic systems. Such systems are likely to become ubiquitous in the not-so-distant future. Concomitantly, the realization that FL, NC, GC, and PR are complementary rather than competitive may put an end to inconclusive debates regarding the superiority of a particular member of the SC consortium over others.

Although *Fuzzy and Neural Approaches in Engineering* is concerned mainly with neurofuzzy systems, the authors address in the last chapters some of the basic aspects of genetic computing and present a succinct and up-to-date account of neurogenetic and fuzzygenetic systems. In this way, their treatise gains in generality and highlights the central role of soft computing in the conception, design, and deployment of intelligent systems.

The organization of the book reflects the basic structure of soft computing. The first six chapters are given over to the exposition of fuzzy logic and its applications. The next six chapters do the same for neurocomputing. The following five chapters present a highly informative and insightful exposition of ways in which fuzzy logic and neurocomputing can be used in combination. The value of these chapters is enhanced by the inclusion of many examples of real-world applications.

What is important to recognize—and what the authors stress—is that the synergism of fuzzy logic and neurocomputing is a two-way street. They do this by devoting a chapter to the discussion of fuzzy methods in neural networks, followed by a chapter on neural methods in fuzzy systems.

An observation which I would like to add is that in many, perhaps most, of the applications of fuzzy logic, the point of departure is a human solution. Thus, fuzzy logic—and, more specifically, the calculus of fuzzy *if/then* rules —is used as a quasi-programming language to express the human solution as a fuzzy rule-set or, more generally, as a fuzzy algorithm. In this sense, fuzzy logic solutions are for the most part descriptive rather than prescriptive.

A case in point is the problem of parking a car. In the fuzzy logic solution of this problem, the starting point is the human knowledge of how to park a car. The next step is to express this knowledge in the language of fuzzy *if/then* rules.

The descriptive approach may fail even though a human solution may exist. For example, one may be able to recognize a person by the way in which that person walks and yet be unable to articulate the fuzzy *if/then* rules that underlie the recognition. The problem of articulating—in the language of fuzzy *if/then* rules—what is subconscious or intuitive is a challenge that has not as yet been fully met.

Although there are many situations in which the problem of articulation remains to be solved, there are many more situations in which articulation is possible, either directly or through the use of rule induction techniques. These issues lie at the center of applications of fuzzy logic, including those applications in which fuzzy methods are used in neural networks.

Fuzzy and Neural Approaches in Engineering makes a major contribution to a better understanding of how fuzzy logic and neurocomputing can be applied, both singly and in combination, to the conception and design of a wide variety of systems. Professors Tsoukalas and Uhrig deserve our thanks and congratulations for producing a text that is informative, insightful, well-written, and forward-looking in both spirit and content.

LOTFI A. ZADEH

PREFACE

Soft computing is the name that is being put forth as an alternative to artificial intelligence for the plethora of advanced information processing technologies that have emerged in the past decade. This new field is characterized by a certain tolerance for imprecision and ambiguity and it includes expert systems, neural networks, genetic algorithms, fuzzy logic, cellular automata, chaotic systems, wavelets, complexity theory, anticipatory systems, and others. Many of these technologies (e.g., neural networks) date back several decades, whereas some (e.g., cellular automata) are still in the early development stages. Neural networks and fuzzy systems individually have reached a degree of maturity where they are each being applied to real-world situations. Researchers often utilize these two technologies in series, using one as the preprocessor or postprocessor for the other. Examples include the use of fuzzy inputs and outputs for neural networks, the use of neural networks to quantify the shape of a fuzzy membership function, and the use of individual neural networks for many sensors mounted on a machine to give individual diagnoses which are then fused using a fuzzy methodology. Although the results clearly suggest that such use of these technologies is synergistic and beneficial, there are indications that even greater benefits may be possible by the integration into a neurofuzzy technology with such concepts as the "fuzzy neuron" and the use of fuzzy logic functions to aggregate weighted inputs of a neuron. It is our perception that neurofuzzy technology (e.g., a technology that combines the feature extraction and modeling capabilities of the neural network with the representation capabilities of fuzzy systems) is at the stage that neural networks and fuzzy logic were at a decade ago.

There are other hybrid combinations of the different elements of soft computing that are also synergistic. Perhaps next in importance is the combination of genetic algorithms with neural networks and/or fuzzy systems. The ability to carry out near-global optimization on any problem for which an objective function can be defined is an incredibly powerful tool that can enhance the capabilities of any technology. The examples cited in this text only hint at the value of integrating genetic algorithms with other soft computing technologies. It is reasonable to expect that optimization of every step in a complex operation could significantly reduce computing time and improve results.

Expert systems offer a framework in which integration of the various soft computing technologies can be carried out. The ability to bring classical logic to bear on the integration process and to seek data and information from whatever sources are available offers the type of environment that could lead to a more flexible, user- and system-adaptive automation. Even more important is the use of fuzzy rules in expert systems so that interactions in complex systems can be represented.

In preparing the manuscript for this book, we were faced with the classical trade-off of breadth versus depth of coverage. We chose to cover the fundamentals of fuzzy systems and neural networks, and to a lesser extent genetic algorithms, in detail while using descriptive material to give perspective to the role of the various technologies involved. The material was originally intended for first-year graduate students, but additional information was included so that it would also be useful for a senior-level course and to practicing engineers. It is our hope that all of these groups will find this text to be useful and that the readers will be motivated to utilize this material in their work.

Many people contributed to the preparation of this manuscript, including many graduate students who used this material in draft form in class. Although it is not possible to acknowledge all of those who contributed, special recognition is due to individuals who read the manuscript in detail and offered constructive comments. Included in this special group, in alphabetical order, are Israel Alguindigue, R. C. Berkan, Mario Fontana, Wesley Hines, Vaclav Hojny, Andreas Ikonomopoulos, and Trent Powers. Graduate students whose research was described in this text are acknowledged by footnotes in the text. Earlier drafts of the manuscript have been used in short courses and seminars in the United States, Europe, and Japan. We are particularly grateful to Professors M. Kitamura and R. Kozma of Tohoku University in Japan, Dr. T. Washio of the Mitsubishi Research Institute, Professors S. Panas and J. Theoharis of Aristotle University in Greece, Professor Elias N. Houstis of Purdue University and Drs. Y. Shinohara, J. Shimazaki, K. Suzuki, K. Hayashi, S. Shinobu, H. Usui, Y. Fujii, K. Watanabe, Y. T. Suzudo, N. Ishikawa, and K. Nabeshima and Ms. S. Tobita (researchers and staff of the Control & AI Laboratory of JAERI in Japan). Their constructive criticism, suggestions, and intellectual support provided much of the inspiration and energy for completing this work. Special thanks are also due to the faculty and staff of the School of Nuclear Engineering at Purdue University and the Department of Nuclear Engineering at the University of Tennessee for their support. Two special individuals merit our gratitude for introducing us to the fields of fuzzy logic and neural networks, Professor M. Ragheb of the University of Illinois and Maureen Caudill of

NeuWorld Services. The word editing and word processing of the manuscript were undertaken by Murray Browne and Lynnetta Holbrook, respectively.

Special thanks are due to Professor S. Haykin, the editor of this series.

We would like to express our appreciation to the John Wiley staff, in particular to George Telecki, who encouraged us to proceed with the book, and to Angioline Loredo, who supervised its production.

Finally, we express our gratitude to our wives, Demetra K. Evangelou and Paula M. Uhrig, whose love, understanding, and patience made it possible for us to write this book.

LEFTERI H. TSOUKALAS
ROBERT E. UHRIG

Purdue University
The University of Tennessee

1

INTRODUCTION TO HYBRID ARTIFICIAL INTELLIGENCE SYSTEMS

As complexity rises, precise statements lose meaning and meaningful statements lose precision.

Lotfi A. Zadeh

1.1 INTRODUCTION

The term "artificial intelligence" (AI), in its broadest sense, encompasses a number of technologies that includes, but is not limited to, expert systems, neural networks, genetic algorithms, fuzzy logic systems, cellular automata, chaotic systems, and anticipatory systems. Interestingly, most of these technologies have their origins in biological or behavioral phenomena related to humans or animals, and many of these technologies are simple analogs of human and animal systems. Hybrid intelligent systems generally involve two, three, or more of these individual AI technologies that are either used in series or integrated in a way to produce advantageous results through synergistic interactions. In this book we have placed emphasis on neural networks and fuzzy systems; to a lesser extent, we have also placed emphasis on genetic algorithms where needed for optimization and expert systems where they are needed to supervise and implement the other three technologies. A major emphasis in this book will be on the integration of fuzzy and neural systems in a synergistic way.

In data and/or information processing, the objective is generally to gain an understanding of the phenomena involved and to evaluate relevant parameters quantitatively. This is usually accomplished through "modeling" of the systems, either experimentally or analytically (using mathematics and physical principles). Most hybrid systems relate experimental data to systems or models. Once we have a model of a system, we can carry out various

1

procedures (e.g., sensitivity analysis, statistical regression, etc.) to gain a better understanding of the system. Such experimentally derived models give insight into the nature of the system behavior that can be used to enhance mathematical and physical models.

There are, however, many situations in which the phenomena involved are very complex and often not well understood and for which first principles models are not possible. Even more often, physical measurements of the pertinent quantities are very difficult and expensive. These difficulties lead us to explore the use of neural networks and fuzzy logic systems as a way of obtaining models based on experimental measurements.

1.2 NEURAL NETWORKS AND FUZZY LOGIC SYSTEMS

In the history of science and technology, new developments often come from observations made from a different perspective. Interrelationships that we take for granted today may not have been so obvious in earlier decades. For instance, we regularly gain insight into the behavior of a dynamic system by viewing it as being in the "time domain" and/or the "frequency domain." However, for the first four decades of the twentieth century, statisticians dealt with autocorrelation and cross-correlation functions (in the time domain) while electrical engineers dealt with power- and cross-spectral densities (in the frequency domain) without either group realizing that these two concepts were related to each other through Fourier transformations.

Both the statisticians and the electrical engineers have found that analysis of the fluctuations in process variables provides useful information about the variables as well as the processes involved. These fluctuations, which result in uncertainties in measured variables, often are caused by some sort of random driving function (i.e., fluid turbulence, rotational unbalance, etc.). Investigation, and the subsequent understanding of these uncertainties (fluctuations), led to the development of the field of "random noise analysis" which spawned such analytical specialties as vibration analysis, seismology, electrocardiography, oceanography, and so on.

Neural networks and fuzzy systems represent two distinct methodologies that deal with uncertainty. Uncertainties that are important include both those in the model or description of the systems involved as well as those in the variables. These uncertainties usually arise from system complexity (often including nonlinearities; we think of complexity as a property of system description—that is, related to the means of computation or language and not merely a system's complicated nature). Neural networks approach the modeling representation by using precise inputs and outputs which are used to "train" a generic model which has sufficient degrees of freedom to formulate a good approximation of the complex relationship between the inputs and the outputs. In fuzzy systems, the reverse situation prevails. The input and output variables are encoded in "fuzzy" representations, while their interrelationships take the form of well-defined *if/then* rules. Zadeh's ingenious observation that the uncritical pursuit of precision may be not only unnecessary but actually a source of error led him to the notion of a fuzzy set.

Each of these approaches has its own advantages and disadvantages. Neural networks can represent (i.e., model) complex nonlinear relationships, and they are very good at classification of phenomena into preselected categories used in the training process. On the other hand, the precision of the outputs is sometimes limited because the variables are effectively treated as analog variables (even when implemented on a digital computer), and "minimization of least squares errors" does not mean "zero error." Furthermore, the time required for proper training a neural network using one of the variations of "backpropagation" training can be substantial (sometimes hours or days). Perhaps the "Achilles heel" of neural networks is the need for substantial data that are representative and cover the entire range over which the different variables are expected to change.

Fuzzy logic systems address the imprecision of the input and output variables directly by defining then with fuzzy numbers (and fuzzy sets) that can be expressed in linguistic terms (e.g., *cold*, *warm*, and *hot*). Furthermore, they allow far greater flexibility in formulating system descriptions at the appropriate level of detail. Fuzziness has a lot to do with the parsimony and hence the accuracy and efficiency of a description. This means that complex process behavior can be described in general terms without precisely defining the complex (usually nonlinear) phenomena involved. Paraphrasing *Occam's Razor*, the philosophical principle holding that more parsimonious descriptions are more representative of nature, we may say that fuzzy descriptions are more parsimonious and hence easier to formulate and modify, more tractable, and perhaps more tolerant of change and even failure.

Neural network and fuzzy logic technologies are quite different, and each has unique capabilities that are useful in information processing. Yet, they often can be used to accomplish the same results in different ways. For instance, they can speed the unraveling and specifying the mathematical relationships among the numerous variables in a complex dynamic process. Both can be used to control nonlinear systems to a degree not possible with conventional linear control systems. They perform mappings with some degree of imprecision. However, their unique capabilities can also be combined in a synergistic way. It is this combination of the two technologies (as well as combinations with other AI technologies) with the goal of gaining the advantages of both that is the focus of this book.

1.3 THE PROGRESS IN SOFT COMPUTING

Soft computing refers to computational tools whose distinguishing characteristic is that they provide approximate solutions to approximately formulated problems (Aminzadeh, 1994). Fuzzy logic, neural networks, probabilistic reasoning, expert systems, and genetic algorithms are some of the constituents of soft computing, all having roots in the field of Artificial Intelligence. Whereas the traditional view of computing considers any imprecision and uncertainty undesirable, in soft computing some tolerance for imprecision and uncertainty is exploited in order to develop more tractable and robust models of systems, at a lower cost and greater economy of communication and computation.

Few of those who attended the historic 1956 Dartmouth Conference to discuss "the potential use of computers and simulation in every aspect of learning and any other feature of intelligence" could have envisioned the evolution and growth of the embryonic artificial intelligence field and the impact it has had on our lives. It was there that the term "artificial intelligence" was coined, perhaps because of the emphasis on learning and simulation. The term "cybernetics" was in vogue at that time with its emphasis on potential control of both man and machines. Vacuum-tube-type analog computers had reached a state of maturity that they (along with high fidelity stereo sound systems) were being marketed as "Heathkits," while the digital "supercomputer" of the time was an IBM-650 with about 2000 words of magnetic drum memory storage that operated at about 2 kHz.

It was in this environment that Frank Rosenblat developed the Perceptron by adding a learning capability to the McCulloch-Pitts model of the neuron, Marvin Minsky built the first "learning machine" (using 40 processing elements, each with six vacuum tubes and a motor/clutch/control system), and Bernard Widrow developed the "Adaline" (adaptive linear element) that even today is used in virtually every high-speed modem and telephone switching system to cancel out the echo of reflected signals. Boolean algebra was standard procedure, and John McCarthy and John von Neuman were putting forth the relative merits of symbolic (LISP) and conventional computer languages. Although there was little in the way of theoretical bases providing an understanding of these systems, work proceeded on an experimental basis that was guided primarily by the genius of the individuals involved.

Today, some 40 years later, the whole world has changed. The computing capacity of that IBM-650 is now encapsulated in a "wristwatch" computer, the Perceptron and Adaline processing elements are instantiated in neural network computing and processing methodologies, learning algorithms are routinely processed on digital computers of all sizes, *Boolean logic* and algebra are being replaced by fuzzy logic concepts, LISP is fading away in favor of object-oriented computer languages for artificial intelligence (e.g., C++), the analog computer has virtually disappeared, and the modern personal computer most of us have on our desks may have more than a gigabyte of memory, operate at a processing rate of 200 MHz or more, and be part of a vast global network of computers capable of sharing on-line information in numerical, textual, visual and audible forms.

The educational, technological, economic, and social impact and significance of the computer as a tool for computation and communication have been continuously discussed and debated in the last few decades. In the 1970's Ralph Lapp, in an interesting book called *The Logarithmic Century*, captured the ever-changing and accelerating trend in the development of technology and economics (Lapp, 1973). Yet, he did not foresee the magnitude of the impact of advanced computer technology, especially the role that communications and information processing would have on society. Perhaps our Japanese colleagues have a better grasp of the issues involved. In a book entitled *The Next Century*, Halberstam (1991) reported a conversation with a retired high official of MITI (Ministry for International Trade and Industry) who in 1987 said "...the (Japanese) educational system is in danger of...producing young people who have the intellectual capacity of computers but who will be inferior to computers in what they can actually do. The computers have caught up."

Of course, the road of technological change is by no means simple. Eloquent critics such as Neil Postman in his evocative book *Technopoly* strongly point out the dangers of subordinating culture and society to an uncritical faith in the machine (Postman, 1993). Indeed, computers cannot magically solve our problems. In today's highly integrated world, however, a diverse world population needs the multiplicity of opportunities provided by the new communications and computer technologies, and soft computing is promising to become a powerful means for obtaining quick, yet accurate and acceptable, solutions to many problems. We, the engineers who work to provide and apply these new soft computing tools, ardently hope that they will be used for the benefit of mankind.

1.4 INTELLIGENT MANAGEMENT OF LARGE COMPLEX SYSTEMS

The real challenge to *soft computing* is the intelligent management of large complex systems—that is, organizations operating on the scale of the global economy and resting on an highly globalized information infrastructure. It is perhaps the most important activity facing industrial, educational, military, and governmental organizations throughout the world today. Management decisions made today will reverberate throughout these organizations for years to come. Management decisions made in the past have shaped these organizations and have made them what they are today. In some cases, large organizations have made the "right decisions" and have been spectacularly successful. However, it is clear that the decisions of other large organizations have not been wise. Multi-hundred million and billion dollar losses, followed by layoffs, restructuring, mergers, and, all too often, bankruptcy are common as these organizations pay the price for past mistakes. Why did these organizations get into trouble or fail? What steps can be taken to ensure that decisions today are better than those in the past? The answers to these questions are as varied as the nature of the organizations. Typical responses given are as follows: incompetent management, too much attention to the next quarterly earnings, lack of vision, fierce new competition, unfair regulatory practices by governments, poor design, failure to keep up with the times, antagonism between labor and management, inadequate research and development, and so on. The list goes on and on. All of these may be valid explanations in individual situations, but correcting these alleged problems will not guarantee that an organization will be successful in meeting its goals in the future. The successful strategies and methodologies of the 1980s may not work in the next century.

Large complex systems, as a general class, are often virtually out of control; indeed they are often deemed to be uncontrollable because of their complexity. The reversal of this situation is absolutely essential in a society in which systems tend to grow without bound because of the perceived benefits of "economy of scale." Indeed, organizations tend to grow until they reach a level of inefficiency that inhibits and impedes their growth. Only an organization with virtually unlimited resources or power (i.e., governmental organizations) can continue to grow under these conditions. The finite resources of the world and of individual nations, as well as the growing population that aspires for improved living conditions, demand improved efficiency. It is absolutely essential for the benefit of mankind, as well as most modern nations that tend to be dominated by large complex systems, that these systems be brought under intelligent control and management. The advances in digital computer technology (both hardware and software) during the past decade, along with the associated development of *soft computing*, appear, for the first time in history, to provide a means of implementing intelligent control of complex systems which are so necessary in delivering the fruits of industrial technology and commerce to global society.

The personal computers or workstations available on the desk of engineers and managers today with its *soft computing* tools has the power of main-frame computers just a few years ago. They provide the capability of keeping track of what is going on in any organization (intelligence), they can provide the tools to examine the data in excruciating detail (analysis), they can provide models of the behavior of complex systems (synthesis) which then permits predictions into the future, at least into the short-term future, and they can provide recommendations for specific actions (intelligent management) that can be communicated to those who have a need to act in a form that they can understand (intelligent communications). To the extent that an organization's management is willing to utilize these tools correctly, significant progress in solving some of these problems by making the "right" decisions will follow.

Unfortunately, making the "right" decision under the circumstance at the time the decision is made does not guarantee success. It may have been the "right" decision at the time, but the consequences may be unpredictable because of the time lag between decision and results in a changing environment. What is needed is a form of anticipatory control as discussed in Chapter 15. In the absence of an ability to predict the future behavior of systems, many conservative organizations have elected a "minimum step" approach—that is, make a decision at the last possible moment that involves the least amount of (financial or resource) commitment and produces results at the earliest possible time. However, this can be a strategy for disaster if the basis on which the decision is made is not valid. All too often, decisions must be made in the absence of complete data, which gives rise to uncertainty in the analysis and a higher probability of an erroneous decision. Even such a "minimum step" approach requires *reliable intelligence, accurate analysis, valid synthesis, intelligent management,* and *intelligent communications,* because there is little margin for error. While a modern digital computer cannot guarantee the availability of these five attributes, they simply would not be available without the modern digital computers and *soft computing.*

Perhaps the single attribute that gives neurofuzzy systems an advantage in addressing the problems of large complex systems is the ability to perform what in mathematical terms would be called *many-to-many mappings.* Such mappings are an inherent part of complex systems, because every single input to a system can influence every single output; i.e., one significant input change may generate significant changes in many outputs. Most approaches to systems analysis can only deal with *one-to-one* or *many-to-one mappings*— that is, with the special class of mathematical mappings that we call *functions,* which have been the premier mathematical relation since the Newtonian revolution of the *Principia.* It is now possible and desirable, however, to effectively compute with more complex mathematical mappings than functions—that is, with *many-to-many relations* (see Section 5.1). This gives us the hope and the expectation that large complex systems can be dealt with in a flexible, reliable, and near-optimal manner.

We do not claim that *neurofuzzy* systems *per se* can bring about the control of large complex systems. It is clear to us that the integration of many technologies in a yet indiscernible manner is an essential step in the right direction. *Neurofuzzy systems* represent an integration of fuzzy logic and neural networks that have capabilities beyond either of these technologies individually (Haykin, 1994; Kartalopoulos, 1996). When we further integrate other technologies, perhaps some not yet discovered, in the decades ahead, we can look forward to tools with sufficient power to tackle problems such as intelligent control of large complex systems.

1.5 STRUCTURE OF THIS BOOK

This book is divided into four parts: Part I, entitled "Fuzzy Systems: Concepts and Fundamentals," explores the fundamentals of fuzzy logic systems and includes the following chapters:

Chapter 2. Foundations of Fuzzy Approaches
Chapter 3. Fuzzy Relations

Chapter 4. Fuzzy Numbers
Chapter 5. Linguistic Descriptions and Their Analytical Forms
Chapter 6. Fuzzy Control

Part II, entitled "Neural Networks: Concepts and Fundamentals," explores the fundamentals of neural networks and includes the following chapters:

Chapter 7. Fundamentals of Neural Networks
Chapter 8. Backpropagation and Related Training Algorithms
Chapter 9. Competitive, Associative, and Other Special Neural Networks
Chapter 10. Dynamic Systems and Neural Control
Chapter 11. Practical Aspects of Using Neural Networks

Part III, entitled "Integrated Neural–Fuzzy Technology," explores the joint use of neural networks and fuzzy logic systems. It includes the following chapters:

Chapter 12. Fuzzy Methods in Neural Networks
Chapter 13. Neural Methods in Fuzzy Systems
Chapter 14. Selected Hybrid Neurofuzzy Applications
Chapter 15. Dynamic Hybrid Neurofuzzy Systems

Part IV, entitled "Other Artificial Intelligence Systems," reviews other artificial intelligence systems that can be used with neural networks and fuzzy systems. It includes the following chapters:

Chapter 16. Expert Systems in Neurofuzzy Systems
Chapter 17. Genetic Algorithms
Chapter 18. Epilogue

1.6 MATLAB[©]1 SUPPLEMENT

In this text, we have included problems for students at the end of most chapters. Generally, these problems are pedagogical in nature and are intended to be simple enough that they can be solved without the aid of computer software. To supplement these exercises, we have enlisted our colleague, Dr. J. Wesley Hines of the University of Tennessee, to prepare a *MATLAB© Supplement for Neural and Fuzzy Approaches in Engineering*, a paperback book of approximately 150 pages published by John Wiley and Sons, in which the Student Edition of MATLAB© (The MathWorks Inc.,

---

[1]MATLAB is copyrighted by MathWorks Inc., of Natick, MA.

1995; Hanselman, 1996) can be used for demonstrations and solving more sophisticated problems. Of course, the Professional Version of MATLAB© can also be used if it is available.

This supplement was written using the MATLAB© Notebook and Microsoft WORD.Version 6.0. The Notebook allows MATLAB© commands to be entered and evaluated while in the WORD environment, which allows the document to both briefly explain the theoretical details and also show MATLAB© implementations. It also allows the user to experiment with changing the MATLAB© code fragments in order to gain a better understanding of their application.

This supplement contains numerous examples that demonstrate the practical implementation of relevant techniques using MATLAB©. Although MATLAB© toolboxes for Fuzzy Logic and Neural Networks are available, they are not required to run the examples given. This supplement should be considered to be a brief introduction to the MATLAB© implementation of neural and fuzzy systems, and we and the author strongly recommend the use of Neural Networks and Fuzzy Logic Toolboxes for a more in-depth study of these information-processing technologies. Many of the m-files and examples are extremely general and portable while other examples will have to be altered significantly for use to solve specific problems.

The content of the *MATLAB© Supplement* is coordinated with *Fuzzy and Neural Approaches in Engineering* so that students can use it to enhance their knowledge of fuzzy systems, neural networks, and neurofuzzy systems. Indeed, it is expected that many instructors will choose to use both this book and the *MATLAB© Supplement* together in their classes. Practicing engineers and scientists in industry who want to use this text to learn about neural, fuzzy, and neurofuzzy systems will find this supplement to be a valuable aid in their self-study.

REFERENCES

Aminzadeh, F., and Jamshidi, M., *Soft Computing*, Prentice-Hall, Englewood Cliffs, NJ, 1994.

Halberstam, D., *The Next Century*, William Morrow and Company, New York, 1991.

Hanselman, D., and Littlefield, B., *Mastering MATLAB*, Prentice-Hall, Englewood Cliffs, NJ, 1996.

Haykin S., *Neural Networks: A Comprehensive Foundation*, IEEE Computer Society Press, Macmillan, New York, 1994.

Kartalopoulos, S., *Understanding Neural Networks and Fuzzy Logic*, IEEE Press, New York, 1996.

Lapp, R., *The Logarithmic Century*, Prentice-Hall, Englewood Cliffs, NJ, 1973.

Mathworks, Inc., *The Student Edition of MATLAB© Users Guide*, Natick, MA, 1995.

Postman, N., *Technopoly*, Vintage Books, New York, 1993.

I

FUZZY SYSTEMS: CONCEPTS AND FUNDAMENTALS

2

FOUNDATIONS OF FUZZY APPROACHES

2.1 FROM CRISP TO FUZZY SETS

The mathematical foundations of fuzzy logic rest in *fuzzy set theory*, which can be thought of as a generalization of *classical set theory*. A familiarity with the novel notions, notations, and operations of fuzzy sets is useful in studying fuzzy logic principles and applications; acquiring it will be our main goal in this chapter.

Fuzziness is a property of language. Its main source is the imprecision involved in defining and using symbols. Consider, for example, the set of chairs in a room. In set theory the set of chairs may be established by pointing to every object in a room asking the question, *Is it a chair?* In classical set theory we are allowed to use only two answers: Yes or No. Let us code Yes as 1 and No as 0. Thus, our answers will be in the pair {0, 1}. If the answer is 1, an element belongs to a set; if the answer is 0, it does not. In the end we collect all the objects whose label is 1 and obtain the *set of chairs in a room*. Suppose, however, that we ask the question, *Which objects in a room may function as a chair?* Again we could point to every object and ask, *Could it function as a chair?* The answer here too could artificially be restricted to {0, 1}. Yet, the set of objects in a room that may *function* as a chair may include not only chairs but also desks, boxes, parts of the floor, and so on. It is a set not uniquely defined. It all depends on what we mean by the word *function*. Words like *function* have many shades of meaning and can be used in many different ways. Their meaning and use may vary with different persons, circumstances, and purposes; it depends on the specifics of a situation. We say therefore that the *set of objects that may function as a chair* is a *fuzzy set*, in the sense that we may not have crisply defined criteria for deciding membership into the set. Objects such as desks, boxes, and part of the floor may function as *chairs*, to a degree. It should be noted, however, that there is nothing fuzzy about the material objects themselves: Chairs, boxes, and desks are what they are. Fuzziness is a feature of their representation in a milieu of symbols and is generally a property of models, computational procedures, and language.

Let us now review some notions of classical set theory. *Classical sets* are crisply defined collections of distinct elements (numbers, symbols, objects, etc.), and for this reason we also call them *crisp sets*. The elements of all the sets under consideration in a given situation belong to an invariable, constant set, called the *universal set* or *universe* or more often the *universe of discourse*.[1] The fact that elements of a set A either belong or do not belong to a crisp set A can be formally indicated by the *characteristic function of A*, defined as

$$\chi_A(x) \equiv \begin{cases} 1 & \text{iff} \quad x \in A \\ 0 & \text{iff} \quad x \notin A \end{cases} \tag{2.1-1}$$

where the symbols $\in$ and $\notin$ denote that x *is* and *is not* a member of A, respectively, and iff is shorthand for "if and only if." The pair of numbers $\{0, 1\}$ is called the *valuation set*. Another way of writing equation (2.1-1) is

$$\chi_A(x): X \rightarrow \{0, 1\} \tag{2.1-2}$$

The notation of equation (2.1-2) is read as follows: *There exists a function $\chi_A(x)$ mapping every element of the set X (our universe of discourse) to the set $\{0, 1\}$.* It emphasizes that the characteristic function is a mechanism for mapping the set X to the valuation set $\{0, 1\}$. Important operations in crisp sets such as *union, intersection,* and *complementation* are familiar to us from elementary mathematics. They are usually represented through Venn diagrams but may also be expressed in terms of the characteristic function.

Fundamentally, sets are *categories*. Defining suitable categories and using operations for manipulating them is a major task of modeling and computation. From image recognition to measurement and control, the notion of *category*, or *set*, is essential in the definition of system variables, parameters, their ranges, and their interactions. The constraint to have a dual degree of membership to a set, an all-or-nothing, is a consequence of a desire to abstract a system description away from the multitude of intricacies and complexities that exist in reality and focus on factors of primary influence. Nevertheless, given our modern-day computational technologies, it may be unduly restrictive. This is particularly the case when it is desired to develop computer models *easily calibrated* to the specifics of a system and endowed with adaptive and self-organizing capabilities (Zadeh, 1973, 1988).

---

[1]The term *universe of discourse* is used in fuzzy logic; it comes from classical logic and describes the complete set of individual elements able to be referred to or quantified.

2.2 FUZZY SETS

As we saw in the previous section, in classical set theory there is a rather strict sense of membership to a set; that is, an element either *belongs* or *does not belong* to the set. In 1965 Lotfi A. Zadeh introduced *fuzzy sets*, where a more flexible sense of membership is possible (Zadeh, 1965). In fuzzy sets many degrees of membership are allowed. The degree of membership to a set is indicated by a number between 0 and 1—that is, a number in the interval [0, 1]. The point of departure for fuzzy sets is simply generalizing the valuation set from the pair of numbers {0, 1} to all numbers found in [0, 1]. By expanding the valuation set we alter the nature of the characteristic function, now called *membership function* and denoted by $\mu_A(x)$. We no longer have *crisp* sets but instead have *fuzzy* sets. Since the interval [0, 1] contains an infinity of numbers, infinite degrees of membership are possible. Thus, in view of equation (2.1-2) we say that *a membership function maps every element of the universe of discourse X to the interval* [0, 1], and we formally write this mapping as

$$\mu_A(x)\colon X \to [0, 1] \qquad\qquad (2.2\text{-}1)$$

Equation (2.2-1) is a generalization of the mapping shown in equation (2.1-2). Membership functions are a simple yet versatile mathematical tool for indicating flexible membership to a set and, as we shall see, for modeling and quantifying the meaning of symbols. A question often asked by people beginning the study of fuzzy sets is, How are membership functions found? Membership functions may represent an individual's (subjective) notion of a vague class—for example, *objects in a room functioning as chairs, tall people, acceptable performance, small contribution to system stability, little improvement, big benefit*, and so on. In designing and operating controllers or automatic decision-making tools, for example, modeling such notions is a very important task. Membership functions may also be determined on the basis of statistical data or through the aid of neural networks. In Part III of this book we will look at the synergistic relation between neural networks and fuzzy logic toward this end (Kosko, 1992). At this point we can simply say that membership functions are primarily subjective in nature; this does not mean that they are assigned arbitrarily, but rather on the basis of application-specific criteria (Kaufmann, 1975; Dubois and Prade, 1980; Zimmermann, 1985).

   There are two commonly used ways of denoting fuzzy sets. If X is a universe of discourse and x is a particular element of X, then a fuzzy set A defined on X may be written as a collection of ordered pairs

$$A = \{(x, \mu_A(x))\}, \qquad x \in X \qquad\qquad (2.2\text{-}2)$$

where each pair $(x, \mu_A(x))$ is called a *singleton* and has x first, followed by its membership in A, $\mu_A(x)$. In crisp sets a singleton is simply the element x by itself. In fuzzy sets a singleton is two things: x and $\mu_A(x)$. For example, the set of *small integers*, A, defined (subjectively) over the universe of discourse of positive integers may be given by the collection of singletons

$$A = \{(1, 1.0), (2, 1.0), (3, 0.75), (4, 0.5), (5, 0.3), (6, 0.3), (7, 0.1), (8, 0.1)\}$$

Thus the fourth singleton from the left tells us that 4 belongs to A to a degree of 0.5. A singleton is also written as $\mu_A(x)/x$—that is, by putting membership first, followed by the marker "/" separating it from x.[2] Singletons whose membership to a fuzzy set is zero may be omitted. The *support set* of a fuzzy set A is the set of its elements that have membership function other than the trivial membership of zero.

An alternative notation, used more often than equation (2.2-2), explicitly indicates a fuzzy as the *union* of all $\mu_A(x)/x$ singletons—that is,

$$A = \sum_{x_i \in X} \mu_A(x_i)/x_i \qquad (2.2\text{-}3)$$

The *summation* sign in equation (2.2-3) indicates the *union* of all singletons (the union operation in set theory is like "addition"). Equation (2.2-3) assumes that we have a *discrete universe of discourse*. In this alternative notation the set of *small integers* above may be written as

$$A = \mu_A(1)/1 + \mu_A(2)/2 + \mu_A(3)/3 + \mu_A(4)/4 + \mu_A(5)/5$$
$$+ \mu_A(6)/6 + \mu_A(7)/7 + \mu_A(8)/8$$
$$= 1.0/1 + 1.0/2 + 0.75/3 + 0.5/4 + 0.3/5 + 0.3/6 + 0.1/7 + 0.1/8$$

For a continuous universe of discourse, we write equation (2.2-3) as

$$A = \int_X \mu_A(x)/x \qquad (2.2\text{-}4)$$

where the *integral* sign in equation (2.2-4) indicates the *union* of all $\mu_A(x)/x$ singletons.[3] Consider, for example, the fuzzy set *small numbers* defined (subjectively) over the set of non-negative real numbers through a continuous

---

[2] It should be noted that "/" does not indicate "division"; it is merely a marker.
[3] Note that the integral sign is not the same as the integral sign of differential and integral calculus. It is used here in the sense that the integral sign is used in set theory—that is, to indicate the *sum* or *union* of individual *singletons*.

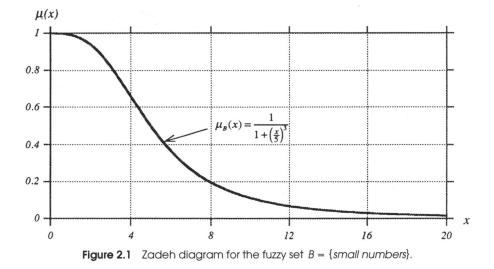

Figure 2.1   Zadeh diagram for the fuzzy set B = {*small numbers*}.

membership function $\mu_B(x)$ given by

$$\mu_B(x) = \frac{1}{1 + \left(\dfrac{x}{5}\right)^3} \tag{2.2-5}$$

Using the form of equation (2.2-4) the fuzzy set B is written as

$$B = \int_{x \geq 0} \mu_B(x)/x = \int_{x \geq 0} \left[\frac{1}{1 + \left(\dfrac{x}{5}\right)^3}\right]/x \tag{2.2-6}$$

The membership function of fuzzy set B is shown in Figure 2.1.[4] A graph like this is called a *Zadeh diagram*.

2.3   BASIC TERMS AND OPERATIONS

Many fuzzy set operations such as *intersection* and *union* are defined through the min ($\wedge$) and max ($\vee$) operators. *Min* and *max* are analogous to *product* ($\cdot$) and *sum* ($+$) in algebra (Dubois and Prade, 1980; Klir and Folger, 1988; Terano et al., 1992). Let us take a look at how they are used.

---

[4]Fuzzy sets are sometimes called *fuzzy subsets*, reflecting the fact that they are subsets of a larger set—that is, the *universe of discourse*. Although the term *fuzzy subsets* is factually correct, we will use the standard term *fuzzy set* for convenience.

First, min ($\wedge$) and max ($\vee$) may be used to select the minimum and maximum of two elements—for example, $2 \wedge 3 = 2$, or $2 \vee 3 = 3$. We also write $\min(2, 3) = 2$, or $\max(2, 3) = 3$. Formally, the minimum of two elements μ_1 and μ_2 denoted either as $\min(\mu_1, \mu_2)$, $\wedge(\mu_1, \mu_2)$, or $\mu_1 \wedge \mu_2$ is defined as

$$\mu_1 \wedge \mu_2 = \min(\mu_1, \mu_2) \equiv \begin{cases} \mu_1 & \text{iff} \quad \mu_1 \le \mu_2 \\ \mu_2 & \text{iff} \quad \mu_1 > \mu_2 \end{cases} \qquad (2.3\text{-}1)$$

where, the "$\equiv$" symbol means "by definition" and iff is shorthand for "if and only if." Similarly the maximum of two elements μ_1 and μ_2, denoted as $\max(\mu_1, \mu_2)$ or $\mu_1 \vee \mu_2$, is defined as

$$\mu_1 \vee \mu_2 = \max(\mu_1, \mu_2) \equiv \begin{cases} \mu_1 & \text{iff} \quad \mu_1 \ge \mu_2 \\ \mu_2 & \text{iff} \quad \mu_1 < \mu_2 \end{cases} \qquad (2.3\text{-}2)$$

Second, min ($\wedge$) and max ($\vee$) may operate on an entire set, selecting the least element (called *infimum* in mathematical analysis) or the greatest element (called *supremum*) of the set. For example, $\wedge (0.01, 0.33, 0.44, 0.999) = 0.01$ and $\vee (0.01, 0.33, 0.44, 0.999) = 0.999$. Formally we write this as

$$\mu = \wedge A = \inf A \qquad (2.3\text{-}3)$$

and

$$\mu = \vee A = \sup A \qquad (2.3\text{-}4)$$

where μ is an element of A—that is, $\mu \in A$.

In addition, min ($\wedge$) and max ($\vee$) may be used as *functions* operating on single elements or on entire sets, for example, to find the smallest element μ out of a list of elements $(\mu_1, \mu_2, \ldots, \mu_m)$—that is,

$$\mu = \wedge(\mu_1, \mu_2, \ldots, \mu_m) \qquad (2.3\text{-}5)$$

which is the same as

$$\mu = \mu_1 \wedge \mu_2 \wedge \cdots \wedge \mu_m \qquad (2.3\text{-}6)$$

We sometimes use a shorthand notation for equations (2.3-5) and (2.3-6) and write them as

$$\mu = \bigwedge_{k=1}^{m} (\mu_k) \qquad (2.3\text{-}7)$$

This notation is analogous to finite *product* notation in algebra (or finite *summation* when $\vee$ is used). There is in fact a more general analogy between min and max and the operations of *multiplication* and *addition*. They both have the same properties of associativity and distributivity, and thus in equations that involve min and max we may employ them in the same manner as *multiplication* ($\cdot$) and *addition* ($+$). We will see an interesting example of these properties in the composition of fuzzy relations (Chapter 3), where we treat composition as *matrix multiplication* with ($\wedge$) and ($\vee$) in place of product ($\cdot$) and sum ($+$).

Min ($\wedge$) and max ($\vee$) can also operate on a collection of sets as for example in

$$A = \wedge (A_1, A_2, \ldots, A_m) \qquad (2.3\text{-}8)$$

which can be succinctly written as

$$A = \bigwedge_{k=1}^{m} (A_k) \qquad (2.3\text{-}9)$$

Using primarily min ($\wedge$) and max ($\vee$), a number of useful notions and operations involving fuzzy sets can be defined.[5]

Empty Fuzzy Set

A fuzzy set A is called *empty* (denoted as $A = \varnothing$) if its membership function is zero everywhere in its universe of discourse X—that is,

$$A \equiv \varnothing \quad \text{if } \mu_A(x) = 0, \forall x \in X \qquad (2.3\text{-}10)$$

where "$\forall x \in X$" is shorthand notation indicating "for any element x in X."

Normal Fuzzy Set

A fuzzy set is called *normal* if there is at least one element x_0 in the universe of discourse where its membership function equals one—that is,

$$\mu_A(x_0) = 1 \qquad (2.3\text{-}11)$$

More than one element in the universe of discourse can satisfy equation (2.3-11).[6]

---

[5]These operations can also be defined in terms of *T-norms* (see Appendix).
[6]It should be noted that the term *normal* does not refer to the area under the curve of the membership function. It simply means what the definition says: At least one point, maybe more, needs to have full membership value.

Equality of Fuzzy Sets

Two fuzzy sets are said to be *equal* if their membership functions are equal everywhere in the universe of discourse—that is,

$$A \equiv B \quad \text{if } \mu_A(x) = \mu_B(x) \tag{2.3-12}$$

Union of Two Fuzzy Sets

The *union* of two fuzzy sets A and B defined over the same universe of discourse X is a new fuzzy set $A \cup B$ also on X, with membership function which is the maximum of the grades of membership of every x to A and B—that is,

$$\mu_{A \cup B}(x) \equiv \mu_A(x) \vee \mu_B(x) \tag{2.3-13}$$

The *union* of two fuzzy sets is related to the logical operation of *disjunction (OR)* in fuzzy logic. Equation (2.3-13) can be generalized to any number of fuzzy sets over the same universe of discourse.

Intersection of Fuzzy Sets

The *intersection* of two fuzzy sets A and B is a new fuzzy set $A \cap B$ with membership function which is the minimum of the grades of every x in X to the sets A and B, i.e.,

$$\mu_{A \cap B}(x) \equiv \mu_A(x) \wedge \mu_B(x) \tag{2.3-14}$$

The *intersection* of two fuzzy sets is related to *conjunction (AND)* in fuzzy logic. The definition of *intersection* in (2.3-14) can be generalized to any number of fuzzy sets over the same universe of discourse.

Complement of a Fuzzy Set

The *complement* of a fuzzy set A is a new fuzzy set, $\bar{A}$, with membership function

$$\mu_{\bar{A}}(x) \equiv 1 - \mu_A(x) \tag{2.3-15}$$

Fuzzy set *complementation* is equivalent to *negation (NOT)* in fuzzy logic.

Product of Two Fuzzy Sets

The product of two fuzzy sets A and B defined on the same universe of discourse X is a new fuzzy set, $A \cdot B$, with membership function that equals the algebraic product of the membership functions of A and B,

$$\mu_{A \cdot B}(x) \equiv \mu_A(x) \cdot \mu_B(x) \tag{2.3-16}$$

The product of two fuzzy sets can be generalized to any number of fuzzy sets on the same universe of discourse.

Multiplying a Fuzzy Set by a Crisp Number

We can multiply the membership function of a fuzzy set A by the crisp number a to obtain a new fuzzy set called *product a · A*. Its membership function is

$$\mu_{aA}(x) \equiv a \cdot \mu_A(x) \tag{2.3-17}$$

The operations of multiplication and raising a fuzzy set to a power that we see next are useful for modifying the meaning of linguistic terms (Zadeh, 1975).

Power of a Fuzzy Set

We can raise fuzzy set A to a power α (positive real number) by raising its membership function to α. The α *power* of A is a new fuzzy set, A^α, with membership function

$$\mu_{A^\alpha}(x) \equiv \left[\mu_A(x) \right]^\alpha \tag{2.3-18}$$

Raising a fuzzy set to the second power is usually taken to be equivalent to linguistically changing it through the modifier *VERY* (Zadeh, 1983) (see Chapter 5). Thus the square of the membership function of $B = \{small$ $numbers\}$ in Figure 2.1 is taken to represent the fuzzy set $B^2 = \{VERY\ small$ $numbers\}$.

Raising a fuzzy set to the second power is a particularly useful operation and therefore has its own name. It is called *concentration* or *CON*. Taking the square root of a fuzzy set is called *dilation* or *DIL* (an operation useful for representing analytically the linguistic modifier *MORE OR LESS*).

Example 2.1   Union, Intersection, and Complement of Fuzzy Sets. Consider the Zadeh diagram of fuzzy sets A and B shown in Figure 2.2*a* and defined by membership functions

$$\mu_A(x) = \frac{1}{1 + 0.3(x - 8)^2} \quad \text{and} \quad \mu_B(x) = \frac{1}{1 + \left(\dfrac{x}{5}\right)^3} \tag{E2.1-1}$$

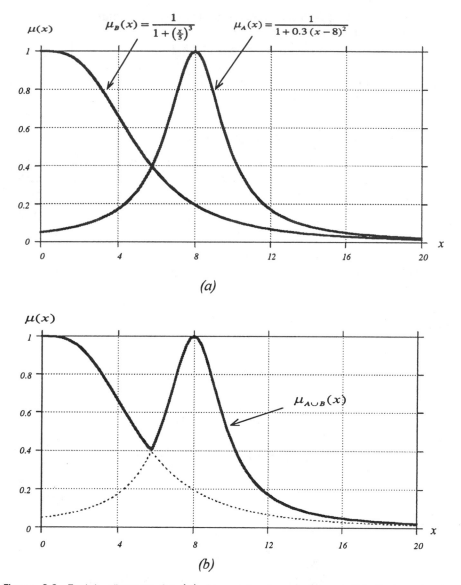

Figure 2.2 Zadeh diagram for (a) fuzzy sets A and B and (b) their *union* in Example 2.1.

Fuzzy set A may be thought of as defining the set of numbers "*about* 8," and fuzzy set B may be thought of as defining "*small numbers.*" We take numbers between 0 and 20 to be the universe of discourse and, would like to find the *union* and *intersection* of A and B and the *complement* of B.

The membership function of the *union* of fuzzy sets A and B is the maximum grade of membership of each element x of the universe of discourse to either A or B in accordance with equation (2.3-13). Figure 2.2b shows the membership function of the *union* $A \cup B$. The interpretation of $A \cup B$ is "*about 8 OR small number.*" Similarly the membership function of the *intersection* of fuzzy sets A and B, shown in Figure 2.3a, represents the new fuzzy set "*about 8 AND small number.*" We observe that although the union of A and B is a *normal* fuzzy set, the intersection shown in Figure 2.3a is not, because fuzzy set $A \cap B$ has no point in the universe of

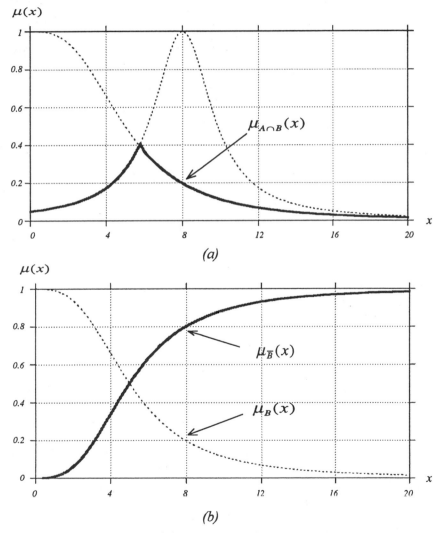

Figure 2.3 Zadeh diagram for (a) the *intersection* of fuzzy sets A and B and (b) the *complement* of B in Example 2.1.

discourse with grade of membership equal to 1. The complement of fuzzy set B is a new fuzzy set with membership function given by equation (2.3-15). Figure 2.3*b* shows the membership function of the *complement* $\overline{B}$. The complement $\overline{B}$ represents the logical negation (*NOT*) of B—that is, the set "*NOT small numbers.*" □ [7]

Concentration

The *concentration* of a fuzzy set A defined over a universe of discourse, X, is denoted as $CON(A)$ and it is a new fuzzy set with membership function given by

$$\mu_{CON(A)}(x) \equiv \left(\mu_A(x) \right)^2 \qquad (2.3\text{-}19)$$

As we said in the previous paragraph, squaring or *concentrating* a fuzzy set is equivalent to linguistically modifying it by the term *VERY*. Figure 2.4 shows the concentration operation applied to the fuzzy set $B = \{small\ numbers\}$. The membership function of the new fuzzy set $CON(B) = B^2 = \{VERY\ small\ numbers\}$ is

$$\mu_{CON(B)}(x) = \left(\mu_B(x) \right)^2 = \cfrac{1}{\left[1 + \left(\dfrac{x}{5} \right)^3 \right]^2}$$

Dilation

The *dilation* of a fuzzy set A, denoted as $DIL(A)$, produces a new fuzzy set in X, with membership function defined as the square root of the membership function of A—that is,

$$\mu_{DIL(A)}(x) \equiv \sqrt{\mu_A(x)} \qquad (2.3\text{-}20)$$

Dilation (*DIL*) and *concentration* (*CON*) are operations with opposing effects. Concentrating a fuzzy set reduces its fuzziness while dilating it increases its fuzziness. The *dilation* operation corresponds to linguistically modifying the meaning of a fuzzy set by the term "*MORE OR LESS.*" Figure 2.4 shows the dilation of $B = \{small\ numbers\}$, resulting in a new fuzzy set $DIL(B) = B^{1/2} = \{MORE\ OR\ LESS\ small\ numbers\}$.

[7]Here and throughout this book, the end of an example is indicated by the symbol "□."

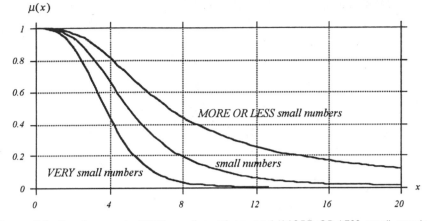

Figure 2.4 The fuzzy sets *VERY small numbers* and *MORE OR LESS small numbers* obtained by *concentrating* and *dilating* the fuzzy set *small numbers*.

Contrast Intensification

In certain applications it is desirable to control the *fuzziness* of a fuzzy set A by modifying the contrast between low and high grades of membership. For instance, we may want to increase the membership function on that part of A where membership values are higher than 0.5, and decrease it for values lower than 0.5. We define the *contrast intensification of A* as

$$\mu_{INT(A)}(x) \equiv 2\left[\mu_A(x)\right]^2 \qquad \text{for} \quad 0 \le \mu_A(x) \le 0.5$$

$$\mu_{INT(A)}(x) \equiv 1 - 2\left[1 - \mu_A(x)\right]^2, \qquad \text{for } 0.5 \le \mu_A(x) \le 1.0$$

$$(2.3\text{-}21)$$

Contrast intensification may be repeatedly applied to a fuzzy set. In the extreme, when the maximum possible contrast is achieved we no longer have a fuzzy set. We are back to a crisp set. The opposite effect—that is, going from a crisp set to fuzzy set—may be achieved through *fuzzification*.

Fuzzyfication

Fuzzification is used to transform a crisp set into a fuzzy set or simply to increase the fuzziness of a fuzzy set. For fuzzification we use a *fuzzyfier function F* that controls the fuzziness of a set. F may be one or more simple parameters. For instance, consider the fuzzy set A that describes *large* *numbers.* We define it (subjectively) through the membership function

$$\mu_{large\ numbers}(x) = \frac{1}{1 + \left(\dfrac{x}{F_2}\right)^{-F_1}} \tag{2.3-22}$$

where x is any positive real number. The membership function in equation (2.3-22) has two fuzzifying parameters: an *exponential fuzzyfier*, F_1, and a *denominational fuzzyfier*, F_2. Through them the fuzzy set $A = \{large\ numbers\}$ can be written as

$$A \equiv \int_X \left[\frac{\text{\includegraphics{}}}{1 + \left(\dfrac{x}{F_2}\right)^{-F_1}}\right] / x \tag{2.3-23}$$

The membership function inside the brackets of equation (2.3-23) can be adjusted when needed in order to better represent the meaning of the term *large numbers.* Consider the case when we fix the value of denominational fuzzifier as $F_2 = 50$ and vary the exponential fuzzyfier F_1. The result is a family of fuzzy sets with decreasing fuzziness as F_1 increases. Figure 2.5 shows membership functions that result from such a variation. Note that when F_1 becomes very large, the set A appears almost like a crisp set. The effect of varying the denominational fuzzyfier F_2 while keeping the exponen-

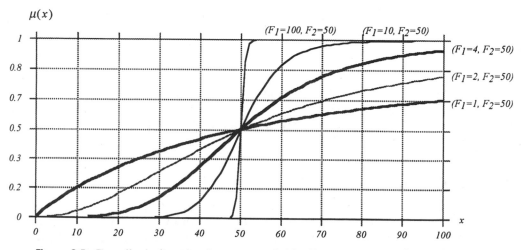

Figure 2.5 The effect of varying the exponential fuzzifier F_1 while keeping the denominational fuzzifier F_2 constant in fuzzifying the set A.

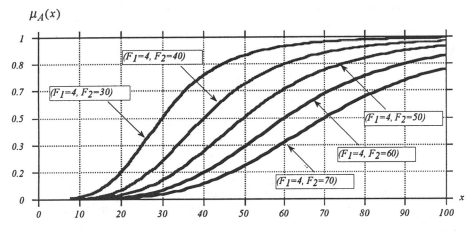

Figure 2.6 The effect of varying the denominational fuzzifier F_2 while keeping the exponential fuzzifier F_1 constant in fuzzifying the set A.

tial fuzzyfier at $F_1 = 4$ is shown in Figure 2.6. Varying F_2 results primarily in translating the membership function left and right, and to a lesser extent it affects the fuzziness of A. Such fuzzifiers are often used in fuzzy pattern recognition and image analysis in defining, for instance, the meaning of the words *vertical, horizontal,* and *oblique lines* (Pal and Majumder, 1986).

Fuzzification may be used more systematically by associating a *fuzzyfier F* with another function, namely a *fuzzy kernel, $K(x)$*, which is the fuzzy set that results from the application of F to a singleton x. This is often done in control applications where the input to an on-line control or diagnostic system comes from sensors and is therefore crisp, usually a real number. In order to use it in fuzzy algorithms (see Chapters 5 and 6), it is often necessary to convert a crisp number to a fuzzy set, a step known as *fuzzification*. As a result of the application of K to a fuzzy set A, we have

$$F(A; K) = \int_X \mu_A(x) \cdot \mu_{K(x)}(x)/x \qquad (2.3\text{-}24)$$

where $F(A; K)$ is a fuzzy set that results from changing the fuzziness of A in accordance with K. The *fuzzy kernel $K(x)$* is simply a fuzzy set imposed on a singleton. It functions as a "mask" that covers the singleton to produce a fuzzy set. For example, suppose that we have the universe of discourse $X = \{1, 2, 3, 4, 5, 6, 7, 8, 9, 10\}$ and a fuzzy kernel $K(x)$ that centers a triangular fuzzy set around 5 given by

$$K(5) = 0.33/3 + 0.67/4 + 1.0/5 + 0.67/6 + 0.33/7 \qquad (2.3\text{-}25)$$

with all other elements of the universe of discourse having trivial (zero) membership. Now suppose that we have the value of 3, which may be a crisp measurement taken at a certain time. We write it as a singleton A given by

$$A = \mu_A(3)/3 = 1.0/3 \qquad (2.3\text{-}26)$$

We fuzzify A using equation (2.3-24) as follows:

$$F(A; K) = \int_X \mu_A(x) \cdot \mu_{K(x)}(x)/x$$

$$= \int_X \left[\mu_A(3) \cdot \mu_{K(3)}(x) \right]/x$$

$$= 0.33/1 + 0.67/2 + 1.0/3 + 0.67/4 + 0.33/5 \qquad (2.3\text{-}27)$$

which results in shifting the fuzzy kernel of (2.3-25) so that its peak is located at the singleton '3'. In other words, the effect of equations (2.3-27) is to *mask* the crisp value '3' by the fuzzy set $K(5)$, shifting its peak from '5' to '3'.

2.4 PROPERTIES OF FUZZY SETS

Fuzzy set properties are useful in performing operations involving membership functions. The properties we list here are valid for crisp and fuzzy sets as well, but some of them are specific to fuzzy sets only; more detailed treatment of properties may be found in Dubois and Prade (1980) and in Klir and Folger (1988). Consider sets A, B, C defined over a common universe of discourse X. We indicate the complement of a set by a bar over it. The following properties are true:

Double Negation Law: $\qquad \overline{(\overline{A})} = A \qquad (2.4\text{-}1)$

Idempotency:
$$A \cup A = A$$
$$A \cap A = A \qquad (2.4\text{-}2)$$

Commutativity:
$$A \cap B = B \cap A$$
$$A \cup B = B \cup A \qquad (2.4\text{-}3)$$

Associative Property:
$$(A \cup B) \cup C = A \cup (B \cup C)$$
$$(A \cap B) \cap C = A \cap (B \cap C) \qquad (2.4\text{-}4)$$

Distributive Property:
$$A \cup (B \cap C) = (A \cup B) \cap (A \cup C)$$
$$A \cap (B \cup C) = (A \cap B) \cup (A \cap C) \qquad (2.4\text{-}5)$$

Absorption:
$$A \cap (A \cup B) = A$$
$$A \cup (A \cap B) = A \qquad (2.4\text{-}6)$$

De Morgan's Laws:
$$\overline{A \cup B} = \overline{A} \cap \overline{B}$$
$$\overline{A \cap B} = \overline{A} \cup \overline{B} \qquad (2.4\text{-}7)$$

In fuzzy sets all these properties can be expressed using the membership function of the sets involved and the definitions of *union*, *intersection*, and *complement*. For example, consider the *associative property* given by equations (2.4-4). In terms of membership functions the associative property is written as

$$(\mu_A(x) \vee \mu_B(x)) \vee \mu_C(x) = \mu_A(x) \vee (\mu_B(x) \vee \mu_C(x))$$

$$(\mu_A(x) \wedge \mu_B(x)) \wedge \mu_C(x) = \mu_A(x) \wedge (\mu_B(x) \wedge \mu_C(x))$$

Similarly, the *distributive property*, equations (2.4-5), in terms of membership functions is written as

$$\mu_A(x) \vee (\mu_B(x) \wedge \mu_C(x)) = (\mu_A(x) \vee \mu_B(x)) \wedge (\mu_A(x) \vee \mu_C(x))$$

$$\mu_A(x) \wedge (\mu_B(x) \vee \mu_C(x)) = (\mu_A(x) \wedge \mu_B(x)) \vee (\mu_A(x) \wedge \mu_C(x))$$

De Morgan's law, equation (2.4-7), is written as

$$\overline{\mu_A(x) \vee \mu_B(x)} = \mu_{\overline{A}}(x) \wedge \mu_{\overline{B}}(x)$$

where the bar over the membership functions indicates that we take the complement. *De Morgan's law* says that the intersection of the complement of two fuzzy sets equals the complement of their union; in terms of membership functions, this is the same as saying that the minimum of two membership functions equals the complement of their maximum. There are also some properties generally not valid for fuzzy sets (although valid in crisp sets), such as the *law of contradiction,*

$$A \cap \overline{A} \neq \emptyset \qquad (2.4\text{-}8)$$

and the *law of the excluded middle,*

$$A \cup \overline{A} \neq X \qquad (2.4\text{-}9)$$

The law of the excluded middle in crisp sets states that the union of a set with its complement results in the universe of discourse. This is generally not true in fuzzy sets. A property unique to fuzzy sets is

$$A \cap \emptyset = \emptyset \qquad (2.4\text{-}10)$$

Equation (2.4-10) says that the intersection of a fuzzy set with the empty set —that is, a set with a membership function equal to zero everywhere on the universe of discourse—is also the empty set. In terms of membership functions equation (2.4-10) is written as

$$\mu_A(x) \wedge 0 = 0$$

Also, the union of a fuzzy set A with the empty set, $\emptyset$, is A itself; that is, $A \cup \emptyset = A$ or, equivalently, $\mu_A(x) \vee 0 = \mu_A(x)$. The intersection of a fuzzy set A with the universe of discourse is the fuzzy set A itself; that is, $A \cap X = A$ or, equivalently, $\mu_A(x) \wedge 1 = \mu_A(x)$. The union of a fuzzy set A with the universe of discourse X is the universe of discourse; that is, $A \cup X = X$, which, in terms of the membership function, is written as $\mu_A(x) \vee 1 = 1$. The universe of discourse may be viewed as a fuzzy set whose membership function equals 1 everywhere; that is, $\mu_X(x) = 1$ for all x in X.

2.5   THE EXTENSION PRINCIPLE

While fuzzification operations such as the ones we saw in Section 2.3 are useful for fuzzifying individual sets or singletons, more general mathematical expressions may also be fuzzified when the quantities they involve are fuzzyfied. For example, the output of arithmetic operations when their arguments are fuzzy sets becomes also a fuzzy quantity. The *extension principle* is a mathematical tool for extending crisp mathematical notions and operations to the milieu of fuzziness. It provides the theoretical warranty that fuzzifying the parameters or arguments of a function results in computable fuzzy sets. It is an important principle, and we will use it on several occasions, particularly in conjunction with fuzzy relations (Chapter 3) and fuzzy arithmetic (Chapter 4). We give here an informal heuristic description of the extension principle; detailed formulations may be found in (Zadeh (1975), and in Dubois and Prade (1980).

Suppose that we have a function f that maps elements $x_1, x_2, \ldots, x_n$ of a universe of discourse X to another universe of discourse Y—that is,

$$y_1 = f(x_1)$$

$$y_2 = f(x_2)$$

$$\ldots \tag{2.5-1}$$

$$y_n = f(x_n)$$

Now suppose that we have a fuzzy set A defined on $x_1, x_2, x_3, \ldots, x_n$ (the input to the function f). A is given by

$$A = \mu_A(x_1)/x_1 + \mu_A(x_2)/x_2 + \cdots + \mu_A(x_n)/x_n \tag{2.5-2}$$

We then ask the question, If the input to our function f becomes fuzzy—for example, the set A of equation (2.5-2)—what happens to the output? Is the output also fuzzy? In other words, is there an output fuzzy set B that can be computed by inputting A to f. Well, the extension principle tells us that there is indeed such an output fuzzy set B and that it is given by

$$B = f(A) = \mu_A(x_1)/f(x_1) + \mu_A(x_2)/f(x_2) + \cdots + \mu_A(x_n)/f(x_n)$$

$$\tag{2.5-3}$$

where every single image of x_i under f—that is, $y_i = f(x_i)$—becomes fuzzy to a degree $\mu_A(x_i)$. Recalling that functions are generally *many-to-one* mappings, it is conceivable that several x's may map to the same y. Thus for a certain y_0 we may have more than one x: Let us say that both x_2 and x_{13} in (2.5-1) are mapping to y_0. Hence, we have to decide which of the two membership values, $\mu_A(x_2)$ or $\mu_A(x_{13})$, we should take as the membership value of y_0. The extension principle says that the *maximum* of the membership values of these elements in the fuzzy set A ought to be chosen as the grade of membership of y_0 to the set B—that is,

$$\mu_B(y_0) = \mu_A(x_2) \vee \mu_A(x_{13}) \tag{2.5-4}$$

If, on the other hand, no element x in X is mapped to y_0—that is, no inverse image of y_0 exists—then the membership value of the set B at y_0 is zero. Having accounted for these two special cases (many x's mapping to the same y and no inverse image for a certain y), we can compute the set B—that is, the grades of membership of elements y in Y produced by the mapping $f(A)$—using equation (2.5-3).

In a more general case where we have several variables, $u, v, \ldots, w$, from different universes of discourse $U, V, \ldots, W$ and m different fuzzy sets $A_1, A_2, \ldots, A_m$ defined on the product space $U \times V \times \cdots \times W$, the multi-variable function, $y = f(u, v, \ldots, w)$, may also be used to fuzzify the space Y through the extension principle. In this case, the grade of membership of any y equals the minimum of the membership values of $u, v, \ldots, w$ in $A_1, A_2, \ldots, A_m$, respectively. The membership function of B is given by

$$\mu_B(y) = \int_{U \times V \times \cdots \times Y} \left[\mu_{A_1}(u) \wedge \mu_{A_2}(v) \wedge \cdots \wedge \mu_{A_m}(w) \right] / f(u, v, \ldots, w) \tag{2.5-5}$$

where there is also a max ($\vee$) operation implicit in the union operation [the integral sign in equation (2.5-5) indicates a union ($\vee$) operation]. The max operation is performed over all $u, v, \ldots, w$ such that $y = f(u, v, \ldots, w)$. This is indicated by the union over the product space $U \times V \times \cdots \times W$ of all the universes on which the m-tuples $u, v, \ldots, w$ are defined under the integral sign. If the inverse image does not exist, then the membership function is simply zero.

In many engineering applications, the interpretation of numerical data may not be precisely known. We consider this type of data to be fuzzy. Using the extension principle, it is quite possible to adapt ordinary algorithms, which are used with precise data, to the case where the data are fuzzy. Example 2.2 is a mathematical illustration of the extension principle.

Example 2.2 Using the Extension Principle. As an illustration of how the extension principle may be used, consider the function f that maps points from the x axis to y axis in the Cartesian plane according to the equation

$$y = f(x) = \sqrt{1 - \frac{x^2}{4}} \qquad \text{(E2.2-1)}$$

Figure 2.7*a* shows the function y of equation (E2.2-1). It is the upper half of an ellipse located on the center of the plane with major axis, $a = 2$, and minor axis (height), $b = 1$. The general equation of the ellipse shown in Figure 2.7*a* is

$$\frac{x^2}{a^2} + \frac{y^2}{b^2} = 1 \qquad \text{(E2.2-2)}$$

In our case with $a = 2$ and $b = 1$, equation (E2.2-2) becomes

$$\frac{x^2}{4} + y^2 = 1 \qquad \text{(E2.2-3)}$$

Equation (E2.2-1) is one of the two solutions of equation (E2.2-3).

Now suppose that we define a fuzzy set A on X as shown in Figure 2.7*b*: We fuzzify the x's of equation (E2.2-1) by specifying a grade of membership $\mu_A(x)$ for each x to fuzzy set A—that is, $\mu_A(x) = \frac{1}{2}|x|$ and

$$A = \int_{-2 \leq x \leq 2} [\tfrac{1}{2}|x|]/x \qquad \text{(E2.2-4)}$$

where $|x|$ is the absolute value of x, and we limit the support of A between -2 and $+2$ as indicated by the limits under the *integration sign* (*union*) of equation (E2.2-4).

Having the x values fuzzyfied by the fuzzy set A, we want to know the effect of fuzzification on y. The extension principle tells us that the fuzziness of A will be extended to y as well. In other words, we will have a fuzzy set B on Y derived by equations (2.5-3) or (2.5-5). To avoid the case where more than one x will map to the same y, we consider first the function f in the first quadrant of the plane (where both x and y are positive). Later we will look at the entire function. The fuzzy set, B, defined on Y is

$$B = f(A) = \int_Y \mu_B(y)/y \qquad \text{(E2.2-5)}$$

We need to find $\mu_B(y)$ in equation (E2.2-5). In terms of the membership function of A and according to the extension principle, equation (2.5-3), the

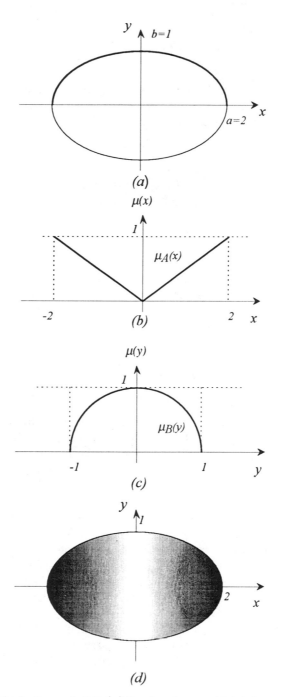

Figure 2.7 Graphs for Example 2.2. (*a*) The function *y*, which is the upper part of the ellipse shown. (*b*) The membership function of the set *A*. (*c*) The membership function of *B*. (*d*) Fuzzifying the interior of the ellipse.

set B will be

$$B = f(A) = \int_Y \mu_A(x)/f(x) \tag{E2.2-6}$$

Of course we want to transform the x variable to y in equation (E2.2-6) since the union (integration) is formed with respect to Y, the universe of discourse for B. We use equation (E2.2-1) to solve for x:

$$x = 2\sqrt{1 - y^2} \tag{E2.2-7}$$

Then we substitute (E2.2-7) in (E2.2-6), noting that $f(x) = y$ and that $\mu_A(x)$ is given by (E2.2-4). Thus we obtain the fuzzy set B:

$$B = \int_{0 \le y \le 1} \sqrt{1 - y^2}/y \tag{E2.2-8}$$

Now if we consider negative values for x as well, we would have to take the maximum of the membership value of A at (x) and $(-x)$ in accordance with equation (2.5-5). Due to the symmetry of the problem these values are actually the same and therefore B is still as derived in (E2.2-8). The membership function of B is,

$$\mu_B(y) = \sqrt{1 - y^2} \tag{E2.2-9}$$

as shown in Figure 2.7c. Figure 2.7d shows the geometric interpretation of fuzzyfying the interior of the ellipse in accordance with the fuzzy sets A and B above. The result is a kind of fuzzy elliptic region, strongest near the x axis and particularly at its $x = \pm 2$ sides and weakest near the origin and the $y = \pm 1$ sides. $\square$

2.6 ALPHA-CUTS

With any fuzzy set A we can associate a collection of crisp sets known as α-cuts (alpha-cuts) or level sets of A. An α-cut is a crisp set consisting of elements of A which belong to the fuzzy set at least to a degree α. As we shall see in the next section, α-cuts offer a method for resolving any fuzzy set in terms of constituent crisp sets (something analogous to resolving a vector into its components). In Chapter 4 we will see that α-cuts are indispensable in performing arithmetic operations with fuzzy sets that represent various qualities of numerical data. It should be noted that α-cuts are crisp, not fuzzy, sets.[8]

[8] Formally, a distinction is made between two types of α-cuts, the strong and the weak α-cut (Dubois and Prade, 1980). We use the weak α-cut, simply calling it α-cut.

The α-cut of a fuzzy set A denoted as A_α is the crisp set comprised of all the elements x of a universe of discourse X for which the membership function of A is *greater than or equal to* α; that is,

$$A_\alpha = \{x \in X|\ \mu_A(x) \geq \alpha\} \tag{2.6-1}$$

where α is a parameter in the range $0 < \alpha \leq 1$; the vertical bar "|" in equation (2.6-1) is shorthand for "such that."

Consider, for example, a fuzzy set A with trapezoidal membership function as shown in Figure 2.8. The 0.5-cut of A is simply the part of its support where its membership function is greater than 0.5. In Figure 2.8 we can see the 0.5-cut of A. Reflecting the fact that the α-cut is a *crisp set*, its membership function appears like a characteristic function. As another example consider the set A of *small integers* given by

$$A = 1.0/1 + 1.0/2 + 0.75/3 + 0.5/4 + 0.3/5 + 0.3/6 + 0.1/7 + 0.1/8$$

The 0.5-cut of A is simply the crisp set $A_{0.5} = \{1, 2, 3, 4\}$.

In the next section we will see that α-cuts provide a useful way both for resolving a membership function in terms of constituent crisp sets as well as for synthesizing a membership function out of crisp sets.

A fuzzy set can have an extensive support since its membership function can be zero or nearly zero, or very small. In order to deal with situations where small degrees of membership are not worthy of consideration, *level*

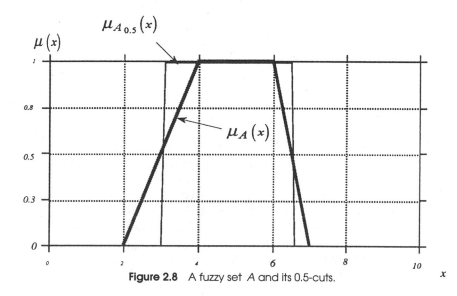

Figure 2.8 A fuzzy set A and its 0.5-cuts.

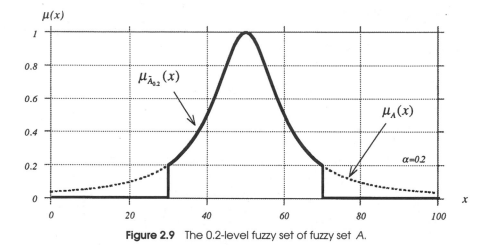

Figure 2.9 The 0.2-level fuzzy set of fuzzy set A.

fuzzy sets were introduced to exclude undesirable grades of membership (Radecki, 1977). We define the level fuzzy sets of a fuzzy set A as fuzzy sets $\tilde{A}_\alpha$ whose membership values are greater than α, where $0 < \alpha < 1$. Formally

$$\tilde{A}_\alpha \equiv \{(x, \mu_A(x)) | x \in A_\alpha\} \tag{2.6-2}$$

where A_α is the α-cut of A. Equation (2.6-2) indicates that for a given α we have a level fuzzy set which is the part of A that has membership greater than α. Let us consider, for example, a fuzzy set A whose membership function is

$$\mu_A(x) = \frac{1}{1 + 0.01(x - 50)^2} \tag{2.6-3}$$

as shown in Figure 2.9 (dotted curve). Suppose that we are not interested in the part of the support that has membership less than 0.2. We obtain the 0.2-level fuzzy set of A by chopping the part of the membership function which is less than 0.2 as shown in the figure. Its membership function $\mu_{\tilde{A}_{0.2}}(x)$ is shown by the solid curve. It is the same as $\mu_A(x)$ between $x = 30$ and $x = 70$ and zero everywhere else. Level fuzzy sets should not be confused with level sets, which is a synonym for α-cuts. Level fuzzy sets are indeed fuzzy sets, whereas α-cuts are crisp sets. They provide a useful way of considering fuzzy sets in the significant part of their support, and hence they save on computing time and storage requirements.

2.7 THE RESOLUTION PRINCIPLE

There are several ways of representing fuzzy sets, and we have already seen a few of them. They all involve two things: identifying a suitable universe of discourse and defining membership functions. One way to represent a fuzzy set would be to list all the elements of the universe of discourse together with the grade of membership of each element (omitting the possibly infinite elements that have zero membership). Alternatively, we can just provide an analytical representation of the membership function. The *resolution principle* offers another way of representing membership to a fuzzy set, namely through its α-cuts. It asserts that the membership function of a fuzzy set A can be expressed in terms of its α-cuts as follows:

$$\mu_A(x) = \bigvee_{0 < \alpha \leq 1} \left[\alpha \cdot \mu_{A_\alpha}(x) \right] \qquad (2.7\text{-}1)$$

where the maximum is taken over all α's. Equation (2.7-1) indicates that the membership function of A is the union (notice the max operator) of all α-cuts, after each one of them has been multiplied by α.

Consider, for example, the fuzzy set A with triangular membership function shown in Figure 2.10. Several α-cuts of A, each multiplied by α, are also shown. Knowing many α's and the α-cuts of A, we can form their products and put them together (in the sense of taking their union) to approximate the function. For example, we multiply the 0.25-cut by 0.25 to get the 0.25-cut pushed down to 0.25, and similarly we multiply the 0.5-cut by 0.5, the 0.75-cut by 0.75, and so on. When put together we have an approximation of the membership function of A as shown in Figure 2.10.

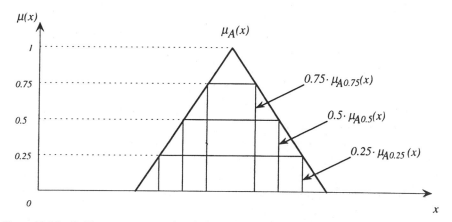

Figure 2.10 Putting many α-cuts of A multiplied by α together approximates the membership function of A.

Thus, a large enough family of α-cuts provides another way of representing a fuzzy set. Although we often know the membership function exactly, in some applications only α-cuts are known and out of them we need to approximate the membership function (see Chapter 4).

2.8 POSSIBILITY THEORY AND FUZZY PROBABILITIES

In the late 1970s Zadeh advanced a theoretical framework for information and knowledge analysis, called *possibility theory*, emphasizing the quantification of the semantic, context-dependent nature of symbols—that is, *meaning* rather than measures of information. The theory of possibility is analogous, and yet conceptually different from the theory of probability. Probability is fundamentally a measure of the frequency of occurrence of an event. Although there are several interpretations of probability (*subjectivistic, axiomatic,* and *frequentistic*), probabilities generally have a physical event basis. They are tied to statistical experiments and are primarily useful for quantifying how frequently a sample occurs in a population. Possibility theory, on the other hand, attempts to quantify how accurately a sample resembles an *ideal* element of a population. The ideal element is a prototypical class or a category of the population which we think of as a fuzzy set. In a sense, possibility theory may be viewed as a generalization of the theory of probability with the *consistency principle*, which we will see later on, providing a heuristic connection between the two. Possibility theory focuses more on the *imprecision* intrinsic in language, whereas probability theory focuses more on events that are *uncertain* in the sense of being random in nature. In natural language processing, automatic speech recognition, knowledge-based diagnosis, image analysis, robotics, analysis of rare events, information retrieval, and related areas, major problems are encountered on quantifying the meaning of events—that is, the efficacious and accurate interpretation of their significance and consequence and not the extent of their occurrence. Let us illustrate with a simple example.

In the field of reliability analysis, probabilistic methods have been the basic instrument for quantifying equipment and human reliability as well. Two very important concepts used are the *failure rate* and the *error rate*. Knowing the failure rate of a component amounts to knowing the duration of time that the component may be trusted to operate safely, and thus a schedule for replacement and maintenance activities can be devised. It is not unusual, however, that after a component is fixed or replaced, the entire system breaks down, a problem particularly acute with electronic components. Indeed, such general failures sometimes cause extremely negative consequences, leading to catastrophic accidents. The problem here is that failure rates are not sufficiently meaningful to account for the complex interactions that a human being, such as a maintenance technician or an operator, may have with a machine. In addition, the correct estimation of failure rate and error rate requires a large amount of data, which is often not practically possible to obtain. Is is obviously impractical to melt nuclear reactors to collect failure rate data. Thus, in practice, the failure rate and error rate are estimated by experts based on their engineering judgment (Onisawa, 1990); from this point of view, fuzzy possibilities and probabilities (which we will examine momentarily) can be used to model such judgments in a flexible and efficient way. Engineering judgment enters many areas of systems and reliability analysis including estimating the effect of environmental factors, operator stress, dependence between functions or units, selection of sequence of events, expressing the degree of uncertainty involved in the formulation of safety criteria, assuming parameter ranges, and so on (Shinohara, 1976). Alternatives to failure and error rates have been developed employing the notion of possibility measures, called *failure* and *error possibilities*, and have been applied to the reliability analysis of nuclear power plants, structural damage assessments, and earthquake engineering. Failure possibilities and error possibilities are essentially fuzzy sets on the interval [0, 1] that employ the notions we examine in this section.

Over the years, two views, or schools of thought, of the definition of fuzziness have emerged. The first view, which we implicitly held in the previous sections, has to do with categorizing or grouping the elements of a universe of discourse into classes or sets whose boundaries of membership are fuzzy. Thus when we defined the set of *small numbers* in Example 2.1 we identified a category of numbers within the universe of all numbers. Implicitly, what we dealt with in the example was the problem of *imprecision*. Our main problem was to find the membership function that most appropriately or accurately described the category of *small numbers*. The other view of fuzziness has to do with the problem of *uncertainty*. Here our main concern is to quantify the certainty of an assertion such as "a number x is a *small number*," where x is an element of the universe of discourse X of numbers (whose location on X is not known in advance) and is therefore called a *nonlocated element*. Possibility theory was advanced in order to address this type of problem. Possibility is more generally known as a *fuzzy measure*, which is a function assigning a value between 0 and 1 to each crisp set of the universe of discourse, signifying the degree of evidence or belief that a particular element belongs to the set. Other types of fuzzy measure are *belief measures*, *plausibility measures*, *necessity measures*, and *probability measures*. The theory of fuzzy measures was advanced in 1974 by Sugeno as part of his Ph.D. dissertation at Tokyo University. Fuzzy measures subsume probability measures as well as belief and plausibility measures used in what is known as the *Dempster–Shafer Theory of Evidence*.

Let us now take a closer look at *possibility*. Possibility is a *fuzzy measure*, which means that possibility is a function with a value between 0 and 1, indicating the degree of evidence or belief that a certain element x belongs to a set (Zadeh, 1978; Dubois and Prade, 1988). A possibility of 0.3 for element x, for example, may indicate a 0.3 degree of evidence or belief that *x* belongs to a certain set. How this belief is distributed to elements other that *x* is quantified through a *possibility distribution*. In possibility theory, the concept of *possibility distribution* is analogous to the notion of *probability distribution* in probability theory. A possibility distribution is viewed as a fuzzy restriction acting as an elastic constraint on the values that may be assigned to a variable. What does this mean? Well, it is best to review the notion of a *variable*, first. Let *A* be a crisp set defined on a universe *X* and let *V* be a variable taking values on some element *x* of *X*, a situation illustrated in Figure 2.11. The crisp set *A* is what in the parlance of probability we call an *event*. Events are comprised out of one or more basic events. Thus, the element *x* may be thought of as a basic event. If *x* is within *A* and *x* occurs, then we say that the event *A* has occurred as well. For example, in reliability analysis, equipment failure and human error are considered to be events whose occurrence is based on the occurrence of basic events known as *initiating events*. To say that *V takes its values in A* is to indicate that any element (basic event) of event *A* could *possibly* be a value of *V* and that any element outside of *A*, the complement of *A*, cannot be a value of *V*. Thus, the statement *V takes its value in A* can be viewed as inducing a possibility, Π over *X*, associating with each value *x* the possibility that *x* is a value of *V*. This can be written as

$$\Pi(V := x) = \pi_V(x) = \chi_A(x) = \begin{cases} 1 & \text{if } x \in A \\ 0 & \text{if } x \notin A \end{cases} \qquad (2.8\text{-}1)$$

where " := " is an assignment symbol indicating that *x* is assigned to the

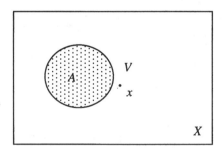

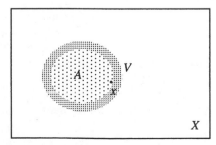

Figure 2.11 The statement about a variable *V*, "*V takes its values in A*," has a different meaning when the set *A* is crisp (*top*) than when the set *A* is fuzzy (*bottom*).

variable V, and $\pi_V(x)$ is the *possibility distribution* associated with V (or the *possibility distribution function* of Π). In equation (2.8-1), $\chi_A(x)$ is the characteristic function of A (see Section 2.1). Mathematically, Π is considered a *measure* which is a special function mapping the universe to the interval $[0, 1]$. Knowing that the values that V may take are members of A is the same as knowing which values of the universe X are restricted to be values of V and which are restricted not to be values of V. We indicated this in equation (2.8-1) by using the *characteristic function* of the crisp set A. We think of the crisp set A as a *restriction* on the values of the variable V, and in view of the nonfuzzy nature of A this type of restriction is called a *crisp restriction*.

Next, suppose that A is a fuzzy set and that its boundary no longer crisp (i.e., does not sharply divide members from nonmembers) but is instead a fuzzy boundary allowing an element x to be a member of A to some degree. As with any fuzzy set, A is uniquely identified by its membership function $\mu_A(x)$. In terms of events we think of A as a fuzzy event, and we can associate with each basic event x a membership function indicating its membership to A. Let us again consider a variable V whose arguments are elements of X[9]. Now suppose that V is constrained to take values on X. The fuzzy set A also restricts the possible values that the variable V may take, but in a fuzzy manner—that is, to a degree. In such a case we consider the fuzzy set A to act as a *fuzzy restriction* on the possible values of V. Generalizing equation (2.8-1) to the fuzzy case we say that the fuzzy set A induces a possibility Π. The associated possibility distribution $\pi_V(x)$ on the values that V may assume is defined to be equal to the membership function of A, $\mu_A(x)$ and is written as

$$\Pi(V := x) = \pi_V(x) = \mu_A(x) \qquad (2.8\text{-}2)$$

Thus, the possibility that V is assigned x—that is, $V := x$, which is sometimes indicated as "V is x"—is postulated to be equal to the membership function of A evaluated at x—that is, $\mu_A(x)$. It is important to observe in equation (2.8-2) that *possibility distributions* are fuzzy sets, while *possibilities* are just numbers between 0 and 1. The possibility Π in (2.8.2) is a measure of the compatibility of a given crisp value x that V may take with an *a priori* defined set A. In this way, V becomes a variable associated with the *possibility distribution* $\pi_V(x)$ in much the same way as a random variable is associated with the probability distribution.

What equation (2.8-2) indicates is that in certain situations, such as in the definition of failure and error possibilities, it is of interest to interpret the membership function $\mu_A(x)$ of a fuzzy set as a *possibility distribution* of a variable V. In this sense the fuzzy set A is viewed as the set of more or less possible values for V.

---

[9]In Chapter 5 the variable V will be generalized to a *fuzzy variable*, which is a variable that takes fuzzy sets as values.

Given a possibility distribution $\pi_V(x)$, the possibility that x may belong to another crisp set B is defined as

$$\Pi(V \subset B) = \bigvee_{x \in B} \pi_V(x) \qquad (2.8\text{-}3)$$

What equation (2.8-3) indicates is that the possibility of B is the possibility of the most possible elementary event x of B. Generalizing this relationship, it can be shown (Dubois and Prade, 1988; Kandel, 1986) that the possibility measure of the union of two crisp sets B and C is the maximum of the possibilities of B and C and can be written as

$$\Pi(B \cup C) = \Pi(B) \vee \Pi(C) \qquad (2.8\text{-}4)$$

Given a fuzzy set A and a possibility distribution function, $\pi_V(x)$, the possibility of A, denoted as $\Pi(A)$, is given by

$$\Pi(A) = \bigvee_{x \in X} [\mu_A(x) \wedge \pi_V(x)] \qquad (2.8\text{-}5)$$

Consider two fuzzy events A and B defined over the universe of discourse X. The possibility of A with respect to B is defined as

$$\Pi(A|B) = \bigvee_{x \in X} [\mu_A(x) \wedge \mu_B(x)] \qquad (2.8\text{-}6)$$

The possibility measure of A with respect to B reflects the extent to which A and B coincide or overlap. Thus, possibility may be viewed as a measure of comparison of fuzzy sets.

Conditional possibilities have been defined in analogy with conditional probabilities; an entire body of theoretical results has been achieved, known generally as *possibility theory*. It is finding an increasing number of applications in the fields of knowledge representation and applied artificial intelligence (Ragheb and Tsoukalas, 1988). A very comprehensive treatment of possibility may be found in the book entitled *Possibility Theory* by Dubois and Prade (1988). The theory of possibility has assumed particular significance in the field of natural language processing due to the inherent fuzziness of natural language. In the late 1970s Zadeh constructed a universal language called *PRUF*, in which the translation of a proposition expressed in natural language takes the form of a procedure for computing the possibility distribution of a set of fuzzy relations in a database. The procedure, then, may be interpreted as a semantic computation transforming the meaning of a proposition to a computed possibility distribution quantifying the information conveyed by the proposition (Zadeh, 1983).

There are certain differences between probability and possibility measures worth pointing out. Possibility measures are "softer" than probability measures, and the interpretation of probability and possibility is quite different. Probability is used to quantify the frequency of occurrence of an event, while possibility (along with fuzzy tools) is used to quantify the *meaning* of an event. Consider the following example offered by Zadeh (1978). Suppose that we have the proposition "*Hans ate V eggs for breakfast*," where $V = \{1, 2, 3, \ldots\}$. A *possibility distribution* and a *probability distribution* may be associated with V, as shown in the following table:

x	1	2	3	4	5	6	7	8	9
$\pi_V(x)$	1	1	1	1	0.8	0.6	0.4	0.2	0.1
$p_V(x)$	0.1	0.8	0.1	0	0	0	0	0	0

The possibility distribution is interpreted as the *degree of ease* with which Hans can eat x eggs, while the probability distribution might have been determined by observing Hans at breakfast for 100 days. Note that the probability distribution function $p_V(x)$ is given a *frequentistic* interpretation and that it sums to '1', while the possibility distribution function $\pi_V(x)$ is imputed with a *situation* or *context-dependent* interpretation and does not have to sum to '1'.

Possibility is an upper bound for probability: A high degree of possibility does not imply a higher degree of probability. If, however, an event is not possible, it is not also probable. This is referred to as the *probability/possibility consistency principle* (Zadeh, 1978). This heuristic principle is useful for drawing a distinction between the *objectivistic* use of probability measures and the *subjectivist* use of possibility or fuzzy measures. When we attempt to use the two to describe a similar thing, we can use the *possibility/probability consistency principle* as a guide. Possibility measures are more flexible measures useful for epistemic (i.e., cognitive) or context-dependent descriptions. In general, according to Zadeh a variable may be associated with both a possibility distribution and a probability distribution, with the weak connection between the two given by the consistency principle (Zadeh, 1978).

In the language of probability theory the set A in Figure 2.11 may be viewed as a *fuzzy event*. Such a fuzzy event induces a distribution on the values of a variable which we called the possibility distribution function and defined in equation (2.8-2). We can also define the *probability of a fuzzy event* A. Suppose that a fuzzy event A is comprised of elementary events x, and with each x we associate a basic probability $p(x)$.

Zadeh defined the *probability of fuzzy event* A as the mathematical expectation (the first moment) of its membership function, that is,

$$P(A) = \frac{\int_X \mu_A(x) p(x) \, dx}{\int_X p(x) \, dx} \tag{2.8-7}$$

where A is a fuzzy event on the universe X, x is an element of X, also called an elementary event, and $p(x)$ is a probability distribution (Zadeh, 1968). When A is not a fuzzy event, equation (2.8-7) reduces back to the usual crisp probability $P(A)$. In equation (2.8-7) we assume that the probability measure on the entire universe of discourse must equal unity—that is, $\int_X p(x)\,dx = 1$.

In addition, given equation (2.8-7) we can define a *fuzzy mean* as

$$m_A = \frac{1}{P(A)} \int_X x \mu_A(x) p(x)\,dx \tag{2.8-8}$$

and a *fuzzy variance* as

$$\sigma_A^2 = \frac{1}{P(A)} \int_X (x - m_A)^2 \mu_A(x) p(x)\,dx \tag{2.8-9}$$

The probability of a fuzzy event as defined in equation (2.8-7) has been an extremely useful notion with wide application in the field of quantification theory (Terano et al., 1992). Quantification methods are useful in analyzing data involving human judgments which are not normally given numerical expression, as well as in interpreting and understanding such data.

Example 2.3   Possibility Measures and Distributions. Let us illustrate the distinction between *possibility measure* or *possibility* and *possibility distribution*. We consider a possibility distribution induced by the proposition "*V is a small integer*" where the possibility distribution is (subjectively) defined as

$$\pi_V(x) = 1.0/1 + 1.0/2 + 0.75/3 + 0.5/4 + 0.3/5$$
$$+ 0.3/6 + 0.1/7 + 0.1/8 \tag{E2.3-1}$$

We also consider the crisp set $A = \{3, 4, 5\}$ which we can write as

$$A = \sum_{x \in X} \mu_A(x)/x = 1/3 + 1/4 + 1/5 \tag{E2.3-2}$$

What is the possibility of A? The *possibility measure* $\Pi(A)$ is found using equation (2.8-5); that is,

$$\Pi(A) = \bigvee_{x \in X} [\mu_A(x) \wedge \pi_V(x)] \tag{E2.3-3}$$

Using equations (E2.3-1) and (E2.3-2) in (E2.3-3), we can obtain the possibility of A:

$$\Pi(A) = 0.75 \vee 0.5 \vee 0.3 = 0.75 \tag{E2.3-4}$$

For another fuzzy set $B = \{$*integers that are not small*$\}$ given by

$$B = 0.2/3 + 0.3/4 + 0.6/5 + 0.8/6 + 1.0/7$$

using equation (E2.3-3), we could obtain that the possibility of B is

$$\Pi(B) = 0.2 \vee 0.3 \vee 0.3 \vee 0.3 \vee 0.1 = 0.3 \qquad \text{(E2.3-5)}$$

It should be noted in equations (E2.3-4) and (E2.3-5) that the possibility is simply a number between 0 and 1, whereas the possibility distribution is a fuzzy set—for example, equation (E2.3-1).

Let us now consider a simple instance of how to generate the possibility distribution itself. Let $C = 1/1 + 1/2 + 0.8/3 + 0.6/4 + 0.4/5 + 0.2/6$ be a fuzzy set that represents *small numbers*. Then the proposition *"V is a small number"* associates with V the possibility distribution, $\pi_V(x)$, taken in view of equation (2.8-2) to be equal to the membership function of C—that is,

$$\pi_V(x) = 1/1 + 1/2 + 0.8/3 + 0.6/4 + 0.4/5 + 0.2/6 \quad \text{(E2.3-6)}$$

In equation (E2.3-6) a singleton such as $0.6/4$ indicates that the possibility that x is 4, given that x is a *small integer*, is 0.6. □

REFERENCES

Dubois, D., and Prade, H., *Fuzzy Sets and Systems: Theory and Applications*, Academic Press, Boston, 1980.

Dubois, D., and Prade, H., *Possibility Theory*, Plenum Press, New York, 1988.

Kandel, A., *Fuzzy Mathematical Techniques with Applications*, Addison-Wesley, Reading, MA, 1986.

Kaufmann, A., *Introduction to the Theory of Fuzzy Subsets*, Volume I, Academic Press, New York, 1975.

Klir, G. J., and Folger, T. A., *Fuzzy Sets, Uncertainty, and Information*, Prentice-Hall, Englewood Cliffs, NJ, 1988.

Kosko, B., *Neural Networks and Fuzzy Systems*, Prentice-Hall, Englewood Cliffs, NJ, 1992.

Onisawa, T., An Application of Fuzzy Concepts to Modelling of Reliability Analysis, *Fuzzy Sets and Systems*, No. 37, pp. 267–286, North-Holland, Amsterdam, 1990.

Pal, S. K., and Majumder, D. K. D., *Fuzzy Mathematical Approach to Pattern Recognition*, John Wiley & Sons, New York, 1986.

Radecki, T., Level Fuzzy Sets, *Journal of Cybernetics*, Vol. 7, pp. 189–198, 1977.

Ragheb, M., and Tsoukalas, L. H., Monitoring Performance of Devices Using a Coupled Probability–Possibility Method, *International Journal of Expert Systems*, Vol. 1, pp. 111–130, 1988.

Shinohara, Y., Fuzzy Set Concepts for Risk Assessment, *International Institute for Applied Systems Analysis*, Report WP-76-2, Laxenburg, Austria, January 1976.

Terano, T., Asai, K., and Sugeno, M., *Fuzzy Systems Theory and Its Applications*, Academic Press, Boston, 1992.

Zadeh, L. A., Fuzzy Sets, *Information and Control*, Vol. 8, pp. 338–353, 1965.

Zadeh, L. A., Probability Measure of Fuzzy Events, *Journal of Mathematical Analysis and Applications*, Vol. 23, pp. 421–427, 1968.

Zadeh, L. A., Outline of a New Approach to the Analysis of Complex Systems and Decision Processes, *IEEE Transactions on Systems, Man and Cybernetics*, SMC-3, pp. 28–44, 1973.

Zadeh, L. A., The Concept of a Linguistic Variable and its Application to Approximate Reasoning, *Information Sciences*, Vol. 8, pp. 199–249, 1975.

Zadeh, L. A., Fuzzy Sets as a Basis for Theory of Possibility, *Fuzzy Sets and Systems*, Vol. 1, pp. 3–28, 1978.

Zadeh, L. A., A Computational Approach to Fuzzy Quantifiers in Natural Languages, *Computer and Mathematics*, Vol. 9, pp. 149–184, 1983.

Zadeh, L. A., Fuzzy Logic, *IEEE Computer*, pp. 83–93, April 1988.

Zimmermann, H. J., *Fuzzy Set Theory and its Applications*, Kluwer-Nijhoff, Boston, 1985.

PROBLEMS

1. What happens to the curves in Figure 2.5 if we set $F_2 = 40$ and vary F_1 as in the figure?

2. In Figure 2.6, what is the significance of the intersection between the $\mu_A = 0.5$ line and the curves?

3. In Example 2.2, substitute $y = \sin x$ for equation (E2.2-1) and utilize the extension principal in the same way as in the example. Choose an appropriate range for x and assume any additional information needed as in the example.

4. The fuzzy variable of Figure 2.9 is given by the equation $\mu_A(x) = 1/[1 + 0.3(x - 50)^2]$. Show that the 0.2 level fuzzy set of fuzzy set A can be represented by α-cuts using the resolution principal.

5. The fuzzy sets A and B are given by

$$A = 0.33/6 + 0.67/7 + 1.00/8 + 0.67/9 + 0.33/10$$
$$B = 0.20/3 + 0.60/4 + 1.00/5 + 0.60/6 + 0.20/7$$

   (a) Write an expression for $A \vee B$.
   (b) Write an expression for $A \wedge B$.

6. Different fuzzy symbols are often used to mean similar things.
   (a) Write all symbols or terms that have the same general meaning as *max* ($\vee$).
   (b) Write all symbols or terms that have the same general meaning as *min* ($\wedge$).

7. Given fuzzy set A, describing *pressure p is higher than* 15 *mPa*, through the membership function:

$$\mu_A(x) = \frac{1}{1 + (x - 15)^{-2}} \quad x > 15,$$
$$= 0 \quad\quad\quad\quad\quad x \le 15,$$

and fuzzy set B, describing *pressure p is approximately equal to* 17 *mPa*, with membership function:

$$\mu_B(x) = \frac{1}{1 + (x - 17)^4}.$$

Find the membership function of the fuzzy set C, describing *pressure p is higher than* 15 *mPa and approximately equal to* 17 *mPa*. Use at least four different norms for interpreting *AND* (see Appendix) and draw all membership functions.

8. Using the data given in Problem 7, find the membership function of the fuzzy set D, describing *pressure p is higher than* 15 *mPa or approximately equal to* 17 *mPa*. Use at least four different norms for interpreting *OR* (see Appendix) and draw all membership functions.

9. Using the data given in Problem 7, find the membership function of the fuzzy set E, describing *pressure p is not higher than* 15 *mPa and approximately equal to* 17 *mPa*. Use four different norms for interpreting *AND* (see Appendix) and draw all membership functions.

10. Determine all α-cuts for the following fuzzy sets, given that $\alpha = 0.0, 0.1, 0.2, \ldots 0.9, 1.0$.

I. $A = 0.1/3 + 0.2/4 + 0.3/5 + 0.4/6 + 0.5/7 + 0.6/8 + 0.7/9 + 0.8/10 + 1.0/11 + 0.8/12$

II. $B = \int_{-\infty < x < +\infty} \left[\dfrac{1}{1 + (x - 15)^{-2}} \right] /x$

Write a MATLAB program that takes a number of α-cuts (minimum 10) and reconstructs the membership function.

11. Let $X = N \times N$, and the fuzzy sets:

$$\mu_A(x) = \frac{1}{1 + 10(x - 2)^2}$$

$$\mu_B(y) = \frac{1}{1 + 2y^2}$$

Let the mappings $z = f(x, y)$, $f: N \times N \rightarrow N$ be the following quadric surfaces

(a) $z = \sqrt{\dfrac{x^2}{4} + \dfrac{y^2}{2}}$, $x \in A$, $y \in B$.

(b) $\dfrac{x^2}{9} + \dfrac{y^2}{15} - \dfrac{z^2}{8} = 1$

(c) $2y^2 + 12z^2 = x^2$

Sketch the surfaces and determine the image $f(A \times B)$ by the extension principle, for each of the above.

3

FUZZY RELATIONS

3.1 INTRODUCTION

In fuzzy approaches, *relations* possess the computational potency and significance that *functions* possess in conventional approaches. Fuzzy *if/then* rules and their aggregations, known as *fuzzy algorithms*, both of central importance in engineering applications, are fuzzy relations in linguistic disguise. Fuzzy relations may be thought of as fuzzy sets defined over high-dimensional universes of discourse. As the name indicates, a relation implies the presence of an association between elements of different sets. If the degree of association is either 0 or 1, we have *crisp relations*. If the degree of association is between 0 and 1, we have *fuzzy relations*; a number between 0 and 1 is taken to indicate partial absence or presence of association. In this chapter we begin by reviewing crisp relations and various ways for representing them. Next, we look at fuzzy relations and properties used to classify them, and finally we come to *composition of fuzzy relations*, a very important tool for approximate reasoning with applications in the fields of expert systems, control, and diagnosis.

On what basis do we associate various elements in a relation? The association may be due to a common property, a quality, a reference, a condition, or a rule, satisfied by pairs of elements (e.g., objects, numbers, words, variables, etc.). For example, the statements *"is greater than"* or *"is a component of"* indicate an association between two elements. The order of the elements is important. For instance, if the relation *"is a component of"* holds for the pair of elements (*u-tube, steam-generator*)—that is, if the statement *"u-tube is a component of steam-generator"* is true—the relation may no longer be true when the elements are interchanged. The relation *"steam generator is a component of u-tube"* is not true. Thus, this is an important point to observe: In relations, *order* is important!

A relation such as *"is a component of"* may also be expressed as an *if/then* rule. We can say *"if an object is a u-tube, then it is a component of a steam generator."* Any ambiguity as to what degree an object is known to be a *u-tube* or a *steam generator*, or any ambiguity as to the degree of truth in such an association, results in a fuzzy relation.

When two elements belong to a relation R, we refer to them as an *ordered pair* denoted as $(a, b) \in R$, or aRb, with element a being distinguished as the first element and b as the second. With two elements in association, we have *binary relations*. With three elements we have *tertiary relations*, and when n elements are in association we have *n-ary relations*. An association of n elements in an *n-ary* relation is called *n-tuple*. A relation is any *set* of ordered *n-tuples*. The keyword here is *"set."* Relations are formed out of sets of elements, and they are sets themselves.

Crisp relations are defined over the *Cartesian product* or *product space* of two or more sets. The *Cartesian product* $X \times Y$ of two sets X and Y is the set of all ordered pairs (x, y) with x in X and y in Y. The product $X \times X$ is often abbreviated as X^2, the product $X^2 \times X$ as X^3, and so on.

We saw that relations are sets where order is important. But relations may also be thought of as *mappings*, with the process of association in mathematics being called a *mapping*. *Functions* are *mappings* as well. Relations, however, are a more general type of mapping. A function performs what is called a *many-to-one mapping*; that is, many elements are associated with one (and only one) element but not vice versa. For example, if the mapping is done between x's and y's in the $X \times Y$ plane, we may have more than one x mapped to the same y but not the other way around. Relations, however, perform *many-to-many* mappings. Many x's can be associated with a single y and vice versa. Many y's can also be associated with a single x. The importance of this abstract-sounding distinction in terms of engineering and computational applications cannot possibly be overstated, as we will see in later chapters. But for the moment let us turn our attention to an example of a crisp relation in order to see some of the ways that relations may be represented.

Example 3.1 A Crisp Relation. Let us consider a *divisibility relation*, R_d, on the set $S = \{1, 2, 3, 4, 6\}$ defined by the statement *"x divides y."* R_d is a *binary* relation because it involves two elements, x and y, drawn from the Cartesian product of the set S with itself—that is, $S \times S$. Furthermore, it is a *crisp* relation since a number either divides another number or not (assuming integer division only). It is easy to list all the pairs of the relation and to see that the relation itself is a set, namely, the crisp set of all the pairs

$$R_d = \{(1,1), (1, 2), (1, 3), (1, 4), (1, 6), (2, 2), (2, 4), (2, 6),$$
$$(3, 3), (3, 6), (4, 4), (6, 6)\} \qquad \text{(E3.1-1)}$$

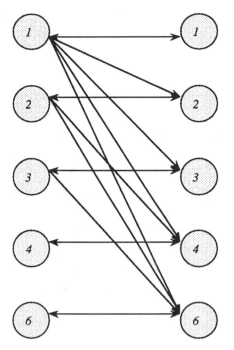

Figure 3.1 The directed graph of the divisibility relation R_d defined on the Cartesian product $S \times S$ of the set $S = \{1, 2, 3, 4, 6\}$.

where the meaning of the elements inside the parentheses is "*1 divides 1*," and so on. The relation R_d can also be represented through a graph as shown in Figure 3.1. The individual elements are represented by circles, called the *vertices* of the graph. If R_d is true for two elements, we connect them by an arrow, with the direction of the arrow indicating the order of the elements in the relation. For example, given that 3 divides 6, there is an arrow going from 3 to 6; and since 6 does not divide 3, there is no arrow going from 6 to 3. Reflecting the fact that the order of elements or the directions of the arrows is important, we call this a *directed graph*.

The binary relation R_d may also be represented by a table or a matrix. Table 3.1 shows the tabular representation of R_d. When a table entry is 1, it indicates that x (row entry) divides the corresponding y (column entry); for example, in the fourth row and fourth column we simply have that the element 4 divides itself. A 0 indicates the absence of such a relation. Should the divisibility relation have been a fuzzy relation, the table entries would be numbers between 0 and 1 as we will see later on.

R_d can also be represented by a matrix obtained from Table 3.1 by removing the column of x's on the side and the row of y's from the top; that is,

$$R_d = \begin{bmatrix} 1 & 1 & 1 & 1 & 1 \\ 0 & 1 & 0 & 1 & 1 \\ 0 & 0 & 1 & 0 & 1 \\ 0 & 0 & 0 & 1 & 0 \\ 0 & 0 & 0 & 0 & 1 \end{bmatrix} \qquad \text{(E3.1-2)}$$

Table 3.1   A tabular representation of the divisibility relation in Example 3.1

R_d:

x \ y	1	2	3	4	6
1	1	1	1	1	1
2	0	1	0	1	1
3	0	0	1	0	1
4	0	0	0	1	0
6	0	0	0	0	1

Thus we have seen five different ways for representing R_d:

1. *Linguistically*, through the statement "*x divides y*"
2. By listing the *set* of all ordered pairs as in equation (E3.1-1)
3. As a *directed graph* (Figure 3.2)
4. As a *table* (Table 3.1)
5. As a *matrix*, equation (E3.1-2)

It should be noted that the last two ways are generally convenient only for *binary* relations. For *tertiary* relations, for example, we would need a three-dimensional table or matrix (for *n-ary* relations *n*-dimensional tables and matrices), and therefore tables and matrices may be conveniently used only with *binary* relations. ☐

3.2   FUZZY RELATIONS

In fuzzy relations we consider *pairs* of elements, and more generally *n-tuples*, that are related to *a degree*. Just as the question of whether some element belongs to a set may be considered a matter of degree, whether some elements are associated may also be a matter of degree (Zadeh, 1971; Dubois and Prade, 1980). For example, suppose we have a diagnosis problem involving vibration data with a set of faults $F = \{f_1, \ldots, f_n\}$ associated to a set of symptoms $S = \{s_1, \ldots, s_m\}$. First we need to establish how symptoms relate to faults—that is, establish a relation from F to S. One of these symptoms, let's say s_i, may be "excessive vibration." Knowing whether a machine vibrates depends on the interpretation of vibration data. If the concept of "excessive vibration" has been crisply defined—that is, it can be readily determined whether the machine vibrates and we can associate a symptom s_i with a fault f_j—we have a crisp relation from F to S. In reality, however, it may be rather difficult to crisply define such associations and hence all faults $F = \{f_1, \ldots, f_n\}$ and all symptoms $S = \{s_1, \ldots, s_m\}$ may be associated to a degree, giving us a fuzzy relation from F to S. What is important in such cases is to compute these degrees. Having established the fuzzy relation from F to S, we can subsequently use it to identify the highest degrees of association given a symptom s_i so that it may be linked to faults f_k, f_j, and so forth (Kaufmann, 1975).

Fuzzy relations are fuzzy sets defined on Cartesian products. Whereas the fuzzy sets we encountered in the previous chapter were defined on a single universe of discourse (e.g., X), fuzzy relations are defined on higher-dimensional universes of discourse (e.g., $X \times X$ or $X \times Y \times Z$). A Cartesian product for us is simply a higher-dimensional universe of discourse. Suppose that we have a binary fuzzy relation R defined on $X \times Y$. As with any fuzzy set, we can list all pairs of the relation explicitly as we did in equation (2.2-2); that is,

$$R = \{((x, y), \mu_R(x, y))\} \tag{3.2-1}$$

where every individual pair (x, y) belongs to the Cartesian product $X \times Y$. Alternatively, we can use the notation of equation (2.2-3) to form the union of all $\mu_R(x, y)/(x, y)$ singletons of $X \times Y$. For a discrete Cartesian product we would have

$$R = \sum_{(x_i, y_j) \in X \times Y} \mu_R(x_i, y_j)/(x_i, y_j) \tag{3.2-2}$$

while for a continuous Cartesian product we have

$$R = \int_{X \times Y} \mu_R(x, y)/(x, y) \tag{3.2-3}$$

The same notation is used for any *n-ary fuzzy relation*.

So much for the fuzzy set nature of fuzzy relations and notation. Let us now take a look at alternative ways of representing them. One of them, which is particularly useful for the composition of relations (see Section 3.5), is to form a matrix of grades of membership in a manner analogous to (E3.1-2), only now we have instead of 0's and 1's various numbers between 0 and 1.

The *membership matrix* of an $n \times m$ binary fuzzy relation has the general form

$$
R = \begin{bmatrix} \mu_R(x_1, y_1) & \mu_R(x_1, y_2) & \cdots & \mu_R(x_1, y_n) \\ \mu_R(x_2, y_1) & \mu_R(x_2, y_2) & \cdots & \mu_R(x_2, y_n) \\ \vdots & & & \\ \mu_R(x_m, y_1) & \mu_R(x_m, y_2) & \cdots & \mu_R(x_m, y_n) \end{bmatrix} \tag{3.2-4}
$$

Let us take a look at some special relations and their membership matrices. The *identity fuzzy relation*, R_I, is a special type of relation which has 1 in all diagonal elements and 0 in all off-diagonal elements—that is,

$$
R_I = \begin{bmatrix} 1 & 0 & & 0 \\ 0 & 1 & & 0 \\ & & \ddots & \\ 0 & 0 & & 1 \end{bmatrix} \tag{3.2-5}
$$

Another special relation is the *universe relation*, R_E, namely a relation with 1 everywhere in its membership matrix—that is,

$$
R_E = \begin{bmatrix} 1 & 1 & \cdots & 1 \\ 1 & 1 & & 1 \\ \vdots & & \ddots & \\ 1 & 1 & & 1 \end{bmatrix} \tag{3.2-6}
$$

The *null relation*, R_0, has a membership matrix with 0 everywhere—that is,

$$
R_0 = \begin{bmatrix} 0 & 0 & \cdots & 0 \\ 0 & 0 & & 0 \\ \vdots & & \ddots & \\ 0 & 0 & & 0 \end{bmatrix} \tag{3.2-7}
$$

The transpose of a membership matrix gives the membership matrix of the *inverse relation* of R denoted by R^{-1} and defined by

$$
\mu_{R^{-1}}(y, x) \equiv \mu_R(x, y) \tag{3.2-8}
$$

Thus the *inverse* of the relation represented by the matrix of equation (3.2-4) has the membership matrix

$$
R^{-1} = \begin{bmatrix} \mu_R(x_1, y_1) & \mu_R(x_2, y_1) & \cdots & \mu_R(x_m, y_1) \\ \mu_R(x_1, y_2) & \mu_R(x_2, y_2) & \cdots & \mu_R(x_m, y_2) \\ \vdots & & & \\ \mu_R(x_1, y_n) & \mu_R(x_2, y_n) & \cdots & \mu_R(x_m, y_n) \end{bmatrix} \tag{3.2-9}
$$

which is the transpose of the matrix found by interchanging the rows of R to produce the columns of R^{-1}, and the columns of R have become the rows of R^{-1} (Klir and Folger, 1988; Terano et al., 1992). The inverse of an inverse relation is the original relation just as the inverse of the inverse of a matrix is the original matrix—that is,

$$(R^{-1})^{-1} = R \qquad (3.2\text{-}10)$$

So far we defined fuzzy relations on crisp Cartesian products. However, fuzzy relations can also be defined on fuzzy Cartesian products (Kandel, 1986; Klir and Folger, 1988). Although fuzzy relations defined over fuzzy sets are of interest, particularly in connection with decision making under uncertainty, we will make no actual use of them in this book. Unless otherwise indicated, fuzzy relations in this book are assumed to be defined over crisp Cartesian products.

Example 3.2 Representing a Fuzzy Relation. Let us take two discrete sets $X = \{x_1, x_2, x_3, x_4\}$ and $Y = \{y_1, y_2, y_3, y_4\}$ and define (subjectively) on their Cartesian product the fuzzy relation $R = $"$x$ *is similar* to y," shown by the directed graph of Figure 3.2. R may be represented through the five different ways we saw in Example 3.1 with regard to crisp relations:

1. Linguistically, for example by the statement "*x is similar to y*"
2. By listing (or taking the union of) all fuzzy singletons
3. As a *directed graph* (Figure 3.2)
4. In *tabular* form
5. As a *matrix*

Let us represent the relation as a *fuzzy set* by taking the union of all singletons—that is, all ordered pairs and their membership values:

$$R = \int_{X \times Y} \mu_R(x, y)/(x, y) \qquad (\text{E3.2-1})$$

Using the data of Figure 3.2, equation (E3.2-1) gives

$$R = 1.0/(x_1, y_1) + 0.3/(x_1, y_2) + 0.9/(x_1, y_3) + 0.0/(x_1, y_4)$$
$$+ 0.3/(x_2, y_1) + 1.0/(x_2, y_2) + 0.8/(x_2, y_3) + 1.0/(x_2, y_4)$$
$$+ 0.9/(x_3, y_1) + 0.8/(x_3, y_2) + 1.0/(x_3, y_3) + 0.8/(x_3, y_4)$$
$$+ 0.0/(x_4, y_1) + 1.0/(x_4, y_2) + 0.8/(x_4, y_3) + 1.0/(x_4, y_4)$$
$$(\text{E3.2-2})$$

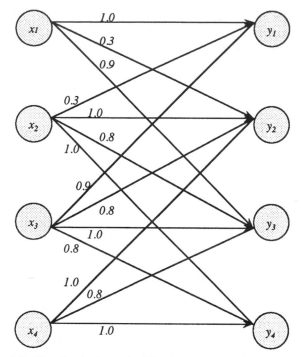

Figure 3.2 The directed graph of the fuzzy relation R in Example 3.2.

The relation R may also be represented in tabular form as

R:

	y_1	y_2	y_3	y_4
x_1	1.0	0.3	0.9	0.0
x_2	0.3	1.0	0.8	1.0
x_3	0.9	0.8	1.0	0.8
x_4	0.0	1.0	0.8	1.0

Note that compared to Table 3.1, where we only used 0's and 1's, in the tabular representation of R we find grades of membership between 0 and 1. Consider the pair (x_3, y_4). From the table of R we see that "x_3 *is similar to* y_4" is true to a 0.8 degree.

In matrix form, R is given by

$$R = \begin{bmatrix} 1.0 & 0.3 & 0.9 & 0.0 \\ 0.3 & 1.0 & 0.8 & 1.0 \\ 0.9 & 0.8 & 1.0 & 0.8 \\ 0.0 & 1.0 & 0.8 & 1.0 \end{bmatrix} \qquad \text{(E3.2-3)}$$

The inverse of R, which we denote as R^{-1}, is the transpose of the membership matrix of equation (E3.2-3), given by

$$R^{-1} = \begin{bmatrix} 1.0 & 0.3 & 0.9 & 0.0 \\ 0.3 & 1.0 & 0.8 & 1.0 \\ 0.9 & 0.8 & 1.0 & 0.8 \\ 0.0 & 1.0 & 0.8 & 1.0 \end{bmatrix} \qquad \text{(E3.2-4)}$$

Of course the inverse fuzzy relation R^{-1} in this case has the same membership matrix due to the fact that R is a symmetric relation (see next section).

□

3.3 PROPERTIES OF RELATIONS

Crisp and fuzzy relations alike are classified on the basis of the mathematical properties they possess. We present here a brief introduction to the subject of properties mostly for the sake of reference. We look first at properties of crisp relations and then examine the properties of fuzzy relations. In fuzzy relations, different properties call for different requirements for the membership function of a relation.

Let S be a Cartesian product (e.g., $S = X \times Y$, with x being an element of X and y being an element of Y) and let R be a relation on S. The relation R could have the following properties:

Reflexive. We say that a relation R is *reflexive* if for any arbitrary element x in S we have that xRx is *valid*—that is, the pair (x, x) also belongs to the relation R.

Antireflexive. A relation R is *antireflexive* if there is no x in S for which xRx is valid.

Symmetric. A relation R is *symmetric* if for all x and y in S, the following is true: If xRy holds, then yRx is valid also.

Asymmetric. A relation R is *asymmetric* if there are no elements x and y in S such that both xRy and yRx are valid.

Antisymmetric. A relation R is *antisymmetric* if for all x and y in S when xRy is valid and yRx is also valid, then $x = y$.

Transitive. A relation R is called *transitive* if the following is true for all x, y, z in S: If xRy is valid and yRz is also valid, then xRz is valid as well.

Connected. A relation R is *connected* when for all x, y in S the following is true: If $x \neq y$, then either xRy is valid or yRx is valid.

Left Unique. A relation R is called *left unique* when for all x, y, z in S the following is true: If xRz is valid and yRz is also valid, then we can infer that $x = y$.

Right Unique. A relation R is *right unique* when for all x, y, z in S the following is true: If xRy and xRz hold true, then $y = z$.

Right Biunique. A relation R which is both *left unique* and *right unique* is called *biunique.*

Relations are classified into different groups on the basis of these properties. For example, an important type of crisp relation is the so-called *equivalence relation.* An equivalence relation is a relation that is *reflexive, symmetric,* and *transitive* (Klir and Folger, 1988). Equivalence relations are found in every corner of mathematics and are particularly useful in engineering fields such as pattern recognition, measurement, and control. Other important relations are the so-called *order relations.* For example, a relation R is called a *partial ordering* if it is *reflexive, transitive,* and *antisymmetric.* If R is also *connected,* then it is called a *total linear ordering.* Order relations are very important in fuzzy arithmetic (Kaufmann and Gupta, 1991).

The properties of fuzzy relations are described in terms of various requirements for their membership function. In a pioneering paper on the subject, Zadeh (1971) showed that most of the important properties of crisp relations stated above are extended to fuzzy relations as well. Let a relation R be a fuzzy relation on the Cartesian product $S = X \times X$. *Reflexivity, symmetry,* and *transitivity* are the three most important properties that help us properly categorize fuzzy relations. R is a *reflexive* relation if for all x in X we have that

$$\mu_R(x, x) = 1 \qquad (3.3\text{-}1)$$

If for at least one x in X but not for all x's, equation (3.3-1) is not true the relation R is called *irreflexive.* If equation (3.3-1) is not satisfied for any x, then R is called *antireflexive.*

A fuzzy relation R is *symmetric* if order is not important—that is, if we can interchange x's and y's. In terms of the membership function of R, this is equivalent to saying that

$$\mu_R(x, y) = \mu_R(y, x) \qquad (3.3\text{-}2)$$

If equation (3.3-2) is not satisfied for some pairs (x, y), then we say that R is *antisymmetric*; if it is not satisfied for all pairs (x, y), then we say that the relation R is *asymmetric.*

A fuzzy relation R on the Cartesian product $X \times X$ is *max–min transitive* if for two pairs (x, y) and (y, z) both in $X \times X$ we have

$$\mu_R(x, y) \geq \bigvee_z [\mu_R(x, z) \wedge \mu_R(z, y)] \qquad (3.3\text{-}3)$$

where all the maxima with respect to z are taken for all the mimima inside the brackets in equation (3.3-3). *Transitivity* can be defined for other operations such as *product* $(\cdot)$ instead of min $(\wedge)$ in equation (3.3-3); in such a case we have what is called *max-product transitivity*. A relation that does not satisfy equation (3.3-3) for all pairs is called *nontransitive*, and if it fails to satisfy (3.3-3) for all pairs, then it is called *antitransitive*.

A fuzzy relation that is *reflexive* and *symmetric* is called a *proximity* or *tolerance relation*. A fuzzy relation that is *reflexive, symmetric,* and *transitive* is called a *similarity* relation, which is the fuzzy generalization of the *equivalence* property of crisp relations (Zadeh, 1971). Similarity relations are very important in fuzzy logic, and together with proximity relations they are crucially important in the field of fuzzy diagnosis. A *fuzzy ordering* is a fuzzy transitive relation. If a fuzzy relation is *reflexive, transitive,* and *antisymmetric,* then we call it a *fuzzy partial ordering*. Fuzzy orderings and similarity relations may be resolved into nonfuzzy partial orderings, in a manner analogous to the way we used the resolution principle in Chapter 2. Let us now look at an example of a fuzzy similarity relation.

Example 3.3 A Similarity Relation. Consider a fuzzy relation R indicating that two points on the $X \times Y$ plane are near the origin. This is a relation we would expect to have a membership function equal to 1 exactly at the origin and to have gradually diminishing membership as we move away from the origin. We can indicate the relation by a statement such as "*x is near the origin with y*" or analytically as a fuzzy set with an appropriately chosen (subjectively) membership function—for example,

$$\mu_R(x, y) = e^{-(x^2 + y^2)} \qquad (E3.3\text{-}1)$$

Thus the relation R is the fuzzy set

$$R = \int_{X \times Y} \mu_R(x, y) / (x, y) \qquad (E3.3\text{-}2)$$

which using equation (E3.3-1) we can write as

$$R = \int_{X \times Y} e^{-(x^2 + y^2)} / (x, y) \qquad (E3.3\text{-}3)$$

The membership function of R is shown in Figure 3.3. It can be shown that R is a *fuzzy similarity relation*; that is, it is *reflexive, symmetric,* and *transitive*.

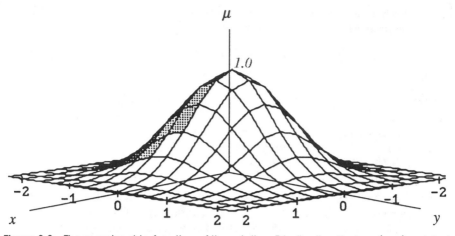

Figure 3.3 The membership function of the relation R indicating that an (x, y) point of the Cartesian plane, $X \times Y$, is close to the origin $(0, 0)$.

Figure 3.3 also illustrates that fuzzy relations are fuzzy sets on high-dimensional universes of discourse. In this case the universe of discourse is the x–y plane—that is, the Cartesian product $X \times Y$. □

3.4 BASIC OPERATIONS WITH FUZZY RELATIONS

Fuzzy relations are fundamentally fuzzy sets defined over higher-dimensional universes of discourse—that is, Cartesian products. All the fuzzy set operations we saw in Chapter 2, such as *union, intersection, α-cuts,* and so on, are also applicable to fuzzy relations. Here we take a look at the *union, intersection, inclusion, α-cuts,* and *resolution* as well as some operations specific to relations such as *projection* and *cylindrical extension* (Dubois and Prades, 1980; Zimmermann, 1985).

   Suppose that we have two fuzzy relations R_1 and R_2. Their *union* is a new relation

$$R_1 \cup R_2 = \int_{X \times Y} \left[\mu_{R_1}(x, y) \vee \mu_{R_2}(x, y) \right] \big/ (x, y) \qquad (3.4\text{-}1)$$

where the membership function of $R_1 \cup R_2$, as indicated in equation (3.4-1), is

$$\mu_{R_1 \cup R_2}(x, y) \equiv \mu_{R_1}(x, y) \vee \mu_{R_2}(x, y) \qquad (3.4\text{-}2)$$

for every (x, y) pair of the Cartesian product.

The intersection of fuzzy relations R_1 and R_2 is a new fuzzy relation whose membership function is the minimum of the membership functions of R_1 and R_2 taken at every point (x, y) of the Cartesian product,

$$R_1 \cap R_2 = \int_{X \times Y} \left[\mu_{R_1}(x, y) \wedge \mu_{R_2}(x, y) \right] / (x, y) \qquad (3.4\text{-}3)$$

where the membership function of $R_1 \cap R_2$ is

$$\mu_{R_1 \cap R_2}(x, y) \equiv \mu_{R_1}(x, y) \wedge \mu_{R_2}(x, y) \qquad (3.4\text{-}4)$$

We define the α-cut of a fuzzy relation in a manner similar to the way we defined in Section 2.6 the α-cuts of one-dimensional fuzzy sets. The *resolution principle* applied to fuzzy relations offers us an alternative way of representing the membership function of a fuzzy relation. It says that the membership function of a fuzzy relation can be represented through its α-cuts. More specifically, the *resolution principle* asserts that the membership function of a fuzzy relation R is expressed in terms of its α-cuts in a manner analogous to equation (2.7-1) as

$$\mu_A(x) = \bigvee_{0 < \alpha \le 1} \left[\alpha \cdot \mu_{R_\alpha}(x, y) \right] \qquad (3.4\text{-}5)$$

where the maximum is taken over all α's and $\mu_{R_\alpha}(x, y)$ is the α-cut of the membership function of the relation R at level α.

We say that a relation R_1 is *included* in R_2 if both are defined over the same product space and we have everywhere

$$\mu_{R_1}(x, y) \le \mu_{R_2}(x, y) \qquad (3.4\text{-}6)$$

Note that the *union* and *intersection* of fuzzy relations are meaningful in the context of relations defined over the same Cartesian product. When the product spaces of two relations are different, these operations have no meaning and instead the important and useful operations become the various *composition operations* which we examine later.

Example 3.4   Union and Intersection of Fuzzy Relations. Suppose that we have the following two relations R_1 and R_2 described by the tables below:

$R_1 =$ "*x is larger than y*":

	y_1	y_2	y_3	y_4
x_1	0.0	0.0	0.1	0.8
x_2	0.0	0.8	0.0	0.0
x_3	0.1	0.8	1.0	0.8

$R_2 = $ "*y is much bigger than x*":

	y_1	y_2	y_3	y_4
x_1	0.4	0.4	0.2	0.1
x_2	0.5	0.0	1.0	1.0
x_3	0.5	0.1	0.2	0.6

The union of the two relations, $R_1 \cup R_2$, is formed by taking the maximum of the two grades of membership for the corresponding elements of the two tables. The table of the new relation is as follows:

$R_1 \cup R_2$:

	y_1	y_2	y_3	y_4
x_1	0.4	0.4	0.2	0.8
x_2	0.5	0.8	1.0	1.0
x_3	0.5	0.8	1.0	0.8

For the intersection, $R_1 \cap R_2$, we take the minimum of the two grades of membership in each cell of the tables of the two relations, and the resulting table is as follows:

$R_1 \cap R_2$:

	y_1	y_2	y_3	y_4
x_1	0.0	0.0	0.1	0.1
x_2	0.0	0.0	0.0	0.0
x_3	0.4	0.1	0.2	0.6

Some caution is needed when we interpret the new relations produced by *union* and *intersection*. For example, the union $R_1 \cup R_2$ can be interpreted as a proposition of the form: "*x is quite different than y.*" The intersection, however, is not very meaningful, since *x* cannot be simultaneously larger than *y* and *y* cannot be larger than *x* (Zimmermann, 1985). □

In relations, when it is desired to go to a space of lower dimension we use *projection*. Starting with a fuzzy relation defined on a two-dimensional space, we can take the *first* and *second projection* and go to one-dimensional universe of discourse, with each projection eliminating the first and second dimension, respectively. The *total projection* takes us to a zero-dimensional singleton, eliminating both dimensions. *Projections* are also called *marginal fuzzy restrictions*. The inverse of projection—that is, going toward higher dimensions—is called *cylindrical extension* (Zadeh, 1971).

Consider the fuzzy relation R defined over the Cartesian product $X \times Y$ —that is,

$$R = \int_{X \times Y} \mu_R(x, y) / (x, y) \qquad (3.4\text{-}7)$$

The first projection is a fuzzy set that results by eliminating the second set Y of the Cartesian product, $X \times Y$, hence projecting the relation on the universe of discourse of the first set X. We write the first projection as

$$R^1 = \int_X \mu_{R^1}(x) / x \qquad (3.4\text{-}8)$$

The membership function of the first projection is defined as

$$\mu_{R^1}(x) \equiv \bigvee_{y} \left[\mu_R(x, y) \right] \qquad (3.4\text{-}9)$$

To obtain $\mu_{R^1}(x)$, equation (3.4-9) indicates that we take the maximum of $\mu_R(x, y)$ with respect to y. Similarly the second projection (projecting on the Y universe of discourse) is a fuzzy set:

$$R^2 = \int_Y \mu_{R^2}(y) / y \qquad (3.4\text{-}10)$$

with membership function defined as

$$\mu_{R^2}(y) \equiv \bigvee_{x} \left[\mu_R(x, y) \right] \qquad (3.4\text{-}11)$$

where we take the maximum of $\mu_R(x, y)$ with respect to x. The *total projection* of R simply identifies the peak point of the relation—that is, a singleton (x_0, y_0) where the membership function of the original relation reaches its highest value.

$$R^T = \bigvee_{x} \bigvee_{y} \mu_R(x_0, y_0) / (x_0, y_0) \qquad (3.4\text{-}12)$$

The opposite of projection is called the *cylindrical extension*. Through cylindrical extension we go from a fuzzy relation defined over a lower-dimensional space to a fuzzy relation on a higher-dimensional space. If a relation R is defined on a subsequence of a product space $X = X_1 \times X_2 \times X_3 \times \cdots \times X_n$, call it $X_{i1} \times X_{i2} \times X_{i3} \times \cdots \times X_{ik}$, then the cylindrical extension of R, denoted as $CE(R)$, is defined as

$$CE(R) \equiv \int_{X_1 \times \cdots \times X_n} \mu_R(x_{i1}, \ldots, x_{ik}) / (x_1, \ldots, x_n) \qquad (3.4\text{-}13)$$

Let us look at an example of projection and cylindrical extension.

Example 3.5   Projection and Cylindrical Extension. Consider the relation R defined over the Cartesian product $X \times Y$ of the sets $X = \{x_1, x_2, x_3\}$ and $Y = \{y_1, y_2, y_3, y_4, y_5, y_6\}$ as shown in Table 3.2. The membership functions for the first and second projection are indicated by the column to the right of the table and the row below the table, respectively. The first projection is what the relation would look like if seen from the direction of the arrow on the left side of the table. Imagine that we look in the direction that the arrow on the left indicates. We see in front of us three rows of the relation and select the highest value in each row. As a result, we obtain the first projection, namely,

$$R^1 = \sum_X \mu_{R^1}(x_i)/x_i = 1.0/x_1 + 0.9/x_2 + 1.0/x_3 \qquad (E3.5\text{-}1)$$

Equation (E3.5-1) indicates that the first projection of the binary fuzzy relation R is simply a fuzzy set on a one-dimensional universe of discourse.

Table 3.2   Fuzzy relation and projections

$$\Downarrow$$

$\mu_{R^1}(x)$

	y_1	y_2	y_3	y_4	y_5	y_6	
x_1	0.1	0.2	0.4	0.8	1.0	0.6	1.0
x_2	0.2	0.4	0.8	0.9	0.8	0.6	0.9
x_3	0.5	0.9	1.0	0.8	0.4	0.2	1.0

$\Rightarrow$

$\mu_{R^2}(y)$ | 0.5 | 0.9 | 1.0 | 0.9 | 1.0 | 0.6 | | 1.0 | μ_{R^T}

The second projection is what the relation would look like if seen from the direction of the arrow on top of the table.

$$R^2 = \sum_Y \mu_{R^2}(y_j)/y_j = 0.5/y_1 + 0.9/y_2 + 1.0/y_3 + 0.9/y_4 + 0.6/y_5 + 0.8/y_6$$

$$(E3.5\text{-}2)$$

The total projection is the single cell in the corner and represents the highest grade of membership that the relation has, namely, 1.

Let us next take a look at the cylindrical extension of the second projection. In a way the cylindrical extension is the *opposite* of projection. We expect therefore to obtain a relation on $X \times Y$ somewhat similar to the original relation R. As equation (E3.5-2) indicates, the second projection is defined on the Y universe of discourse. The generalization of this to the $X \times Y$ two-dimensional space is given by the *cylindrical extension*. Using equation (3.4-7) we obtain that the cylindrical extension of the second projection of the relation R^2 is simply the fuzzy set of the second projection extended in one more dimension, namely,

$CE(R^2)$:

	y_1	y_2	y_3	y_4	y_5	y_6
x_1	0.5	0.9	1.0	0.9	1.0	0.6
x_2	0.5	0.9	1.0	0.9	1.0	0.6
x_3	0.5	0.9	1.0	0.9	1.0	0.6

Note that although the cylindrical extension of the second projection R^2 results in a relation of higher dimensionality, it did not recover the original relation R. Some information was lost through the operation of the cylindrical extension. □

3.5   COMPOSITION OF FUZZY RELATIONS

Fuzzy relations defined on different Cartesian products can be combined with each other in a number of different ways through *composition*. Composition may be thought of metaphorically as a bridge that allows us to connect one product space to another, provided that there is a common boundary. Figure 3.4 illustrates the notion. Given two fuzzy relations—one in $X \times Y$ and another on $Y \times Z$—we want to associate directly elements of X with elements of Z. The set Y is the common boundary. Composition results in a

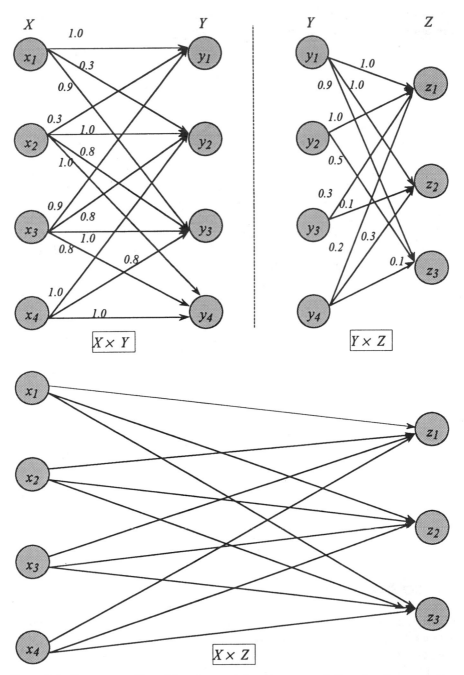

Figure 3.4 The composition of two fuzzy relations is a new relation directly associating elements from X and Z.

new relation shown at the bottom of Figure 3.4 that directly relates X to Z. Our main task in composition is to compute the grades of membership of the pairs (x, z) in the composed relation, namely, $\mu(x, z)$ (not shown in Figure 3.4).

Composition is very important for inferencing procedures used in linguistic descriptions of systems and is particularly useful in fuzzy controllers and expert systems (Klir and Folger, 1988). As we shall see in Chapters 5 and 6, collections of fuzzy *if/then* rules or *fuzzy algorithms* are mathematically equivalent to fuzzy relations, and the problem of *inferencing* or (*evaluating them with specific inputs*) is mathematically equivalent to *composition*. There are several types of composition. By far the most common in engineering applications is *max–min composition*, but we will also look at *max-star*, *max-product*, and *max-average*. In general, different types of composition result in different composed relations.

Max–Min Composition

The max–min composition of two fuzzy relations uses the familiar operators of fuzzy sets, max ($\vee$) and min ($\wedge$) (see Section 2.3). Suppose that we have two fuzzy relations $R_1(x, y)$ and $R_2(y, z)$ defined over the Cartesian products $X \times Y$ and $Y \times Z$, respectively. The max–min composition of R_1 and R_2 is a new relation $R_1 \circ R_2$ defined on $X \times Z$ as

$$R_1 \circ R_2 \equiv \int_{X \times Z} \bigvee_y \left[\mu_{R_1}(x, y) \wedge \mu_{R_2}(y, z) \right] / (x, z) \qquad (3.5\text{-}1)$$

where the symbol "$\circ$" stands for max–min composition of relations R_1 and R_2. When the Cartesian product $X \times Y$ is discrete, then the integral (union) sign in (3.5-1) is replaced by summation. From equation (3.5-1) we see that the grade of membership of each (x, z) pair in the new relation is

$$\mu_{R_1 \circ R_2}(x, z) = \bigvee_y \left[\mu_{R_1}(x, y) \wedge \mu_{R_2}(y, z) \right] \qquad (3.5\text{-}2)$$

where the outer maximum is taken with respect to the elements y of the common boundary. The operation on the right-hand side of equation (3.5-2) is actually very similar to matrix multiplication, with max ($\vee$) being analogous to *summation* ($+$) and min ($\wedge$) being analogous to *multiplication* ($\cdot$), as we will see in the examples that follow. Interchanging min and max in (3.5-1) is known as the *min–max composition*. In this book, however, we will mostly use max–min composition and compositions where the final (outer) operand is max ($\vee$). Max–min composition is used extensively in diagnostic and control applications of fuzzy logic.

Max–Star Composition

We can use *multiplication, summation,* or some other *binary operation* ($*$) in place of min ($\wedge$) in equations (3.5-1) and (3.5-2) while still performing maximization with respect to y. This type of composition of two fuzzy relations is generally known as the *"max-star"* or *"max-$*$ composition."*[1]

Suppose that we have two fuzzy relations R_1 and R_2 defined over the Cartesian products $X \times Y$ and $Y \times Z$, respectively. The max-$*$ composition of R_1 and R_2 is the new relation

$$R_1 * R_2 \equiv \int_{X \times Z} \bigvee_y \left[\mu_{R_1}(x, y) * \mu_{R_2}(y, z) \right] / (x, z) \qquad (3.5\text{-}3)$$

We see from equation (3.5-3) that the membership function of the new relation is

$$\mu_{R_1 * R_2}(x, z) = \bigvee_y \left[\mu_{R_1}(x, y) * \mu_{R_2}(y, z) \right] \qquad (3.5\text{-}4)$$

When the Cartesian product is discrete the integral sign in equation (3.5-3) is replaced by summation. Again as we shall see in the examples that follow this is essentially a computational procedure very similar to matrix multiplication. Two special cases of the max-star composition are the *max-product* (or *max-prod*) and the *max-average* composition.

Max-Product Composition

In max-product composition we use product ($\cdot$) in place of ($*$) in equations (3.5-3) and (3.5-4). Thus the max-product composition of two relations R_1 and R_2 is

$$R_1 \cdot R_2 \equiv \int_{X \times Z} \bigvee_y \left[\mu_{R_1}(x, y) \cdot \mu_{R_2}(y, z) \right] / (x, z) \qquad (3.5\text{-}5)$$

For discrete product spaces we use the summation sign in equation (3.5-5). The membership function of the composed relation is given by

$$\mu_{R_1 \cdot R_2}(x, z) = \bigvee_y \left[\mu_{R_1}(x, y) \cdot \mu_{R_2}(y, z) \right] \qquad (3.5\text{-}6)$$

---

[1]The name "star" refers to the star symbol that stands for a number of operations such as *average* and *product.*

Max–Average Composition

In the max-average composition of fuzzy relations we use the arithmetic sum $(+)$ divided by 2 in place of $(*)$ in equations (3.5-3) and (3.5-4). Thus the max-average composition of R_1 with R_2 is a new relation $R_1\langle + \rangle R_2$ given by

$$R_1\langle + \rangle R_2 \equiv \int_{X \times Z} \bigvee_{y} \left[\tfrac{1}{2}\left(\mu_{R_1}(x, y) + \mu_{R_2}(y, z) \right) \right] \big/ (x, z) \quad (3.5\text{-}7)$$

with membership function

$$\mu_{R_1\langle + \rangle R_2}(x, z) = \bigvee_{y} \left[\tfrac{1}{2}\left(\mu_{R_1}(x, y) + \mu_{R_2}(y, z) \right) \right] \quad (3.5\text{-}8)$$

Let us take a look at a few examples of composition.

Example 3.6   Max–Min Composition of Fuzzy Relations. Let's use max–min composition with the two relations shown in the upper part of Figure 3.4. The membership matrices of the relations R_1 on $X \times Y$ and R_2 on $Y \times Z$ are

$$R_1 = \begin{bmatrix} \mu_{R_1}(x_1, y_1) & \mu_{R_1}(x_1, y_2) & \mu_{R_1}(x_1, y_3) & \mu_{R_1}(x_1, y_4) \\ \mu_{R_1}(x_2, y_1) & \mu_{R_1}(x_2, y_2) & \mu_{R_1}(x_2, y_3) & \mu_{R_1}(x_2, y_4) \\ \mu_{R_1}(x_3, y_1) & \mu_{R_1}(x_3, y_2) & \mu_{R_1}(x_3, y_3) & \mu_{R_1}(x_3, y_4) \\ \mu_{R_1}(x_4, y_1) & \mu_{R_1}(x_4, y_2) & \mu_{R_1}(x_4, y_3) & \mu_{R_1}(x_4, y_4) \end{bmatrix}$$

$$= \begin{bmatrix} 1.0 & 0.3 & 0.9 & 0.0 \\ 0.3 & 1.0 & 0.8 & 1.0 \\ 0.9 & 0.8 & 1.0 & 0.8 \\ 0.0 & 1.0 & 0.8 & 1.0 \end{bmatrix} \quad (E3.6\text{-}1)$$

and

$$R_2 = \begin{bmatrix} \mu_{R_2}(y_1, z_1) & \mu_{R_2}(y_1, z_2) & \mu_{R_2}(y_1, z_3) \\ \mu_{R_2}(y_2, z_1) & \mu_{R_2}(y_2, z_2) & \mu_{R_2}(y_2, z_3) \\ \mu_{R_2}(y_3, z_1) & \mu_{R_2}(y_3, z_2) & \mu_{R_2}(y_3, z_3) \\ \mu_{R_2}(y_4, z_1) & \mu_{R_2}(y_4, z_2) & \mu_{R_2}(y_4, z_3) \end{bmatrix}$$

$$= \begin{bmatrix} 1.0 & 1.0 & 0.9 \\ 1.0 & 0.0 & 0.5 \\ 0.3 & 0.1 & 0.0 \\ 0.2 & 0.3 & 0.1 \end{bmatrix} \quad (E3.6\text{-}2)$$

We want to compute the membership matrix of the max–min composition of R_1 and R_2. We can use equation (3.5-2) to obtain the membership function of the composed relation. The operations in (3.5-2) are similar to matrix multiplication, with ($\vee$) being treated like *summation* ($+$) and ($\wedge$) being treated like *multiplication* ($\cdot$). With this in mind, instead of using

$$\mu_{R_1 \circ R_2}(x, z) = \bigvee_y \left[\mu_{R_1}(x, y) \wedge \mu_{R_2}(y, z) \right]$$

we can use the matrix form of max–min composition, namely,

$$R_1 \circ R_2 = \begin{bmatrix} 1.0 & 0.3 & 0.9 & 0.0 \\ 0.3 & 1.0 & 0.8 & 1.0 \\ 0.9 & 0.8 & 1.0 & 0.8 \\ 0.0 & 1.0 & 0.8 & 1.0 \end{bmatrix} \circ \begin{bmatrix} 1.0 & 1.0 & 0.9 \\ 1.0 & 0.0 & 0.5 \\ 0.3 & 0.1 & 0.0 \\ 0.2 & 0.3 & 0.1 \end{bmatrix} \quad \text{(E3.6-3)}$$

To evaluate equation (E3.6-3) we proceed, like in matrix multiplication, by forming the pairs of minima of each element in the first row of membership matrix R_1 with every element in the first column of membership matrix R_2. For example, to obtain the first element, (x_1, z_1), of the composition we perform the following operations:

$$\begin{bmatrix} 1.0 & 0.3 & 0.9 & 0.0 \end{bmatrix} \circ \begin{bmatrix} 1.0 \\ 1.0 \\ 0.3 \\ 0.2 \end{bmatrix}$$

$$= [1.0 \wedge 1.0] \vee [0.3 \wedge 1.0] \vee [0.9 \wedge 0.3] \vee [0.0 \wedge 0.2]$$
$$= 1.0 \vee 0.3 \vee 0.3 \vee 0.0$$
$$= 1.0$$

We repeat this procedure for all rows and columns and the result is the membership matrix of the composed relation $R_1 \circ R_2$ given by

$$R_1 \circ R_2 = \begin{bmatrix} 1.0 & 1.0 & 0.9 \\ 1.0 & 0.3 & 0.5 \\ 0.9 & 0.9 & 0.9 \\ 1.0 & 0.3 & 0.5 \end{bmatrix} \quad \text{(E3.6-4)}$$

The new relation is a fuzzy set over the Cartesian product $X \times Z$ which may also be written as

$$R_1 \circ R_2 = 1.0/(x_1, z_1) + 1.0/(x_1, z_2) + 0.9/(x_1, z_3)$$
$$+ 1.0/(x_2, z_1) + 0.3/(x_2, z_2) + 0.5/(x_2, z_3)$$
$$+ 0.9/(x_3, z_1) + 0.9/(x_3, z_2) + 0.9/(x_3, z_3)$$
$$+ 1.0/(x_4, z_1) + 0.3/(x_4, z_2) + 0.5/(x_4, z_3) \quad \text{(E3.6-5)}$$

$\square$

Example 3.7   Max–Min, Max-Product, and Max-Average Composition of Fuzzy Relations. Suppose we have the two relations R_1 and R_2, shown below, and we want to compute a new relation which is the max–min composition of the two, $R = R_1 \circ R_2$. We will also find the max-product and max-average compositions. We perform max–min composition using the tabular representation of the relations and the definition of max–composition given in equations (3.5-1) or (3.5-2). The relations to be composed are described by the following membership tables:

R_1:

	y_1	y_2	y_3	y_4	y_5
x_1	0.1	0.2	0.0	1.0	0.7
x_2	0.3	0.5	0.0	0.2	1.0
x_3	0.8	0.0	1.0	0.4	0.3

R_2:

	z_1	z_2	z_3	z_4
y_1	0.9	0.0	0.3	0.4
y_2	0.2	1.0	0.8	0.0
y_3	0.8	0.0	0.7	1.0
y_4	0.4	0.2	0.3	0.0
y_5	0.0	1.0	0.0	0.8

To find the new relation $R = R_1 \circ R_2$ we use equation (3.5-2), the definition of max–min composition, namely,

$$\mu_{R_1 \circ R_2}(x, z) = \bigvee_{y} \left[\mu_{R_1}(x, y) \wedge \mu_{R_2}(y, z) \right] \qquad \text{(E3.7-1)}$$

To use (E3.7-1) we proceed in the following manner. First, we fix x and z—for example, $x = x_1$ and $z = z_1$—and vary y. Next, we evaluate the following pairs of minima, using the numbers from the shaded cells in the tables of the two relations:

$$\mu_{R_1}(x_1, y_1) \wedge \mu_{R_2}(y_1, z_1) = 0.1 \wedge 0.9 = 0.1$$

$$\mu_{R_1}(x_1, y_2) \wedge \mu_{R_2}(y_2, z_1) = 0.2 \wedge 0.2 = 0.2$$

$$\mu_{R_1}(x_1, y_3) \wedge \mu_{R_2}(y_3, z_1) = 0.0 \wedge 0.8 = 0.0 \qquad \text{(E3.7-2)}$$

$$\mu_{R_1}(x_1, y_4) \wedge \mu_{R_2}(y_4, z_1) = 1.0 \wedge 0.4 = 0.4$$

$$\mu_{R_1}(x_1, y_5) \wedge \mu_{R_2}(y_5, z_1) = 0.7 \wedge 0.0 = 0.0$$

We take the maximum of all these terms and obtain the value of the (x_1, z_1) element of the relation, namely,

$$\mu_{R_1 \circ R_2}(x_1, z_1) = 0.1 \vee 0.2 \vee 0.0 \vee 0.4 = 0.4 \qquad \text{(E3.7-3)}$$

This is the value in the shaded cell in the table of the composed relation shown below. In a similar manner, we determine the grades of membership for all other pairs and finally we have

$$R = R_1 \circ R_2:$$

	z_1	z_2	z_3	z_4
x_1	0.4	0.7	0.3	0.7
x_2	0.3	1.0	0.5	0.8
x_3	0.8	0.3	0.7	1.0

Let us now compose these two relations using max-product composition as defined by equation (3.5-6)—that is,

$$\mu_{R_1 \cdot R_2}(x, z) = \bigvee_y \left[\mu_{R_1}(x, y) \cdot \mu_{R_2}(y, z) \right] \qquad \text{(E3.7-4)}$$

Again we fix x and z and vary y—for example, $x = x_1$, $z = z_1$, and $y = y_i$ for $i = 1, \ldots, 5$. We form and evaluate the products of the shaded cells in the relation tables—that is,

$$\mu_{R_1}(x_1, y_1) \cdot \mu_{R_2}(y_1, z_1) = 0.1 \times 0.9 = 0.09$$

$$\mu_{R_1}(x_1, y_2) \cdot \mu_{R_2}(y_2, z_1) = 0.2 \times 0.2 = 0.04$$

$$\mu_{R_1}(x_1, y_3) \cdot \mu_{R_2}(y_3, z_1) = 0.0 \times 0.8 = 0.0 \qquad \text{(E3.7-5)}$$

$$\mu_{R_1}(x_1, y_4) \cdot \mu_{R_2}(y_4, z_1) = 1.0 \times 0.4 = 0.4$$

$$\mu_{R_1}(x_1, y_5) \cdot \mu_{R_2}(y_5, z_1) = 0.7 \times 0.0 = 0.0$$

Taking the maximum of these terms, we obtain the grade of membership of the (x_1, z_1) pair in the composed relation, namely,

$$\mu_{R_1 \cdot R_2}(x_1, z_1) = 0.09 \vee 0.04 \vee 0.0 \vee 0.4 \vee 0.0 \qquad \text{(E3.7-6)}$$

which coincidentally evaluates also to $\mu_{R_1 \cdot R_2}(x_1, z_1) = 0.4$. This is the number in the shaded cell of the table below. Similarly, we obtain the membership of all other pairs and finally we get the membership table of the composition as

$R_1 \cdot R_2$:

	z_1	z_2	z_3	z_4
x_1	0.4	0.7	0.3	0.56
x_2	0.27	1.0	0.4	0.8
x_3	0.8	0.3	0.7	1.0

For the max-average composition of the two relations, again we fix x and z and vary y in order to find the max with respect to y in equation (3.5-8) for each (x, z) pair. Thus first we form and evaluate the sums of the shaded cells as before:

$$\mu_{R_1}(x_1, y_1) + \mu_{R_2}(y_1, z_1) = 0.1 + 0.9 = 1.0$$

$$\mu_{R_1}(x_1, y_2) + \mu_{R_2}(y_2, z_1) = 0.2 + 0.2 = 0.4$$

$$\mu_{R_1}(x_1, y_3) + \mu_{R_2}(y_3, z_1) = 0.0 + 0.8 = 0.8 \qquad \text{(E3.7-7)}$$

$$\mu_{R_1}(x_1, y_4) + \mu_{R_2}(y_4, z_1) = 1.0 + 0.4 = 1.4$$

$$\mu_{R_1}(x_1, y_5) + \mu_{R_2}(y_5, z_1) = 0.7 + 0.0 = 0.0$$

Thus, using equation (3.5-8), the grade of membership of the (x_1, z_1) pair is

$$\mu_{R_1\langle+\rangle R_2}(x_1, z_1) = \tfrac{1}{2}[1.0 \vee 0.4 \vee 0.8 \vee 1.4 \vee 0.0] = 0.7 \quad (E3.7\text{-}8)$$

This is the grade of membership of the shaded cell in the table shown below. In a similar manner the membership function for each pair is computed, and finally we get the max-average composition of the two relations in the table:

$R_1\langle+\rangle R_2$:

	z_1	z_2	z_3	z_4
x_1	0.7	0.85	0.65	0.75
x_2	0.6	1.0	0.65	0.9
x_3	0.9	0.65	0.85	1.0

We observe from the tables of the composed relations that max–min, max-product, and max-average compositions of R_1 and R_2 may result in different relations. □

REFERENCES

Dubois, D., and Prade, H., *Fuzzy Sets and Systems: Theory and Applications*, Academic Press, Boston, 1980.

Kandel, A., *Fuzzy Mathematical Techniques with Applications*, Addison-Wesley, Reading, MA, 1986.

Kaufmann, A., *Introduction to the Theory of Fuzzy Subsets*, Vol. I, Academic Press, New York, 1975.

Kaufmann, A., and Gupta, M. M., *Introduction to Fuzzy Arithmetic*, Van Nostrand Reinhold, New York, 1991.

Klir, G. J., and Folger, T. A., *Fuzzy Sets, Uncertainty, and Information*, Prentice-Hall, Englewood Cliffs, NJ, 1988.

Terano, T., Asai, K., and Sugeno, M., *Fuzzy Systems Theory and its Applications*, Academic Press, Boston, 1992.

Zadeh, L., Similarity Relations and Fuzzy Orderings, *Information Sciences*, Vol. 3, Elsevier, Amsterdam, 1971, pp. 177–200.

Zimmermann, H. J., *Fuzzy Set Theory and its Applications*, Kluwer-Nijhoff, Boston, 1985.

PROBLEMS

1. A fuzzy "diagnostic relation" R_d for an automobile relates the system set S to the fault set F. These sets are given below.

$$S = [x_1 \text{ (low gas mileage)}, x_2 \text{ (excessive vibration)}, x_3 \text{ (loud noise)},$$

$$x_4 \text{ (high collant temperature)}, x_5 \text{ (steering instability)}]$$

$$F = [y_1 \text{ (bad spark plugs)}, y_2 \text{ (wheel imbalance)}, y_3 \text{ (bad muffler)},$$

$$y_4 \text{ (thermostat stuck closed)}]$$

Assume reasonable numerical values ($0 \rightarrow 1$) for membership values relating members of sets S and F and use them. Give all five representations of this fuzzy diagnostic relationship R_d in terms of x_i and y_j.

2. Give the max–min composition, max-star composition, and the max-average composition of the relation fuzzy "diagnostic relation" of Problem 1.

3. Repeat Example 3.3 for a fuzzy relation R indicating that "x is near the perimeter of a circle having a radius 1 with y".

4. In Example 3.4, give a table for $[R_1 \cap R_2] \cup [R_1 \cap R_2]$.

5. Find the first, second, and total projection as well as the cylindrical extension of the fuzzy relation R given by Equation (E3.2-2).

6. Find the max-product and max-average composition of relations R_1 and R_2 given by Equations (E3.6-1) and (E3.6-2), respectively.

7. Find the max–min composition of relations R_1 and R_2 given in Example 3.7.

8. **a.** Show that the max–min composition of fuzzy relations is *associative*. Illustrate with an example of your own.

   b. Consider the max-min composition and a relation R which is *reflexive*. Show that:

$$R \circ R = R.$$

9. Suppose that we have three relations involved in max–min composition

$$P \circ Q = R$$

When two of the components in the above equation are given and the other is unknown we have a set of equations known as *fuzzy relation equations*. Solve the following fuzzy relation equations:

(a) $P \circ \begin{bmatrix} .9 & .6 & 1 \\ .8 & .8 & .5 \\ .6 & .4 & .6 \end{bmatrix} = [.6 \quad .6 \quad .5]$

(b) $P \circ \begin{bmatrix} .2 & .4 & .5 & .7 \\ .3 & .1 & .6 & .8 \\ .1 & .4 & .6 & .7 \\ 0 & .3 & 0 & 1 \end{bmatrix} = \begin{bmatrix} .2 & .4 & .6 & .7 \\ .1 & .1 & .2 & .2 \end{bmatrix}$

10. Consider two probability distributions that are independent and described by

$$dP(x_1) = e^{-x_1}\, dx_1 \text{ and } dP(x_2) = x_2 e^{-x_2}\, dx_2, \; x_1, x_2 \geq 0$$

How can we model the *similarity* of x_1, x_2 through a fuzzy set and what would be the probability of occurrence of such a set?

4

FUZZY NUMBERS

4.1 INTRODUCTION

Fuzzy numbers are fuzzy sets used in connection with applications where an explicit representation of the ambiguity and uncertainty found in numerical data is desirable. In an intuitive sense, they are fuzzy sets representing the meaning of statements such as *"about 3"* or *"nearly five and a half."* In other words, fuzzy numbers take into account the "about," "almost," and "not quite" qualities of numerical labels. Fuzzy set operations such as *union* and *intersection*, as well as the notions of *α-cuts*, *resolution*, and the *extension principle* (Chapter 2), are all applicable to fuzzy numbers. In addition, a set of operations very similar to the familiar operations of arithmetic, *addition*, *subtraction*, *multiplication*, and *division* can be defined for fuzzy numbers as well. In this chapter we look at such operations and examples of their use. Fuzzy numbers have been successfully applied in expert systems, fuzzy regression, and fuzzy data analysis methodologies (Kaufmann and Gupta, 1991; Terano et al., 1992). Fuzzy numbers have also been used in connection with fuzzy equations, and alternative operations of fuzzy arithmetic have been introduced for the purpose of reducing fuzziness in successive computations (Sanchez, 1993).

The universe of discourse on which fuzzy numbers are defined is the set of real numbers and its subsets (e.g., integers or natural numbers), and their membership functions ought to be *normal* and *convex*. We recall from Section 2.3 that a fuzzy set is called *normal* if there is at least one point in the universe of discourse where the membership function reaches unity [equation (2.3-11)]. But what is a "convex" fuzzy set? The intuitive meaning

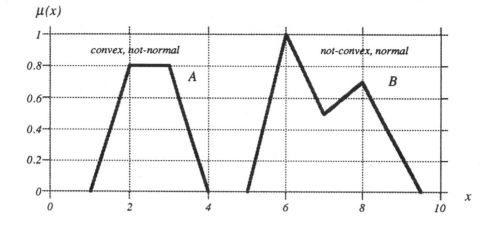

(a)

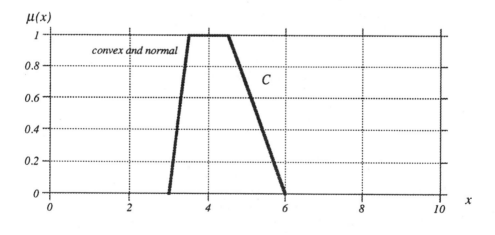

(b)

Figure 4.1 (*a*) Two fuzzy sets that *cannot* be used as fuzzy numbers. (*b*) A fuzzy set that may be used as fuzzy number.

of *convexity*[1] is that the membership function of a convex fuzzy set does not go "up-and-down" more than once. Consider, for example, the fuzzy sets *A* and *B* shown in Figure 4.1*a*. Fuzzy set *A* is *convex* but *not normal* since nowhere in the universe of discourse does its membership function reach unity. Therefore it is not a fuzzy number. Fuzzy set *B* is *normal* but *not convex* since its membership function goes "up-and-down" twice, and hence it is also not a fuzzy number. On the other hand, consider the set *C* shown in Figure 4.1*b*. It is both *normal* and *convex* and therefore may be considered a fuzzy number. We will see in following sections that changing the shape of a membership function results in a different number. "Shape" is what fuzzy numbers are all about, and fuzzy arithmetic may be thought of as a way of computing with "shapes" (areas) instead of "points" (we consider crisp numbers as "points").

Fuzzy numbers may also be defined on a multidimensional universe of discourse that is a Cartesian product. Such fuzzy numbers are used, for example, in connection with scene analysis and robotics to define the meaning of a region in space, or a domain on the *x–y* plane, and also to add, subtract, and multiply regions (Pal and Majumder, 1986). In this chapter, however, we consider fuzzy numbers defined on a simple, one-dimensional universe of discourse. A very comprehensive treatment of fuzzy numbers, including multidimensional ones, may be found in the book entitled *Introduction to Fuzzy Arithmetic* by Kaufmann and Gupta (1991).

4.2 REPRESENTING FUZZY NUMBERS

We denote fuzzy numbers by boldfaced italics—for example, *3* or *A*—or by referring to their membership function. As we said earlier, fuzzy numbers are fuzzy sets used to represent the *"about,"* *"almost,"* or *"nearly"* qualities of numerical data. We observe, however, that there are many possible meanings to a statement such as *"about* 3." Therefore, several different sets may be used to represent *"about* 3." In the context of fuzzy arithmetic operations, however, at any given time we use only one meaning, chosen on the basis of application-specific criteria and needs. Figure 4.2*a* shows a triangular membership function representing the fuzzy number *3*. Another possible representation is the bell-shaped membership function in Figure 4.2*b*. These are two different *3*'s. If we start a computation using the triangular *3*, we cannot halfway through switch to the bell-shaped *3*. Note that on both instances the shape of the membership function meets our *normal* and *convex* require-

---

[1] The notion of convexity is derived through references to geometrical objects. A body Ω in Euclidean space is called *convex* if the line segment joining any two points of Ω lies in Ω. Examples of convex bodies in three-dimensional space are the sphere, the ellipsoid, a cylinder, a cube, and a cone.

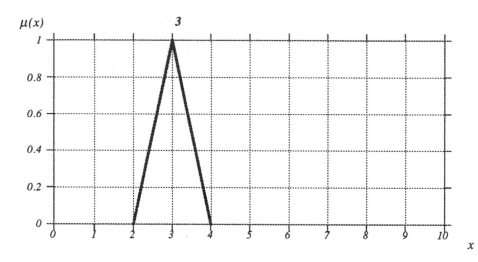

(a)

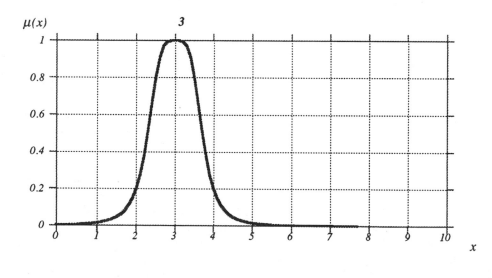

(b)

Figure 4.2 Two different fuzzy numbers: (a) triangular **3** and (b) bell-shaped **3**.

Table 4.1   Tabular representation of a fuzzy number 3

	0.4	0.7	1	0.7	0.4	0.2	0.1	0	0
α=1			1						
α=0.9			1						
α=0.8			1						
α=0.7		1	1	1					
α=0.6		1	1	1					
α=0.5		1	1	1					
α=0.4	1	1	1	1	1				
α=0.3	1	1	1	1	1				
α=0.2	1	1	1	1	1	1			
α=0.1	1	1	1	1	1	1			
α=0	1	1	1	1	1	1	1	1	1
	1	2	3	4	5	6	7	8	9

ments. Another possible fuzzy number *3* is shown in Table 4.1, where the shaded cells, the 1's, indicate the shape of the number. Here *3* is defined over the universe of natural numbers shown at the bottom of the table. In the leftmost column we list the values of a parameter, α, ranging between 0 and 1, used to parametrize the shape of the function (Kaufmann and Gupta, 1991). In fact, this is the same α we saw in connection with α-cuts (see Section 2.6). The α-cuts of fuzzy numbers are very useful in fuzzy arithmetic operations. Looking at Table 4.1 we see that the grade of membership of crisp number 4 to the fuzzy number *3* is 0.7, and the grade of membership of crisp 3 is 1.0. Although the fuzzy numbers shown in Figure 4.2 and Table 4.1 are all different, we designate them with the same symbol (i.e., *3*) since they all peak at crisp 3 (Zimmermann, 1985; Kandel, 1986).

Fuzzy numbers, like any fuzzy set, may be represented by its α-cuts. We saw in Chapter 2 that a membership function may be parameterized by a parameter α in a manner similar to the tabular representation of number *3* shown in Table 4.1. The parameter α is a number between 0 and 1 (i.e., in the interval [0, 1]). Parameterizing the shape of a fuzzy number by α offers a convenient way for computing with fuzzy numbers because it essentially transforms fuzzy arithmetic operations into operations of interval arithmetic. It is easy to see what we are talking about by looking at Table 4.1. At each level α we have a horizontal "slice," or interval of the membership function, which is its α-cut. For example, at $\alpha = 0.5$, the α-cut is the interval from 2 to 4, and at $\alpha = 0.2$ it is the interval from 1 to 6. The tabular representation exemplifies the length of each α-cut; that is, it shows the number of cells and thus the length of the membership function at level α.

Consider the fuzzy number A shown in Figure 4.3. The membership function of A is parameterized by the parameter α. With each α we identify an interval $[a_1^{(\alpha)}, a_2^{(\alpha)}]$. As may be seen from the figure, we indicate by $a_1^{(\alpha)}$ the left endpoint of the interval ("left" is denoted by the subscript "1") and by $a_2^{(\alpha)}$ the right endpoint of the interval ("right" is denoted by the subscript "2"). Requiring that the membership function of a fuzzy number be *convex* and *normal* is another way of saying that the intervals that comprise the interval representation of A should be nested into one another as we move from the bottom of the membership function to the top (Klir and Folger, 1988; Terano et al., 1992). In other words, when $\alpha_1 < \alpha_2$, as shown in Figure 4.3, we have

$$\left[a_1^{(\alpha_2)}, a_2^{(\alpha_2)} \right] \subset \left[a_1^{(\alpha_1)}, a_2^{(\alpha_1)} \right] \qquad (4.2\text{-}1)$$

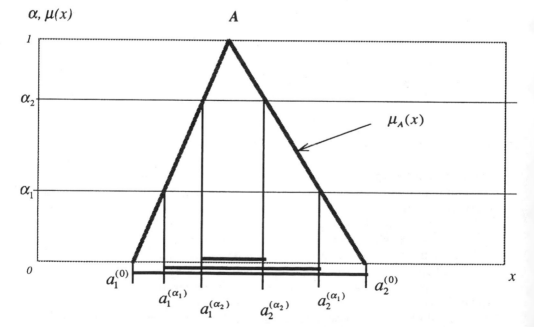

Figure 4.3 Nested intervals (α-cuts) associated with a fuzzy number **A**.

where the symbol $\subset$ denotes that the interval $[a_1^{(\alpha_2)}, a_2^{(\alpha_2)}]$ is contained within the interval $[a_1^{(\alpha_1)}, a_2^{(\alpha_1)}]$.

We can uniquely describe two fuzzy numbers A and B as two collection of intervals i.e., $[a_1^{(\alpha)}, a_2^{(\alpha)}]$ and $[b_1^{(\alpha)}, b_2^{(\alpha)}]$ respectively. We recall that the α-cuts of A and B (Section 2.6) were defined as the crisp sets

$$A_\alpha = \{x | \mu_A(x) \geq \alpha\} \tag{4.2-2}$$

and

$$B_\alpha = \{x | \mu_B(x) \geq \alpha\} \tag{4.2-3}$$

The α-cuts in equations (4.2-2) and (4.2-3) are simply intervals on the x axis, and hence for each α we have

$$A_\alpha = \left[a_1^{(\alpha)}, a_2^{(\alpha)}\right] \tag{4.2-4}$$

and

$$B_\alpha = \left[b_1^{(\alpha)}, b_2^{(\alpha)}\right] \tag{4.2-5}$$

Thus the fuzzy numbers A and B can be described (using the resolution principle—see Section 2.7) as collections of intervals, that is,

$$A = \bigvee_{0 \leq a \leq 1} \alpha \cdot A_\alpha = \bigvee_{0 \leq a \leq 1} \alpha \cdot \left[a_1^{(\alpha)}, a_2^{(\alpha)}\right] \tag{4.2-6}$$

and

$$B = \bigvee_{0 \leq a \leq 1} \alpha \cdot B_\alpha = \bigvee_{0 \leq a \leq 1} \alpha \cdot \left[b_1^{(\alpha)}, b_2^{(\alpha)}\right] \tag{4.2-7}$$

To simplify matters, we will not use the rather awkward representation of the two numbers given by equations (4.2-6) and (4.2-7) but will, instead, use equations (4.2-4) and (4.2-5), which we call the α-cut or *interval representation* of A and B (with the understanding that the number is the collection of all "slices," all α-cuts as α varies from 0 to 1).

Having two different ways of representing fuzzy numbers, through *membership functions* and through α-cuts or *intervals*, gives us the choice of defining arithmetic operations either through the extension principle (i.e., through a fuzzification of arithmetic operations on crisp numbers) or, equivalently, through the operations of interval arithmetic. This last approach is often more practical and straightforward as we will see in several examples.

Let us go next to the definition of *addition, subtraction, multiplication,* and *division with fuzzy numbers.* Although we will define operations for two numbers A and B, they are generally true for more than two numbers. A word of caution: Some of the properties of crisp numbers—for example,

$(7 \div 3) \times 3 = 7$—may not be valid for arithmetic operations involving fuzzy numbers. We will see that usually when fuzzy numbers are involved we have that $(7 \div 3) \times 3$ may not equal 7.

4.3   ADDITION

When *adding* two fuzzy numbers A and B we seek to compute a new fuzzy number $C = A + B$. The new number C is uniquely described when we obtain its membership function, $\mu_C(z) \equiv \mu_{A+B}(z)$, with z being the crisp sum of x and y, the elements of the universe of discourse of A and B. The addition of A and B may be defined in terms of addition of the α-cuts of the two numbers as follows:

$$A + B \equiv \left[a_1^{(\alpha)}, a_2^{(\alpha)} \right] + \left[b_1^{(\alpha)}, b_2^{(\alpha)} \right] \qquad (4.3\text{-}1)$$

where $[a_1^{(\alpha)}, a_2^{(\alpha)}]$ is the collection of intervals representing the fuzzy number A, and $[b_1^{(\alpha)}, b_2^{(\alpha)}]$ is the collection of intervals representing the fuzzy number B. Intervals are added by adding their corresponding left and right endpoints, and therefore equation (4.3-1) becomes

$$A + B = \left[a_1^{(\alpha)} + b_1^{(\alpha)}, a_2^{(\alpha)} + b_2^{(\alpha)} \right] \qquad (4.3\text{-}2)$$

Equation (4.3-2) indicates that the new number is also a collection of intervals with endpoints obtained from the endpoints of A and B.

Another way of defining fuzzy addition is through the *extension principle* (Section 2.5). We give here a cursory description of how this is done; more detailed treatments may be found in Dubois and Prade (1980) and in Terano et al. (1992). Suppose we want to add two crisp numbers x and y. The result is another crisp number $z = x + y$. Now, if x and y are variables, obviously their sum may be thought of as a function of x and y; that is,

$$z(x, y) = x + y \qquad (4.3\text{-}3)$$

Fuzzifying x and y—that is, defining fuzzy sets on x and y—results in a fuzzified function, $z = f(x, y)$. We saw in Section 2.5 how we can use the extension principle to obtain the fuzzy set C on $z = f(x, y)$. Suppose that we have two fuzzy numbers, A and B, defined over x and y (the universe of discourse of real numbers). According to the extension principle, their sum is a fuzzy set on z denoted as C, whose membership function is

$$\mu_C(z) \equiv \bigvee_{z=x+y} \left[\mu_A(x) \wedge \mu_B(y) \right] \qquad (4.3\text{-}4)$$

Equation (4.3-4) tells us that to compute the grade of membership of a certain crisp number z to the fuzzy number C, we take the maximum of the minima of the grades of membership of all pairs x and y which add up to z. How equation (4.3-4) works will be seen in Example 4.2, where a rather simple tabular way of carrying out the max–min operations will be presented.

Example 4.1   Addition of Discrete Fuzzy Numbers. Let us compute the sum C of two fuzzy numbers $A = 3$ and $B = 7$ defined as

$$A = 3 = 0.3/1 + 0.7/2 + 1.0/3 + 0.7/4 + 0.3/5 + 0/6 \quad \text{(E4.1-1)}$$

$$B = 7 = 0.2/5 + 0.6/6 + 1.0/7 + 0.6/8 + 0.2/9 + 0/10 \quad \text{(E4.1-2)}$$

and seen in Table 4.2. We compute C by adding the α-cuts of A, B in accordance with equation (4.3-2). We see from Table 4.2 that when $\alpha = 0.4$, for example, the 0.4-cuts of A and B are

$$A_{0.4} = \left[a_1^{(0.4)}, a_2^{(0.4)} \right] = [2, 4] \quad \text{(E4.1-3)}$$

and

$$B_{0.4} = \left[b_1^{(0.4)}, b_2^{(0.4)} \right] = [6, 8] \quad \text{(E4.1-4)}$$

The intervals in equations (E4.1-3) and (E4.1-4) are shown as shaded "slices" of cells in Table 4.2. According to equation (4.3-1) the 0.4-cut of C is the sum of the two intervals given by (E4.1-3) and (E4.1-4)—that is,

$$\begin{aligned} C_{0.4} &= \left[a_1^{(0.4)}, a_2^{(0.4)} \right] + \left[b_1^{(0.4)}, b_2^{(0.4)} \right] \\ &= \left[a_1^{(0.4)} + b_1^{(0.4)}, a_2^{(0.4)} + b_2^{(0.4)} \right] \\ &= [2 + 6, 4 + 8] \\ &= [8, 12] \end{aligned} \quad \text{(E4.1-5)}$$

We can obtain the same result from Table 4.2 simply by adding the endpoints of the shaded rows. We repeat this for each α to compute the entire sum. We start from the bottom of the table and go up in a *row-by-row* manner identifying the corresponding intervals of the two numbers and adding them up. The result is the number shown in Table 4.3. The 0.4-cut of C is indicated as a shaded group of cells in the table. As seen from the table, the new fuzzy number reaches unity at crisp number 10 (in the universe of discourse shown at the bottom) and therefore we think of it as a fuzzy number **10**. Thus we see that the sum is $7 + 3 = 10$, as would also be the case with crisp numbers. $\square$

Table 4.2   Fuzzy numbers 3 and 7 in Example 4.1

3:

	0.3	0.7	1	0.7	0.3	0	0	0	0
$\alpha=1.0$			1						
$\alpha=0.9$			1						
$\alpha=0.8$			1						
$\alpha=0.7$		1	1	1					
$\alpha=0.6$		1	1	1					
$\alpha=0.5$		1	1	1					
$\alpha=0.4$		1	1	1					
$\alpha=0.3$	1	1	1	1	1				
$\alpha=0.2$	1	1	1	1	1				
$\alpha=0.1$	1	1	1	1	1				
$\alpha=0.0$	1	1	1	1	1	1	1	1	1
	1	2	3	4	5	6	7	8	9

7:

	0	0	0	0	0.2	0.6	1.0	0.6	0.2	0	0
$\alpha=1.0$							1				
$\alpha=0.9$							1				
$\alpha=0.8$							1				
$\alpha=0.7$							1				
$\alpha=0.6$						1	1	1			
$\alpha=0.5$						1	1	1			
$\alpha=0.4$						1	1	1			
$\alpha=0.3$						1	1	1			
$\alpha=0.2$					1	1	1	1	1		
$\alpha=0.1$					1	1	1	1	1		
$\alpha=0.0$	1	1	1	1	1	1	1	1	1	1	1
	1	2	3	4	5	6	7	8	9	10	11

Table 4.3 Sum of fuzzy numbers 3 and 7 in Example 4.1

10:

	0	0	0	0	0	0.2	0.3	0.6	0.7	1.0	0.7	0.6	0.3	0.2	0	0
α=1.0										1						
α=0.9										1						
α=0.8										1						
α=0.7									1	1	1					
α=0.6								1	1	1	1	1				
α=0.5								1	1	1	1	1				
α=0.4								1	1	1	1	1				
α=0.3							1	1	1	1	1	1	1			
α=0.2						1	1	1	1	1	1	1	1	1		
α=0.1						1	1	1	1	1	1	1	1			
α=0.0	1	1	1	1	1	1	1	1	1	1	1	1	1	1	1	1
	1	*2*	*3*	*4*	*5*	*6*	*7*	*8*	*9*	*10*	*11*	*12*	*13*	*14*	*15*	*16*

Example 4.2 Addition of Fuzzy Numbers Through the Extension Principle.
In this example we compute the sum of the two numbers A and B of
Example 4.1 using the alternative definition of addition through the extension principle, namely, equation (4.3-4). At first glance, equation (4.3-4) looks
somewhat esoteric. We present here a rather simple technique for using it.
The same technique may be used with other fuzzy arithmetic operations as
well (Kaufmann and Gupta, 1991). Let's repeat equation (4.3-4) here:

$$\mu_{A+B}(z) \equiv \bigvee_{z=x+y} [\mu_A(x) \wedge \mu_B(y)] \qquad \text{(E4.2-1)}$$

A convenient way to compute the sum according to equation (E4.2-1) is to
create a table as shown in Table 4.4. We take the *support* of B and make as
many columns in the table as there are elements in the support; and similarly
we take the *support* of A and make as many rows in the table as there are
elements in the support of A. We recall that the support is the part of the
universe of discourse that has nonzero membership. A and B can be

Table 4.4   Adding fuzzy numbers through the extension principle

	$S\ u\ p\ p\ o\ r\ t$	$o\ f$	B							
	y=1	y=2	y=3	y=4	y=5	y=6	y=7	y=8	y=9	y=10
x=1	0.0 / 0.3	0.0 / 0.3	0.0 / 0.3	0.0 / 0.3	0.2 / 0.3	0.6 / 0.3	1.0 / 0.3	0.6 / 0.3	0.2 / 0.3	0.0 / 0.3
x=2	0.0 / 0.7	0.0 / 0.7	0.0 / 0.7	0.0 / 0.7	0.2 / 0.7	0.6 / 0.7	1.0 / 0.7	0.6 / 0.7	0.2 / 0.7	0.0 / 0.7
x=3	0.0 / 1.0	0.0 / 1.0	0.0 / 1.0	0.0 / 1.0	0.2 / 1.0	0.6 / 1.0	1.0 / 1.0	0.6 / 1.0	0.2 / 1.0	0.0 / 1.0
x=4	0.0 / 0.7	0.0 / 0.7	0.0 / 0.7	0.0 / 0.7	0.2 / 0.7	0.6 / 0.7	1.0 / 0.7	0.6 / 0.7	0.2 / 0.7	0.0 / 0.7
x=5	0.0 / 0.3	0.0 / 0.3	0.0 / 0.3	0.0 / 0.3	0.2 / 0.3	0.6 / 0.3	1.0 / 0.3	0.6 / 0.3	0.2 / 0.3	0.0 / 0.3
x=6	0.0 / 0.0	0.0 / 0.0	0.0 / 0.0	0.0 / 0.0	0.2 / 0.0	0.6 / 0.0	1.0 / 0.0	0.6 / 0.0	0.2 / 0.0	0.0 / 0.0
x=7	0.0 / 0.0	0.0 / 0.0	0.0 / 0.0	0.0 / 0.0	0.2 / 0.0	0.6 / 0.0	1.0 / 0.0	0.6 / 0.0	0.2 / 0.0	0.0 / 0.0
x=8	0.0 / 0.0	0.0 / 0.0	0.0 / 0.0	0.0 / 0.0	0.2 / 0.0	0.6 / 0.0	1.0 / 0.0	0.6 / 0.0	0.2 / 0.0	0.0 / 0.0
x=9	0.0 / 0.0	0.0 / 0.0	0.0 / 0.0	0.0 / 0.0	0.2 / 0.0	0.6 / 0.0	1.0 / 0.0	0.6 / 0.0	0.2 / 0.0	0.0 / 0.0
x=10	0.0 / 0.0	0.0 / 0.0	0.0 / 0.0	0.0 / 0.0	0.2 / 0.0	0.6 / 0.0	1.0 / 0.0	0.6 / 0.0	0.2 / 0.0	0.0 / 0.0

(Row labels B above and A at left; left margin reads "S u p p o r t   o f   A". In each cell the upper-right value is $\mu_B(y)$ and the lower-left value is $\mu_A(x)$.)

interchanged in terms or rows and columns, but for the moment let's make columns from the support of A and make rows from the support of A. In every cell of the table we put at the lower left corner the grade of membership of x to A and put in the upper right corner the grade of membership of y to B. Thus we have $\mu_A(x)$ in the lower left corner and $\mu_B(y)$ in the upper right corner as shown in Table 4.4. Now, let's take another look in the equation above. It calls for taking the maximum of pairs of singletons that add up to a certain z. For example, suppose that have $z = 9$. There are three different ways to get $z = 9$: adding $y = 8$ and $x = 1$, adding $y = 7$ and

$x = 2$, adding $y = 6$ and $x = 3$ and so on. Both elements for each addition are found inside a cell. These are the shaded cells shown in Table 4.4. Equation (E4.2-1) says that for $z = 9$ we need to take the maximum of the minima of the three pairs of grades of membership inside the shaded cells. First we find the minimum of the grades of membership inside each cell—that is,

$$\mu_A(1) \wedge \mu_B(8) = 0.3 \wedge 0.6 = 0.3$$
$$\mu_A(2) \wedge \mu_B(7) = 0.7 \wedge 1.0 = 0.7$$
$$\mu_A(3) \wedge \mu_B(6) = 1.0 \wedge 0.6 = 0.6$$
$$\mu_A(4) \wedge \mu_B(5) = 0.7 \wedge 0.2 = 0.2$$
$$\mu_A(5) \wedge \mu_B(4) = 0.3 \wedge 0 = 0$$
$$\mu_A(6) \wedge \mu_B(3) = 0 \wedge 0 = 0$$
$$\mu_A(7) \wedge \mu_B(2) = 0 \wedge 0 = 0$$
$$\mu_A(8) \wedge \mu_B(1) = 0 \wedge 0 = 0 \qquad \text{(E4.2-2)}$$

Now if we look only at the shaded part of the table, we can replace the contents of each cell with the minima found in equations (E4.2-2)—that is,

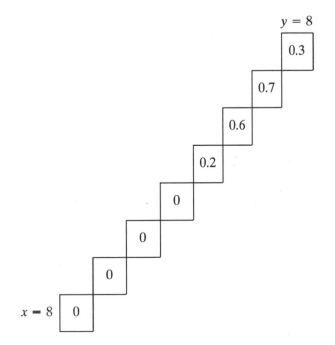

Next, we take the maximum of these numbers, which in this case is 0.7; this is the maximum with respect to $z = 9$ in equation (E4.2-1). At this point we have completed the entire operation on equation (E4.2-1) for $z = 4$—that is,

$$\mu_{A+B}(9) = [(0.3) \vee (0.7) \vee (0.6) \vee (0.2) \vee (0) \vee (0) \vee (0) \vee (0)]$$

$$= 0.7 \qquad \qquad \text{(E4.2-3)}$$

This is the grade of membership of $z = 9$ to the sum $C = A + B$. We repeat this procedure for all other cells to obtain the membership function of C. The result is

$$C = 0/5 + 0.2/6 + 0.3/7 + 0.6/8 + 0.7/9 + 1.0/10$$
$$+ 0.7/11 + 0.6/12 + 0.3/13 + 0.2/14 + 0/15$$

which is the same number as the one we found by the interval approach in Example 4.1—that is, the number shown in Table 4.3. □

4.4 SUBTRACTION

The difference C of two fuzzy numbers A, B may be defined either through interval subtraction utilizing the *α-cut representation* of the two numbers or through the *extension principle*. Using α-cuts we subtract them as follows

$$A - B \equiv \left[a_1^{(\alpha)}, a_2^{(\alpha)}\right] - \left[b_1^{(\alpha)}, b_2^{(\alpha)}\right] \qquad (4.4-1)$$

where $[a_1^{(\alpha)}, a_2^{(\alpha)}]$ is the collection of closed intervals representing A, and $[b_1^{(\alpha)}, b_2^{(\alpha)}]$ is the collection of closed intervals representing B. Two intervals are subtracted by subtracting their left and right endpoints, and thus equation (4.4-1) becomes

$$A - B = \left[a_1^{(\alpha)} - b_2^{(\alpha)}, a_2^{(\alpha)} - b_1^{(\alpha)}\right] \qquad (4.4-2)$$

The alternative way to define the difference of fuzzy numbers A and B is through the extension principle—that is, by fuzzifying a function $z = x - y$. Fuzzification means that we define fuzzy sets on the universes of discourse where the crisp elements x and y are found. As a result, z gets fuzzified as well; that is, there is a fuzzy set C over the universe of discourse of the z's, which is the result of fuzzifying the function $z = f(x, y) = x - y$. The membership function of $C = A - B$ can be computed from

$$\mu_{A-B}(z) \equiv \bigvee_{z=x-y} \left[\mu_A(x) \wedge \mu_B(y)\right] \qquad (4.4-3)$$

Equation (4.4-3) gives, of course, the same number C obtained through (4.4-2).

Example 4.3   Subtracting Fuzzy Numbers as Intervals. Let us compute a fuzzy number $C = 7 - 3$, where the fuzzy numbers 7 and 3 are as defined in Table 4.2 (Example 4.1):

$$A = 3 = 0.3/1 + 0.7/2 + 1.0/3 + 0.7/4 + 0.3/5 + 0/6 \quad \text{(E4.3-1)}$$

$$B = 7 = 0.2/5 + 0.6/6 + 1.0/7 + 0.6/8 + 0.2/9 + 0/10 \quad \text{(E4.3-2)}$$

Subtracting the two numbers is the same as interval subtraction at each α. From Table 4.2 we see that when $\alpha = 0.3$, for example, the 0.3-cuts of the two numbers are

$$A_{0.3} = \left[a_1^{(0.3)}, a_2^{(0.3)} \right] = [1, 5] \quad \text{(E4.3-3)}$$

and

$$B_{0.3} = \left[b_1^{(0.3)}, b_2^{(0.3)} \right] = [6, 8] \quad \text{(E4.3-4)}$$

The α-cut of C at $\alpha = 0.3$ is the difference of the α-cuts in by (E4.3-3) and (E4.3-4)

$$\begin{aligned}
C_{0.3} &= \left[b_1^{(0.3)}, b_2^{(0.3)} \right] - \left[a_1^{(0.3)}, a_2^{(0.3)} \right] \\
&= \left[b_1^{(0.3)} - a_2^{(0.3)}, b_2^{(0.3)} - a_1^{(0.3)} \right] \\
&= [6 - 5, 8 - 1] \\
&= [1, 7] \quad \text{(E4.3-5)}
\end{aligned}$$

shown as a "slice" of shaded cells in Table 4.5. In a similar manner we compute the α-cuts of C at the other levels of α and obtain the fuzzy number

$$C = 0.2/0 + 0.3/1 + 0.6/2 + 0.7/3 + 1.0/4$$
$$+ 0.7/5 + 0.6/6 + 0.3/7 + 0.2/8$$

which is also shown in Table 4.5. As may be seen from Table 4.5, C can be considered a fuzzy 4. $\square$

Example 4.4   Subtracting Fuzzy Numbers with Continuous Membership Functions. Consider the two triangular fuzzy numbers A and B shown in Figure 4.4. We want to compute their difference—that is, find a fuzzy number $C = A - B$. When continuous (or piecewise continuous) membership functions are used, we subtract them by parameterizing their membership functions by α and subtracting their α-cuts. The membership functions of

Table 4.5 Difference of fuzzy numbers 7 and 3 in Example 4.3

4:

	0.2	0.3	0.6	0.7	1	0.7	0.6	0.3	0.2	0
α=1.0					1					
α=0.9					1					
α=0.8					1					
α=0.7				1	1	1				
α=0.6			1	1	1	1	1			
α=0.5			1	1	1	1	1			
α=0.4			1	1	1	1	1			
α=0.3		1	1	1	1	1	1	1		
α=0.2	1	1	1	1	1	1	1	1	1	
α=0.1	1	1	1	1	1	1	1	1	1	
α=0.0	1	1	1	1	1	1	1	1	1	1
	0	1	2	3	4	5	6	7	8	9

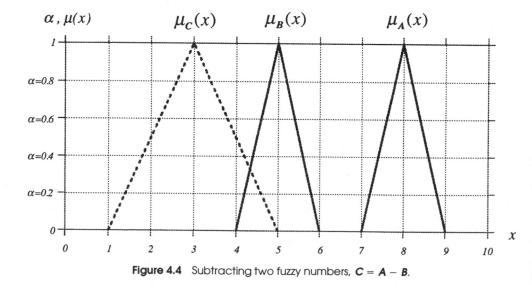

Figure 4.4 Subtracting two fuzzy numbers, **C = A − B**.

A, *B* are

$$\mu_A(x) = 0, \qquad\qquad x \le 7$$
$$= x - 7, \qquad 7 \le x \le 8$$
$$= -x + 9, \qquad 8 \le x \le 9$$
$$= 0, \qquad\qquad x \ge 9 \qquad\qquad \text{(E4.4-1)}$$

and

$$\mu_B(x) = 0, \qquad\qquad x \le 4$$
$$= x - 4, \qquad 4 \le x \le 5$$
$$= -x + 6, \qquad 5 \le x \le 6$$
$$= 0, \qquad\qquad x \ge 6 \qquad\qquad \text{(E4.4-2)}$$

Let us parameterize them by α. To simplify matters, consider the left and right side of each membership function separately. There is one equation for the left side and another for the right side of the membership function of *A*, and likewise for *B*. Thus, we have a total of four equations to parameterize. From equations (E4.4-1) we take the part that describes the left side of *A*, $\mu_A^L(x) = x - 7$, and write it in terms of a. We note that the value of α is the same as the value of the membership function at the left endpoint $a_1^{(\alpha)}$ of an α-cut, and $a_1^{(\alpha)}$ is the value of x at that point. Thus we have for the left side of *A*,

$$\alpha = a_1^{(\alpha)} - 7 \Rightarrow a_1^{(\alpha)} = \alpha + 7 \qquad\qquad \text{(E4.4-3)}$$

where $a_1^{(\alpha)}$ is the left endpoint of the "slice" of *A* at level α.

Similarly for the right side of *A* we parameterize the right endpoint $a_2^{(\alpha)}$ of each α-cut in terms of α as

$$\alpha = -a_2^{(\alpha)} + 9 \Rightarrow a_2^{(\alpha)} = -\alpha + 9 \qquad\qquad \text{(E4.4-4)}$$

Using equations (E4.4-3) and (E4.4-4) the α-cut representation of *A* is written as

$$A = \left[a_1^{(\alpha)}, a_2^{(\alpha)} \right] = \left[\alpha + 7, -\alpha + 9 \right] \qquad\qquad \text{(E4.4-5)}$$

The membership function of the number *B* is parameterized in terms of α in a similar fashion. We express the left endpoint $b_1^{(\alpha)}$ in terms of α by

$$\alpha = b_1^{(\alpha)} - 4 \Rightarrow b_1^{(\alpha)} = \alpha + 4 \qquad\qquad \text{(E4.4-6)}$$

The right endpoint $b_2^{(\alpha)}$ is given as a function of α by

$$\alpha = -b_2^{(\alpha)} + 6 \Rightarrow b_2^{(\alpha)} = -\alpha + 6 \qquad \text{(E4.4-7)}$$

From equations (E4.4-6) and (E4.4-7) the interval representation of B is

$$B = \left[b_1^{(\alpha)}, b_2^{(\alpha)} \right] = \left[\alpha + 4, -\alpha + 6 \right] \qquad \text{(E4.4-8)}$$

From the α-cut representations of A and B (equations (E4.4-5) and (E4.4-8)), we find their difference by subtracting their corresponding intervals at each α, that is,

$$\begin{aligned} C = A - B &= \left[a_1^{(\alpha)} - b_2^{(\alpha)}, a_2^{(\alpha)} - b_1^{(\alpha)} \right] \\ &= \left[(\alpha + 7) - (-\alpha + 6), (-\alpha + 9) - (\alpha + 4) \right] \\ &= \left[2\alpha + 1, -2\alpha + 5 \right] \end{aligned} \qquad \text{(E4.4-9)}$$

Therefore, C is

$$C = \left[c_1^{(\alpha)}, c_2^{(\alpha)} \right] = \left[2\alpha + 1, -2\alpha + 5 \right] \qquad \text{(E4.4-10)}$$

We note that the left and right endpoints of C are functions of α. To express the fuzzy number C in terms of a membership function, we derive equations for the left and right side of C. The left endpoint $c_1^{(\alpha)}$ in equation (E4.4-10) is equal to the value of x when the left-side membership function's value is α. Similarly, the right endpoint $c_2^{(\alpha)}$ is equal to the value of x when the right-side membership function is α. Thus the equation of the left side is obtained by setting $c_1^{(\alpha)} = x$ and recalling that $\alpha = \mu_C^L(x)$, where $\mu_C^L(x)$ is the left-side membership function for C. We have

$$x = 2\mu_C^L(x) + 1 \Rightarrow \mu_C^L(x) = \tfrac{1}{2}(x - 1) \qquad \text{(E4.4-11)}$$

In a similar manner we obtain an equation for $\mu_C^R(x)$, the right side of the membership function of C, and solve it to obtain the membership function of the right side—that is,

$$x = -2\mu_C^R(x) + 5 \Rightarrow \mu_C^R(x) = -\tfrac{1}{2}(x - 5) \qquad \text{(E4.4-12)}$$

From equations (E4.4-11) and (E4.4-12) we obtain

$$\begin{aligned} \mu_C(x) &= 0, & x &\leq 1 \\ &= \tfrac{1}{2}(x - 1), & 1 &\leq x \leq 3 \\ &= -\tfrac{1}{2}(x - 5), & 3 &\leq x \leq 5 \\ &= 0, & x &\geq 5 \end{aligned} \qquad \text{(E4.4-13)}$$

The number C described by equations (E4.4-11) is shown in Figure 4.4. Note that C has its peak at crisp 3, and therefore it can be considered as a fuzzy number *3* (as expected since, *8 − 5 = 3*). □

4.5 MULTIPLICATION

As in the case of addition and subtraction, fuzzy number multiplication may be defined either as α-cut multiplication or through the extension principle. Using the α-cut representation of two numbers A, B, their product is defined as

$$A \cdot B \equiv \left[a_1^{(\alpha)}, a_2^{(\alpha)} \right] \cdot \left[b_1^{(\alpha)}, b_2^{(\alpha)} \right] \tag{4.5-1}$$

In general, the product of two intervals is a new interval whose left endpoint is the product of the left endpoints of the two intervals and the right endpoint is the product of the right endpoints of the two intervals. Thus, equation (4.5-1) is

$$A \cdot B = \left[a_1^{(\alpha)} \cdot b_1^{(\alpha)}, a_2^{(\alpha)} \cdot b_2^{(\alpha)} \right] \tag{4.5-2}$$

Alternatively, we define the product of A and B through the extension principle by fuzzifying the function $z(x, y) = x \cdot y$. The extension principle tells us that their product is a fuzzy set on z, denoted as $A \cdot B$, whose membership function is

$$\mu_{A \cdot B}(z) \equiv \bigvee_{z = x \cdot y} \left[\mu_A(x) \wedge \mu_B(y) \right] \tag{4.5-3}$$

Of course, equations (4.5-2) and (4.5-3) are equivalent in that they give us the same number $C = A \cdot B$.

A special case of fuzzy multiplication is the product of fuzzy number by crisp number. Let k be a crisp positive real number and A a fuzzy number defined over the universe of discourse of positive real numbers also. We define the product of k with A either as interval multiplication or through the extension principle. Crisp number k may be viewed as an interval also, a trivial interval whose left and right endpoints are the same—that is, $k = [k, k]$. We use equations (4.5-1) and (4.5-2) to obtain the product of k with A as

$$k \cdot A = [k, k] \cdot \left[a_1^{(\alpha)}, a_2^{(\alpha)} \right]$$

$$= \left[k a_1^{(\alpha)}, k a_2^{(\alpha)} \right] \tag{4.5-4}$$

Alternatively, we define the product of fuzzy number A with a crisp number k, $k \cdot A$, through the extension principle. It may be shown using equation

(4.5-3) that the membership function of $k \cdot A$ is

$$\mu_{k \cdot A}(x) = \mu_A\left(\frac{x}{k}\right)$$ (4-5-5)

where equations (4.5-4) and (4.5-5) give the same result.

Example 4.5   Multiplication of Two Fuzzy Numbers. Consider the triangular fuzzy numbers $A = 8$ and $B = 2$ defined over the positive real numbers as shown in Figure 4.5 (since both numbers are defined over the same universe of discourse we simply use x to indicate an element of the universe of discourse, instead of x, y, etc.). We want to compute a fuzzy number C which is the product of A and B—that is, $C = A \cdot B$. Let us do this through α-cut multiplication—that is, by parameterizing their membership functions and multiplying their α-cuts in the manner indicated by equation (4.5-2).

First, we write the analytical expressions for the membership functions of A and B:

$$\mu_A(x) = 0, \qquad\qquad x \le 4$$

$$= \tfrac{1}{4}x - 1, \qquad 4 \le x \le 8$$

$$= -\tfrac{1}{4}x + 3, \qquad 8 \le x \le 12$$

$$= 0, \qquad\qquad x \ge 12 \qquad\qquad \text{(E4.5-1)}$$

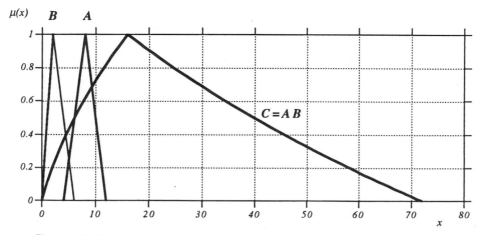

Figure 4.5 The product $C = A \cdot B$ of numbers $A = 8$ and $B = 2$ in Example 4.5.

and

$$\mu_B(x) = 0, \qquad\qquad x \le 0$$
$$= \tfrac{1}{2}x, \qquad\qquad 0 \le x \le 2$$
$$= -\tfrac{1}{4}x + \tfrac{3}{2}, \qquad 2 \le x \le 6$$
$$= 0 \qquad\qquad\quad x \ge 6 \qquad\qquad \text{(E4.5-2)}$$

Next, we parameterize the membership functions in equations (E4.5-1) and (E4.5-2) in terms α (a procedure of renaming the left and right side of the membership functions and thus the endpoints of all intervals in terms of α). Let us take the left and right side of each membership function separately and rewrite it in terms of α. It should be noted that a given value of α is the same as the value of the membership function at that level. From equation (E4.5-1) we have that the left and right endpoints of A are

$$\alpha = \tfrac{1}{4}a_1^{(\alpha)} - 1 \Rightarrow a_1^{(\alpha)} = 4(\alpha + 1) \qquad\qquad \text{(E4.5-3)}$$

and

$$\alpha = -\tfrac{1}{4}a_2^{(\alpha)} + 3 \Rightarrow a_2^{(\alpha)} = -4(\alpha - 3) \qquad\qquad \text{(E4.5-4)}$$

Using equations (E4.5-3) and (E4.5-4) we obtain A as

$$A = \left[a_1^{(\alpha)}, a_2^{(\alpha)} \right] = [4(\alpha + 1), -4(\alpha - 3)] \qquad \text{(E4.5-5)}$$

Similarly, we parameterize the membership function of B and write its left and right endpoints at each α as

$$\alpha = \tfrac{1}{2}b_1^{(\alpha)} \Rightarrow b_1^{(\alpha)} = 2\alpha \qquad\qquad \text{(E4.5-6)}$$

and

$$\alpha = -\tfrac{1}{4}b_2^{(\alpha)} + \tfrac{3}{2} \Rightarrow b_2^{(\alpha)} = -4\left(\alpha - \tfrac{3}{2}\right) \qquad \text{(E4.5-7)}$$

Thus, from equations (E4.5-6) and (E4.5-7) the interval representation of B is

$$B = \left[b_1^{(\alpha)}, b_2^{(\alpha)} \right] = \left[2\alpha, -4\left(\alpha - \tfrac{3}{2}\right) \right] \qquad \text{(E4.5-8)}$$

Having the endpoints of A and B in terms of α, we multiply the two numbers using equation (4.5-2) and obtain

$$C = A \cdot B = \left[a_1^{(\alpha)} \cdot b_1^{(\alpha)}, a_2^{(\alpha)} \cdot b_2^{(\alpha)} \right]$$
$$= \left[4(\alpha + 1) \cdot 2\alpha, -4(\alpha - 3) \cdot \left(-4\left(\alpha - \tfrac{3}{2}\right)\right) \right]$$
$$= [8\alpha^2 + 8\alpha, 16\alpha^2 - 72\alpha + 72] \qquad \text{(E4.5-9)}$$

The interval representation of C is

$$C = \left[c_1^{(\alpha)}, c_2^{(\alpha)} \right] = [8\alpha^2 + 8\alpha, 16\alpha^2 - 72\alpha + 72] \quad \text{(E4.5-10)}$$

where the left and right endpoints in equation (E4.5-10) are functions for α. We can obtain the membership function of C as well. Equation (E4.5-10) provides us with left and right endpoints of each α-cut. The equation for the left-side membership function $\mu_C^L(x)$ is obtained by setting $c_1^{(\alpha)} = x$ and recalling that $\alpha = \mu_C^L(x)$. Thus, we obtain an equation involving $\mu_C^L(x)$, which is

$$8\left(\mu_C^L(x) \right)^2 + 8\mu_C^L(x) - x = 0 \quad \text{(E4.5-11)}$$

Solving quadratic equation (E4.5-11) for $\mu_C^L(x)$, we obtain two solutions and accept only the value of $\mu_C^L(x)$ in [0, 1], ignoring the other one. The result is

$$\mu_C^L(x) = -\tfrac{1}{2} + \tfrac{1}{2}\sqrt{1 + \tfrac{1}{2}x} \quad \text{(E4.5-12)}$$

Similarly we obtain an equation for $\mu_C^R(x)$, the right side of the membership function of C, and solve it, keeping the solution which is within [0, 1]. The result is

$$\mu_C^R(x) = \tfrac{1}{2}\left(4.5 - \sqrt{(4.5)^2 - 4(4.5 - \tfrac{1}{16}x)} \right) \quad \text{(E4.5-13)}$$

The membership function of C is

$$\mu_C(x) = 0, \qquad\qquad\qquad\qquad\qquad\qquad x \leq 0$$

$$= -\tfrac{1}{2} + \tfrac{1}{2}\sqrt{1 + \tfrac{1}{2}x}, \qquad\qquad\quad 0 \leq x \leq 16$$

$$= \tfrac{1}{2}\left(4.5 - \sqrt{(4.5)^2 - 4(4.5 - \tfrac{1}{16}x)} \right), \quad 16 \leq x \leq 72$$

$$= 0, \qquad\qquad\qquad\qquad\qquad\qquad x \geq 72 \quad \text{(E4.5-14)}$$

as shown in Figure 4.5. It should be noted that C has its peak point at crisp 16 and therefore may be considered a fuzzy number *16*. It should also be noted that multiplying two fuzzy numbers results in a new number whose shape has been considerably changed, no longer having a triangular membership function with linear sides but in this case parabolic sides. Multiplication in general has the effect of "fattening" the lower part of the membership functions involved. □

4.6   DIVISION

We can find the *quotient* of two fuzzy numbers A and B either through interval division or by the extension principle. In terms of their α-cut representation, we write the *quotient* of the two numbers as

$$A \div B \equiv \left[a_1^{(\alpha)}, a_2^{(\alpha)} \right] \div \left[b_1^{(\alpha)}, b_2^{(\alpha)} \right] \qquad (4.6\text{-}1)$$

In general, the *quotient* of two intervals is a new interval given by

$$\left[a_1^{(\alpha)}, a_2^{(\alpha)} \right] \div \left[b_1^{(\alpha)}, b_2^{(\alpha)} \right] \equiv \left[\frac{a_1^{(\alpha)}}{b_2^{(\alpha)}}, \frac{a_2^{(\alpha)}}{b_1^{(\alpha)}} \right]$$

Hence, provided that $b_2^{(\alpha)} \neq 0$ and $b_1^{(\alpha)} \neq 0$, the *quotient* of A, B is

$$A \div B = \left[\frac{a_1^{(\alpha)}}{b_2^{(\alpha)}}, \frac{a_2^{(\alpha)}}{b_1^{(\alpha)}} \right] \qquad (4.6\text{-}2)$$

Alternatively, we find the *quotient* of A and B through the extension principle by fuzzifying the function $z(x, y) = x \div y$, where x and y are crisp elements of the universe of discourse of A and B. The extension principle tells us that $A \div B$ is a fuzzy set with membership function

$$\mu_{A \div B}(z) \equiv \bigvee_{z = x \div y} \left[\mu_A(x) \wedge \mu_B(y) \right] \qquad (4.6\text{-}3)$$

The results obtained through equations (4.6-4) and (4.6-2) are of course the same. Equation (4.6-3) may be used in the manner shown in Example 4.2. We construct a table such as Table 4.4 and proceed as outlined in the example. A word of caution: Fuzzy number division is not the reverse of multiplication; that is, generally it is not true that $(A \div B) \times C = A$.

Example 4.6   Division of Fuzzy Numbers. Consider the triangular fuzzy numbers $A = 8$ and $B = 2$ used in Example 4.5. Let us find $C = A \div B$ using interval division. The analytical expressions for the membership functions of A and B are given in Example 4.5 [equations (E4.5-1) and (E4.5-2)], and their parameterized interval representation is found in equations (E4.5-5) and (E4.5-8), which for convenience we repeat here:

$$A = \left[a_1^{(\alpha)}, a_2^{(\alpha)} \right] = \left[4(\alpha + 1), -4(\alpha - 3) \right] \qquad (E4.6\text{-}1)$$

$$B = \left[b_1^{(\alpha)}, b_2^{(\alpha)} \right] = \left[2\alpha, -4(\alpha - \tfrac{3}{2}) \right] \qquad (E4.6\text{-}2)$$

Thus their *quotient* $C = A \div B$ is obtained using equation (4.6-2):

$$C = A \div B = \left[\frac{a_1^{(\alpha)}}{b_2^{(\alpha)}}, \frac{a_2^{(\alpha)}}{b_1^{(\alpha)}} \right]$$

$$= \left[\frac{4(\alpha + 1)}{(-4(\alpha - \frac{3}{2}))}, \frac{-4(\alpha - 3)}{2\alpha} \right] \qquad \text{(E4.6-3)}$$

The α-cut representation of C is

$$C = \left[c_1^{(\alpha)}, c_2^{(\alpha)} \right] = \left[-\frac{(\alpha + 1)}{(\alpha - \frac{3}{2})}, -\frac{2(\alpha - 3)}{\alpha} \right] \qquad \text{(E4.6-4)}$$

where the left and right endpoints are functions of α. We may also express C in terms of a membership function by deriving equations for the left and right sides of the membership function as we did in Example 4.5. Equation (E4.6-4) gives us the endpoints of the interval of each α-cut. The equation of the left side is obtained by setting $c_1^{(\alpha)} = x$ and recalling that $\alpha = \mu_C^L(x)$, where, $\mu_C^L(x)$ is the left side membership function for C. The result is

$$\mu_C^L(x) = \frac{\frac{3}{2}x - 1}{x + 1} \qquad \text{(E4.6-5)}$$

Similarly we obtain an equation for $\mu_C^R(x)$, the right side of the membership function of C, and solve it to obtain

$$\mu_C^R(x) = \frac{6}{x + 2} \qquad \text{(E4.6-6)}$$

The quotient is shown in Figure 4.6, and the analytical description of the

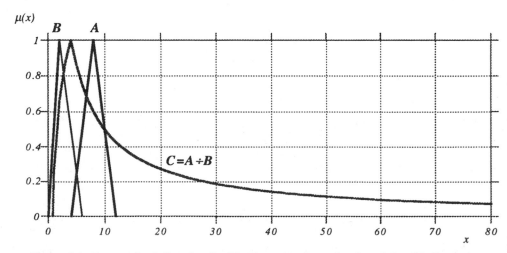

Figure 4.6 The quotient $C = A \div B$ of the fuzzy numbers $A = 8$ and $B = 2$ in Example 4.6.

membership function of C is

$$\mu_C(x) = 0, \qquad\qquad x \leq 0$$

$$= \frac{\frac{3}{2}x - 1}{x + 1}, \qquad 0 \leq x \leq 4$$

$$= \frac{6}{x + 2}, \qquad 4 \leq x \leq 72$$

$$= 0, \qquad\qquad x \geq 72 \qquad\qquad \text{(E4.6-7)}$$

It should be noted from equation (E4.6-7) that the quotient is a new fuzzy number that no longer has a triangular shape with linear sides. As may be seen from the figure, the fuzzy number C only asymptotically reaches zero and hence we may consider the use of a *level fuzzy set* (Chapter 2) in order to limit and exclude trivially small grades of membership—for example, less than 0.2. □

4.7 MINIMUM AND MAXIMUM

The minimum and maximum of two fuzzy numbers A, B result in finding the smallest and the biggest one, respectively, and may be defined either through their interval representation or by the extension principle. In interval arithmetic the minimum of two intervals is a new interval whose left endpoint is the minimum of the left endpoints of the original intervals and whose right endpoint is the minimum of the right endpoints of the two intervals. Thus the minimum of A, B is a new number, $A \wedge B$, given by

$$A \wedge B \equiv \left[a_1^{(\alpha)}, a_2^{(\alpha)}\right] \wedge \left[b_1^{(\alpha)}, b_2^{(\alpha)}\right]$$

$$= \left[a_1^{(\alpha)} \wedge b_1^{(\alpha)}, a_2^{(\alpha)} \wedge b_2^{(\alpha)}\right] \qquad (4.7\text{-}1)$$

Alternatively, the minimum of two fuzzy numbers may be obtained through the extension principle. The membership function of $A \wedge B$ is

$$\mu_{A \wedge B}(z) \equiv \bigvee_{z = x \wedge y} \left[\mu_A(x) \wedge \mu_B(y)\right] \qquad (4.7\text{-}2)$$

In an analogous manner we define the maximum of two fuzzy numbers A and B, recalling that in interval arithmetic the maximum of two intervals is a new interval whose left endpoint is the maximum of the left endpoints of the original intervals and whose right endpoint is the maximum of the right endpoints of the two intervals. Thus the maximum $A \vee B$ is given by

$$A \vee B \equiv \left[a_1^{(\alpha)}, a_2^{(\alpha)}\right] \vee \left[b_1^{(\alpha)}, b_2^{(\alpha)}\right]$$

$$= \left[a_1^{(\alpha)} \vee b_1^{(\alpha)}, a_2^{(\alpha)} \vee b_2^{(\alpha)}\right] \qquad (4.7\text{-}3)$$

Alternatively, by the extension principle the membership function of the

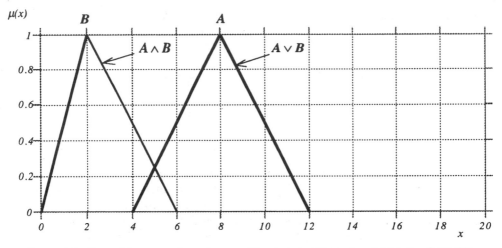

Figure 4.7 The minimum and maximum of the two numbers **A** = **8** and **B** = **2** used in Example 4.6.

maximum of the two numbers A and B is

$$\mu_{A \vee B}(z) \equiv \bigvee_{z=x \vee y} \left[\mu_A(x) \wedge \mu_B(y) \right] \tag{4.7-4}$$

It should be noted that the *maximum* and *minimum* of two fuzzy numbers are different than the maximum and minimum of membership functions used in connection with the *union* and *intersection* of two fuzzy sets. Let us illustrate this by finding the minimum of the numbers $A = 8$ and $B = 2$ used in Examples 4.5 and 4.6 and redrawn in Figure 4.7. Equations (4.7-1) or (4.7-2) do not give us the little wedge between A and B, which is the *intersection* of A and B. They will simply give us the number $B = 2$ itself, which is the smallest of the two fuzzy numbers. Similarly the largest of the numbers is found by using the maximum operation of either equation (4.7-3) or (4.7-4), which is simply the number $A = 8$, as shown in Figure 4.7. For more intricately overlapping membership functions the maximum or minimum may not simply be a number with the membership function of either A or B, but may have a totally new shape (Kaufmann and Gupta, 1991).

REFERENCES

Dubois, D., and Prade, H., *Fuzzy Sets and Systems*: *Theory and Applications*, Academic Press, Boston, 1980.

Kandel, A., *Fuzzy Mathematical Techniques with Applications*, Addison-Wesley, Reading, MA, 1986.

Kaufmann, A., and Gupta, M. M., *Introduction to Fuzzy Arithmetic*, Van Nostrand Reinhold, New York, 1991.

Klir, G. J., and Folger, T. A., *Fuzzy Sets, Uncertainty, and Information*, Prentice-Hall, Englewood Cliffs, NJ, 1988.

Pal, S. K., and Majumder, D. K. D, *Fuzzy Mathematical Approach to Pattern Recognition*, John Wiley & Sons, New York, 1986.

Sanchez, E., Non Standard Fuzzy Arithmetic, in *Between Mind and Computer-Fuzzy Science and Engineering*, P-Z. Wang, and K-F. Loe, eds. World Scientific, Singapore, 1993, pp. 271–282.

Terano, T., Asai, K., and Sugeno, M., *Fuzzy Systems Theory and Its Applications*, Academic Press, Boston, 1992.

Zimmermann, H. J., *Fuzzy Set Theory and its Applications*, Kluwer-Nijhoff, Boston, 1985.

PROBLEMS

1. The fuzzy numbers A and B are given by

$$A = 0.33/6 + 0.67/7 + 1.00/8 + 0.67/9 + 0.33/10$$
$$B = 0.33/1 + 0.67/2 + 1.00/3 + 0.67/4 + 0.33/5$$

Subtract B from A to give fuzzy number C. Draw a sketch of C.

2. Multiply fuzzy numbers A and B of Problem 1. Draw a sketch of C.

3. Divide fuzzy number A by fuzzy number B where the fuzzy numbers are defined in Problem 1. Draw a sketch of C.

4. Modify Example 4.2 to subtract the two fuzzy numbers using the extension principle.

5. Consider the fuzzy numbers A and B described by the membership functions:

$$\mu_A(x) = 0, \qquad\qquad x \le 8,$$
$$= \frac{1}{10}x - \frac{8}{10}, \qquad 8 \le x \le 18,$$
$$= -\frac{1}{14}x + \frac{32}{14}, \qquad 18 \le x \le 32,$$
$$= 0, \qquad\qquad x > 32,$$
$$\mu_B(x) = 0, \qquad\qquad x \le -3,$$
$$= \frac{1}{9}x - \frac{1}{3}, \qquad -3 \le x \le 6,$$
$$= -\frac{1}{18}x + \frac{4}{3}, \qquad 6 \le x \le 24,$$
$$= 0, \qquad\qquad x > 24$$

Compute:

(a) $A\,(+)\,B$,

(b) $A\,(-)\,B$,

(c) $A\,(\div)\,B$.

6. Repeat the computations in Problem 5 for the fuzzy numbers A and B given below, and using C state and show the distributivity property (with respect to addition and multiplication)

$$A = 0.6/1 + 0.8/2 + 1.0/3 + 0.6/4$$
$$B = 0.5/0 + 0.7/1 + 0.9/2 + 1.0/3 + 0.4/4$$
$$C = 0.7/1 + 0.8/2 + 1.0/3 + 0.3/4$$

5

LINGUISTIC DESCRIPTIONS AND THEIR ANALYTICAL FORMS

5.1 FUZZY LINGUISTIC DESCRIPTIONS

Fuzzy linguistic descriptions (often called *fuzzy systems* or simply *linguistic descriptions*) are formal representations of systems made through fuzzy *if/then* rules. They offer an alternative and often complementary language to conventional (analytic) approaches to modeling systems (involving differential or difference equations). Informal linguistic descriptions used by humans in daily life as well as in the performance of skilled tasks, such as control of industrial facilities, troubleshooting, aircraft landing, and so on, are usually the starting point for the development of fuzzy linguistic descriptions. Although fuzzy linguistic descriptions are formulated in a human-like language, they have rigorous mathematical foundations involving fuzzy sets and relations (Zadeh, 1988). They encode knowledge about a system in statements of the form

if (a set of conditions are satisfied)

then (a set of consequences can be inferred)

For example, in process control the desirable behavior of a system may be formulated as a collection of rules combined by the connective *ELSE* such as

if   error is   *ZERO*   *AND* Δerror is ZERO   *then* Δu is ZERO   ELSE

if   error is   *PS*   *AND* Δerror is ZERO   *then* Δu is NS   ELSE

...

if   error is   *SMALL*   *AND* Δerror is NS   *then* Δu is BIG

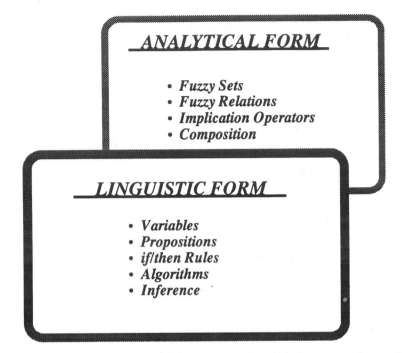

Figure 5.1 Fuzzy linguistic descriptions possess a linguistic form as well as a background analytical form involving fuzzy set operations.

where *error* and Δ*error* (change in error) are linguistic variables describing the input to a controller and Δ*u* is a linguistic variable describing the change in output. A *linguistic variable* is a variable whose arguments are fuzzy numbers (and more generally words modeled by fuzzy sets), which we refer to as *fuzzy values*. For example, in the rules above the *fuzzy values* of the linguistic variable *error* are *ZERO*, *PS* (*positive small*), and *SMALL*, the values of Δ*error* are *ZERO* and *NS* (*negative small*) and the values of Δ*u* are *ZERO*, *NS*, and *BIG*.[1] A specific evaluation of a fuzzy variable—for example, "*error is ZERO*"—is called *fuzzy proposition*. Individual fuzzy propositions on either *left-* (*LHS*) or *right-hand side* (*RHS*) of a rule maybe connected by connectives such as *AND* and *OR*— for example, "*error is PS AND* Δ*error is ZERO*."[2] Individual *if/then* rules are connected with the connective *ELSE* to form a *fuzzy algorithm*. Propositions and *if/then* rules in classical logic are supposed to be either true or false. In fuzzy logic they can be true or false to a degree.

---

[1] The convention we follow is to use lowercase italics for linguistic variables and capital italics for fuzzy values, unless otherwise specified or implied by the context.
[2] These are also called *antecedent* (*LHS*) and *consequent* (*RHS*) propositions. We find alternative designations for the *LHS* and *RHS* of a rule in different application areas. In process control, for example, the *if part* is often referred to as the *situation side* and the *then part* is often referred to as the *action* side.

Figure 5.1 shows schematically what is involved in linguistic descriptions. In the front end we find *linguistic forms* representing a system in a human-like manner. In the background we have rigorously defined *analytical forms* involving fuzzy set operations, relations, and composition procedures such as the ones we saw in Chapters 2 and 3.

Despite the difference in appearance, linguistic and conventional (analytic) descriptions are in fact equivalent to each other. Both can be used to describe the same system. However, the computational costs incurred using one or the other may be significantly different. Consider, for example, a function $y = f(x)$ shown in Figure 5.2, describing analytically a specific relation between x's and y's.[3] The same relation may be described by listing all possible, or at least a sufficiently large number of, (x, y) pairs or *points* of $f(x)$, indicating (for example), that when $x = a_1$ the value of the function is $y = b_1$, when $x = a_2$ the value of the function is $y = b_2$, when $x = a_i$ the value of the function is $y = b_i$, and so on. Knowing n such points we may alternatively represent $y = f(x)$ by listing the pairs

$$(a_1, b_1)$$

$$(a_2, b_2)$$

$$\cdots$$

$$(a_i, b_i) \qquad\qquad (5.1\text{-}1)$$

$$\cdots$$

$$(a_n, b_n)$$

Of course this representation is an acceptable approximation of the analytic representation only when n becomes sufficiently large, with the precision of the approximation being controlled by choosing an appropriate n. A point (a_i, b_i) can also be thought of as a crisp *if/then* rule of the form, "*if x is a_i then y is b_i*." Obviously, the pairs of (5.1-1) may be expressed linguistically as crisp rules:

$$\begin{aligned}
&if \quad x\ is\ a_1 \quad then \quad y\ is\ b_1 \\
&if \quad x\ is\ a_2 \quad then \quad y\ is\ b_2 \\
&\quad\cdots \\
&if \quad x\ is\ a_i \quad then \quad y\ is\ b_i \qquad\qquad (5.1\text{-}2)\\
&\quad\cdots \\
&if \quad x\ is\ a_n \quad then \quad y\ is\ b_n
\end{aligned}$$

[3]As we saw in Chapter 3, *functions* are a particular kind of *relation* allowing one and only one value of y for each x. This is also referred to as a *many-to-one mapping*.

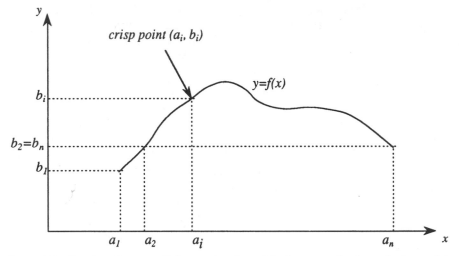

Figure 5.2 The function $y = f(x)$ may be thought of as a collection of crisp points (a_i, b_i), and each point may also be articulated as a crisp *if / then* rule.

Every representation has a cost. We can think of it as related to the number of symbols used and the complexity of operations involved, but actually it involves much more—for example, the cost of extracting the knowledge used, its realization in a machine, the cost of updating and maintaining it, and so on. When we use several crisp rules to represent $y = f(x)$ in the manner of (5.1-2), we are obviously using a more costly representation in a computational sense. By comparison, the analytical description $y = f(x)$ offers a more economical way of describing the function. In this sense the analytical description $y = f(x)$ is said to be a more *parsimonious* description than (5.1-2), in reference to the reduced cost of representation.

Intuitively we expect the crisp linguistic rendition of $y = f(x)$ to become more accurate with increasing number of rules. Having 1000 crisp rules for $f(x)$ is preferable to, say, 10 rules. However, the number of crisp *if/then* rules needed to describe a function such as the one shown in Figure 5.2 actually depends on the specific nature of $f(x)$ as well as our tolerance for approximation error. Take, for instance, a linear function, a straight line going through the origin. In this case, one crisp *if/then* rule may suffice since an additional point on the x–y plane outside the origin uniquely identifies a straight line. On the other hand, a very "noisy" function with many "spikes" and slope changes will require considerably more rules. In practical terms, however, an approximate description of $y = f(x)$ may be acceptable, sometimes even preferable. We are often interested in associations such as *if x is "about a_i," then y is "about b_i"*; that is to say, we are interested not in a crisp point of $f(x)$ but in an area or neighborhood around a point. This is illustrated in Figure 5.3, where instead of crisp point (a_i, b_i) we consider the circled area around (a_i, b_i) which may be thought of as an *area-cum-point*, an

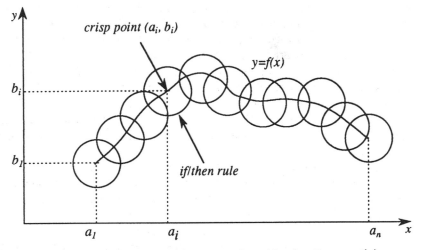

Figure 5.3 Building a linguistic description of the function $y = f(x)$.

area obtained from a point. Such an *area-cum-point* may be described by a fuzzy *if/then* rule. Let us consider "*about a_i,*" to be a fuzzy number A_i on the universe of discourse of the x's and consider "*about b_i,*" to be a fuzzy number B_i on the universe of discourse of the y's. As we will see later on (Section 5.2), we can define a linguistic variable x whose arguments are fuzzy numbers on the x-axis, such as A_i, and a linguistic variable y whose arguments are fuzzy numbers on the y-axis, such as B_i. Hence the *area-cum-point* "about (a_i, b_i)" can be described by a fuzzy *if/then* rule of the form

$$\text{if } x \text{ is } A_i \quad \text{then} \quad y \text{ is } B_i \tag{5.1-3}$$

The analytical form of rule (5.1-3) is a fuzzy relation $R_i(x, y)$ called the *implication relation* of the rule. How we obtain this implication relation is a rather complicated issue which we will examine in more detail in Section 5.3. For the moment we assume that each fuzzy *if/then* rule has an *implication relation*.

The function $y = f(x)$ may be approximated by collecting several fuzzy *if/then* rules—for example,

$$\text{if} \quad x \text{ is } A_1 \quad \text{then} \quad y \text{ is } B_1 \quad ELSE$$

$$\text{if} \quad x \text{ is } A_2 \quad \text{then} \quad y \text{ is } B_2 \quad ELSE$$

$$\cdots$$

$$\text{if} \quad x \text{ is } A_i \quad \text{then} \quad y \text{ is } B_i \quad ELSE \tag{5.1-4}$$

$$\cdots$$

$$\text{if} \quad x \text{ is } A_n \quad \text{then} \quad y \text{ is } B_n$$

where $A_1, A_2, \ldots, A_i, \ldots, A_n$ are fuzzy numbers on the x axis and $B_1, B_2, \ldots, B_i, \ldots, B_n$ are fuzzy numbers on the y axis. The rules of (5.1-4) are combined by the connective *ELSE*, which could be analytically modeled as either *intersection* or *union* [and more generally as *T norms* or *S norms* (see Appendix A)] depending on the *implication relation* of the individual rules (we will have more on *ELSE* in Section 5.5). The collection of *if/then* rules in (5.1-4) is called a *fuzzy algorithm*, and its analytical form is a relation $R_\alpha(x, y)$ between the x's and the y's, called the *algorithmic relation*. As may be expected, the *algorithmic relation* depends on the *implication relation* of constituent rules.

The transition from conventional descriptions, such as $y = f(x)$, to linguistic descriptions addresses the fact that functions are often mathematical idealizations. In most real-world problems, we do not have a curve such as the one shown in Figure 5.2 but rather something like the region shown in Figure 5.4. For example, suppose that the function $y = f(x)$ is viewed as a control policy—that is, a prescription recommending a control action y—for each state x. In many applications, the control system changes with time (*time-varying*) and in general manifests nonlinear and complex behaviors. Hence, the control policy may actually be a more general relation $R_\alpha(x, y)$ as shown in Figure 5.4. Figure 5.2 could in fact be an idealization of the real-world control policy shown in Figure 5.4. We recall (see also Chapter 3) that a function is a special kind of relation that associates a unique y with each x. A function performs what is called a *many-to-one mapping*; that is, several values of x may have the same value of y but not vice versa. Most real-world applications, however, involve *many-to-many mappings*. Situations like the one shown in Figure 5.4—that is, relations that are *many-to-many*

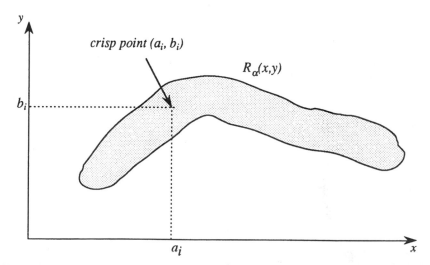

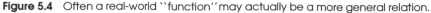

Figure 5.4   Often a real-world ''function'' may actually be a more general relation.

mappings—are far more common in complex engineering systems than usu-ally. Sometimes conventional descriptions, being overly idealized models of complex systems, may suffer from lack of robustness and exhibit undesirable side effects.

Let us look again in Figure 5.3. We note that the transition from *points* to *area-cum-points* reduces the number of *if/then* rules needed to describe $y = f(x)$. For example, we could approximate $f(x)$ with only 11 fuzzy *if/then* rules (circled areas) as shown in Figure 5.3. The rules are overlapping as are the various fuzzy numbers on the x and y axes. Yet, we no longer have a function (a *many-to-one mapping*) but a more general relation $R_\alpha(x, y)$ (a *many-to-many mapping*), and the obvious question is: How do we use such a relation? In conventional descriptions we evaluate functions by inputting a crisp value of x to $f(x)$ and obtain a unique crisp value of y as output. Something similar can be done with linguistic descriptions as well. The process of evaluating a fuzzy linguistic description is called *fuzzy inference*. There are two important problems in fuzzy inference. First, given a fuzzy number A' as input to a linguistic description, we want to obtain a fuzzy number B' as its output; and, second, given B', we want to obtain A' (the inverse problem). The first problem is addressed with an inferencing procedure called *generalized modus ponens* (GMP), and the second is ad-dressed with another inferencing procedure called *generalized modus tollens* (GMT). Both GMP and GMT have their origin in the field of logic and approximate reasoning (Section 5.4), and analytically they involve composi-tion of fuzzy relations (Chapter 3).

In GMP, when an *if/then* rule and its antecedent are approximately matched, a consequent may be inferred. For simplicity let us consider only a generic rule of (5.1-4) having an implication relation $R(x, y)$. GMP is formally stated as

$$if \quad x \, is \, A \quad then \quad y \, is \, B$$

$$\underline{\quad x \, is \, A' \quad}$$

$$y \, is \, B'$$

(5.1-5)

where A' is an input value matching the antecedent A to a degree (including totally perfect and totally imperfect match). The implication relation of the rule $R(x, y)$ and the input A' above the line are considered known, whereas what is below the line—in other words B'—is considered unknown. B' is what we want to find. Analytically, GMP (5.1-5) is performed by composing A' with the implication relation $R(x, y)$ as in the max–min composition (see Chapter 3)

$$B' = A' \circ R(x, y)$$

(5.1-6)

We will see how this is done in detail in Section 5.4. For the moment let us simply keep in mind that we can evaluate linguistic descriptions just as we can evaluate functions and that the procedure of evaluation involves composition of fuzzy relations. GMP is related to forward-chaining or data-driven inference and is the main inferencing procedure in fuzzy control. When $A' = A$ and $B' = B$, GMP (5.1-5) reduces to an inferencing procedure of classical logic known as *modus ponens* (depending on the implication relation).

In GMT a rule and its consequent are approximately matched and from that we can obtain an antecedent. GMT is formally stated as

$$\text{if} \quad x \, is \, A \quad \text{then} \quad y \, is \, B$$

$$\frac{y \, is \, B'}{x \, is \, A'} \tag{5.1-7}$$

Again, everything above the line is known and we want to find out what is below the line—that is, A'. The analytical problem involved in GMT is addressed by composing the implication relation $R(x, y)$ with fuzzy number B' as

$$A' = R(x, y) \circ B' \tag{5.1-8}$$

GMT is closely related to backward-chaining or goal-driven inference, which is the main form of inference used in diagnostic expert systems. When $A' = NOT \ A$ and $B' = NOT \ B$, GMT reduces to classical *modus tollens* (depending on the implication relation used).

In general, fuzzy linguistic descriptions offer convenient tools for controlling the *granularity* of a description,[4] in the sense that they facilitate the choice of appropriate precision levels—that is, levels that application-specific considerations call for. In terms of our example, when we use fuzzy numbers and fuzzy *if/then* rules to describe $y = f(x)$, we have at our disposal a mechanism for reducing the number of rules needed and, hence, for controlling the *granularity* of this particular description and the overall cost of computation (Zadeh, 1979). In addition, the technology for computing with *if/then* rules has already advanced to the point where fuzzy microprocessors, called *fuzzy chips*, are widely available (Yamakawa, 1987; Isik, 1988; Hirota and Ozawa, 1988; Huertas et al., 1992; Shimizu et al., 1992). Fuzzy chips encoding knowledge in the form of linguistic descriptions can function as "mounted devices"—that is, dedicated processors fine-tuned to the specifics of a component and its environment, performing domain-specific computations. Such processors are already deployed in several control and robotics applications with remarkable successes (Yamakawa, 1988; Pin et al., 1992). Of course, software is a commonly used medium for the implementation of fuzzy algorithms on a variety of different computers. However, the advent of fuzzy logic hardware and the development of fuzzy computers may have a

---

[4]By *granularity* we roughly mean the coarseness of a description, the level of precision necessary to effectively represent a given system.

profound impact on the design and operation of engineering systems (Yamakawa, 1988). Fuzzy linguistic descriptions are of growing importance in many areas of engineering ranging from expert systems and artificial intelligence applications to process control, pattern recognition, signal analysis, reliability engineering, and machine learning (Ray and Majumder, 1988). The basic ideas, however, are rather similar and rest on the mathematics of fuzzy sets. Describing a system through a linguistic description, no matter for what purpose, involves specifying in some way *linguistic variables, if/then rules*, and evaluation procedures known as *fuzzy inference*.

5.2 LINGUISTIC VARIABLES AND VALUES

As we saw in the previous section, a linguistic variable is a variable whose arguments are fuzzy numbers and more generally words represented by fuzzy sets. For example, the arguments of the linguistic variable *temperature* may be *LOW*, *MEDIUM*, and *HIGH*. We call such arguments *fuzzy values*. Each and every one of them is modeled by its own membership function. The fuzzy values *LOW*, *MEDIUM*, and *HIGH* may be modeled as shown in Figure 5.5 or Figure 5.6. In Figure 5.5 we have three discrete fuzzy values, while in Figure 5.6 we have three (piecewise) continuous membership functions— $\mu_{LOW}(T)$, $\mu_{MEDIUM}(T)$, and $\mu_{HIGH}(T)$—modeling the words *LOW*,

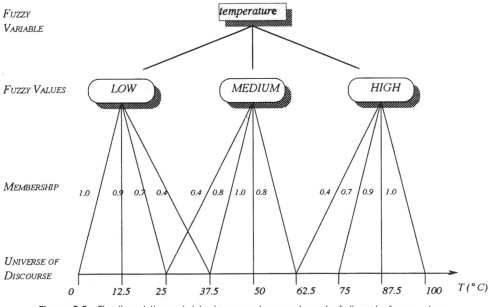

Figure 5.5 The linguistic variable *temperature* and a set of discrete fuzzy values.

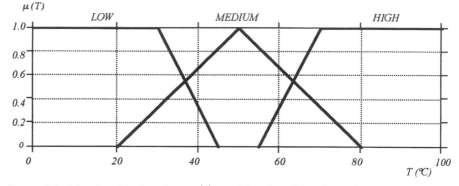

Figure 5.6 Membership functions $\mu(T)$ used for describing the primary values, *LOW*, *MEDIUM*, and *HIGH*, of the linguistic variable *temperature*.

MEDIUM, and *HIGH*, respectively. Any crisp value of temperature e.g., 60°C) has a unique degree of membership to each fuzzy value of *temperature*. In Figure 5.6, for example, crisp temperature 60°C is *LOW* to a degree zero, *MEDIUM* to a degree 0.65, and *HIGH* to a degree 0.35.

We distinguish four different levels in the definition of a linguistic variable as shown in Figure 5.5. At the top level we have the name of the variable (e.g., *temperature*). At the level below it we have the labels of *fuzzy values* (starting with an initial set of values called *primary values* or *term set*[5]). Further down we have *membership functions*, and at the bottom we have the *universe of discourse*. All four levels are indispensable in the definition of a variable. It is important to observe that linguistic variables have a dual nature; at higher levels we have a symbolic linguistic form, and at lower levels we have a well-defined quantitative analytical form—that is, the membership function. This double identity is a general feature of fuzzy linguistic descriptions rendering them convenient for performing both symbolic (qualitative) and numerical (quantitative) computations (Zadeh, 1975).

Generally, the values of a linguistic variable may be *compound values*—that is, values constructed through the use of *primary values* and linguistic modifiers such as *NOT, VERY, RATHER, ALMOST*, and *MORE OR LESS*. For example, out of the initial set of primary values *LOW, MEDIUM*, and *HIGH* for *temperature*, compound values such as *NOT LOW, VERY LOW, RATHER MEDIUM*, and *ALMOST HIGH* may be formed.

Fuzzy values are essentially *aggregations* or *categories* of crisp values. In fuzzy logic the flexibility of adjusting membership functions is useful for categorizing the parameters of a domain in accordance with the domain's own unique features. If temperature is considered in a conventional

---

[5]The *values* of a linguistic variable are referred to by a variety of names in the literature. Often they are called *fuzzy variables, primary terms, set of prototypes*, or the *term set*. We will mostly use the name *fuzzy value* or simply *value*. Linguistic variables are also called *fuzzy variables*.

sense—that is, as a numerical variable—its arguments are simply the crisp numbers of a universe of discourse (e.g., natural numbers between 1°C and 100°C). We may think of each number as a crisp category of temperature; in this case we could have 100 different categories. For certain applications this may be an acceptable categorization of the values of temperature. For others we may need 1000 categories, and still for others 3 categories may suffice. Fuzzy values provide this kind of flexibility. They allow for adjustable categories and explicitly acknowledge the ambiguous and application-dependent nature of this or the other categorization.

Primary Values

The words which function as the initial values of a linguistic variable are called *primary values*. They are the principal categorization of a universe of discourse—for example, the values *LOW*, *MEDIUM*, and *HIGH* shown in Figures 5.5 and 5.6. To model them we often use functions whose shape is adjusted through a finite set of parameters. For example, the function

$$\mu(x) = \frac{1}{1 + a(x - c)^b} \tag{5.2-1}$$

has parameters a, b, and c which may be used to adjust the overall form of $\mu(x)$. Parameter a adjusts the width of the membership function, b determines the extent of fuzziness, and c describes the location of the "peak" of the membership function. This is the point in the universe of discourse where $\mu(x) = 1$. Consider the primary values of *temperature*, *SMALL*, *MEDIUM*, and *LARGE* shown in Figure 5.7. Their membership functions are of the form of Equation (5.2-1) with $a = 0.0005$, $b = 3$, and $c = 20$, 50, and 80,

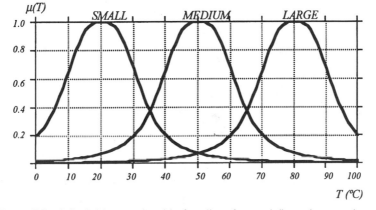

Figure 5.7 Adjustable membership functions for modeling primary values.

respectively—that is,

$$\mu_{SMALL}(T) = \frac{1}{1 + 0.0005(|T - 20|)^3}$$

$$\mu_{MEDIUM}(T) = \frac{1}{1 + 0.0005(|T - 50|)^3} \qquad (5.2\text{-}2)$$

$$\mu_{LARGE}(T) = \frac{1}{1 + 0.0005(|T - 80|)^3}$$

In many control applications, continuous membership functions such as the trapezoidal/triangular functions of Figure 5.6 are used. Fuzzy values defined through trapezoidal/triangular membership functions have adjustable parameters as well, namely the "corners" of the function—that is, the points where the monotonicity changes. We recall their use in Chapter 4 in connection with fuzzy numbers. In fuzzy arithmetic, however, we required that fuzzy sets be normalized—that is, that there be at least one point of the universe of discourse where the membership function reaches unity, whereas in fuzzy linguistic descriptions this requirement is relaxed. Fuzzy values ought to be convex, just as fuzzy numbers, but not necessarily normal.

Primary values can also be modeled through *S-shaped* and *Π-shaped* functions named by their general form (Zimmermann, 1985; Kandel, 1986). *S*-shaped and *Π*-shaped membership functions may be adjusted to suit various application needs merely by altering a limited number of parameters as in the case of trapezoidal and triangular membership functions. *S*-shaped functions are defined through three parameters α, β, and γ as follows:

$$S(x; \alpha, \beta, \gamma) = 0 \qquad \text{for } x \leq \alpha$$

$$S(x; \alpha, \beta, \gamma) = 2\left(\frac{x - \alpha}{\gamma - \alpha}\right)^2 \qquad \text{for } \alpha \leq x \leq \beta$$

$$\qquad (5.2\text{-}3)$$

$$S(x; \alpha, \beta, \gamma) = 1 - 2\left(\frac{x - \gamma}{\gamma - \alpha}\right)^2 \qquad \text{for } \beta \leq x \leq \gamma$$

$$S(x; \alpha, \beta, \gamma) = 1 \qquad \text{for } x \geq \gamma$$

where x is any real number and α, β, and γ are appropriately chosen parameters. For continuity of slope at $x = \beta$, the two intervals $(\beta - \alpha)$ and $(\gamma - \beta)$ must be equal.

A Π-shaped function may be thought of as two S-shaped functions put together "back-to-back" and can be expressed as

$$\Pi(x; \delta, \gamma) = S\left(x; \gamma - \delta, \frac{\gamma - \delta}{2}, \gamma\right) \qquad \text{for } x \le \gamma$$

$$\Pi(x; \delta, \gamma) = 1 - S\left(x; \gamma, \frac{\gamma + \delta}{2}, \gamma + \delta\right) \qquad \text{for } x \ge \gamma$$

(5.2-4)

The parameter δ in Π-shaped functions is called the *bandwidth.* It is the distance between the crossover (inflection) points—that is, the points where the function equals 0.5. The parameter γ is the point where the Π-shaped function reaches unity. Fuzzy values modeled by S-shaped and Π-shaped functions are more often encountered in software than in hardware realizations of fuzzy linguistic descriptions. Triangular/trapezoidal membership functions are the preferred shapes for fuzzy values used in hardware realizations.

Compound Values

Using the connectives *AND* and *OR* and a collection of linguistic modifiers such as *NOT, VERY, MORE OR LESS, RATHER,* and so on, we can generate compound values from primary values. Modifiers and connectives are modeled by fuzzy set operations as well. For example, *AND* and *OR* are modeled by the fuzzy set operations of *intersection* and *union,* respectively, while *NOT* is modeled by *complementation.* More generally they are modeled by T and S norms (see Appendix A). Through linguistic modifiers we may easily construct a larger, potentially infinite set of values from a relatively small and finite set of primary values. Some modifiers are also called *linguistic hedges* due to the property of semantically constraining (*hedging*) the general meaning of a word by operating on the fuzzy set that represents it (Zadeh, 1983).

The connective *OR* generates a compound value with membership function equal to the max ($\vee$) of the membership functions of other values. Consider the values A and B defined over the same universe of discourse X as

$$A = \int_X \mu_A(x)/x, \qquad B = \int_X \mu_B(x)/x$$

The compound value "$A \ OR \ B$" is defined as

$$A \ OR \ B \equiv \int_X [\mu_A(x) \vee \mu_B(x)]/x \qquad (5.2\text{-}5)$$

The connective *AND* uses the min operator ($\wedge$) to generate the membership function of the compound value out of the membership functions of two (or more) other values. The compound value constructed through the connective *AND* is defined as

$$A \; AND \; B \equiv \int_X [\, \mu_A(x) \wedge \mu_B(x)]/x \qquad (5.2\text{-}6)$$

The *AND* connective has to be used with caution when generating compound values because it may lead to nonsensical words such as in the proposition "*temperature* is (*HIGH AND LOW*)." As shown in Figure 5.8a, this compound value has zero membership function and may be thought of as meaningless. The connective *AND* can produce correct compound values when used with the complement of primary values as, for example, in the proposition "*temperature* is ((*NOT LOW*) *AND* (*NOT HIGH*))," whose membership function can be seen in Figure 5.8b.

The membership function of a compound value produced by negating another value is the complement of the membership function of the original value—that is,

$$NOT \; A \equiv \int_X [1 - \mu_A(x)]/x \qquad (5.2\text{-}7)$$

The semantics of the modifier *NOT* are fairly straightforward, and it may be used very much as negation is used in natural language—for example, "*temperature* is (*NOT HIGH*)."

Every linguistic modifier is associated with a corresponding fuzzy set operation involving membership functions. Table 5.1 lists some of these associations. The *PLUS* and *MINUS* modifiers in Table 5.1 offer a smaller degree of concentration and dilation than do the concentration *CON* and dilation *DIL* operations which we saw in Chapter 2. Modifiers may be connected in series in order to form larger compound values. Suppose, for example, that we start with the primary values *SMALL* and *LARGE*. We form compound values such as (*VERY SMALL*) and (*NOT VERY SMALL*) by logically multiplying *SMALL* by *VERY* and *NOT*. We can go on in this manner obtaining more compound values—for example, *C* = ((*NOT VERY SMALL*) *AND* (*NOT LARGE*)). Using the operations in Table 5.1 we model *C* by the following membership function:

$$\mu_C(x) = [1 - \mu_{SMALL}^2(x)] \wedge [1 - \mu_{LARGE}(x)] \qquad (5.2\text{-}8)$$

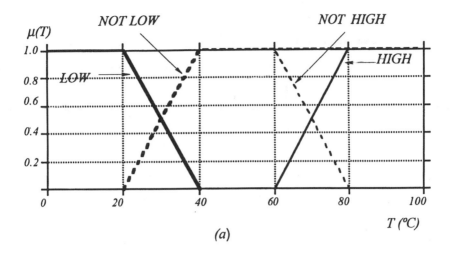

(a)

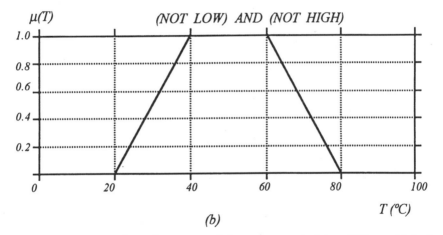

(b)

Figure 5.8 The semantics of compound terms generated by *AND* ought to be carefully examined. In (*a*) the compound term *LOW AND HIGH* has trivial membership function, while in (*b*) the compound term (*NOT LOW*) *AND* (*NOT HIGH*) is well defined.

It should be noted that compound fuzzy values may not be arbitrarily generated. We need to examine their semantics—that is, their meaning in the context of a specific application. An interesting quantitative guide to the semantics of compound values is provided by their membership function. When the new membership function becomes uniformly 1 or 0 we may have a semantically suspect compound value.

Table 5.1   Translation of linguistic modifiers into fuzzy set operations

MODIFIER	*MEMBERSHIP FUNCTION OPERATION*
VERY A	$\mu_{CON(A)}(x) \equiv \left[\mu_A(x) \right]^2$
MORE OR LESS A	$\mu_{DIL(A)}(x) \equiv \left[\mu_A(x) \right]^{1/2}$
INDEED A	$\mu_{INT(A)}(x)$ [see Equation (2.3–21)]
PLUS A	$\left[\mu_A(x) \right]^{1.25}$
MINUS A	$\left[\mu_A(x) \right]^{0.75}$
OVER A	$1 - \mu_A(x), \quad x \geq x_{max}$ $0, \qquad\quad x < x_{max}$
UNDER A	$1 - \mu_A(x), \quad x \leq x_{min}$ $0, \qquad\quad x > x_{min}$

5.3   IMPLICATION RELATIONS

Fuzzy *if/then* rules are conditional statements that describe the dependence of one (or more) linguistic variable on another. As we already alluded to earlier, the underlying analytical form of an *if/then* rule is a fuzzy relation called the *implication relation*. There are over 40 different forms of implication relations reported in the literature (Lee, 1990a, b). Implication relations are obtained through different *fuzzy implication operators* ϕ. Information from the left- (LHS) and right-hand side (RHS) of a rule is inputted to ϕ, and it outputs an implication relation. The choice of implication operator is a rather significant step in the overall development of a fuzzy linguistic description. It reflects application-specific criteria, as well as logical and intuitive considerations focusing on the interpretation of the connectives *AND*, *OR*, and *ELSE*. An extensive discussion of different implication relations may be found in Mizumoto (1988), Lee (1990a, b), and Ruan and Kerre (1993).[6] We will examine here the most common implication operators used in engineering applications, particularly in fuzzy control (Chapter 6). Our focus will be on the implication relation of a simple *if/then* rule and on how to obtain it from LHS and RHS membership functions.

Let us consider a generic *if/then* rule involving two linguistic variables, one on each side of the rule—for example,

$$if \quad x \text{ is } A \quad then \quad y \text{ is } B \tag{5.3-1}$$

where linguistic variables x and y take the values A and B, respectively. The underlying analytical form of rule (5.3-1) is the *implication relation*

$$R(x, y) = \int_{(x, y)} \mu(x, y)/(x, y) \tag{5.3-2}$$

where $\mu(x, y)$ is the membership function of the implication relation, the thing we want to obtain. When the linguistic variables in (5.3-1) are defined over discrete universes of discourse, an implication relation is written as

$$R(x_i, y_j) = \sum_{(x_i, y_j)} \mu(x_i, y_j)/(x_i, y_j) \tag{5.3-3}$$

There are several options for obtaining the membership function of the implication relation. We explore them through the implication operator notion. For the rule of (5.3-1) an implication operator ϕ takes as input the membership functions of the antecedent and consequent parts, namely, $\mu_A(x)$ and $\mu_B(y)$, and takes as outputs $\mu(x, y)$, namely

$$\mu(x, y) = \phi[\mu_A(x), \mu_B(y)] \tag{5.3-4}$$

---

[6] Implication operators can also be expressed through T and S norms (see Appendix). It should be noted that the term "implication" is somewhat of a misnomer (since strictly speaking there is no logical implication in a rule); nonetheless it is widely used in the literature.

We distinguish the following implication operators:

Zadeh Max–Min Implication Operator

The *Zadeh max–min* implication operator (Zadeh, 1973) is

$$\phi_m[\,\mu_A(x), \mu_B(y)\,] \equiv (\,\mu_A(x) \wedge \mu_B(y)) \vee (1 - \mu_A(x)) \quad (5.3\text{-}5)$$

Thus the membership function of the implication relation (5.3-2) is

$$\mu(x, y) = (\,\mu_A(x) \wedge \mu_B(y)) \vee (1 - \mu_A(x))$$

Mamdani Min Implication Operator

The *Mamdani min* implication operator is a simplified version of Zadeh max–min proposed by Mamdani in the 1970s in connection with fuzzy control (Mamdani, 1977) and is defined as

$$\phi_c[\,\mu_A(x), \mu_B(y)\,] \equiv \mu_A(x) \wedge \mu_B(y) \quad (5.3\text{-}6)$$

Larsen Product Implication Operator

The *Larsen product* implication operator uses arithmetic product (Larsen, 1980) and is defined as

$$\phi_p[\,\mu_A(x), \mu_B(y)\,] \equiv \mu_A(x) \cdot \mu_B(y) \quad (5.3\text{-}7)$$

Arithmetic Implication Operator

The *arithmetic* implication operator is based in multivalued logic (Zadeh, 1975) and is defined as

$$\phi_a[\,\mu_A(x), \mu_B(y)\,] \equiv 1 \wedge (1 - \mu_A(x) + \mu_B(y)) \quad (5.3\text{-}8)$$

Boolean Implication Operator

The *Boolean* implication operator is based on classical logic and has been used in control and decision-making applications. It is defined as

$$\phi_b[\,\mu_A(x), \mu_B(y)\,] \equiv (1 - \mu_A(x)) \vee \mu_B(y) \quad (5.3\text{-}9)$$

The Bounded Product Implication Operator

The *bounded product* fuzzy implication operator has been used in fuzzy control and is defined as

$$\phi_{bp}[\,\mu_A(x), \mu_B(y)\,] \equiv 0 \vee (\,\mu_A(x) + \mu_B(y) - 1) \quad (5.3\text{-}10)$$

The Drastic Product Implication Operator

The *drastic product* implication operator has also been used in the field of control. As the name implies, it involves a more drastic (crisp) decision as to the form of the implication relation and is defined as

$$\phi_{dp}[\,\mu_A(x), \mu_B(y)\,] \equiv \begin{cases} \mu_A(x), & \mu_B(y) = 1 \\ \mu_B(y), & \mu_A(x) = 1 \\ 0, & \mu_A(x) < 1, \mu_B(y) < 1 \end{cases} \qquad (5.3\text{-}11)$$

The Standard Sequence Implication Operator

The *standard sequence* implication operator has crisp logic features. It is defined as

$$\phi_s[\,\mu_A(x), \mu_B(y)\,] \equiv \begin{cases} 1, & \mu_A(x) \le \mu_B(y) \\ 0, & \mu_A(x) > \mu_B(y) \end{cases} \qquad (5.3\text{-}12)$$

Gougen Implication Operator

The *Gougen* implication operator considers the fuzzy implication relation to be strong, reaching unity, if the membership function of the antecedent $\mu_A(x)$ is smaller than the membership function of the consequent $\mu_B(x)$. Otherwise, the greater $\mu_A(x)$ becomes, relative to $\mu_B(y)$, the more the membership function of the implication relation $\mu(x, y)$ comes to resemble that of the consequent. The Gougen implication relation is in a way a more tempered version of the standard sequence operator. It is formally defined as

$$\phi_\Delta[\,\mu_A(x), \mu_B(y)\,] \equiv \begin{cases} 1, & \mu_A(x) \le \mu_B(y) \\ \dfrac{\mu_B(y)}{\mu_A(x)}, & \mu_A(x) > \mu_B(y) \end{cases} \qquad (5.3\text{-}13)$$

Gödelian Implication Operator

The *Gödelian* implication operator is defined as

$$\phi_g[\,\mu_A(x), \mu_B(y)\,] \equiv \begin{cases} 1, & \mu_A(x) \le \mu_B(y) \\ \mu_B(y), & \mu_A(x) > \mu_B(y) \end{cases} \qquad (5.3\text{-}14)$$

These fuzzy implication operators are listed in Table 5.2. They are frequently encountered in engineering applications particularly in fuzzy control (Chapter 6). One interesting issue arises in connection with whether or not some of

Table 5.2 Some fuzzy implication operators

NAME	*IMPLICATION OPERATOR* $\phi[\mu_A(x), \mu_B(y)] =$
ϕ_m, *Zadeh Max-Min*	$(\mu_A(x) \wedge \mu_B(y)) \vee (1 - \mu_A(x))$
ϕ_c, *Mamdani min*	$\mu_A(x) \wedge \mu_B(y)$
ϕ_p, *Larsen Product*	$\mu_A(x) \cdot \mu_B(y)$
ϕ_a, *Arithmetic*	$1 \wedge (1 - \mu_A(x) + \mu_B(y))$
ϕ_b, *Boolean*	$(1 - \mu_A(x)) \vee \mu_B(y)$
ϕ_{bp}, *Bounded Product*	$0 \vee (\mu_A(x) + \mu_B(y) - 1)$
ϕ_{dp}, *Drastic Product*	$\mu_A(x),\quad if\quad \mu_B(y) = 1$ $\mu_B(x),\quad if\quad \mu_A(y) = 1$ $0,\quad if\quad \mu_A(y) < 1, \mu_B(y) < 1$
ϕ_s, *Standard Sequence*	$1,\quad if\quad \mu_A(x) \leq \mu_B(y)$ $0,\quad if\quad \mu_A(x) > \mu_B(y)$
ϕ_Δ, *Gougen*	$1,\qquad if\quad \mu_A(x) \leq \mu_B(y)$ $\frac{\mu_B(y)}{\mu_A(x)},\quad if\quad \mu_A(x) > \mu_B(y)$
ϕ_g, *Gödelian*	$1,\qquad if\quad \mu_A(x) \leq \mu_B(y)$ $\mu_B(y),\quad if\quad \mu_A(x) > \mu_B(y)$

these operators satisfy *classical modus ponens* and *modus tollens*. Another issue has to do with the manner that they satisfy certain intuitive criteria about inferencing such as, for example, the expectation that evaluating *"if x is A then y is B"* by *"x is VERY A"* ought to result in *"y is VERY B."* A good discussion of these issues is found in Mizumoto (1988) and Lee (1990a, b).

5.4   FUZZY INFERENCE AND COMPOSITION

Fuzzy inference refers to computational procedures used for evaluating fuzzy linguistic descriptions. There are two important inferencing procedures: *generalized modus ponens* (GMP) and *generalized modus tollens* (GMT). For simplicity let us consider a linguistic description involving only a simple *if/then* rule with known implication relation $R(x, y)$ and a fuzzy value A' approximately matching the antecedent of the rule. GMP allows us to compute (infer) the consequent B'. It is formally stated as

$$\begin{array}{c} if \quad x \text{ is } A \quad then \quad y \text{ is } B \\ \underline{x \text{ is } A'} \\ y \text{ is } B' \end{array} \qquad (5.4\text{-}1)$$

where everything above the line is analytically known, and what is below is analytically unknown. Suppose, for example, that we have the rule *"if temperature is HIGH then humidity is ZERO."* Given that *"temperature is VERY HIGH,"* GMP allows us to evaluate the rule and infer a value for *humidity*. The inferred value B' is computed through the composition of A' with the implication relation $R(x, y)$. Let us look at what is involved analytically in (5.4-1). We know the implication relation $R(x, y)$ of the rule *"if x is A then y is B"* (obtained by using one of the operators shown in Table 5.2) and the membership function of A'. To compute the membership function of B' in (5.4-1), we use *max–min composition* of fuzzy set A' with $R(x, y)$—that is,

$$B' = A' \circ R(x, y) \qquad (5.4\text{-}2)$$

In terms of membership functions, equation (5.4-2) is (see Chapter 3)

$$\mu_{B'}(y) = \bigvee_{x} \left[\mu_{A'}(x) \wedge \mu(x, y) \right] \qquad (5.4\text{-}3)$$

where $\mu_{A'}(x)$ is the membership function of A', $\mu(x, y)$ is the membership function of the implication relation, and $\mu_{B'}(y)$ the membership function of B'. We recall from Chapter 3 that max–min composition ($\circ$) is analogous to matrix multiplication with max ($\vee$) and min ($\wedge$) in place of *addition* ($+$) and *multiplication* ($\times$).

In GMT a rule and a fuzzy value approximately matching its consequent are given and it is desired to infer the antecedent—that is,

$$if \quad x \; is \; A \quad then \quad y \; is \; B$$

$$\underline{\hspace{2cm} y \; is \; B' \hspace{2cm}} \qquad (5.4\text{-}4)$$

$$x \; is \; A'$$

In GMT we know $R(x, y)$ and the consequent B'. To compute the membership function of A' in (5.4-2), we can use max–min composition of $R(x, y)$ with fuzzy set B'—that is,

$$A' = R(x, y) \circ B' \qquad (5.4\text{-}5)$$

In terms of membership functions, equation (5.4-5) is (see Chapter 3)

$$\mu_{A'}(x) = \bigvee_{y} [\mu(x, y) \wedge \mu_{B'}(y)] \qquad (5.4\text{-}6)$$

Of course, other compositions may be used in place of max–min. For example, using max-product composition the membership function of B' in (5.4-2) is given by

$$\mu_{B'}(y) = \bigvee_{x} [\mu_{A'}(x) \cdot \mu_{R}(x, y)] \qquad (5.4\text{-}7)$$

where we take the maximum with respect to x of all the products of the pairs inside the brackets (see Example 5.2). In general max-* composition may be used to infer the membership function of B':

$$\mu_{B'}(y) = \bigvee_{x} [\mu_{A'}(x) * \mu_{R}(x, y)] \qquad (5.4\text{-}8)$$

Using composition of relations to infer consequents—that is, to draw conclusions on the basis of imprecise premises—is known as the *compositional rule of inference*, since logical inferencing such as GMP is performed analytically through composition. As shown in Figure 5.9, GMP works in a manner analogous to *evaluating* a function and GMT is analogous to finding the *inverse* (Pappis and Sugeno, 1985). When a fuzzy value A' is given as input to a linguistic description (single rule or fuzzy algorithm) we can obtain B' through GMP; conversely, if we know B' we can obtain A' through GMT. Generally, we have several overlapping rules, and more than one may contribute a nontrivial B' (or A'). The *union* or *intersection* (depending on the implication operator used as we will see in the next section) of all contributions is the output of the linguistic description for a given A' (or B'). Often the fuzzy values used are not symmetric or of the same form, and

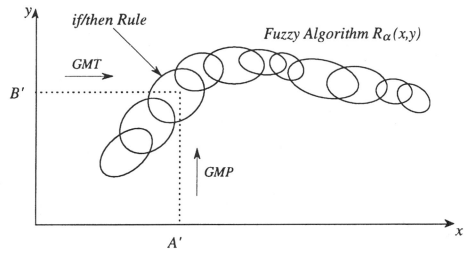

Figure 5.9 GMP and GMT are procedures for evaluating fuzzy linguistic descriptions.

hence we may not have circular *area-cum-points* as in Figure 5.3 but instead have the more general shapes shown in Figure 5.9. Of course in order to use composition we must have available *implication* and *algorithmic relations*.

Logical operations other than GMP or GMT may also be performed analytically through composition—for example, by combining two or more rules in a *syllogism* (Zimmermann, 1985). Consider the following rules:

$$
\begin{aligned}
&\text{if}\quad x \text{ is } A \quad \text{then } y \text{ is } B\\
&\text{if}\quad y \text{ is } B \quad \text{then } z \text{ is } C
\end{aligned}
\tag{5.4-9}
$$

from which we can infer another rule: "*If x is A then z is C*" through *syllogism*. Each rule in (5.4-9) is analytically described by a fuzzy relation, the first by $R_1(x, y)$ and the second by $R_2(y, z)$. From these relations we may infer a new relation $R_{12}(x, z)$ for the rule "*if x is A then z is C*" using max−min composition of $R_1(x, y)$ and $R_2(y, z)$—that is, $R_{12}(x, z) = R_1(x, y) \circ R_2(y, z)$. Again, max−min, max-product, or max-$*$ composition may also be used to obtain $R_{12}(x, z)$.

Example 5.1   GMP and Mamdani Min Implication. In this example we use GMP to evaluate a linguistic description comprised of a single rule "*if x is A then y is B*" with LHS and RHS membership functions $\mu_A(x)$ and $\mu_B(y)$, as shown in Figures 5.10a and 5.10b. The implication relation of the rule is modeled through Mamdani min implication operator. Fuzzy number A' (a singleton) shown in Figure 5.10c is the input to the rule. From Figure 5.10

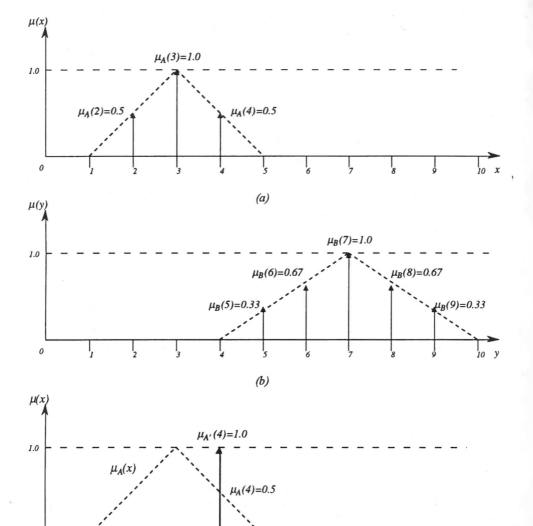

Figure 5.10 (*a*) The membership function of the antecedent *A*. (*b*) The fuzzy value *B* of the consequent. (*c*) A fuzzy *A'* that approximately matches the antecedent in Example 5.1.

we have

$$A = \sum_{i=0}^{10} \mu_A(x_i)/x_i$$

$$= 0.5/2 + 1.0/3 + 0.5/4 \qquad (E5.1\text{-}1)$$

$$B = \sum_{i=0}^{10} \mu_B(y_i)/y_i$$

$$= 0.33/5 + 0.67/6 + 1.0/7 + 0.67/8 + 0.33/9 \qquad (E5.1\text{-}2)$$

$$A' = \sum_{i=0}^{10} \mu_A(x_i)/x_i = 1.0/4 \qquad (E5.1\text{-}3)$$

All variables are defined over the same universe of discourse, the set of integers from 0 to 10; and, as is customary, zero membership singletons are omitted.

First, let us compute the membership function $\mu(x_i, y_j)$ of the implication relation $R(x_i, y_j)$ that analytically describes the rule using the Mamdani min implication operator. Having discrete fuzzy values we use discretized fuzzy relations as well. From Table 5.2 we see that the membership function of the implication relation is given by

$$\mu(x_i, y_j) = \phi_c\big[\,\mu_A(x_i), \mu_B(y_j)\big]$$

$$= \mu_A(x_i) \wedge \mu_B(y_j) \qquad (E5.1\text{-}4)$$

Thus the analytical form of the rule is given by the implication relation (5.3-3)—that is,

$$R(x_i, y_j) = \sum_{(x_i, y_j)} \mu(x_i, y_j)/(x_i, y_j)$$

$$= 0.33/(2,5) + 0.5/(2,6) + 0.5/(2,7) + 0.5/(2,8)$$

$$+ 0.33/(2,9) + 0.33/(3,5) + 0.67/(3,6) + 1.0(3,7)$$

$$+ 0.67/(3,8) + 0.33/(3,9) + 0.33/(4,5) + 0.5/(4,6)$$

$$+ 0.5/(4,7) + 0.5/(4,8) + 0.33/(4,9) \qquad (E5.1\text{-}5)$$

where the membership function is computed using (E5.1-4). The implication relation $R(x_i, y_j)$ of (E5.1-5) is shown in Table 5.3. We note that the full relation is defined over the Cartesian product of the discrete universe of discourse of LHS and RHS variables. Since each universe of discourse is the set of integers from 0 to 10, the Cartesian product is the 11×11 product

Table 5.3 The fuzzy implication relation in Example 5.1

y_j x_i	0	1	2	3	4	5	6	7	8	9	10
0	0	0	0	0	0	0	0	0	0	0	0
1	0	0	0	0	0	0	0	0	0	0	0
2	0	0	0	0	0	0.33	0.50	0.5	0.5	0.33	0
3	0	0	0	0	0	0.33	0.67	1.0	0.67	0.33	0
4	0	0	0	0	0	0.33	0.5	0.5	0.5	0.33	0
5	0	0	0	0	0	0	0	0	0	0	0
6	0	0	0	0	0	0	0	0	0	0	0
7	0	0	0	0	0	0	0	0	0	0	0
8	0	0	0	0	0	0	0	0	0	0	0
9	0	0	0	0	0	0	0	0	0	0	0
10	0	0	0	0	0	0	0	0	0	0	0

space shown in Table 5.3. The nontrivial part of the relation is found in the shaded cells of Table 5.3.

To find B' we compose A' with $R(x_i, y_j)$ in accordance with equation (5.4-2). It is sufficient to consider the nonzero part of the relation—that is, the shaded part of Table 5.3. We use matrix notation and remind ourselves (see Chapter 3) that max–min composition ($\circ$) is analogous to matrix multiplication with max ($\vee$) and min ($\wedge$) in place of addition ($+$) and multiplication ($\times$), respectively. From Equation (5.4-2) we have

$$B'(y_j) = A'(x_i) \circ R(x_i, y_j)$$

$$= \begin{bmatrix} 0 & 0 & 1 \end{bmatrix} \circ \begin{bmatrix} 0.33 & 0.50 & 0.50 & 0.50 & 0.33 \\ 0.33 & 0.67 & 1.00 & 0.66 & 0.33 \\ 0.33 & 0.50 & 0.50 & 0.50 & 0.33 \end{bmatrix} \quad \text{(E5.1-6)}$$

where the column vector for A' ranges from $x = 2$ to $x = 4$ (see Figure 5.10) which is the same as the row range of the implication matrix. The columns of the implication matrix range from $y = 5$ to $y = 9$ (see Table 5.3). From equation (5.4-3) the membership function of the first element of the consequent—that is, at $y = 5$—is computed as follows:

$$\mu_{B'}(5) = \bigvee_x [0 \wedge 0.33, 0 \wedge 0.33, 1 \wedge 0.33]$$

$$= \bigvee_x [0, 0, 0.33]$$

$$= 0.33 \qquad\qquad\qquad\qquad (E5.1\text{-}7)$$

Similarly we compute the rest of B'. The result is

$$B' = 0.33/5 + 0.50/6 + 0.50/7 + 0.5/8 + 0.33/9 \qquad (E5.1\text{-}8)$$

as shown in Figure 5.11. It should be noted in Figure 5.11 that the membership function of B' is essentially the membership function of B clipped at a height equal to the degree that A' matches A (see Figure 5.10c). This value is called the *degree of fulfillment* (DOF) of the rule. It is a measure of the degree of similarity between the input A' and the antecedent of the rule A. In the present case we have that

$$DOF = 0.5 \qquad\qquad\qquad\qquad (E5.1\text{-}9)$$

Clipping the membership function of the consequent by DOF is a feature of ϕ_c, the Mamdani min implication operator. Whenever we use ϕ_c to model the implication relation involved in GMP we get such a clipping transformation of the consequent. The situation is shown in general in Figure 5.12. We shall encounter clipping in Chapter 6 when dealing with control applications of linguistic descriptions. We should keep in mind that clipping depends on the Mamdani min implication operator (not to be confused with max–min composition). Using different implication operators to model the implication relation leads to different shape transformations of the RHS of a rule evaluated under GMP.   □

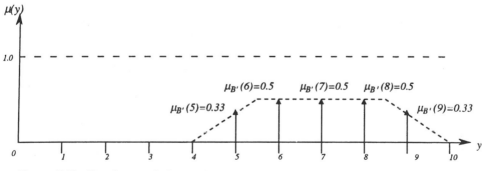

Figure 5.11 The fuzzy set B' produced by evaluating the linguistic description of Example 5.1.

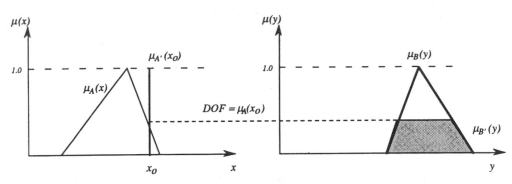

Figure 5.12 When the Mamdani min implication operator is used to model an implication relation, GMP clips the membership function of the consequent by the DOF of the rule.

Example 5.2 GMP with Larsen Product Implication. In this example we evaluate a fuzzy *if/then* rule, whose implication relation is modeled by the Larsen product fuzzy implication operator ϕ_p (see Table 5.2) using GMP. The antecedent and consequent variables of rule *if x is A then y is B* are shown in Figures 5.13a and 5.13b. The membership function of the input value A' is shown in Figure 5.13c. From Figure 5.14 we have

$$A = \sum_{i=-5}^{5} \mu_A(x_i)/x_i$$

$$= 0.33/(-1) + 0.67/0 + 1.0/1 + 0.75/2 + 0.5/3 + 0.25/4 \quad \text{(E5.2-1)}$$

$$B = \sum_{i=-5}^{5} \mu_B(y_i)/y_i$$

$$= 0.50/(-4) + 1.0/(-3) + 0.67/(-2) + 0.33/(-1) \quad \text{(E5.2-2)}$$

$$A' = \sum_{i=-5}^{5} \mu_A(x_i)/x_i = 1.0/3 \quad \text{(E5.2-3)}$$

Using equations (5.4-2) and (5.4-3) for GMP we can compute the membership function of value B'. First, however, we have to obtain the membership function of the implication relation, $\mu(x, y)$, using the Larsen product fuzzy implication operator ϕ_p (see Table 5.2). The implication relation has the membership function

$$\mu(x_i, y_j) = \phi_p[\mu_A(x_i), \mu_B(y_j)] = \mu_A(x_i) \cdot \mu_B(y_j) \quad \text{(E5.2-4)}$$

and plugging in numbers from equations (E5.2-1) to (E5.2-3) we obtain the

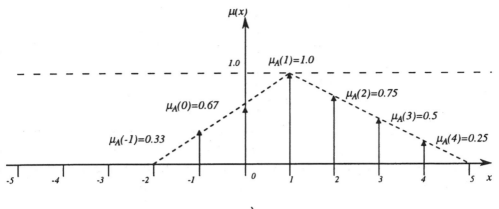

(a)

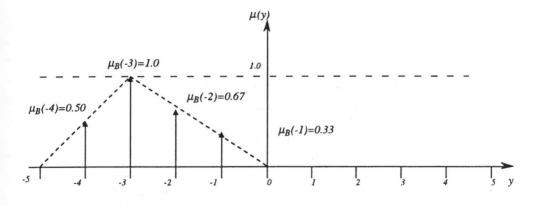

(b)

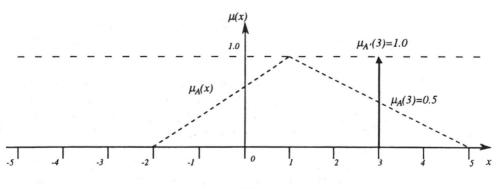

(c)

Figure 5.13 (a) The fuzzy value A of the antecedent. (b) The fuzzy value B of the consequent. (c) The singleton A' that approximately matches the antecedent in Example 5.2.

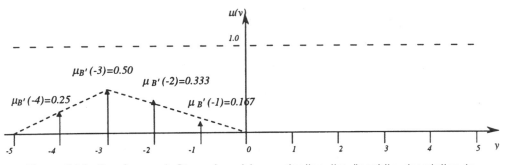

Figure 5.14 The fuzzy set B' produced by evaluating the linguistic description in Example 5.2.

implication relation

$$R(x_i, y_j) = \sum_{(x_i, y_j)} \mu(x_i, y_j)/(x_i, y_j)$$

$$= 0.167/(-1, -4) + 0.333/(-1, -3) + 0.222/(-1, -2)$$

$$+ 0.111/(-1, -1) + 0.333/(0, -4) + 0.667/(0, -3)$$

$$+ 0.445/(0, -2) + 0.222/(0, -1) + 0.500/(1, -4)$$

$$+ 1.000/(1, -3) + 0.667/(1, -2) + 0.333/(1, -1)$$

$$+ 0.375/(2, -4) + 0.750/(2, -3) + 0.500/(2, -2)$$

$$+ 0.250/(2, -1) + 0.250/(3, -4) + 0.500/(3, -3)$$

$$+ 0.333/(3, -2) + 0.167/(3, -1) + 0.125/(4, -4)$$

$$+ 0.250/(4, -3) + 0.167/(4, -2) + 0.083/(4, -1)$$

$$(E5.2\text{-}5)$$

The implication relation of (E5.2-5) can also be seen as the shaded part of Table 5.4, where we use a similarly scaled discrete universe of discourse for both antecedent and consequent variables, namely, integers from -5 to $+5$. Thus, the implication relation is taking values on an 11×11 Cartesian product space as shown.

We find B' through GMP—that is, max–min composition of A' with $R(x_i, y_j)$. Again we need only consider the nonzero part of the relation—that is, the shaded part of Table 5.4. We use matrix notation and remind ourselves (see Chapter 3) that max–min composition is analogous to matrix multiplication with the max ($\vee$) and min ($\wedge$) in the role of addition ($+$) and

Table 5.4 Implication relation in Example 5.2

y_j x_i	-5	-4	-3	-2	-1	0	1	2	3	4	5
-5	0	0	0	0	0	0	0	0	0	0	0
-4	0	0	0	0	0	0	0	0	0	0	0
-3	0	0	0	0	0	0	0	0	0	0	0
-2	0	0	0	0	0	0	0	0	0	0	0
-1	0	0.165	0.333	0.222	0.111	0	0	0	0	0	0
0	0	0.333	0.667	0.445	0.222	0	0	0	0	0	0
1	0	0.500	1.00	0.667	0.333	0	0	0	0	0	0
2	0	0.375	0.750	0.495	0.250	0	0	0	0	0	0
3	0	0.250	0.50	0.333	0.166	0	0	0	0	0	0
4	0	0.125	1.00	0.167	0.083	0	0	0	0	0	0
5	0	0	0	0	0	0	0	0	0	0	0

multiplication ($\times$). Thus we have

$$B' = A' \circ R = \begin{bmatrix} 0 & 0 & 0 & 0 & 1.0 & 0 \end{bmatrix} \circ \begin{bmatrix} 0.165 & 0.333 & 0.222 & 0.111 \\ 0.333 & 0.667 & 0.444 & 0.222 \\ 0.50 & 1.000 & 0.667 & 0.333 \\ 0.375 & 0.750 & 0.500 & 0.250 \\ 0.250 & 0.500 & 0.333 & 0.167 \\ 0.125 & 0.250 & 0.167 & 0.083 \end{bmatrix}$$

(E5.2-6)

The column vector to the left—that is, the discrete membership function of A'—ranges from $x = -1$ to $x = 4$, matching the x dimension (which is the row dimension) of the matrix. The y dimension (the columns) of the matrix ranges from $y = -4$ to $y = -1$ as shown in Table 5.4. The result is the fuzzy value

$$B' = 0.25/(-4) + 0.50/(-3) + 0.33/(-2) + 0.167/(-1) \quad \text{(E5.2-7)}$$

B' is shown in Figure 5.14. Its membership function is essentially the

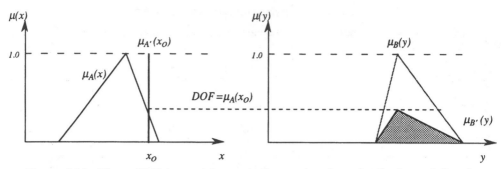

Figure 5.15 When GMP is used to evaluate a rule whose implication relation is modeled by the Larsen product, the membership function of the consequent is scaled by the DOF.

membership function of B scaled (multiplied) by the degree that A' matches the membership function of A at $x = 5$—that is, the DOF of the rule by A'. Scaling the membership function of the consequent by DOF is a feature of the Larsen product fuzzy implication operator ϕ_p. Schematically this property of ϕ_p is shown in Figure 5.15. Other fuzzy implication operators (Table 5.2) result in different shape transformations of the consequent. $\square$

5.5 FUZZY ALGORITHMS

A *fuzzy algorithm* is a procedure for performing a task formulated as a collection of fuzzy *if/then* rules. The rules are defined over the same product space and are connected by the connective *ELSE* which may be interpreted either as *union* or *intersection* depending on the implication operator used for the individual rules.[7] Consider for example the algorithm

$$
\begin{aligned}
&\textit{if} \quad x \textit{ is } A_1 \quad \textit{then} \quad y \textit{ is } B_1 \quad ELSE \\
&\textit{if} \quad x \textit{ is } A_2 \quad \textit{then} \quad y \textit{ is } B_2 \quad ELSE \\
&\quad \ldots \\
&\textit{if} \quad x \textit{ is } A_n \quad \textit{then} \quad y \textit{ is } B_n
\end{aligned}
\tag{5.5-1}
$$

We recall that analytically each rule in (5.5-1) is represented by an implication relation $R(x, y)$ and that the form of $R(x, y)$ depends on the implication operator used (see Table 5.2). Table 5.5 lists the most common interpretation

---

[7]*ELSE* can also be interpreted as *arithmetic sum* and *product* (as well as other T and S norms), which we do not use in this book.

Table 5.5   Interpretation of *ELSE* under various implications

IMPLICATION	INTERPRETATION OF ELSE
ϕ_m, Zadeh Max-Min	AND (∧)
ϕ_c, Mamdani Min	OR (∨)
ϕ_p, Larsen Product	OR (∨)
ϕ_a, Arithmetic	AND (∧)
ϕ_b, Boolean	AND (∧)
ϕ_{bp}, Bounded Product	OR (∨)
ϕ_{dp}, Drastic Product	OR (∨)
ϕ_s, Standard Sequence	AND (∧)
ϕ_Δ, Gougen	AND (∧)
ϕ_g, Gödelian	AND (∧)

of the connective *ELSE* for the implication operators shown in Table 5.2 (in the next chapter we will see more on this). The relation of the entire collection of rules (5.5-1) is called the *algorithmic relation*

$$R_\alpha(x, y) = \int_{(x, y)} \mu_\alpha(x, y)/(x, y) \qquad (5.5\text{-}2)$$

and is either the *union* (∨) or the *intersection* (∧) of the implication relations of the individual rules. A fuzzy algorithm is a linguistic description evaluated analytically using composition operations just as we did in the case of single-rule linguistic descriptions. Given a new fuzzy value A' we evaluate

(5.5-1) through GMP formally stated as

$$
\begin{array}{lll}
\textit{if} & x \text{ is } A_1 & \textit{then} \quad y \text{ is } B_1 \quad \textit{ELSE} \\
\textit{if} & x \text{ is } A_2 & \textit{then} \quad y \text{ is } B_2 \quad \textit{ELSE} \\
& \cdots \\
\textit{if} & x \text{ is } A_n & \textit{then} \quad y \text{ is } B_n \\
& x \text{ is } A'
\end{array}
$$

$$\overline{}$$

$$y \text{ is } B'$$

(5.5-3)

The output value B' in (5.5-3) is computed by max–min composition (and more generally max-$*$) of A' and $R_\alpha(x, y)$—that is,

$$B' = A' \circ R_\alpha(x, y) \tag{5.5-4}$$

The membership function of B' is

$$\mu_{B'}(y) = \bigvee_x \left[\mu_{A'}(x) \wedge \mu_\alpha(x, y) \right] \tag{5.5-5}$$

Then inverse problem is solved through GMT, stated as

$$
\begin{array}{lll}
\textit{if} & x \text{ is } A_1 & \textit{then} \quad y \text{ is } B_1 \quad \textit{ELSE} \\
\textit{if} & x \text{ is } A_2 & \textit{then} \quad y \text{ is } B_2 \quad \textit{ELSE} \\
& \cdots \\
\textit{if} & x \text{ is } A_n & \textit{then} \quad y \text{ is } B_n \\
& y \text{ is } B'
\end{array}
$$

$$\overline{}$$

$$x \text{ is } A'$$

(5.5-6)

The membership function of A' in (5.5-4) can be computed by max–min composition (and more generally max-$*$) of $R_\alpha(x, y)$ and B'—that is,

$$A' = R_\alpha(x, y) \circ B' \tag{5.5-7}$$

with the membership function of A' given by

$$\mu_{A'}(x) = \bigvee_x \left[\mu_\alpha(x, y) \wedge \mu_{B'}(y) \right] \tag{5.5-8}$$

In the elementary fuzzy algorithm of (5.5-1) there is only one variable in the antecedent side of each implication and one on the consequent side. Generally, we are interested in linguistic descriptions that may have more than one variable in either side, which we refer to as *multivariate fuzzy algorithms*. The interpretations of the connective *ELSE* are the same as for the elementary algorithm of (5.5-1). Consider an *if/then* rule of the form

$$if \quad x_1 \ is \ A_1 \quad AND \quad x_2 \ is \ A_2 \quad AND \ \cdots \ AND \quad x_m \ is \ A_m \quad then \quad y \ is \ B$$

$$(5.5\text{-}9)$$

where $x_1, \ldots, x_m$ are antecedent linguistic variables with $A_1, \ldots, A_m$ their respective fuzzy values and y is the consequent linguistic variable with B its fuzzy value. The connective *AND* in the LHS of rule (5.5-9) can be analytically modeled either as min or as arithmetic product. In such cases we can combine the propositions in the LHS either through min ($\wedge$) or through product ($\cdot$) and use an appropriate implication operator ϕ (Table 5.2) to obtain the membership function of implication relation of (5.5-9). Thus we have

$$\mu(x_1, x_2, \ldots, x_m, y) = \phi\left[\mu_{A_1}(x_1) \wedge \mu_{A_2}(x_2) \wedge \cdots \wedge \mu_{A_m}(x_m), \mu_B(y)\right]$$

$$(5.5\text{-}10)$$

In case *AND* is analytically modeled as product, the implication relation has membership function

$$\mu(x_1, x_2, \ldots, x_m, y) = \phi\left[\mu_{A_1}(x_1) \cdot \mu_{A_2}(x_2) \cdot \cdots \cdot \mu_{A_m}(x_m), \mu_B(y)\right]$$

$$(5.5\text{-}11)$$

where ϕ is an appropriate implication operator from Table 5.2. In a similar manner the connective *OR* can be interpreted as max ($\vee$) or as sum ($+$) or other S norms (see Appendix A)).

Less frequently we encounter *multivariate fuzzy implications* involving m nested fuzzy implications, each having one antecedent variable, of the form

$$if \quad x_1 \ is \ A_1 \quad then \ (if \ x_2 \ is \ A_2 \ then \ \cdots \ (if \ x_m \ is \ A_m \ then \ y \ is \ B) \ \cdots) \quad (5.5\text{-}12)$$

The membership function of a multivariate fuzzy implication of equation (5.5-12) is obtained through repeated application of an implication operator (see Table 5.2), once for each nested *if/then* rule:

$$\mu(x_1, x_2, \ldots, x_m, y) = \phi\left[\mu_{A_1}(x_1), \phi\left[\mu_{A_2}(x_2), \ldots, \phi\left[\mu_{A_m}(x_m), \mu_B(y)\right]\right]\right]$$

$$(5.5\text{-}13)$$

When we have several rules of the form of (5.5-9) or (5.5-12) the overall algorithmic relation depends upon the implication operator used and the related interpretation of the connective *ELSE*.

Let us consider a fuzzy algorithm consisting of n multivariate fuzzy implications of the form shown in (5.5-9). We have m variables $x_1, \ldots, x_m$ on the antecedent side of the jth *if/then* rule taking values $A_{1j}, \ldots, A_{mj}$ ($j = 1, \ldots, n$) and only one consequent linguistic variable y, taking values $B_1, B_2, \ldots, B_n$. Our fuzzy algorithm is the collection of rules

if x_1 *is* A_{11} *AND* x_2 *is* A_{21} *AND* $\cdots$ *AND* x_{1m} *is* A_{m1} *then y is* B_1 *ELSE*

if x_1 *is* A_{12} *AND* x_2 *is* A_{22} *AND* $\cdots$ *AND* x_m *is* A_{m2} *then y is* B_2 *ELSE*

$\cdots$

if x_1 *is* A_{1n} *AND* x_2 *is* A_{2n} *AND* $\cdots$ *AND* x_m *is* A_{mn} *then y is* B_n

$$(5.5\text{-}14)$$

The fuzzy algorithm of (5.5-14) is analytically described by an algorithmic relation of the form

$$R_\alpha(x_1, x_2, \ldots, x_m, y)$$

$$= \int_{(x_1, x_2, \ldots, x_m, y)} \mu_\alpha(x_1, x_2, \ldots, x_m, y) / (x_1, x_2, \ldots, x_m, y) \quad (5.5\text{-}15)$$

and when discrete fuzzy sets are used we obtain

$$R_\alpha(x_{1i}, x_{2i}, \ldots, x_{mi}, y_j)$$

$$= \sum_{(x_{1i}, x_{2i}, \ldots, x_{mi}, y_j)} \mu_\alpha(x_{1i}, x_{2i}, \ldots, x_{mi}, y_j) / (x_{1i}, x_{2i}, \ldots, x_{mi}, y_j)$$

$$(5.5\text{-}16)$$

The membership function in (5.5-15) or (5.5-16) can be obtained from the implication relation of the individual rules and appropriate interpretation of the connectives *AND* and *ELSE*. Once the algorithmic relation is known, GMP may be used to obtain an output B' given inputs $A'_1, A'_2, \ldots, A'_m$— that is,

if x_1 *is* A_{11} *AND* x_2 *is* A_{21} *AND* $\cdots$ *AND* x_{1m} *is* A_{m1} *then y is* B_1 *ELSE*

if x_1 *is* A_{12} *AND* x_2 *is* A_{22} *AND* $\cdots$ *AND* x_m *is* A_{m2} *then y is* B_2 *ELSE*

$\cdots$

if x_1 *is* A_{1n} *AND* x_2 *is* A_{2n} *AND* $\cdots$ *AND* x_m *is* A_{mn} *then y is* B_n

x_1 *is* A'_1 x_2 *is* A'_2 $\cdots$ x_m *is* A'_m

---

y is B'

$$(5.5\text{-}17)$$

Let A'_1 be a new input to (5.5-17). The membership function of B' is given by max–min composition of the fuzzy set $A'_1 = A'(x_1)$ and $R_\alpha(x_1, x_2, \ldots, x_m, y)$— that is,

$$B'(y) = A'_1 \circ R_\alpha(x_1, x_2, \ldots, x_m, y) \tag{5.5-18}$$

When m inputs are offered to the algorithm and the connective *AND* in the LHS of each rule is interpreted as min, GMP will give an output value

$$B'(y) = \left(\bigwedge_{j=1}^{m} A'(x_j) \right) \circ R_\alpha(x_1, x_2, \ldots, x_m, y) \tag{5.5-19}$$

with membership function

$$\mu_{B'}(y) = \bigvee_{x_1} \bigvee_{x_2} \cdots \bigvee_{x_m} \left[\left(\bigwedge_{j=1}^{m} \mu_{A'}(x_j) \right) \wedge \mu_\alpha(x_1, x_2, \ldots, x_m, y) \right] \tag{5.5-20}$$

Other compositions may be used as well, such as the max-product or, more generally max-∗, to obtain the membership function of the new consequent B' (see Chapter 3).

REFERENCES

Hirota, K., and Ozawa, K., Fuzzy Flip-Flop as a Basis of Fuzzy Memory Modules, in *Fuzzy Computing Theory, Hardware, and Applications*, M. M. Gupta and T. Yamakawa, eds., Elsevier/North-Holland, Amsterdam, 1988, pp. 173–183.

Huertas, J. L., Sanchez-Solano, S., Barriga, A., and Baturone, I., Serial Architecture for Fuzzy Controllers: Hardware Implementation Using Analog/Digital VLSI Techniques, in *Proceedings of the 2nd International Conference on Fuzzy Logic and Neural Networks, IIZUKA '92*, Iizuka, Japan, July 17–22, 1992, pp. 535–539.

Isik, C., Inference Hardware for Fuzzy Rule-Based Systems, in *Fuzzy Computing Theory, Hardware, and Applications*, M. M. Gupta and T. Yamakawa, eds., Elsevier/North-Holland, Amsterdam, 1988, pp. 185–194.

Kandel, A., *Fuzzy Mathematical Techniques with Applications*, Addison-Wesley, Reading, MA, 1986.

Larsen, P. M., Industrial Applications of Fuzzy Logic Control, *International Journal of Man–Machine Studies*, Vol. 12, No. 1, 1980, pp. 3–10.

Lee, C. C., Fuzzy Logic in Control Systems: Fuzzy Logic Controller—Part-I, *IEEE Transactions on Systems, Man and Cybernetics*, Vol. 20, No. 2, pp. 404–418, 1990a.

Lee, C. C., Fuzzy Logic in Control Systems: Fuzzy Logic Controller—Part-II, *IEEE Transactions on Systems, Man and Cybernetics*, Vol. 20, No. 2, pp. 419–435, March/April 1990b.

Mamdani, E. H., Applications of Fuzzy Set Theory to Control Systems: A Survey, in *Fuzzy Automata and Decision Processes*, M. M. Gupta, G. N. Saridis and B. R. Gaines, eds., North-Holland, New York, 1977, pp. 1–13.

Mizumoto, M., Fuzzy Controls Under Various Reasoning Methods, *Information Sciences*, Vol. 45, pp. 129–141, 1988.

Pappis, C. P., and Sugeno, M., Fuzzy Relational Equations and the Inverse Problem, *Fuzzy Sets and Systems*, Vol. 15, pp. 79–90, 1985.

Pin, G. F., Watanabe, H., Symon, J., Pattay, R. S., Autonomous Navigation of Mobile Robot Using Custom-Designed Qualitative Reasoning VLSI Chips and Boards, *Proceedings of the 1992 IEEE International Conference on Robotics and Automation*, Nice, France, May 1992, pp. 123–128.

Ray, S. K., and Majumder, D. D., Fuzzy Rule Based Approach to Image Segmentation, in *Fuzzy Computing Theory, Hardware, and Applications*, M. M. Gupta and T. Yamakawa, eds., Elsevier/North-Holland, Amsterdam, 1988, pp. 375–397.

Ruan, D., and Kerre, E. E., Fuzzy Implication Operators and Generalized Fuzzy Method of Cases, *Fuzzy Sets and Systems*, Vol. 54, pp. 23–37, 1993.

Shimizu, K., Osumi, M., and Imae, F., Digital Fuzzy Processor FP-5000, *Proceedings of the 2nd International Conference on Fuzzy Logic and Neural Networks, IIZUKA '92*, pp. 539–542, Iizuka, Japan, July 17–22, 1992.

Yamakawa, T., Fuzzy Hardware Systems of Tomorrow, in *Approximate Reasoning in Intelligent Systems, Decision and Control*, E. Sanchez, and L. A. Zadeh, eds., Pergamon Press, Oxford, 1987, pp. 1–20.

Yamakawa, T., Intrinsic Fuzzy Electronic Circuits for Sixth Generation Computers, in *Fuzzy Computing Theory, Hardware, and Applications*, M. M. Gupta and T. Yamakawa, eds., Elsevier/North-Holland, Amsterdam, 1988, pp. 157–173.

Zadeh, L. A., Outline of a New Approach to the Analysis of Complex Systems and Decision Processes, *IEEE Transactions on Systems, Man and Cybernetics*, Vol. 1, pp. 28–44, 1973.

Zadeh, L. A., The Concept of a Linguistic Variable and Its Application to Approximate Reasoning, *Information Sciences*, Vol. 8, pp. 199–249, 1975.

Zadeh, L. A., Fuzzy Sets and Information Granularity, *Advances in Fuzzy Set Theory and Applications*, M. M. Gupta, R. K. Ragade, and R. R. Yager, eds., North-Holland, Amsterdam, 1979, pp. 3–18.

Zadeh, L. A., A Computational Approach to Fuzzy Quantifiers in Natural Languages, *Computer and Mathematics*, Vol. 9, pp. 149–184, 1983.

Zadeh, L. A., Fuzzy Logic, *IEEE Computer*, pp. 83–93, April 1988.

Zimmermann, H. J., *Fuzzy Set Theory and Its Applications*, Kluwer-Nijhoff, Boston, 1985.

PROBLEMS

1. The Mamdani min implication operator given by Equation (5.3-6) is alleged to be a simplification of the Zadeh max–min implication operator given by Equation (5.3-5). Explain what simplifications were made and discuss how these influence implication operations in fuzzy operations such as control. Illustrate your discussion with sketches.

2. A linguistic description is comprised of a single rule

$$\text{if} \quad x \text{ is } A \quad \text{then } y \text{ is } B$$

where A and B are the fuzzy numbers

$$A = 0.33/6 + 0.67/7 + 1.00/8 + 0.67/9 + 0.33/10$$
$$B = 0.33/1 + 0.67/2 + 1.00/3 + 0.67/4 + 0.33/5$$

The implication relation of the rule is modeled through the Larsen product implication operator. If a fuzzy number $x = A'$ is a premise, use generalized modus ponens to infer a fuzzy number $y = B'$ as the consequent. A' is defined by

$$A' = 0.5/5 + 1.00/6 + 0.5/7$$

3. Using the data given in Problem 2, Mamdani min implication operator, and generalized modus ponens, evaluate the rule.

4. Using the data given in Problem 2, arithmetic implication operator, and generalized modus ponens, evaluate the rule.

5. Using the data given in Problem 2, Boolean implication operator, and generalized modus ponens, evaluate the rule.

6. Using the data given in Problem 2, bounded product implication operator, and generalized modus ponens, evaluate the rule.

7. Using the data given in Problem 2, Zadeh max−min implication operator and generalized modus ponens, evaluate the rule.

8. Given the rule and fuzzy values for A and B as well as the B' that you found in Problem 2, use generalized modus tollens to infer an A'.

9. What happens if you repeat Problem 8, having used bounded product implication operator to model the rule?

10. Which of the fuzzy implication operators given in Table 5.2 reduce to classical modus ponens under max−min composition? Examine each operator and show an example of what happens using the data found in Example 5.1.

11. This problem requires an investigation on your part of the concept of fuzzy functions. Generally, a fuzzy function can be understood as a mapping between fuzzy sets and the extension principle can serve as a tool for generalizing ordinary mappings. Depending on where fuzziness

occurs one gets different types of fuzzy functions. The problem is this: Set up a fuzzy function that will take as input ambient temperatures and will produce as output energy demand to a power plant. There are no unique solutions, but rather, different approaches to formulating the solution. State clearly, what could be fuzzy in this problem; what assumptions you need to make; what crisp function, if any, you start with. Also, give the functional form and test it. Does it make sense? Could you get higher energy demand for lower temperatures from your model?

12. Given the assumptions made in Problem 11, find a fuzzy algorithm that describes the same general relation as the fuzzy function you developed in Problem 11.

6

FUZZY CONTROL

6.1 INTRODUCTION

Fuzzy control primarily refers to the control of processes through fuzzy linguistic descriptions. Since 1974, when E. H. Mamdani and S. Assilian (Mamdani, 1974) demonstrated that fuzzy *if/then* rules could regulate a model steam engine, a great number of fuzzy control applications have been successfully deployed. The list is very long and growing and includes cement kilns, subway trains, unmanned helicopters, autonomous mobile robots, process heat exchangers, and blast furnaces (Mamdani, 1977; Ostergaard, 1982; Yasunobu and Miyamoto, 1985; King and Karonis, 1988).[1] In the 1970s and early 1980s most applications were minicomputer-based, often found in the process industry in areas where automatic control was rather difficult to realize and hence left in the hands of human operators. More recently, with the advent of fuzzy microprocessors, a growing number of fuzzy control applications have emerged in consumer electronics and home appliances such as hand-held cameras, vacuum cleaners, air conditioners, and washing machines (Hirota, 1993; Yamakawa, 1989; Schwartz, 1992; Terano et al., 1992).

In this chapter we begin by reviewing conventional process control in order to establish the relevant context and proceed to fuzzy control, a subject we view primarily as an application of fuzzy linguistic descriptions (Chapter 5). Of course, the appropriate choice of controller in engineering applications

[1]There are a number of excellent books available on fuzzy control. The interested reader may want to consult, for example, Driankov et al., 1993; Pedrycz, 1993; Harris et al., 1993; Yager and Filev, 1994; and Wang, 1994.

is made not as much by a commitment to a particular methodology or technology as by careful examination of the needs and features of a given application. In fact, some of the most successful applications of fuzzy control have been in conjunction with conventional controllers such as the *proportional integral derivative* (PID) controller (Lee, 1990a, b). In fuzzy control we are concerned with two broad questions: How can we implement a control strategy as a fuzzy linguistic description? and What are the crucial factors involved in fuzzy algorithmic synthesis and analysis? Although fuzzy linguistic descriptions are a subject of wider interest than the replacement or enhancement of PID controllers, their application to control serves to illustrate some of the basic ideas we encountered in earlier chapters.

Consider the simple process system shown in Figure 6.1. Here, a tank is filled with liquid flowing from a pipe at the top (inlet flow). Liquid leaves the tank through a pipe at the bottom (outlet flow). The upper pipe is fitted with control valve *A*, used to adjust inlet flow, and the bottom pipe with valve *B* is assumed to remain at a preset position. A *controller* maintains the liquid in the tank at the desired level. By *process* here we mean the tank, the liquid, the pipes, and the valves. The term *process control system* refers to the

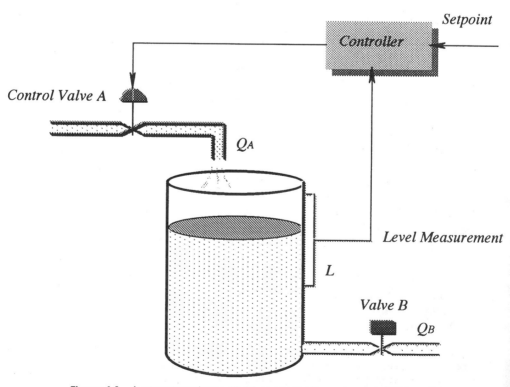

Figure 6.1 A process system with level control through control valve *A*.

process plus the controller and any required components for measurement and actuation.

The purpose of any process control system is to regulate some *dynamic variable* or *variables* of the process. In the liquid level process control system shown in Figure 6.1, the dynamic variable is the liquid level L, a process parameter that depends on other parameters and thus suffers changes from many different inputs. We select one of these other parameters to be our *controlling parameter*—in this case control valve A, the adjustment of which leads to control of flow rate, Q_A. Liquid level depends on flow rates via control valve A and valve B, ambient temperature T_A (not shown), liquid temperature T_l (also not shown), and the physical condition of valves A and B. This dependence may be described by a process relation of the form

$$L = f(Q_A, Q_B, T_a, T_l) \qquad (6.1\text{-}1)$$

where Q_A is the flow rate through control valve A, Q_B is the flow rate through valve B, T_a is the ambient temperature, and T_l is the liquid temperature. In many cases the relationship of equation (6.1-1) is not analytically known and actually may not be a function (a *many-to-one mapping*) but instead a more general relation (a *many-to-many mapping*) as we discussed in Chapter 5.

The input to the controller is usually not L itself but instead the error e between a measured indication of L, denoted as y, and a *setpoint* or *reference* value r representing the desired value of the dynamic variable. The controller's *output* or *manipulated variable* is denoted by u and is a signal representing action to be taken when the measured value of the dynamic variable y deviates from reference r. Thus, the output of the controller u serves as input to the process. The error $e = r - y$ is actually *smoothed* and *scaled* before input to the controller. Smoothing is performed in sampled systems in order to avoid the instantaneous changes during sampling that misled the general direction of change for the variable. Such a smoothing function may be defined recursively as $(e_k) \equiv 0.9e_{k-1} + 0.1e_k$, where e_k is the error value at time $t = k$. Scaling is required in order to transform instrument values to a predetermined interval or transform them to a range of numbers that correspond to natural magnitudes.

The most common controller in the process industry is the *PID controller*, where the control relation associated with equation (6.1-1) takes the form

$$u(t) = K_P e(t) + K_P K_I \int_{t=0}^{t} e(t)\, dt + K_P K_D \frac{de(t)}{dt} + u(0) \quad (6.1\text{-}2)$$

where K_P is the *controller gain* representing a proportionality constant between error and controller output (dimensionless), K_I is the *reset constant* relating the rate to the error in units of $[\%/(\% - \text{sec})]$, K_D is the *rate constant* (or *derivative gain constant*) in units of $[(\% - \text{sec})/\%]$, and $u(0)$ is the controller output at $t = 0$ (when a deviation from setpoint starts).

The first term in equation (6.1-2) is called the *proportional term*, and if it was the only term in the equation it would represent a mode of control where the output of the controller $u(t)$ is changed in proportion to the error $e(t)$, which is the percent deviation from the setpoint. The second term is called the *integral term* and represents a mode of control where the present controller output depends on the history of errors from when observations started at $t = 0$. The amount of corrective action due to integral mode is directly proportional to the length of time that the error has existed. The reset constant K_I expresses the scaling between error and controller output. A large value of K_I means that a small error produces a large rate of change of u and vice versa. If this term alone was used in equation (6.1-2), in addition to the constant $u(0)$, then we would have a mode of control called *integral mode*. The third term in equation (6.1-2) represents the *derivative mode of control*. This mode provides that the controller output depends on the rate of change of error. Derivative mode tends to minimize oscillation of the system and prevent overshooting. Since derivative control is based solely on the rate of change of error, the controlled variable can stabilize at a value different from r, a condition termed "offset." In pure derivative mode the output depends upon the rate at which the error is changed and not on the value of the error. Integral control is used to address situations when permanent offset or slow returns to desired values cannot be tolerated. The combination of these three modes is called *proportional integral derivative* or (PID) control. PID is a powerful composite mode of control that has been used for virtually any linear process condition.

The process of adjusting the coefficients of each mode of control in equation (6.1-2) is called *tuning*. There are several methods for determining the optimum value of these gains such as *frequency response methods* and the *Ziegler–Nichols method* (Johnson, 1977). Fuzzy and neural approaches with adaptive characteristics have also been used for PID tuning and more generally for emulating and enhancing PID controllers (Matia et al., 1992; Maeda and Murakami, 1992; Shoureshi and Rahmani, 1992; He et al., 1993).

Example 6.1   PID Level Control. Consider the process control system shown in Figure 6.1. Suppose that we control the liquid level in the tank by adjusting control valve A (inlet flow) through a PID controller. The output of the controller $u(t)$ is based on the error $e(t)$—that is, the difference between a reference value r and the measured value of level y. The output of the controller is given by equation (6.1-2) as

$$u(t) = K_P e(t) + K_P K_I \int_{t=0}^{t} e(t)\, dt + K_P K_D \frac{de(t)}{dt} + u(0) \quad \text{(E6.1-1)}$$

with the following values for the various gains and initial controller output: $K_P = -1.3$, $K_I = 0.5[\%/(\% - \text{min})]$, $K_D = 1.9[(\% - \text{min})/\%]$, and $u(0) = 50\%$.

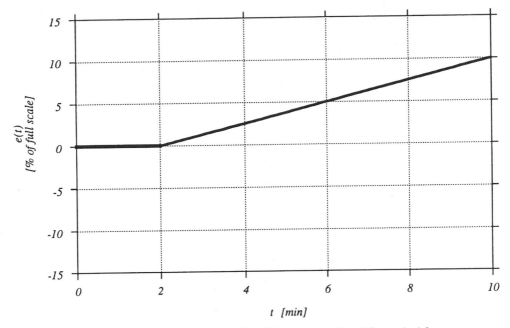

Figure 6.2 Error introduced to the PID level controller of Example 6.1.

Suppose that the error shown in Figure 6.2 is introduced to the system at $t = 2$ min. Such an error may be due to any change in process parameters—for example, an unforeseen change in the position of valve B—since valve B is not under control. The equation of error as function of time is

$$e(t) = 1.25t - 2.5 \qquad (E6.1\text{-}2)$$

Using equation (E6.1-2) in equation (E6.1-1) the output of the controller after $t = 2$ is given by

$$u(t) = -1.3[1.25t - 2.5] - 1.3(0.5 \text{ min}^{-1})\int_{t=2}^{t}[1.25t - 2.5]\,dt$$

$$- 1.3(1.9 \text{ min})\frac{d}{dt}[1.25t - 2.5] + 50\% \qquad (E6.1\text{-}3)$$

The first term in equation (6.1-3) represents the *proportional mode* of the controller, the second term the *integral mode*, and the third term the *derivative mode*. Let us call them $u_P(t)$, $u_I(t)$, and $u_I(t)$, respectively. Figure 6.3 shows the response due to each mode and the total response of the controller $u(t)$, which is the sum of the three terms plus the initial output of the controller, in this case 50%. Looking at Figure 6.3 we note that at the

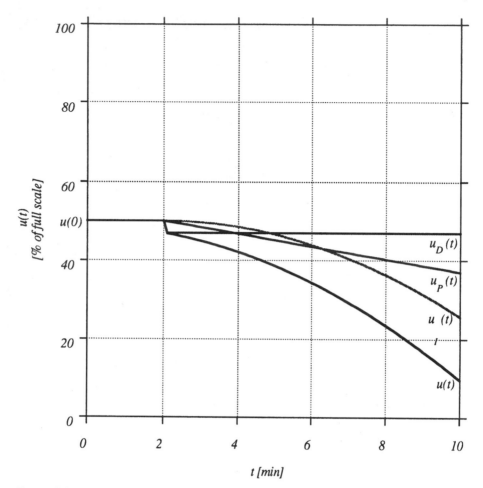

Figure 6.3 Proportional, integral, derivative, and total response of the PID level controller in Example 6.1.

end of the 10-min interval the controller sends a signal to control valve A, which is about 10% of its full scale. This does not necessarily mean that the valve itself allows at that time 10% of full flow to the tank. Different valves have different characteristics, often nonlinear. The relation between a dynamic variable and its transduced equivalent, although desired to be linear for many transducers, is almost always actually nonlinear. As an example, suppose that control valve A is an *equal-percentage* valve. In such valves a given percent change in the valve's stem position (which is what actually the controller controls) produces an equivalent change in flow, hence the name *equal-percentage*. Generally this type of valve does not shut off flow completely in its limit of stem travel. Let Q denote the flow rate through the valve (in m^3/sec), Q_{min} the minimum flow when the stem of the valve is at the lowest limit of its travel, and Q_{max} the flow rate when the valve is fully open. The ratio $R = (Q_{max}/Q_{min})$ is called the *rangeability* of the valve; it is a parameter specific to a given valve. The actual flow at any given time varies nonlinearly with rangeability and is often given by an exponential expression of the form

$$Q = Q_{min} R^{u/u_{max}} \qquad \text{(E6.1-4)}$$

where u/u_{max} is the ratio of the actual to the maximum control signal sent to the actuator (actually the valve stem position at any given time divided by the maximum position of valve stem). Suppose that control valve A has rangeability $R = 30$. Thus when the control signal is 10% of full range—that is, $(u/u_{max}) = 0.1$—the flow rate according to equation (E6.1-4) becomes

$$Q = Q_{min}(30)^{0.1} = 1.4 Q_{min} \qquad \text{(E6.1-5)}$$

We can see from equation (E6.1-5) that at the end of the 10-min interval the controller output is 10% of maximum value even though the flow through control valve A is 1.4 times the minimum flow through the valve. Actuators in general have such nonlinear characteristics, and furthermore their characteristics change due to aging or other environmental factors. Some of the difficulties in the field utilization of control algorithms, such as the PID level controller here, arise from the collective impact of such changes. Their extent and nature may not be fully known when the controller is designed, tested, and initially deployed. In the course of time the control engineer has to make various judgments about the overall performance of the process control system and, in collaboration with operations and maintenance personnel, intervene to retune gains, repair or replace equipment, revise procedures for operation, and so on. An objective of linguistic control is to make this entire process somewhat easier. It may therefore be seen as irony in the choice of words, but indeed a benefit of *fuzzy* control is introducing even more *clarity* to the development, evaluation, and maintenance of control systems.   □

6.2   FUZZY LINGUISTIC CONTROLLERS

The core of a fuzzy controller is a linguistic description prescribing appropriate action for a given state. As we saw in Chapter 5, fuzzy linguistic descriptions involve associations of fuzzy variables and procedures for inferencing. Whereas in a conventional PID controller what is modeled is the physical system or process being controlled, in fuzzy controllers the aim is to incorporate expert human knowledge in the control algorithm. In this sense, a fuzzy controller may be viewed as a real-time expert system—that is, a model of the thinking processes an expert might go through in the course of manipulating the process.

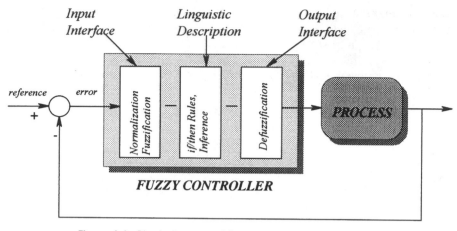

Figure 6.4   Block diagram of fuzzy process control system.

The basic structure of a fuzzy controller is outlined in Figure 6.4. The fact that measuring devices give crisp measurements and that actuators require crisp inputs calls for two additional considerations when linguistic descriptions are employed for control purposes: *fuzzifying* the input of the controller and *defuzzifying* its output. *Fuzzification* can be achieved through a fuzzifier kernel as we saw in Section 2.3, and *defuzzification* can be achieved through special procedures that select a crisp value representative of the fuzzy output (see Section 6.3). Many controllers, however, use directly crisp inputs. Figure 6.4 shows that in addition to a set of *if/then* rules,[2] a fuzzy controller has an *input interface* and an *output interface* handling fuzzification and defuzzification as well as various signal manipulations such as *normalization, scaling, smoothing,* and *quantization.* Scaling maps the range of values of the controlled variables into predefined universes of discourse, and quantization procedures assist in the mapping when discrete membership functions are used (Larkin, 1985; Efstathiou, 1987; Yager and Filev, 1994).

Fuzzy controllers operate in discrete time intervals. The rules are evaluated at regular intervals in the same way as in conventional digital control, with several rules being executed together (in *parallel*) within the same time interval. This parallel feature makes it possible to develop highly dispersed fuzzy algorithms as we will see later on. We use the subscript k to indicate a specific moment in time—that is, when $t = t_k$. The choice of sampling interval depends on the process being controlled and is usually selected so that at least several significant control actions are made during the process settling time (King and Mamdani, 1977).

Let us look at typical *input* or *left-hand side* (LHS) and *output* or *right-hand side* (RHS) fuzzy variables used in the knowledge base of fuzzy

---

[2]The set of *if/then* rules is also referred to as the controller's *knowledge base.*

controllers. Many fuzzy controllers use *error*, *change of error*, and *sum of errors* in the LHS,[3] based on measured process variables and setpoint values, and any process variable that can be manipulated directly in the RHS.

Input Variables

The most common LHS variable in fuzzy control is the *error*, or *e*. It is usually defined on the universe of discourse of crisp error *e*, which is the deviation of some measured variable *y* from a setpoint or reference *r*. At any time $t = k$ crisp error is defined as

$$e(k) \equiv r - y(k) \tag{6.2-1}$$

The change in error, Δe or $\Delta error$, between two successive time steps is also commonly used as an LHS variable. It is defined on the universe of discourse of crisp changes in error. At time $t = k$ the crisp change in error is the difference between present error and error in the previous time step $t = k - 1$, namely,

$$\Delta e(k) \equiv e(k) - e(k - 1) \tag{6.2-2}$$

Fuzzy variables can also be defined for the *rate of change in error* $\Delta^2 e(k) \equiv \Delta e(k) - \Delta e(k - 1)$, and so on. The *sum of errors* $\bar{e}(k)$ may be used as an LHS fuzzy variable also. It takes into account the integrated effect of all past errors and is defined as

$$\bar{e}(k) \equiv \sum_{i=1}^{k} e_i \tag{6.2-3}$$

In some cases, actual state variables may be used (instead of error, etc.) depending on the availability of parameter and structure estimation knowledge. It is even possible to use variables not directly measurable, such as *performance* or *reliability*, provided that they can be estimated in a timely and reliable manner (Tsoukalas, 1991).

Output Variables

RHS variables may be any directly manipulated variable. An RHS fuzzy variable *u* can be defined on the universe of discourse of a crisp manipulated variable. Actually the change in output Δu is more often used as the RHS variable. Δu indicates the extent of change of the control variable *u* at time $t = k$—that is, the *change in action*. Hence, if the defuzzified output at

---

[3] Using these variables, one can write *if/then* rules emulating PID modes of control.

time k is $\Delta u^*(k)$, the overall crisp output of the controller will be

$$u(k) = u(k - 1) + \Delta u^*(k) \qquad (6.2\text{-}4)$$

Using Δu is preferable, since it requires a smaller number of data points in the output universe of discourse in order for the controller to operate with reasonable accuracy.

if / then Rules and Inference

Often, but not always, LHS and RHS variables are scaled to the same universe of discourse and possess fuzzy values that have the same form. Scaling to a common universe of discourse with a common set of values for all variables may offer considerable savings in memory and speed as far as the computer implementation of a fuzzy algorithm is concerned. In addition, it may be helpful in analyzing the behavior of the controller itself, as we will see later in this chapter. With the advent of fuzzy microprocessors and fuzzy development shells, it is no longer necessary for a user to do scaling because it is done by the system automatically. Nonetheless, scaling helps to simplify algorithmic development and investigate factors involved in synthesis and analysis. As an example, consider the fuzzy values for the variables *error*, $\Delta error$, and Δu shown in Figure 6.5 in connection with a fuzzy controller that emulates the derivative mode of a conventional controller (Sugeno, 1985;

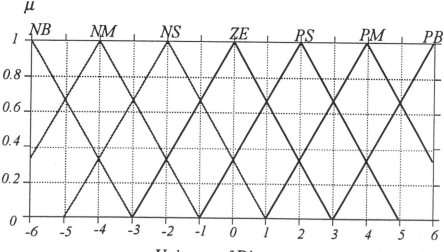

Figure 6.5 Common fuzzy values for the *error*, $\Delta error$, and Δu variables, scaled to the same universe of discourse.

Mizumoto, 1988). The common fuzzy values are as follows:

$$NB \equiv negative\,big, \qquad PS \equiv positive\,small$$
$$NM \equiv negative\,medium, \qquad PM \equiv positive\,medium$$
$$NS \equiv negative\,small, \qquad PB \equiv positive\,big$$
$$ZE \equiv zero$$

All variables share the same universe of discourse ranging between -6 and $+6$ as shown in Figure 6.5. In computer implementations, fuzzy values are usually quantized and stored in memory in the form of a look-up table as shown in Table 6.1. In this case the fuzzy values are stored in a 7×13 table, with every row in the table representing a quantized fuzzy value. The fuzzy algorithm of a controller that emulates a derivative mode is comprised of the following *if/then* rules:

R_1:   *if error is NB AND Δerror is ZE then Δu is PB*   ELSE

R_2:   *if error is NM AND Δerror is ZE then Δu is PM*   ELSE

R_3:   *if error is NS AND Δerror is ZE then Δu is PS*   ELSE

R_4:   *if error is ZE AND Δerror is ZE then Δu is ZE*   ELSE

R_5:   *if error is PS AND Δerror is ZE then Δu is NS*   ELSE

R_6:   *if error is PM AND Δerror is ZE then Δu is NM*   ELSE

R_7:   *if error is PB AND Δerror is ZE then Δu is NB*   ELSE   (6.2-5)

R_8:   *if error is ZE AND Δerror is NB then Δu is PB*   ELSE

R_9:   *if error is ZE AND Δerror is NM then Δu is PM*   ELSE

R_{10}:   *if error is ZE AND Δerror is NS then Δu is PS*   ELSE

R_{11}:   *if error is ZE AND Δerror is PS then Δu is NS*   ELSE

R_{12}:   *if error is ZE AND Δerror is PM then Δu is NM*   ELSE

R_{13}:   *if error is ZE AND Δerror is PB then Δu is NB*

When two LHS and one RHS variables are used as in (6.2-5), the algorithm can be visualized in the form of a table as shown in Table 6.2. Such an arrangement is sometimes called a "fuzzy associative memory (FAM) matrix." Blank items in the table indicate that there is no rule present for the particular combination of LHS variables. Obviously for algorithms with more than two LHS variables a tabular representation requires additional dimensions.

Table 6.1   Table of fuzzy values

	-6	-5	-4	-3	-2	-1	0	1	2	3	4	5	6
NB	1	0.67	0.33	0	0	0	0	0	0	0	0	0	0
NM	0.33	0.67	1	0.67	0.33	0	0	0	0	0	0	0	0
NS	0	0	0.33	0.67	1	0.67	0.33	0	0	0	0	0	0
ZE	0	0	0	0	0.33	0.67	1	0.67	0.33	0	0	0	0
PS	0	0	0	0	0	0	0.33	0.67	1	0.67	0.33	0	0
PM	0	0	0	0	0	0	0	0	0.33	0.67	1	0.67	0.33
PB	0	0	0	0	0	0	0	0	0	0	0.33	0.67	1

Table 6.2   A fuzzy algorithm in tabular form

Δerror error	NB	NM	NS	ZE	PS	PM	PB
NB				PB			
NM				PM			
NS				PS			
ZE	PB	PM	PS	ZE	NS	NM	NB
PS				NS			
PM				NM			
PB				NB			

Fuzzy control algorithms are evaluated using *generalized modus ponens* (GMP). We recall from Chapter 5 that GMP is a data-driven inferencing procedure that analytically involves the composition of fuzzy relations, usually *max–min composition*. We also saw that max–min composition under a given implication operator affects the RHS in a specific manner—for example, by *clipping* (when *Mamdani min*, ϕ_c, is used) or *scaling* (when *Larsen product*, ϕ_p, is used). In general, GMP can be thought of as a transformation of the RHS by a degree commensurate with the degree of fulfillment (DOF) of the rule and in a manner dictated by the implication operator chosen (see Examples 5.1 and 5.2). In this chapter, instead of explicitly using composition operations, we will mostly focus on such transformations as is often done, for the sake of convenience, in many fuzzy control applications. As far as the entire algorithm is concerned, the connective *ELSE* is analytically modeled as either *OR* ($\vee$) or *AND* ($\wedge$), again depending on the implication operator used for the individual *if/then* rules. For example, when the Mamdani min implication is used, the connective *ELSE* is interpreted as *OR* (see Table 5.5).

Fuzzy controller inputs are usually crisp numbers. Fuzzy inputs may also be considered in the case of uncertain or noisy measurements and crisp numbers may be fuzzified (see Section 2.3). Consider the situation shown in Figure 6.6 involving rules R_3, R_4, and R_{11} of (6.2-5). When at time $t = k$ crisp error e' and crisp change in error $\Delta e'$ as shown in Figure 6.6 are given to these rules we say that the rules have "fired," provided that their DOF is not zero. For example, in rule R_3 the crisp error e' shown has a 0.8 degree of membership to *NS* while the crisp change in error $\Delta e'$ has a 0.6 degree of membership to *ZE*. Thus the degree of fulfillment of rule R_3 at this particular time is

$$\text{DOF}_3 = \mu_{NS}(e') \wedge \mu_{ZE}(\Delta e') = 0.8 \wedge 0.6 = 0.6 \qquad (6.2\text{-}6)$$

Provided that we interpreted the LHS connective *AND* as min ($\wedge$) [a common alternative is *product* ($\cdot$)], the RHS value *PS* will be transformed in accordance with DOF_3 in equation (6.2-6). The nature of the transformation depends on the implication used as we saw in Chapter 5. When Mamdani min is used the transformation amounts to clipping *PS* at the height of DOF_3 as shown in Figure 6.6. Thus R_3 contributes $\mu_{PS'}(\Delta u)$, the shaded part of the RHS value, to the total fuzzy output. Similarly rules R_4 and R_{11} have degrees of fulfillment

$$\text{DOF}_4 = \mu_{ZE}(e') \wedge \mu_{ZE}(\Delta e') = 0.4 \wedge 0.6 = 0.4 \qquad (6.2\text{-}7)$$

$$\text{DOF}_{11} = \mu_{ZE}(e') \wedge \mu_{PS}(\Delta e') = 0.4 \wedge 1.0 = 0.4 \qquad (6.2\text{-}8)$$

and they contribute $\mu_{ZE'}(\Delta u)$ and $\mu_{NS'}(\Delta u)$, shown as shaded parts of the RHS values in Figure 6.6. The rest of the rules of algorithm (6.2-5) do not fire, that is, they contribute a zero output. The total fuzzy output is the *union* of the three outputs since we interpret the connective *ELSE* in (6.2-5) as *OR* ($\vee$)—that is,

$$\mu_{OUT}(\Delta u) = \mu_{PS'}(\Delta u) \vee \mu_{ZE'}(\Delta u) \vee \mu_{NS'}(\Delta u) \qquad (6.2\text{-}9)$$

$\mu_{OUT}(\Delta u)$ is shown at the lower part of Figure 6.6. At this point we need to defuzzify $\mu_{OUT}(\Delta u)$ and obtain a crisp value Δu_k^* representative of $\mu_{OUT}(\Delta u)$ to be used as input to the process. In the next section we will look at different methods for defuzzification.

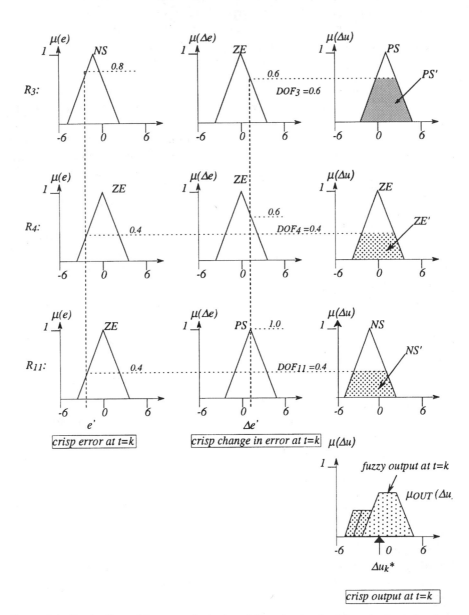

Figure 6.6 Evaluation of three control rules at time $t = k$ using Mamdani min implication and min interpretation of *AND* (DOF).

If *Larsen product* is used as the fuzzy implication operator for the individual rules of (6.2-5), the membership function of the RHS value is *scaled* by the degree of fulfillment of each rule as shown in Figure 6.7 (see Example 5.2). Since the connective *ELSE* is interpreted as *OR* ($\vee$) when Larsen product implication is used (see Table 5.5), the total output $\mu_{OUT}(\Delta u)$ is also the *union* of the three individual outputs. Out of that we need to select a representative crisp value as input to the process. We note in Figure 6.7 that $\mu_{OUT}(\Delta u)$ looks quite different from the total fuzzy output obtained using Mamdani min shown in Figure 6.6. Other fuzzy implication operators (see Table 5.2) would produce different transformations in the shape of the RHS fuzzy value and, hence, a different $\mu_{OUT}(\Delta u)$.

Interpretations of *AND* other than min ($\wedge$) may be used in the *AND* connective found in the LHS of the rules, hence obtaining different degrees of fulfillment. Arithmetic product has been used (particularly in conjunction with max-product implication) and more generally *T-norms* (Zimmermann, 1985; Fuller and Zimmermann, 1992).[4] Using arithmetic product the degree of fulfillment for the rules of (6.2-5) that fire would be evaluated in the manner shown in Figure 6.8. The degree of fulfillment for R_3, R_4, and R_{11} are

$$\text{DOF}_3 = \mu_{NS}(e') \cdot \mu_{ZE}(\Delta e') = 0.8 \cdot 0.6 = 0.48$$

$$\text{DOF}_4 = \mu_{ZE}(e') \cdot \mu_{ZE}(\Delta e') = 0.4 \cdot 0.6 = 0.24$$

$$\text{DOF}_{11} = \mu_{ZE}(e') \cdot \mu_{PS}(\Delta e') = 0.4 \cdot 1.0 = 0.4 \qquad (6.2\text{-}10)$$

Comparing equations (6.2-10) with equations (6.2-6)–(6.2-8) we see that generally the two different interpretations of *AND* lead to different results under the same fuzzy implication operators as we can also see by comparing Figures 6.7 and 6.8.

After we defuzzify $\mu_{OUT}(\Delta u)$ by one of the methods discussed in the next section, we obtain a crisp value Δu_k^*, which in the case of the algorithm of (6.2-5) would be an integer between -6 and $+6$. Values greater than the extremes of the universe of discourse are set to the extreme values, in this case -6 or $+6$. This value is then multiplied by a scaling factor that maps it into the appropriate range of the manipulated variable before using it to actuate a device (Larkin, 1985).

Since so much of actual process control knowledge has historically been obtained through PID controllers, it is often convenient to emulate various modes and combinations of the PID controller by fuzzy rules. Thus a fuzzy controller emulating a conventional PD mode of control controller would

---

[4]See the Appendix for an introduction to *T norms* and their co-norms, called *S norms*.

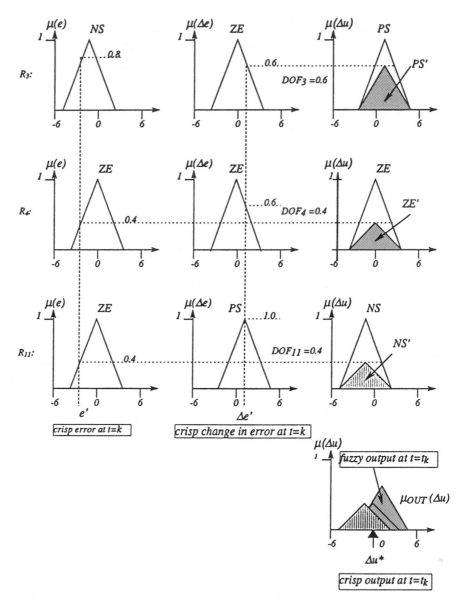

Figure 6.7 Evaluation of three control rules at time $t = k$ Lusing Larsen product implication and min DOF.

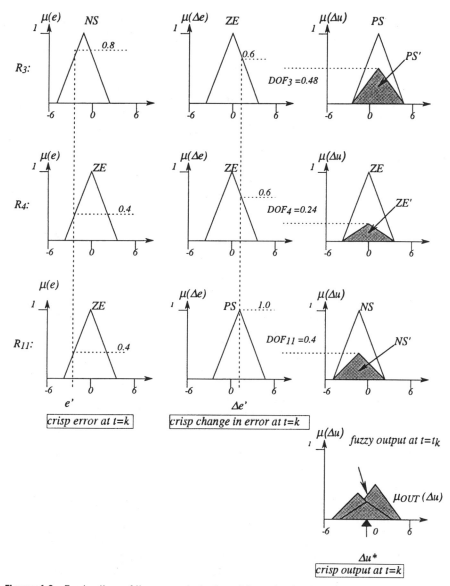

Figure 6.8 Evaluation of three control rules at time $t = k$ using Larsen product implication and product DOF.

consist of rules having the form

$$if \quad e \ is \ A \quad AND \quad \Delta e \ is \ B \quad then \ u \ is \ C \qquad (6.2\text{-}11)$$

where e is the error and Δe is the change in error. A PI-like fuzzy controller would have rules of the form

$$if \quad e \ is \ A \quad AND \quad \Delta e \ is \ B \quad then \ \Delta u \ is \ C \qquad (6.2\text{-}12)$$

while a P-like controller would have rules

$$if \quad e \ is \ A \quad then \quad u \ is \ C \qquad (6.2\text{-}13)$$

The rule form of a PID-like fuzzy controller is

$$if \quad e \ is \ A \quad AND \quad \Delta e \ is \ B \quad AND \quad \bar{e} \ is \ C \quad then \quad u \ is \ D \qquad (6.2\text{-}14)$$

where $\bar{e}$ is the sum of errors.

Although we have formulated fuzzy algorithms in terms of rules involving fuzzy values on their RHS (such rules are referred to as *Mamdani rules*), there are advantages to consider crisp or special shape membership functions as well. Several fuzzy controllers use rules where the output variable is given in terms of a functional relation of the inputs. This is known as the *Sugeno* or *TSK*[5] *form* of fuzzy rules. Such rules are typically written as

$$if \quad x_1 \ is \ A_1 \quad AND \quad x_2 \ is \ A_2 \ldots then \quad u = f(x_1, \ldots, x_n) \qquad (6.2\text{-}15)$$

where f is a function of the inputs $x_1, \ldots, x_n$. When $f(x_1, \ldots, x_n)$ is a constant, rules of the form (6.1-15) constitute a *zero-order Sugeno controller*. When $f(x_1, \ldots, x_n)$ is a first-order polynomial we have what is called a *first-order Sugeno controller*. For example, we may describe a PI controller of (6.2-12) by rules of the form

$$if \quad e \ is \ LARGE \quad AND \quad \Delta e \ is \ MEDIUM \quad then \ u = 2e + 3\Delta e \qquad (6.2\text{-}16)$$

An interesting application of Sugeno rules is when a PID controller is put directly in the RHS of (6.2-15). The result is a fuzzy "supervisor" changing the parameters of a PID controller [see Tzafestas and Papanikolopoulos (1990)]. Sugeno fuzzy models are well suited for modeling nonlinear systems by interpolating multiple linear models and are also well suited to mathematical analysis and lend themselves to adaptive techniques, whereas Mamdani rules are more intuitive and better suited to human. [See Jang and Sun (1995) and Jang and Gulley (1995) for a review of different controllers.]

---

[5]After the *Takagi, Sugeno,* and *Kang* who first proposed it in 1985; also referred to as the *Sugeno fuzzy model.*

An alternative to either Sugeno or Mamdani rules is to consider rules whose consequents employ monotonical membership functions. This form of rules is known as the *Tsukamoto fuzzy model* (Tsukamoto, 1979). In Tsukamoto rules the inferred output of each rule is a crisp value equal to the rule's *degree of fulfillment*, with the overall output being taken as the weighted average of all outputs (a crisp value).

Fuzzy algorithms such as (6.2-5) are inherently parallel in the sense that individual *if/then* rules are fired independent of each other, with a specific input being processed by several rules each contributing to a collective result, namely, $\mu_{OUT}(\Delta u)$. Actual process systems, however, may have many inputs and outputs, and hence they are referred to as *many-input-many-output* (MIMO) systems. The question then arises of how relevant are the rather simple *if/then* rules we have seen thus far, such as the algorithm of (6.2-5), to the control of such systems and what happens to parallelism at a higher level of system complexity.

Generally the control strategies of complicated process systems may be organized in such a manner that relatively simple *if/then* rules are used (Terano, 1992; Yager, 1994). This is achieved by partitioning the knowledge base of the controller into rule clusters. In each cluster there are *if/then* rules that may have several LHS variables but only one RHS variable. Suppose that we have p input variables $x_1, x_2, \ldots, x_p$ and r manipulated variables $u_1, u_2, \ldots, u_r$. The algorithmic development generally proceeds from some general and maybe complicated *if/then* rules that form the *a priori* knowledge prescribing what has to be done under a set of hypothetical situations. Often, but not always, it is possible to reduce these initial rules to simpler rules with one control variable in the RHS. Rules that have the same RHS variable are collected together to form a rule cluster. In the end we have one cluster of rules whose RHS is used to manipulate variable u_1, another for variable u_2, and so on. Thus, a complicated process control system may be decomposed into a number of *many-input-single-output* controllers. Such rule clusters may be executed independently, hence maintaining the overall parallel characteristics of fuzzy systems. Of course, more elaborate architectures can be devised that may include *metarules*. The developers of fuzzy control algorithms exercise considerable creativity in setting up special variables and rules for the interaction of these clusters. In principle, however, rule clusters can be noninteractive, in which case they can be executed in parallel, achieving considerable speed and computational efficiency.

6.3 DEFUZZIFICATION METHODS

After the input to the controller has been processed by the control algorithm the result is a fuzzy output $\mu_{OUT}(u)$. Selecting a crisp number u^* representative of $\mu_{OUT}(u)$ is a process known as *defuzzification*. Over the years several defuzzification techniques have been suggested (Terano et al., 1992; Pedrycz, 1993; Yager and Filev, 1994). The choice of defuzzification method may have a significant impact on the speed and accuracy of a fuzzy controller.[6] The most frequently used ones are the *centroid* or *center of area* (COA), the *center of sums* (COS), and *mean of maxima* (MOM).

Center of Area (COA) Defuzzification

In COA defuzzification[7] the crisp value u^* is taken to be the geometrical center of the output fuzzy value $\mu_{OUT}(u)$, where $\mu_{OUT}(u)$ is formed by taking the union of all the contributions of rules whose DOF > 0.[8] The center is the point which splits the area under the $\mu_{OUT}(u)$ curve in two equal parts. Let us assume we have a discretized universe of discourse. The defuzzified output is defined as

$$u^* = \frac{\sum_{i=1}^{N} u_i \, \mu_{OUT}(u_i)}{\sum_{i=1}^{N} \mu_{OUT}(u_i)} \tag{6.3-1}$$

where the summation (integration) is carried over (discrete) values of the universe of discourse u_i sampled at N points. COA is a well known and often used defuzzification method. Some potential drawbacks of COA are that it favors "central" values in the universe of discourse and that, due to its complexity, it may lead to rather slow inference cycles. COA defuzzification takes into account the area of the resultant membership function $\mu_{OUT}(u)$ as a whole. If the areas of two or more contributing rules overlap, equation (6.3-1) does not take into account the overlapping area only once [since we take the union to form $\mu_{OUT}(u)$, the resultant membership function].

When $\mu_{OUT}(u) = 0$ we simply set the crisp output to a pre-agreed value (in order to avoid dividing by zero), typically $u^* = 0$. The crisp output value may also be computed in terms of the DOF of each contributing rule as

$$u^* = \frac{\sum_{k=1}^{n} \mathrm{DOF}_k \cdot M_k}{\sum_{k=1}^{n} \mathrm{DOF}_k \cdot B_k'} \tag{6.3-2}$$

where B_k' is the contribution due to the firing of rule k,[9] M_k is the moment of B_k', and DOF_k is the *degree of fulfillment* of the kth rule ($k = 1, \ldots, n$).

---

[6]Certain defuzzification methods may introduce nonlinearities and discontinuities in the control hypersurface [see Jager (1995)].
[7]Also known as *center of gravity* defuzzification, a name more appropriate for multidimensional fuzzy output.
[8]For convenience we use the control signal u as the output variable. The control signal change Δu or any other output variable may be used as well.
[9]The subscript k is used to indicate the kth rule and should not be confused with the letter k used earlier to indicate the $t = k$ time step.

We recall that the moment of B'_k is the product of B'_k and the distance of its center of gravity from the μ axis (the moment about zero).

Center of Sums (COS) Defuzzification

To address the problems associated with COA and take into account the overlapping areas of multiple rules more than once, a variant of COA called *center of sums* (COS) is used. As shown in Figure 6.9, COS builds the resultant membership function by taking the sum (not just the union) of output from each contributing rule. Hence overlapping areas are counted more than once. COS is actually the most commonly used defuzzification method. It can be implemented easily and leads to rather fast inference cycles. It is given by

$$u^* = \frac{\sum_{i=1}^{N} u_i \cdot \sum_{k=1}^{n} \mu_{B'_k}(u_i)}{\sum_{i=1}^{N} \sum_{k=1}^{n} \mu_{B'_k}(u_i)} \tag{6.3-3}$$

where $\mu_{B'_k}(u_i)$ is the membership function (at point u_i of the universe of discourse) resulting from the firing of the kth rule.

Mean of Maxima (MOM) Defuzzification

One simple way to defuzzify the output is to take the crisp value with the highest degree of membership in $\mu_{OUT}(u)$. Oftentimes, however, there may be more than one element in the universe of discourse having the maximum value, as may be seen in the $\mu_{OUT}(u)$ of Figure 6.6. In such cases we can randomly select one of them or, even better, take the mean value of the maxima. Suppose that we have M such maxima in a discrete universe of discourse. The crisp output can be obtained by

$$u^* = \sum_{m=1}^{M} \frac{u_m}{M} \tag{6.3-4}$$

where u_m is the mth element in the universe of discourse where the membership function of $\mu_{OUT}(u)$ is at the maximum value, and M is the total number of such elements.

MOM defuzzification is faster than COA, and furthermore it allows the controller to reach values near the edges of the universe of discourse. A disadvantage of this method, however, is that it does not consider the overall shape of the fuzzy output $\mu_{OUT}(u)$. On the other hand, with COA the extreme values of the universe of discourse cannot be reached—for example, near ± 6 in Figure 6.5. Both methods have been used in control applications; several variants of them exist, such as the *indexed center of gravity method*, where a threshold level is used to eliminate elements with degrees of membership lower than a threshold in the computation of $\mu_{OUT}(\Delta u)$

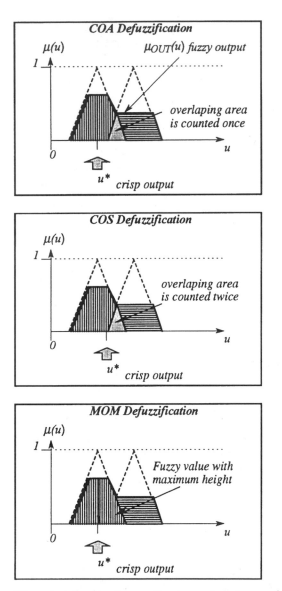

Figure 6.9 Three different defuzzification methods: center of area (COA, center of sums (COS), and mean of maxima (MOM).

(Pedrycz, 1993). It is also possible to employ defuzzification methods in an adaptive manner (Yager and Filev, 1993, 1994).

Example 6.2   A Simple Fuzzy Controller for Level Control. Consider the process system shown in Figure 6.1 (and controlled by a PID controller in

Example 6.1). We want to develop a fuzzy controller to control the flow of liquid in the tank and maintain its level at a reference value. The simplest possible fuzzy controller would have only one LHS variable and one RHS variable. We define fuzzy variables *error* and *output* for the LHS and RHS, respectively, as shown in Figure 6.10. It should be noted that the simplicity of the problem does not call for scaling our variables to a common universe of discourse. The universe of discourse for *error* is made of crisp percent error values from -20% to 20%. The fuzzy values of the variable *error* are: *NB* (*negative big*), *NS* (*negative small*), *Z* (*zero*), *PS* (*positive small*), and *PB* (*positive big*). The universe of discourse for *output* is the crisp values of output ranging from 0% to 100%, with fuzzy values: *VL* (*very low*), *LOW*, *MED* (*medium*), *HIGH*, and *VH* (*very high*). The fuzzy algorithm is com-

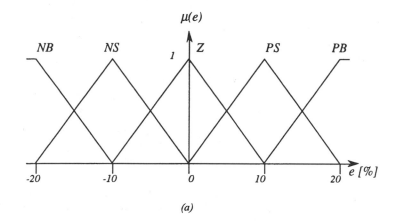

(a)

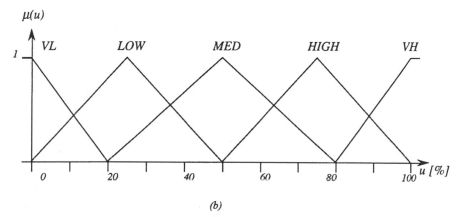

(b)

Figure 6.10 Fuzzy values for (*a*) *error* and (*b*) *output* fuzzy variables used in the rules of Example 6.2

prised of the following rules:

R_1:   *if error is NB then output is VH*   *ELSE*

R_2:   *if error is NS then output is HIGH ELSE*

R_3:   *if error is Z then output is MED*   *ELSE*   (E6.2-1)

R_4:   *if error is PS then output is LOW*   *ELSE*

R_5:   *if error is PB then output is VL*

We use Mamdani min for fuzzy implication and hence interpret the connective *ELSE* in (E6.2-1) as *OR* (see Table 5.5). Suppose that at time $t = 0$ min the output of our controller was at 50% of its full range and 2 min later we introduce the error shown in Figure 6.2. What would be the output according to the algorithm of (E6.2-1)? Let us look at what happens at $t = 3$ min. From Figure 6.2 we see that the input to the algorithm at this time is a crisp error $e_k = 1.25\%$, which belongs to the Z value of *error* to a degree of 0.87 and to PS to a degree of 0.12. The degree of membership to other fuzzy values is zero, as can be seen in Figure 6.10. Thus, rules R_3 and R_4 of algorithm (E6.2-1) will fire, since they are the only rules involving the Z and PS values. The situation is shown in Figure 6.11, where we see a schematic (geometrical) rendition of the evaluation of the control algorithm under GMP at time $t = 3$ min. The *degree of fulfillment* of R_3 is $DOF_3 = 0.87$ and for R_4 we have $DOF_4 = 0.12$. All other rules have DOF = 0.0. Using Mamdani min implication the result of evaluating rules R_3 and R_4 under GMP is to *clip* the RHS values of rules R_3 and R_4—that is, *MED* and *LOW*—at $\mu_{LOW}(u) = 0.87$ and $\mu_{MED}(u) = 0.12$, respectively. In other words, *MED* is clipped at 0.12. Out of all rules, only R_3 and R_4 contribute at $t = 3$ min, and their contributions are $\mu_{LOW'}(u)$ and $\mu_{MED'}(u)$ as shown in Figure 6.11. Since we interpret the connective *ELSE* as *OR*, the total fuzzy output of the entire algorithm at $t = 3$ min is the union of these two values—that is,

$$\mu_{OUT}(u) = \mu_{LOW'}(u) \vee \mu_{MED'}(u) \qquad (E6.2-2)$$

$\mu_{OUT}(u)$ is shown in the lower part of Figure 6.11. We use the COA defuzzification—that is, equation (6.3-2)—to defuzzify $\mu_{OUT}(u)$. The result is $u_3^* = 47\%$. If MOM is used, the average value of the maximum values of $\mu_{OUT}(u)$ is at the middle of the plateau of $\mu'_{MED}(u)$—that is, at about 50%. Hence, the two methods give somewhat different results. In MOM the contribution of $\mu'_{LOW}(u)$ is totally ignored since only the values where $\mu_{OUT}(u)$ is at a maximum are taken into account. The above procedure is repeated in subsequent time steps. The defuzzified output of the controller using COA is shown in Figure 6.12. Comparing with the three different modes of PID control shown in Figure 6.3, we see that our fuzzy controller

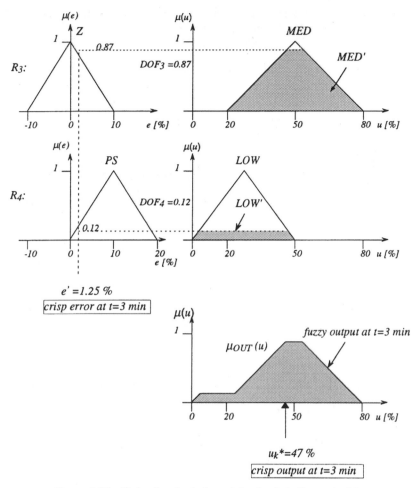

Figure 6.11   Output calculation at $t = 3$ min in Example 6.2.

behaves like a proportional controller, since the form (not the magnitude) of the output is similar to $u_p(t)$.   □

Example 6.3   A Two-Input Fuzzy Controller for Level Control. Consider the process system shown in Figure 6.1, addressed by a PID controller in Example 6.1 and a simple fuzzy controller in Example 6.2. Let us now develop a fuzzy controller with two LHS and one RHS variables. We use *error* and change in error, $\Delta error$, in the LHS of the rules and use *output* in the RHS. The fuzzy values of these variables are shown in Figure 6.13. $\Delta error$ express the direction of change in error; that is, *increasing* is described by the fuzzy value P (*positive*), *decreasing* is described by N (*negative*), and *no* *change* is described by *ZE* (*zero*). We use Mamdani min implication and COA defuzzification. The fuzzy algorithm is:

R_1: *if error is NB AND Δerror is N then output is HIGH* *ELSE*

R_2: *if error is NB AND Δerror is ZE then output is VH* *ELSE*

R_3: *if error is NB AND Δerror is P then output is VH* *ELSE*

R_4: *if error is NS AND Δerror is N then output is HIGH* *ELSE*

R_5: *if error is NS AND Δerror is ZE then output is HIGH ELSE*

R_6: *if error is NS AND Δerror is P then output is MED* *ELSE*

R_7: *if error is Z   AND Δerror is N then output is MED* *ELSE*

R_8: *if error is Z   AND Δerror is ZE then output is MED* *ELSE*

R_9: *if error is Z   AND Δerror is P then output is MED* *ELSE*

R_{10}: *if error is PS AND Δerror is N then output is MED* *ELSE*

R_{11}: *if error is PS AND Δerror is ZE then output is LOW* *ELSE*

R_{12}: *if error is PS AND Δerror is P then output is LOW* *ELSE*

R_{13}: *if error is PB AND Δerror is N then output is LOW* *ELSE*

R_{14}: *if error is PB AND Δerror is ZE then output is VL* *ELSE*

R_{15}: *if error is PB AND Δerror is P then output is VL*

$$(E6.3-1)$$

We recall that with Mamdani min the connective *ELSE* in (E6.3-1) is interpreted as *OR* and therefore the total fuzzy output will be the *union* of individual rule contributions (see Table 5.5). It is customary in the control literature to refer to the fuzzy relations (E6.3-1) as *control surfaces* (or *hypersurfaces*). Figure 6.14 is a graphical representation of the control surface indicating hypersurface dependence on the rules. In Figure 6.14*a*, no rules exist in our algorithm; hence the control hypersurface is a flat plane at $u = 0$. If the control algorithm in (E6.3-1) was comprised only of the two rules R_1 and R_2 (the rest did not exist), then the control hypersurface would look like what is shown in Figure 6.14*b*. If only the first eight rules of the algorithm are present, the control hypersurface looks like Figure 6.14*c*, while if the first 13 rules are present, the control hypersurface would look like Figure 6.14*d*. Finally, if all 15 rules are present, the control hypersurface looks like

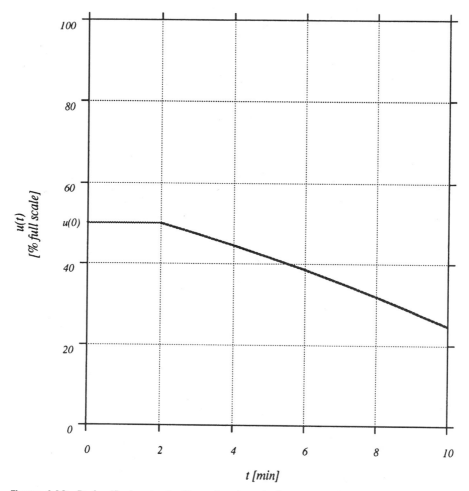

Figure 6.12 Defuzzified output of fuzzy level controller with one input and one output.

Figure 6.14*e*. Plotting the control hypersurface helps to visualize the manner in which a fuzzy controller covers the control space. Unfortunately it is not convenient to use when more than three variables are present.

At $t = 2$ min we introduce the error shown in Figure 6.2 (same as in Examples 6.1 and 6.2). Figure 6.15 shows a schematic representation of the fuzzy inference or *generalized modus ponens* (GMP) at $t = 5$ min. Crisp inputs $e' = 3.75\%$ and $\Delta e' = 1.25\%$ are presented to the algorithm (E6.3-1) at this time. Crisp error $e' = 3.75\%$ belongs to fuzzy value Z to degree of 0.75 and to fuzzy value PS to a degree of 0.4. Similarly, crisp change-in-error $\Delta e' = 1.25\%$ belongs to fuzzy value ZE to a degree of 0.6 and to fuzzy value P to a degree of 0.4. Hence the only rules that will have DOF greater than

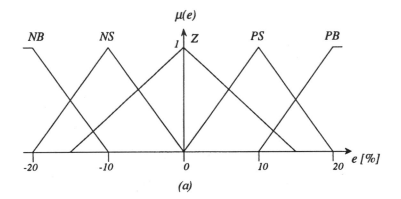

(a)

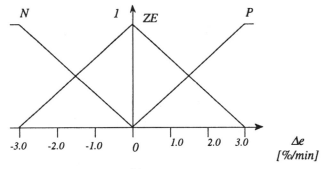

(b)

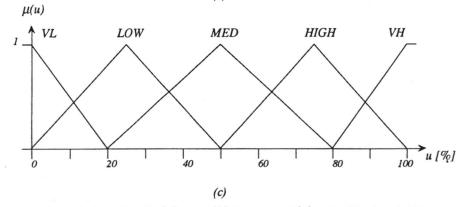

(c)

Figure 6.13 Fuzzy values for (*a*) *error*, (*b*) Δ*error*, and (*c*) *output* fuzzy variables used in Example 6.3.

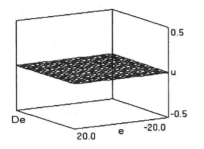

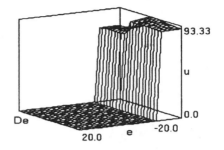

(a)   *(b)*

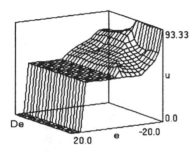

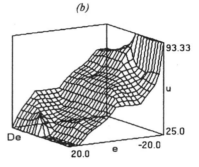

(c)   *(d)*

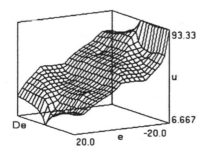

(e)

Figure 6.14 *(a)* The control hypersurface when there are no rules present, *(b)* with rules R_1 and R_2 only, *(c)* with rules R_1 through R_8, *(d)* with rules R_1 through R_{13}, and *(e)* with all 15 rules.

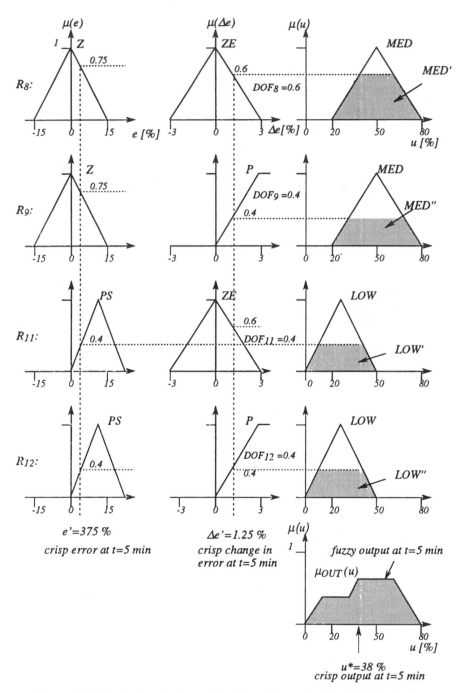

Figure 6.15 Evaluating the fuzzy algorithm of Example 6.3 at time $t = 5$ min.

zero in (E6.3-1)—that is, the rules that fire—will be R_8, R_9, R_{11}, and R_{12}. Using the min form of DOF (i.e., the min ($\wedge$) interpretation of *AND*), each rule contributes the shaded part of the RHS value shown in Figure 6.14. We recall that GMP with Mamdani min implication clips the RHS at the height of DOF, as shown in Figure 6.15. Rule R_8 contributes *MED'*, R_9 contributes *MED''*, R_{11} contributes *LOW'*, and R_{12} contributes *LOW''*. The fuzzy output $\mu_{out}(u)$ is the union (max) of these four contributions (shaded parts); that is,

$$\mu_{out}(u) = \mu_{MED'}(u) \vee \mu_{MED''}(u) \vee \mu_{LOW'}(u) \vee \mu_{LOW''}(u) \quad (E6.3-2)$$

$\mu_{out}(u)$ is shown at the lower part of Figure 6.15. Using COA defuzzification, we obtain the crisp output $u^* = 38\%$. The procedure is repeated for other time steps. The crisp output of the controller for the duration of the problem is shown in Figure 6.16. Comparing with Figure 6.13, we see that introducing

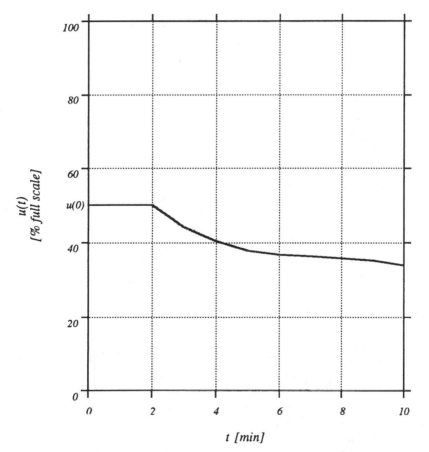

Figure 6.16 Defuzzified output of fuzzy controller with two inputs (*error* and Δ *error*) under ramp input.

one more variable, namely the change in error $\Delta error$, makes a significant difference in the control action. To obtain a desired response by our controller, we can now modify the shape and position of constituent membership functions or the rules, the implications used, and so on. A number of factors contribute to different outputs, such as the knowledge encoded in (E6.3-1), the fuzzy values used, the interpretation of *AND* (affecting DOF), the implication operator, and the defuzzification method used. The role and significance of these factors will be examined in the next section.   □

6.4   ISSUES INVOLVED IN DESIGNING FUZZY CONTROLLERS

Although there are automatic ways of identifying the rules and membership functions involved in a fuzzy controller,[10] in many ways the development of a good fuzzy controller reflects the maturity of knowledge about a process. The choice of fuzzy variables and values and the rules themselves are intimately related to the knowledge a developer has about the entire process control system. The knowledge can be extracted by interviewing skilled operators or analyzing records of system responses to prototypes of input sequences (Dubois and Prade, 1980; Bernard, 1988). In addition, important decisions need to be made about the algorithm itself, such as what kind of implication to use, the appropriate defuzzification method, and implementation-related issues such as how to store the fuzzy relation of the algorithm, how to quantize membership functions, and so on. A difficult issue in fuzzy control arises in connection with determining the stability characteristics of the system. Stability itself can be thought of as a fuzzy variable and can be included in a description, with various degrees of stability (not just *stable* or *unstable*) being considered. Generally though, stability questions are hard to answer exclusively within fuzzy linguistic descriptions (Kiszka et al., 1985; Jianqin and Laijiu, 1993).

Once an algorithm has been developed, its quality can be assessed by examining the shape of the fuzzy output. Consider the situation shown in Figure 6.17 (King and Mamdani, 1977). Here we have three different general shapes for the membership function of the fuzzy output at some particular time step. They reveal three different instances of algorithmic quality. In situation A, a well-peaked fuzzy output indicates presence of strong firing rules. In situation B, the output points to two different and opposite areas of the universe of discourse, and hence we identify the presence of some contradictory rules or groups of rules, at the same time suggesting an output toward -3 and toward $+3$. An algorithm that points its output in opposite directions at the same time needs some further refinement to remove this kind of contradiction. In situation C we have the presence of an *unsatisfactory* set of rules since there is no representative output. In general, low

---

[10] Often referred to as *structure* and *parameter identification*.

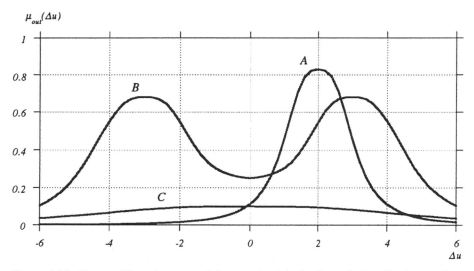

Figure 6.17 Three different cases of fuzzy output indicative of algorithmic quality: (A) *dominant* rule, (B) *contradictory* rules, and (C) *no satisfactory* rule.

plateaus like what is shown in situation C indicate that the knowledge encoded in the algorithm is incomplete and that additional rules are needed.

Let us now turn our attention to the various factors involved in the development of a fuzzy algorithm other than the quality of the encoded knowledge. Fuzzy algorithms are linguistic descriptions of the desirable behavior of a system. As such they have an analytical form involving fuzzy variables, relations, implication operators, and inferencing procedures. In order to examine the factors involved in algorithmic synthesis and analysis, let us look at the general analytical description of a fuzzy algorithm. Suppose that we have a control algorithm with linguistic form:

$$
\begin{array}{llll}
if & x\ is\ A_1 & AND & y\ is\ B_1 & then\ u\ is\ C_1 & ELSE \\
if & x\ is\ A_2 & AND & y\ is\ B_2 & then\ u\ is\ C_2 & ELSE \\
& \ldots \\
if & x\ is\ A_j & AND & y\ is\ B_j & then\ u\ is\ C_j & ELSE \\
& \ldots \\
if & x\ is\ A_n & AND & y\ is\ B_n & then\ u\ is\ C_n
\end{array}
$$

(6.4-1)

At time $t = k$, crisp inputs x' and y' are given to algorithm (6.4-1), and through GMP (see Chapters 3 and 5) we determine the output membership function. Analytically the operation of inferring a fuzzy output at any given time step may be written as

$$\mu_{C'}(u) = \phi\big(\mathrm{DOF}_1(k), \mu_{C_1}(u)\big)$$

$$\vee\ \phi\big(\mathrm{DOF}_2(k), \mu_{C_2}(u)\big)$$

$$\cdots$$

$$\vee\ \phi\big(\mathrm{DOF}_j(k), \mu_{C_j}(u)\big)$$

$$\cdots$$

$$\vee\ \phi\big(\mathrm{DOF}_n(k), \mu_{C_n}(u)\big) \tag{6.4-2}$$

for implication operators ϕ where the connective *ELSE* is interpreted as *union* (see Table 5.5). For implication operators interpreted as *intersection* we change ($\vee$) in equation (6.4-2) to ($\wedge$). Equation (6.4-2) tells us that the collective output of the controller depends on aggregating the outputs of individual rules with the output of each rule depending, in turn, on the degree of fulfillment plus the consequent membership function of the rule.

The degree of fulfillment of the *j*th rule DOF_j in (6.4-2) depends on the interpretation of the connective *AND* (generally thought of as a T norm (see Appendix)). If *AND* is analytically described as min ($\wedge$), the *degree of fulfillment* at timestep k is

$$\mathrm{DOF}_j(k) = \mu_{A_j}(x') \wedge \mu_{B_j}(y') \tag{6.4-3}$$

where x' and y' are the measured input values at a given time k. If *AND* is analytically described as *product* ($\cdot$), the DOF is

$$\mathrm{DOF}_j(k) = \mu_{A_j}(x') \cdot \mu_{B_j}(y') \tag{6.4-4}$$

It should be noted that the DOF is a function of time as different input values activate the rules to different degrees at different times. Equation (6.4-2) gives the fuzzy output (before defuzzification) of the controller in a general form and helps us to identify choices the developer needs to make such as the appropriate fuzzy implication operator ϕ and the associated interpretation of the connective *ELSE*, the form of DOF, and the defuzzification method.

In the design of fuzzy systems it is important to adequately cover the state space of the problem. Generally the development of a rule set that is both complete and correct is one of the most difficult problems in fuzzy control. Although various approaches have been suggested for learning a control algorithm on-line and adapting it to changing process conditions (Graham and Newell, 1988; Cox, 1993), this is still a rather heuristic process, and a good understanding of the various factors influencing the output of the controller is very helpful in its development and evaluation. Generally, which rules and to what extent will contribute toward an output at any given time depends primarily on the *form of the degree of fulfillment* (min or product), the *defuzzification method*, and the *implication operator*.

Let us consider the *j*th rule of (6.4-1) where triangular membership functions are used as shown in Figure 6.18. We assume min ($\wedge$) form for DOF as in equation (6.4-3). We also assume common quantized universe of discourse for all variables, of the type shown in Table 6.1. In Figure 6.18 we see the part of the Cartesian product of LHS variables covered by the *j*th rule. The $x \times y$ plane is the *state space* of our system. The state space covered by the *j*th rule is a square of six units edge, centered at (x_{jc}, y_{jc}) as shown in the figure. At time $t = k$, crisp inputs (x', y') are given to the rule. Let us first see what happens when the point (x', y') is located within the innermost square centered at (x_{jc}, y_{jc}) that has an edge of 2 units as shown in Figure 6.18. In such a case the degree of fulfillment DOF_j of the *j*th rule will be the same regardless of the exact location of point (x', y') so long as it remains within this particular square, since we have that

$$\mu_{A_j}(x_{jc} + 1) \wedge \mu_{B_j}(y_{jc} + 1) = 0.67 \wedge 0.67 = 0.67$$

$$\mu_{A_j}(x_{jc}) \wedge \mu_{B_j}(y_{jc} - 1) = 0.67 \wedge 0.67 = 0.67$$

$$\mu_{A_j}(x_{jc}) \wedge \mu_{B_j}(y_{jc} + 1) = 0.67 \wedge 0.67 = 0.67$$

$$\mu_{A_j}(x_{jc} - 1) \wedge \mu_{B_j}(y_{jc} - 1) = 0.67 \wedge 0.67 = 0.67 \qquad (6.4\text{-}5)$$

$$\mu_{A_j}(x_{jc} - 1) \wedge \mu_{B_j}(y_{jc}) = 0.67 \wedge 0.67 = 0.67$$

$$\mu_{A_j}(x_{jc} - 1) \wedge \mu_{B_j}(y_{jc} + 1) = 0.67 \wedge 0.67 = 0.67$$

Thus the DOF is 0.67 everywhere within this innermost square. Similar considerations lead us to the conclusion that if the point (x', y') falls within a square of edge 4, the DOF is 0.33 everywhere, whereas if it falls outside, the DOF is 0. Thus the distribution of the DOFs of a rule centered at (x_{jc}, y_{jc}) is as shown in Table 6.3.

When at time $t = k$ the crisp inputs (x', y') are given to the controller, the DOF of individual rules depends linearly on the distance between the input and the center (or peak) of the rule (x_{jc}, y_{jc}). Obviously the number of rules that will influence and contribute to the collective fuzzy output at any given time are only those within a distance d from input (x', y') (see Figure 6.18). Thus in a control algorithm, only the part of state space a distance d from a crisp input needs to be considered for rules that may be "fired." The rest have DOF = 0. The distance d is taken to be half of the support of a fuzzy value (considering the support to be where the membership function is not trivial). As shown in Figure 6.18, the edge of a square with (x_{jc}, x_{jc}) at its

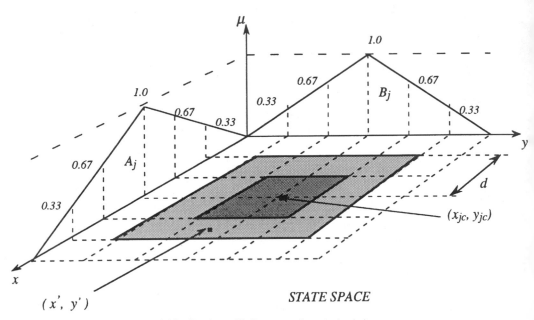

Figure 6.18 Region of influence of a rule in state space.

Table 6.3 Distribution of DOFs around rule center (rule with min DOF)

0	0	0	0	0	0	0
0	0.33	0.33	0.33	0.33	0.33	0
0	0.33	0.67	0.67	0.67	0.33	0
0	0.33	0.67	1.0	0.67	0.33	0
0	0.33	0.67	0.67	0.67	0.33	0
0	0.33	0.33	0.33	0.33	0.33	0
0	0	0	0	0	0	0

center is $2d$. We assumed of course that the two fuzzy values A_j and B_j have supports of the same length. When the support sets do not have the same length, then instead of a square we have a parallelogram and the distance of (x', y') from (x_{jc}, x_{jc}) will vary directionally as we move in different locations of state space.

The number of rules contributing to the fuzzy output depends also on the defuzzification method. When MOM is used, only the rules that are very close to the input (x', y') contribute maximum values to the output and therefore they are the only ones that need to be taken into account. We recall that with MOM, only the maximum values of the various contributions to the fuzzy output are used; and hence, only rules with high DOF and therefore small distance from (x', y') will influence the output. When COA is used, all the rules within a distance d from (x', y') need to be taken into account. Of course their contribution is in proportion to their distance from the input. Those which are the closest have the highest degree of fulfillment and therefore contribute more than those far away. Nonetheless, all rules within a distance d from (x', y') need to be taken into account.

On the other hand, if product is used in the DOF—that is, $\text{DOF}_j = \mu_{A_j}(x_k) \cdot \mu_{B_j}(y_k)$—the distribution of the DOF for input values in the vicinity of the jth rule would be as shown in Table 6.4. We see that DOF is varying with distance from (x_{jc}, y_{jc}) in a nonlinear manner. Again, if COA defuzzification is used, all the rules within a distance d need to be taken into account. When MOM is used, the rule peaking at (x_{jc}, y_{jc}) will have less influence than the earlier situation when the degree of fulfillment was defined through min. In the present case, it will influence the rule in a directional manner. For this reason with product DOF, COA defuzzification is more appropriate. Sometimes we may have very low DOFs, and therefore a cutoff number ought to be used to limit the number of rules that need to be considered. Thus we may choose $\text{DOF}_j = \alpha$ and ignore rules below whose DOF is less than α.

When continuous instead of discrete fuzzy values are used, their membership functions can be defined by various functions such as *S-shaped* and *Π-shaped* functions (see Sections 2.6.3, 2.6.4, and 2.6.5). Similar considerations hold for such cases as for discretized membership functions. With MOM, tremendous accuracy is not required since only the relative size of the membership values influences the final result and not the precise magnitude. Thus in order to take advantage of the fact that we have more precise membership values with continuous membership functions, it is best to utilize the COA (or COS) method of defuzzification.

We turn our attention now to the influence of the shape of membership functions describing the antecedent and consequent fuzzy values. The support of the fuzzy values of the antecedents (e.g., $2d$) determines the area of influence of every rule and hence plays a crucial role in the calculation of the control value. Generally the *shape* of the membership functions of LHS values has a substantial impact on the computation of the control action at any given time since it affects the DOF of each rule. Obviously the shape of

Table 6.4   Distribution of DOFs around rule center
(rule with product DOF)

0	0	0	0	0	0	0
0	0.09	0.21	0.33	0.21	0.09	0
0	0.21	0.49	0.67	0.49	0.21	0
0	0.33	0.67	1.0	0.67	0.33	0
0	0.21	0.49	0.67	0.49	0.21	0
0	0.09	0.21	0.33	0.21	0.09	0
0	0	0	0	0	0	0

the RHS membership function affects directly the contribution of the rule to the overall fuzzy output.

When MOM is used, the exact shape of the LHS membership functions does not play a major role provided that it is in the general shape of a "hill, and symmetric with respect to a normal point." MOM defuzzification effectively distinguishes the rules with the highest priority (highest DOF) that is, the rules closest to the input (x', y'). Thus with MOM, the DOF suggests the distance from (x', y') and therefore the absolute values of the membership functions are not crucial, just their magnitude in relation to the membership functions of other rules. Similarly the exact shape of RHS membership functions does not play a crucial role in the calculation of the crisp output. When the support is not symmetric, the peaks in the membership function of the antecedents move relative to the support set and thus offer different DOF for the same inputs to the controller and different nonsymmetric shapes of "hills."

When COA defuzzification is used, the exact shape of the membership functions of antecedent as well as consequent plays an important role, even when symmetric membership functions are used. This happens because COA defuzzification takes into account the area under the curve of the total fuzzy output at any given time. This area is influenced directly by the shape of the consequent membership functions of the contributing rules, and indirectly through the DOF (the shape of the consequent membership functions of the contributing rules). The above-mentioned influence on the crisp controller output is emphasized (accentuated) even more if nonsymmetric membership functions are used, as well as if different membership functions are used for the different variables.

Let us now examine the influence of fuzzy implication operator ϕ on the computations of the controller output at a time $t = k$. We consider a hypothetical case where only one rule exists in the vicinity of (x', y'); that is, only one rule fires. Suppose that we use Mamdani min implication operator ϕ_c and fuzzy sets defined through symmetric triangular functions of the form shown in Table 6.1 (also Figure 6.5). Figure 6.19 (top) shows what happens to the RHS value for different degrees of fulfillment of a rule. The consequent membership function remains the same for DOF = 1 and is gradually clipped, finally becoming zero when DOF = 0. The defuzzified output is the same with either COA or MOM methods. The situation is similar when Larsen product implication operator ϕ_p is used as can be seen in Figure 6.18.

On the other hand, if the *Boolean* implication operator ϕ_b is used, a "plateau" is created that grows, as DOF is getting smaller, until it covers the entire universe of discourse when DOF = 0. While this is exactly the opposite of what happens when the Mamdani min implication operator is used, it is counterbalanced by interpreting *ELSE* as intersection (min) when a number of rules are connected in order to compute the total fuzzy output of the controller. In fact, this is the reason for using min for the connective *ELSE* with this implication operator (see Table 5.5). It should be noted from Figure 6.19 that COA and MOM defuzzification may give different crisp outputs when Boolean implication is used.

If the *arithmetic* implication operator ϕ_a is used, a "plateau" is also formed as it happens with Boolean implication. The peak of the function is *clipped* as with Mamdani min implication. When DOF = 0, the "plateau" covers the entire universe of discourse. Again it should be noted that COA and MOM defuzzification may give different crisp outputs.

As can be seen in Figure 6.19, when Mamdani min and Larsen product implications are used, both COA and MOM defuzzification methods give similar results. In Boolean and arithmetic implications, on the other hand, the two defuzzification methods will give rather different results due to the developing plateaus. When plateaus appear, MOM defuzzification is better because COA considers the peak of the rule together with the developing plateau, and hence it shifts the final crisp output away from the location that is suggested by the peak of the rule. This is undesirable since "plateaus" do not contain useful information. They can be interpreted as a fuzzy value "*unknown*." In Figure 6.19 we note that when DOF = 0, Mamdani min and Larsen product implications give "nothing" as the output of the controller.

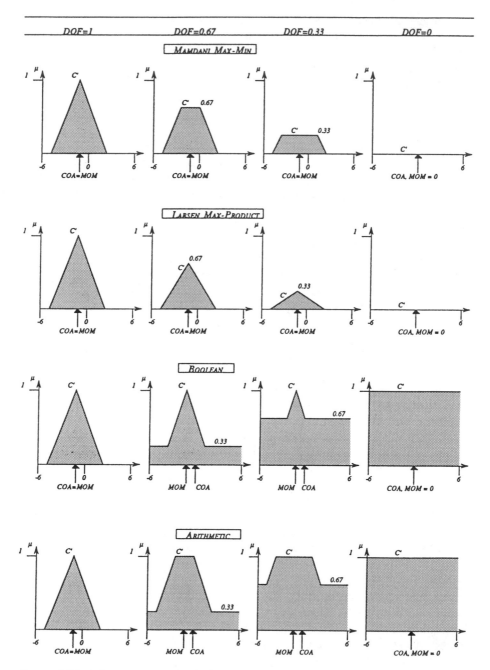

Figure 6.19 Influence on output of various implications under different degrees of fulfillment.

On the other hand, using the Boolean and arithmetic implications produces *unknown* as output. In all cases we can set the output to crisp zero.

In addition, with Mamdani min and Larsen product implications we may effectively use either method of defuzzification since the total fuzzy output contains contributions from all (or many) rules. Generally, if a rule or several rules are an equal distance away from the point (x', y'), we have the same results with either defuzzification method. With Boolean and arithmetic implications we do not need to use COA defuzzification (which is actually computationally more demanding) since in many cases the total fuzzy output does not effectively represent the contribution of the individual rules (due to the "plateau" or "flattening" effect). In general, it is preferable (but not required) to use MOM in conjunction with Boolean and arithmetic implications.

REFERENCES

Bernard, J. A., Use of a Rule-Based System for Process Control, *IEEE Control Systems Magazine*, pp. 3–13, October 1988.

Cox, E., Adaptive Fuzzy Systems, *IEEE Spectrum*, pp. 27–31, February 1993.

Driankov, D., Hellendoorn, H., and Reinfrank, M., *An Introduction to Fuzzy Control*, Springer-Verlag, Berlin, 1993.

Dubois, D., and Prade, H., *Fuzzy Sets and Systems: Theory and Applications*, Academic Press, Boston, 1980.

Efstathiou, J., Rule-Based Process Control Using Fuzzy Logic, in *Approximate Reasoning in Intelligent Systems, Decision and Control*, E. Sanchez and L. A. Zadeh, eds., Pergamon Press, Oxford, 1987, pp. 145–158.

Fuller, R., and Zimmermann, H.-J., On Computation of the Compositional Rule of Inference under Triangular Norms, *Fuzzy Sets and Systems*, Vol. 51, pp. 267–275, 1992.

Graham, B. P., and Newell, R. B., Fuzzy Identification and Control of a Liquid Level Rig, *Fuzzy Sets and Systems*, Vol. 26, pp. 255–273, 1988.

Harris, C. J., Moore, C. G., and Brown, M., *Intelligent Control—Aspects of Fuzzy Logic and Neural Nets*, World Scientific, Singapore, 1993.

He, S-Z., Tan, S., and Xu, F-L., Fuzzy Self-Tuning of PID Controllers, *Fuzzy Sets and Systems*, Vol. 56, pp. 37–46, 1993.

Hirota, K., *Industrial Applications of Fuzzy Control*, Springer-Verlag, Tokyo, 1993.

Jager, R., *Fuzzy Logic in Control*, Unpublished Ph.D. Dissertation, University of Delft, The Netherlands, 1995.

Jang, J-S., and Gulley, N., *Fuzzy Logic Toolbox User's Guide*, Mathworks, 1995.

Jang, J-S., and Sun, C-T., Neuro-Fuzzy Modeling and Control, *Proceedings of the IEEE*, Vol. 83, No. 3, pp. 378–406, 1995.

Jiangin, C., and Laijiu, C., Study on Stability of Fuzzy Closed-Loop Control Systems, *Fuzzy Sets and Systems*, Vol. 57, pp. 159–168, 1993.

Johnson, C. D., *Process Control Instrumentation Technology*, John Wiley & Sons, New York, 1977.

King, P. J., and Mamdani, E. H., The Application of Fuzzy Control Systems to Industrial Processes, in *Fuzzy Automata and Decision Processes*, M. M. Gupta, G. N. Saridis, and B. R. Gaines, eds., North-Holland, New York, 1977, pp. 321–330.

King, R. E., and Karonis, F. C., Multi-Level Expert Control of a Large-Scale Industrial Process, in *Fuzzy Computing Theory, Hardware, and Applications*, M. M. Gupta and T. Yamakawa, eds., Elsevier/North-Holland, Amsterdam, 1988, pp. 323–340.

Kiszka, J. B., Gupta, M. M., and Nikiforuk, P. N., Some Properties of Expert Control Systems, in *Approximate Reasoning in Expert Systems*, M. M. Gupta, A. Kandel, W. Bandler, and J. B. Kiszka, eds., Elsevier/North-Holland, Amsterdam, 1985, pp. 283–306.

Larkin, L. I., A Fuzzy Logic Controller for Aircraft Flight Control, in *Industrial Applications of Fuzzy Control*, M. Sugeno, ed., North-Holland, New York, 1985, pp. 87–103.

Lee, C. C., Fuzzy Logic in Control Systems: Fuzzy Logic Controller—Part-I, *IEEE Transactions on Systems, Man and Cybernetics*, Vol. 20, No. 2, pp. 404–418, March/April 1990a.

Lee, C. C., Fuzzy Logic in Control Systems: Fuzzy Logic Controller—Part-II, *IEEE Transactions on Systems, Man and Cybernetics*, Vol. 20, No. 2, pp. 419–435, March/April 1990b.

Maeda, M., and Murakami, S., A Self-Tuning Fuzzy Controller, *Fuzzy Sets and Systems*, Vol. 51, pp. 29–40, 1992.

Mamdani, E. H., Application of Fuzzy Algorithms for Control of Simple Dynamic Plants, *Proceedings of IEE*, Vol. 121, No. 12, pp. 1585–1588, 1974.

Mamdani, E. H., Advances in the Linguistic Synthesis of Fuzzy Controllers, *International Journal of Man-Machine Studies*, Vol. 8, pp. 669–678, 1976.

Mamdani, E. H., Applications of Fuzzy Set Theory to Control Systems: A Survey, in *Fuzzy Automata and Decision Processes*, M. M. Gupta, G. N. Saridis, and B. R. Gaines, eds., North-Holland, New York, 1977, pp. 1–13.

Matia, F., Jimenez, A., Galan, R., and Sanz, R., Fuzzy Controllers: Lifting the Linear–Nonlinear Frontier, *Fuzzy Sets and Systems*, Vol. 52, pp. 113–128, 1992.

Mizumoto, M., Fuzzy Controls Under Various Reasoning Methods, *Information Sciences*, Vol. 45, pp. 129–141, 1988.

Ostergaard, J. J., and Holmblad, L. P., Control of Cement Kiln by Fuzzy Logic, in *Fuzzy Information and Decision Processes*, M. M. Gupta and E. Sanchez, eds., North-Holland, Amsterdam, 1982, pp. 389–399.

Pedrycz, W., *Fuzzy Control and Fuzzy Systems*, second (extended) edition, John Wiley & Sons, New York, 1993.

Schwartz, D. G., Fuzzy Logic Flowers in Japan, *IEEE Spectrum*, pp. 32–35, July 1992.

Shoureshi, R., and Rahmani, K., Derivation and Application of an Expert Fuzzy Optimal Control System, *Fuzzy Sets and Systems*, Vol. 49, pp. 93–101, 1992.

Sugeno, M., An Introductory Survey of Fuzzy Control, *Information Sciences*, Vol. 36, pp. 59–83, 1985.

Terano, T., Asai, K., and Sugeno, M., *Fuzzy Systems Theory and Its Applications*, Academic Press, Boston, 1992.

Tsukamoto, Y., An Approach to Fuzzy Reasoning Method, *Advances in Fuzzy Set Theory and Applications*, M. M. Gupta, R. K. Ragade, and R. R. Yager, eds., North-Holland, Amsterdam, 1979, pp. 134–149.

Tzafestas, S. G., and Papanikolopoulos, N. P., Incremental Fuzzy Expert PID Control, *IEEE Transactions on Industrial Electronics*, Vol. 37, No. 5, pp. 365–371, 1990.

Wang, Li-Xin, *Adaptive Fuzzy Systems and Control*, Prentice-Hall, Englewood Cliffs, NJ, 1994.

Yager, R. R., and Filev, D. P., SLIDE: A Simple Adaptive Defuzzification Method, *IEEE Transactions on Fuzzy Systems*, Vol. 1, No. 1, pp. 69–78, 1993.

Yager, R. R., and Filev, D. P., *Essentials of Fuzzy Modeling and Control*, John Wiley & Sons, New York, 1994.

Yamakawa, T., Fuzzy Hardware Systems of Tomorrow, in *Approximate Reasoning in Intelligent Systems, Decision and Control*, E. Sanchez and L. A. Zadeh, eds., Pergamon Press, Oxford, 1987, pp. 1–20.

Yamakawa, T., Stabilization of an Inverted Pendulum by a High-Speed Fuzzy Logic Controller Hardware System, *Fuzzy Sets and Systems*, Vol. 32, pp. 161–180, 1989.

Yasunobu, S., and Miyamoto, S., Automatic Train Operation System by Predictive Fuzzy Control, in *Industrial Applications of Fuzzy Control*, M. Sugeno, ed., North-Holland, Amsterdam, 1985, pp. 1–18.

Zimmermann, H.-J., *Fuzzy Set Theory and Its Applications*, Kluwer-Nijhoff, Boston, 1985.

PROBLEMS

1. A fuzzy control system used inputs of error e and change in error Δe to control an output variable u. Their fuzzy membership functions have the following characteristics:

Variable	Range		Description of membership function
e Error (%)	−20 to +20	N (negative)	Straight line from 1 at −20% to 0 at 0%
		Z (zero)	Straight line from 0 at −20% to 1 at 0% and another straight line from 1 at 0% to 0 at +20%
		P (positive)	Straight line from 0 at 0% to 1 at +20%
Δe Change in	−10 to +10	N (negative)	Straight line from 1 at −10%/min to 0 at +10%/min
error (%/min)		P (positive)	Straight line from 0 at −10%/min to 1 at +10%/min
u Output (%)	−25 to +25	N (negative)	Straight line from 1 at −25% to 0 at 0%
		Z (zero)	Straight line from 0 at −25% to 1 at 0% and another straight line from 1 at 0% to 0 at +25%
		P (positive)	Straight line from 0 at 0% to 1 at +25%

The fuzzy algorithm is given below. Determine the output u for $e' = +16\%$ and $\Delta e' = -2\%/\text{min}$ using the Mamdani min implication operator and max–min composition (as well as min interpretation for *AND* in the degree of fulfillment). Use the Center of area method to defuzzify the answer. (Sketch the various membership functions involved and show how you obtained your solution.)

FUZZY ALGORITHM

R_1 *if e is N   AND Δe is N   then u is P   ELSE*

R_2 *if e is N   AND Δe is P   then u is P   ELSE*

R_3 *if e is Z   AND Δe is N   then u is Z   ELSE*

R_4 *if e is Z   AND Δe is P   then u is Z   ELSE*

R_5 *if e is P   AND Δe is N   then u is N   ELSE*

R_6 *if e is P   AND Δe is P   then u is N*

2. Repeat Problem 1 using the Larsen product implication, max–min composition, and product for the degree of fulfillment.

3. In Problem 1, the error starts at a value of $+16\%$ at time 0 and decreases at a rate of $2\%/min$ for 4 minutes. Determine the output u at times $t = 0, 1, 2, 3,$ and 4.

4. Analyze the fuzzy controller given in Problem 1 using the criteria given in Section 4. Are there contradictions within the rule set? Is there a dominant rule? Are the rules covering the state space in a satisfactory manner?

5. Using MATLAB, draw the control hypersurface for the fuzzy controller given in Problem 1. Simulate the controller for the range of all possible inputs and answer the questions posed in Problem 4.

6. Show what the different interpretations for *ELSE* could be for the fuzzy controller of Problem 1 and the implication operators given in Table 5.2.

II

NEURAL NETWORKS: CONCEPTS AND FUNDAMENTALS

7

FUNDAMENTALS
OF NEURAL NETWORKS

7.1 INTRODUCTION

In 1956, the Rockefeller Foundation sponsored a conference at Dartmouth
College that had as its scope

> The potential use of computers and simulation in every aspect of learning and
> any other feature of intelligence.

It was at this conference that the term "artificial intelligence" came into
common use. Artificial intelligence can be broadly defined as

> Computer processes that attempt to emulate the human thought processes that
> are associated with activities that require the use of intelligence.

Generally, this definition included the fields of automatic learning, understanding natural language, vision-image recognition, voice recognition, game
playing, mathematical problem solving, robotics, and expert systems. In
recent years, some researchers have included neural networks and other
related technologies as constituents of artificial intelligence, while others,
pointing to their origin in biological sciences, have sought to avoid this
association. In this text we accept neural networks as a legitimate field of
artificial intelligence. Furthermore, we include genetic algorithms, fuzzy logic
or fuzzy systems, wavelets, cellular automata, and chaotic systems as being
within the general field of artificial intelligence.

7.2 BIOLOGICAL BASIS OF NEURAL NETWORKS

The human brain is a very complex system capable of thinking, remembering, and problem solving. There have been many attempts to emulate brain functions with computer models, and although there have been some rather spectacular achievements coming from these efforts, all of the models developed to date pale into oblivion when compared with the complex functioning of the human brain.

A *neuron* is the fundamental cellular unit of the brain's nervous system. It is a simple processing element that receives and combines signals from other neurons through input paths called *dendrites*. If the combined input signal is strong enough, the neuron "fires," producing an output signal along the axon that connects to the dendrites of many other neurons. Figure 7.1 is a sketch of a neuron showing the various components. Each signal coming into a neuron along a dendrite passes through a *synapse* or *synaptic junction*. This junction is an infinitesimal gap in the dendrite that is filled with neurotransmitter fluid that either accelerates or retards the flow of electrical charges. The fundamental actions of the neuron are chemical in nature, and this neurotransmitter fluid produces electrical signals that go to the nucleus or *soma* of the neuron. The adjustment of the impedance or conductance of the synaptic gap is a critically important process. Indeed, these adjustments lead to memory and learning. As the synaptic strengths of the neurons are adjusted, the brain "learns" and stores information.

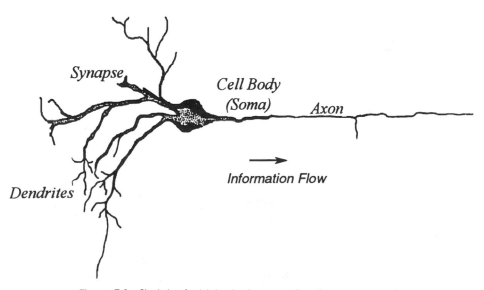

Figure 7.1 Sketch of a biological neuron showing components.

When a person is born, the cerebral cortex portion of his or her brain contains approximately 100 billion neurons. The outputs of each of these neurons are connected through their respective *axons* (output paths) to about 1000 other neurons. Each of these 1000 paths contains a synaptic junction by which the flow of electrical charges can be controlled by a neurochemical process. Hence, there are about 100 trillion synaptic junctions that are capable of having influence on the behavior of the brain. It is readily apparent that in our attempts to emulate the processes of the human brain, we cannot think of billions of neurons and trillions of synaptic junctions. Indeed, the largest of our neural networks typically contain a few thousand artificial neurons and less than a million artificial synaptic junctions.

The one area in which artificial neural networks may have an advantage is speed. When a person walks into a room, it typically takes another person about half a second to recognize them. We are told that this recognition process involves about 200–250 individual separate operations within the brain. As a benchmark for speed, this means that the human brain operates at about 400–500 hertz (Hz). Modern digital computers typically operate at clock speeds between 100 and 200 megahertz (MHz), which means that they have a very large speed advantage over the brain. However, this advantage is dramatically reduced because digital computers operate in a serial mode whereas the brain operates in a parallel mode. However, neural network chips have been developed in recent years that enable neural computers to operate in a parallel mode.

The nomenclature in the neural network field is still not standardized. You will find books and technical articles that refer to artificial neural networks as *connectionist systems* and artificial neurons as *processing elements* (PEs), *neurodes*, *nodes*, or simply *neurons*. In this text we shall use the terms *neurons* and *neural networks*, except in situations where an alternate designation would be more descriptive. Often we drop the adjective "artificial," because we deal only with artificial neurons in this text.

7.3 ARTIFICIAL NEURONS

An artificial neuron is a model whose components have direct analogs to components of an actual neuron. Figure 7.2 shows the schematic representation of an artificial neuron. The input signals are represented by $x_0, x_1, x_2, \ldots, x_n$. These signals are continuous variables, not the discrete electrical pulses that occur in the brain. Each of these inputs is modified by a *weight* (sometimes called the *synaptic weight*) whose function is analogous to that of the synaptic junction in a biological neuron. These weights can be either positive or negative, corresponding to acceleration or inhibition of the flow of electrical signals. This processing element consists of two parts. The first part simply aggregates (sums) the weighted inputs resulting in a quantity

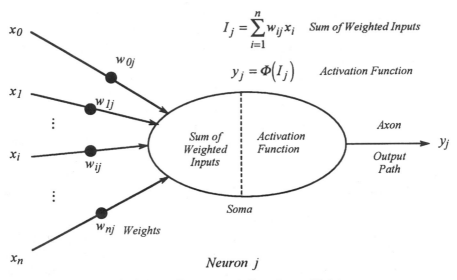

$$I_j = \sum_{i=1}^{n} w_{ij} x_i \quad \textit{Sum of Weighted Inputs}$$

$$y_j = \Phi(I_j) \quad \textit{Activation Function}$$

Figure 7.2   Schematic representation of an artificial neuron.

I; the second part is effectively a nonlinear filter, usually called the *activation function*,[1] through which the combined signal flows.

Figure 7.3 shows several possible activation functions. It may be a threshold function as shown in Figure 7.3*a* that passes information (usually a +1 signal) only when the output I of the first part of the artificial neuron exceeds the threshold T. It can be the signum function (sometimes called a quantizer function) shown in Figure 7.3*b* that passes negative information when the output is less than the threshold T and positive information when the output is greater than the threshold T. More commonly, the activation function is a continuous function that varies gradually between two asymptotic values, typically 0 and 1, or -1 and $+1$, called the *sigmoidal function*. The most widely used activation function is the logistic function, one of the sigmoidal activation functions, which is shown in Figure 7.3*c* and is represented by the equation

$$\Phi(I) = \frac{1}{1 + e^{-\alpha \cdot I}} \tag{7.3-1}$$

where α is a coefficient that adjusts the abruptness of this function as it changes between the two asymptotic values.

A more descriptive term for the activation function is "squashing function," which indicates that this function squashes or limits the values of the output

---

[1]A rather common name used in many books for the activation function is "transfer function." We will avoid the use of this term in the text to avoid confusion, because this term is commonly used in engineering—to describe the input–output behavior of linear systems.

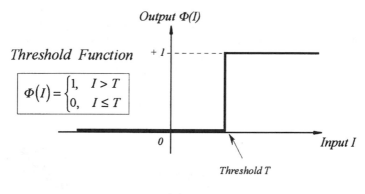

Threshold Function

$$\Phi(I) = \begin{cases} 1, & I > T \\ 0, & I \le T \end{cases}$$

Output $\Phi(I)$

$+1$

0

Threshold T

Input I

(a)

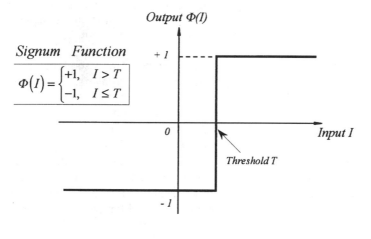

Signum Function

$$\Phi(I) = \begin{cases} +1, & I > T \\ -1, & I \le T \end{cases}$$

Output $\Phi(I)$

$+1$

0

-1

Threshold T

Input I

(b)

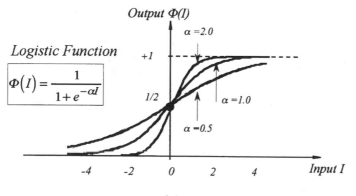

Logistic Function

$$\Phi(I) = \frac{1}{1 + e^{-\alpha I}}$$

Output $\Phi(I)$

$\alpha = 2.0$

$+1$

$1/2$

$\alpha = 1.0$

$\alpha = 0.5$

-4 -2 0 2 4 Input I

(c)

Figure 7.3 Transfer functions for neurons: (*a*) Threshold activation function (when $T = 0$, this is called a binary activation function). (*b*) Signum activation function (sometimes called a quantizer'). (*c*) Logistic activation functions for $\alpha = 0, 5, 1$ and 2.

of an artificial neuron to values between the two asymptotes. This limitation is very useful in keeping the output of the processing elements within a reasonable dynamic range. However, there are certain situations in which a linear relation, sometimes only in the right half-plane, is used for the activation function. It should be noted, however, that the use of a linear activation function removes the nonlinearity from the artificial neuron. Without nonlinearities, a neural network cannot model nonlinear phenomena.

7.4 ARTIFICIAL NEURAL NETWORKS

An artificial neural network can be defined as

A data processing system consisting of a large number of simple, highly interconnected processing elements (artificial neurons) in an architecture inspired by the structure of the cerebral cortex of the brain.

These processing elements are usually organized into a sequence of layers or slabs with full or random connections between the layers. This arrangement is shown in Figure 7.4, where the input layer is a buffer that presents data to the network. This input layer is *not* a neural computing layer because the nodes have no input weights and no activation functions. (Some authors do not count this layer in describing neural networks. We will count it, but we will use different symbols for the nodes in this layer where there is a need to distinguish between the different kinds of neurons.) The top layer is the output layer which presents the output response to a given input. The other layer (or layers) is called the intermediate or hidden layer because it usually

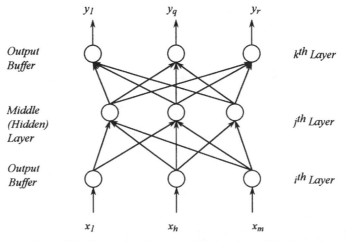

Figure 7.4 Example of an neural network architecture.

has no connections to the outside world. Typically the input, hidden, and output layers are designated the ith, jth, and kth layers, respectively.

Two general kinds of neural networks are in use: the *heteroassociative* neural network in which the output vector is different than the input vector, and the *autoassociative* neural network in which the output is identical to the input. Unless otherwise indicated, all neural networks in this book are heteroassociative.

A typical neural network is "fully connected," which means that there is a connection between each of the neurons in any given layer with each of the neurons in the next layer as shown in Figure 7.5. When there are no lateral connections between neurons in a given layer and none back to previous layers, the network is said to be a feedforward network. Neural networks with feedback connections (i.e., networks with connections from one layer back to a previous layer) are also useful and are discussed in the following chapters. Lateral connections between neurons in the same layer are also called feedback connections. In certain cases, a neuron has feedback from its

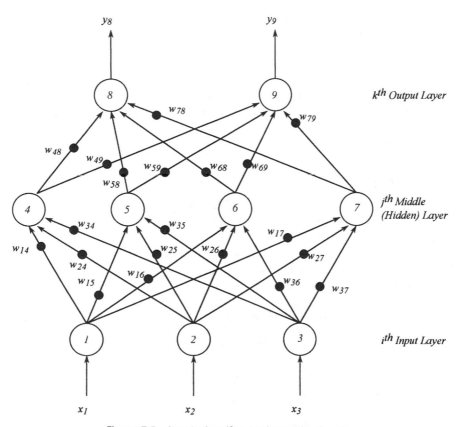

Figure 7.5 Simple feedforward neural network.

output to its own input. In all cases, these connections have weights that must be trained.

Each of the connections between neurons has an adjustable weight as shown in Figure 7.5. This simple neural network is a fully connected, feedforward network with three neurons in the input layer, four in the middle or hidden layer, and two in the output layer. The individual weights are shown as solid dots on the connection and are designated by symbols such as w_{ij}. For instance, the symbol w_{37} indicates a weight on the connection between neurons 3 and 7.

Let us consider the neural network of Figure 7.5, which has an input vector $\mathbf{X}$ consisting of components x_1, x_2, and x_3 and an output vector $\mathbf{Y}$ having components y_8, and y_9. When a signal x_1 is applied to neuron 1 in the input layer, the output x_1 goes to each of the artificial neurons in the middle or hidden layer, passing through weights w_{14}, w_{15}, w_{16}, and w_{17}. The input signal x_2 and x_3 behave in a similar manner, sending signals to neurons 4, 5, 6, and 7 through the appropriate weights, as shown in Figure 7.5.

Now let us consider the behavior of neuron 4. It has three inputs from the three neurons in the input layer that have been modified by the connection weights w_{14}, w_{24}, and w_{34}. The first part of this neuron simply sums up these three weighted inputs. Then this summation is passed to the second part of the neuron, which is a nonlinear function—typically a logistic curve between 0 and 1 as shown in Figure 7.3c. The output of this activation function or squashing function is then sent to neurons 8 and 9 through weights w_{48} and w_{49}. Neurons 5, 6, and 7 behave in a similar manner. Neurons 8 and 9 collect the weighted inputs from neurons 4, 5, 6 and 7, sum them, and pass the sums through the activation functions to produce y_8 and y_9, the components of the output vector $\mathbf{Y}$.

Vector and Matrix Notation

It is convenient to utilize vector and matrix notation in dealing with the inputs, outputs, and weights. Let us cut the neural network in Figure 7.5, just above the hidden layer as shown in Figure 7.6. The outputs of neurons 4, 5, 6, and 7 are shown to be the vector $\mathbf{V}_j$, which has components v_4, v_5, v_6, and v_7. If we limit the activation functions to linear functions, the mathematical relationships described in the previous section can be written in matrix form; that is, the column vector $\mathbf{V}_j$ is equal to the dot product of the weight matrix $\mathbf{W}_{ji}$ and the input vector $\mathbf{X}_i$. This relationship is given by

$$\begin{bmatrix} v_4 \\ v_5 \\ v_6 \\ v_7 \end{bmatrix} = \begin{bmatrix} w_{14} & w_{24} & w_{34} \\ w_{15} & w_{25} & w_{35} \\ w_{16} & w_{26} & w_{36} \\ w_{17} & w_{27} & w_{37} \end{bmatrix} \cdot \begin{bmatrix} x_1 \\ x_2 \\ x_3 \end{bmatrix} \qquad (7.4\text{-}1)$$

$$\mathbf{V}_j = \mathbf{W}_{ij} \cdot \mathbf{X}_i \qquad (7.4\text{-}2)$$

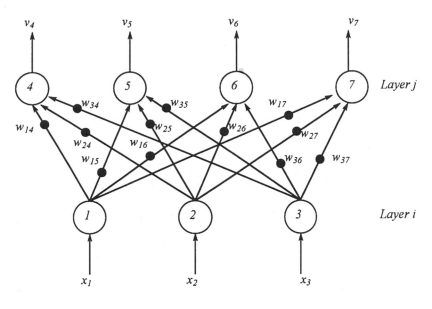

$$V = W_{ij} \cdot X$$

Figure 7.6 Lower portion of neural network cut above the hidden layer.

In a similar manner using the upper half of the artificial neural network shown in Figure 7.7, it can be shown that the output vector Y_k is equal to the dot product of the weight matrix W_{jk} and the input vector V_j. This relationship is given by

$$\begin{bmatrix} y_8 \\ y_9 \end{bmatrix} = \begin{bmatrix} w_{48} & w_{58} & w_{68} & w_{78} \\ w_{49} & w_{59} & w_{69} & w_{79} \end{bmatrix} \cdot \begin{bmatrix} v_4 \\ v_5 \\ v_6 \\ v_7 \end{bmatrix} \tag{7.4-3}$$

or

$$Y_k = W_{jk} \cdot V_i \tag{7.4-4}$$

By combining equations (7.4-1) and (7.4-3), it is apparent that the output vector Y_k is equal to the dot product of the two matrices and the input vector X_i.

$$\begin{bmatrix} y_8 \\ y_9 \end{bmatrix} = \begin{bmatrix} w_{48} & w_{58} & w_{68} & w_{78} \\ w_{49} & w_{59} & w_{69} & w_{79} \end{bmatrix} \cdot \begin{bmatrix} w_{14} & w_{24} & w_{34} \\ w_{15} & w_{25} & w_{35} \\ w_{16} & w_{26} & w_{36} \\ w_{17} & w_{27} & w_{37} \end{bmatrix} \cdot \begin{bmatrix} x_1 \\ x_2 \\ x_3 \end{bmatrix} \tag{7.4-5}$$

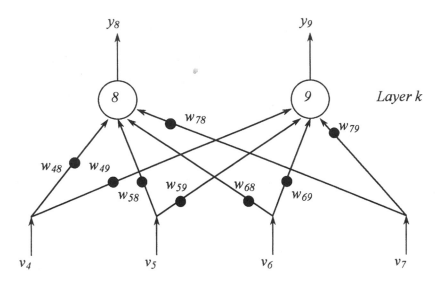

$$Y = W_{jk} \cdot V$$

Figure 7.7   Upper portion of neural network cut above the hidden layer.

Since the two matrices can then be reduced to a single matrix, W_{ijk}, it follows that the output vector Y is equal to the dot product of the combined matrix and the input vector:

$$Y_k = W_{ij} \cdot W_{jk} \cdot X_j = W_{ijk} \cdot X_i \qquad (7.4\text{-}6)$$

The limitation of linear activation functions means that the relationships given in equations (7.4-1) through (7.4-6) are severely limited. Indeed, this indicates that a three-layer perceptron with linear activation functions in the middle and output layers can be replaced with a two-layer network with a linear activation function in the output layer. Nevertheless, this process has introduced the concept of the weight matrices, which is very useful in many situations.

Neural Networks and Feedback

The feedforward neural network shown in Figure 7.5 operates in a simple straightforward manner. When the vector X_i is applied to the input layer, the calculations for weighting inputs, as well as summing and filtering, are rapidly carried out for each neuron as the process moves from the input to the middle layer and on to the output layer. However, when there are feedback connections, either between neurons in the same layer or from one layer to an earlier layer, the process is much more complicated. In a neural network where the mathematical operations are performed almost instantaneously, information reverberates around the network, across layers and within layers, until some sort of convergence status is reached. When the mathematical operations are implemented serially, the process is more complicated. The outputs for the feedforward connections are performed first, then the calculations for the feedback connections are performed, then the calculations for the feedforward connections are again performed using the results of the previous calculations, and this process continues until equilibrium values are reached. Under many circumstances, artificial neural networks with feedback connections can be very useful. However, about 80% of the neural network applications today utilize feedforward neural networks.

Neural Networks in Perspective

Neural networks have profound strengths and weaknesses, and these must be recognized if they are to be used properly. Although neural networks are sometimes called neural computers, they are in fact not computers; but rather, they are basically memories that memorize results, just as the human brain memorizes certain results. For instance, a person memorizes the fact that the product of four times six is twenty-four, and this fact is stored in the person's memory for life. On the other hand, the cheapest digital calculator actually calculates the product every time the numbers are entered.

Neural networks use memory-based storage of information in ways that are different and more flexible than simple storage in a look-up table. In the neural network, as in the brain, the storage of information is distributed throughout the network. Although this makes it hard to keep things separate that should be kept separate, it does give rise to the networks' ability to make generalizations that are so important to the practical applications of neural networks. Furthermore, the loss of a few neurons (real or artificial) does not materially affect the information stored.

Linear Associator Neural Network

The most elementary neural network is a "linear associator" that, along with its learning rules, can be used to demonstrate the abilities and limitations of neural networks. We start with the fundamental assumption that information is stored by a pattern or a set of activities of many neurons that is often represented as a "state vector." Hence, the output of the network is the result of the interaction of many neurons (sometimes called neural computing), not just the response of a single neuron. As discussed earlier, the fundamental neuron sums the weighted inputs and then subjects this sum to a nonlinear activation function, typically a sigmoidal function, to keep the output of the neuron within a reasonable range.

The architecture of a linear associator is a set of input neurons that are connected to a set of output neurons (i.e., a two-layer neural network). Any particular output y_j (one component of the output state vector **Y**) can be computed from the activities of all the various inputs x_i and the strengths of the weights on the connections. Mathematically, the output from the summing unit is equal to the *inner product* (dot product) between the weight matrix and the input vector. In the linear associator, the activation function is a linear function. While this simplifies the network considerably, care must be taken to ensure that the outputs do not exceed the range of the output neuron.

In simple terms, the operation of a linear associator involves the input of a pattern that then produces the output pattern that we want (i.e., the "right" answer). For this to happen, we have to train the weights of the linear associator to give the desired pattern. This can be accomplished by presenting the network with training vector pairs (inputs and desired outputs) and utilizing an appropriate training rule. Any of the different training rules discussed later can be used to perform this training. In theory, the initial weights can have any values, but experience indicates that starting with small randomized weights is advantageous.

Suppose that we have one set of neurons projecting to another set through modifiable weights as shown in Figure 7.8. When the activation functions of the neurons are linear, this network is a linear associator. What this means is

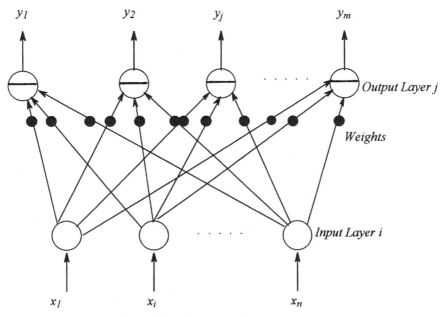

Figure 7.8 Sketch of a linear associator.

that after the neural network is trained, presentation of an input pattern to the input layer will produce the desired (associated) output pattern. This is represented mathematically by the following equation:

$$\mathbf{Y}_i = \mathbf{W} \cdot \mathbf{X}_i \qquad (7.4\text{-}7)$$

where $\mathbf{W}$ is the trained weight matrix, and $\mathbf{X}_i$ and $\mathbf{Y}_i$ are the ith input and output vectors, respectively.

One of the unique and advantageous features of the linear associator is its ability to store more than one relationship simultaneously. This is discussed and demonstrated in a later section. The problems with the linear associator is that it is not very accurate, especially if too many items are stored in the associator. Second, simple networks that use *Hebbian learning*[2] cannot compute some functions that may be desired. This leads to the concept of comparing the output with the desired output and using the difference (error) as a basis for adjusting the weights, such as is the case in *Widrow–Hoff learning*. In effect, this procedure constitutes a form of "supervised" learning that is discussed in the next chapter.

7.5 LEARNING AND RECALL

Neural networks perform two major functions: *learning* and *recall*. Learning is the process of adapting the connection weights in an artificial neural network to produce the desired output vector in response to a stimulus vector presented to the input buffer. Recall is the process of accepting an input stimulus and producing an output response in accordance with the network weight structure. Recall occurs when a neural network globally processes the stimulus presented at its input buffer and creates a response at the output buffer. Recall is an integral part of the learning process since a desired response to the network must be compared to the actual output to create an error function.

The learning rules of neural computation indicate how connection weights are adjusted in response to a learning example. In *supervised learning*, the artificial neural network is trained to give the desired response to a specific input stimulus. In *graded learning*, the output is "graded" as good or bad on a numerical scale, and the connection weights are adjusted in accordance with the grade.

In *unsupervised learning* there is no specific response sought, but rather the response is based on the networks ability to organize itself. Only the input stimuli are applied to the input buffers of the network. The network then organizes itself internally so that each hidden neuron responds strongly to a different set of input stimuli. These sets of input stimuli represent

[2]*Hebbian learning* as well as other learning paradigms are presented in Chapter 9.

clusters in the input space (which often represent distinct real-world concepts or features).

The vast majority of learning in engineering applications involves supervised learning. In this case a stimulus is presented at the input buffer representing the input vector, and another stimulus is presented at the output buffer representing the desired response to the given input. This desired response must be provided by a knowledgeable teacher. The difference between actual output and desired response constitutes an error, which is used to adjust the connection weights. In other cases, the weights are adjusted in accordance with criteria that are prescribed by the nature of the learning process, as in competitive learning or in Hebbian learning.

There are a number of common supervised learning algorithms utilized in neural networks. Perhaps the oldest is Hebbian learning, named after Donald Hebb, who proposed a model for biological learning (Hebb, 1949) where a connection weight is incremented if both the input and the desired output are large. This type of learning comes from the biological world, where a neural pathway is strengthened each time it is used. "Delta rule" learning takes place when the error (i.e., the difference between the desired output response and the actual output response) is minimized, usually by a least squares process. Competitive learning, on the other hand, occurs when the artificial neurons compete among themselves, and only the one that yields the largest response to a given input modifies its weight to become more like the input. There is also random learning in which random incremental changes are introduced into the weights, and then either retained or dropped, depending upon whether the output is improved or not (based on whatever criteria the user specifies).

In the recall process, a neural network accepts the signal presented at the input buffer and then produces a response at the output buffer that is determined by the "training" of the network. The simplest form of recall occurs when there are no feedback connections from one layer to another or within a layer (i.e., the signals flow from the input buffer to the output buffer in a "feedforward" manner). In a feedforward network the response is produced in one cycle of calculations by the computer.

Supervised Learning

In order to demonstrate supervised learning, let us modify the neural network shown in Figure 7.5, to include a desired output pattern, a comparator, and a weight adjusting algorithm. This arrangement is shown in Figure 7.9, where the desired output is represented by the vector $\mathbf{Z}$ with components z_8 and z_9. The inputs to the comparator are the desired output pattern $\mathbf{Z}$ and the actual output pattern $\mathbf{Y}$. The error coming from the comparator—that is, the difference between $\mathbf{Y}$ and $\mathbf{Z}$—is then utilized in the weight-adjusting algorithm to determine the amount of the adjustment to be made in the weights in both layers.

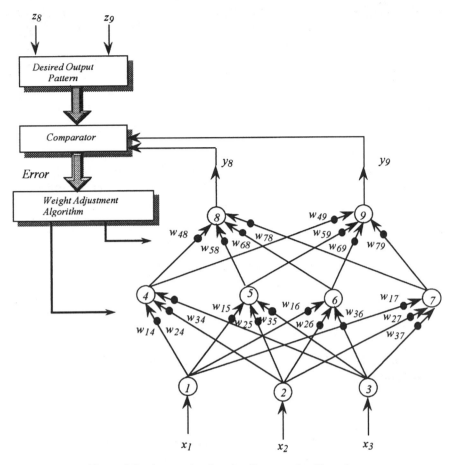

Figure 7.9 A neural network with supervised learning.

In order to start the process, let us randomly adjust all the weights in the neural network in Figure 7.9 to small random values, and then consider the training pair **X** and **Z** with components x_1, x_2, x_3 and z_8, z_9, respectively. When the vector **X** is applied to the neural network, it produces an output vector **Y**, which is compared with the vector **Z** to produce the error. The weight-adjusting algorithm then modifies the weights in the direction that reduces this error. When the input vector **X** is again applied, it produces a new **Y**, which is compared with **Z**, and the error is applied to the weight-adjusting algorithm again to adjust the weights. This process is repeated over and over until the error is reduced to some specified value or an irreducible small quantity. At that point the output vector **Y** and the desired output vector **Z** are substantially equivalent, and the neural network is said to have been trained to map input vector **X** into the desired output vector **Z**. This is the essence of supervised training. Of course, we must specify and explain the mechanism by which the weights are adjusted before there is a complete understanding of this process.

Example 7.1 Mapping the Alphabet to a Five-Bit Code. In order to understand how a neural network training process works, let us consider the arrangements shown in Figure 7.10. On the left we have a 7×5 matrix array of inputs that are restricted to either 0 or 1. In the center we have a neural network, with the input layer on the left having 35 input artificial neurons. Each of these 35 neurons is connected to one of the inputs from the 7×5 matrix array. On the right we have a 5×1 matrix array of outputs, each of them connected to one of the neurons in the output layer of the neural network in the center. The hidden layer in the neural network in this case has 20 artificial neurons, a number that was chosen arbitrarily. The input vector on the left, **X**, has 35 components $(x_1, x_2, \ldots, x_{35})$ and the output vector **Y** on the right has five components $(y_1, y_2, y_3, y_4, y_5)$. In effect, we are going to map the pattern contained in the 7×5 matrix on the left into a pattern on the right contained by the 5×1 matrix. In a sense, this is a form of data compression where the data contained in the 35-bit matrix on the left is mapped into the five-bit matrix on the right. The compression ratio in this case is $7:1$.

Let us introduce a pattern to represent an uppercase letter A in the 7×5 matrix on the left, where the shaded areas in Figure 7.10 represent 1s and the unshaded areas represent 0s. Suppose we want to map this pattern into the five-bit pattern in the matrix on the right, which is shown to be $(1, 0, 1, 0, 1)$. The artificial neural network has an input pattern representing the A, and the desired output pattern is represented by the five-bit matrix on the right. To carry out this mapping, we must adjust the weights in both the

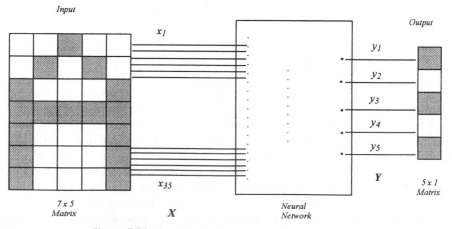

Figure 7.10 Input–output mapping of the letter *A*.

connections between the input and hidden layers and the connection between the hidden and output layers.

If the neural network is fully connected, we have 700 (35 × 20) connections with 700 weights between the input and hidden layers. In the connections between the hidden and output layers, we have another 100 (20 × 5) weights, giving a total of 800 weights that must be adjusted. In effect, we can think of this arrangement as having 800 degrees of freedom because of the 800 adjustable weights. It is very clear that we do not need 800 degrees of freedom to map a 35-bit input into a five-bit output. What this means is that there are hundreds, if not thousands, of different combinations of these 800 weights that will permit this neural network to carry out this mapping.

In order to start the training for this mapping, all the weights in the neural network are set to small random values, usually between −0.3 and +0.3. Then the training process is started. This involves applying the pattern from the 35-bit matrix on the left to the input layer, multiplying these inputs by 700 connection weights between the input and hidden layer, and then summing the 35 weighted inputs going into each of the 20 neurons in the hidden layer. These 20 sums then pass through the nonlinear activation function to produce the 20 outputs that go to each of the five neurons in the output layer. Each of these 20 outputs is multiplied by the appropriate weights, summed by each output neuron and passed through the nonlinear activation function to produce the five outputs. (Note that these outputs are not 0s and 1s, but rather numerical values between 0 and 1. Therefore, an interpretation of the outputs is needed. For instance, an output greater than 0.9 could be considered as a 1; an output less than 0.1 could be considered as a 0; and any value in between 0.1 and 0.9 could be considered as indeterminate.) These outputs are then compared with the desired output shown in the 5 × 1 matrix on the right. The difference between the actual output of the neural network and the desired output becomes the error vector that is then used to adjust both layers of weights in such a way that the overall error is reduced. Then the process is repeated over and over again until eventually, every time an A is applied to the input, the desired output is produced by the neural network within limits prescribed by some specific criteria. At this point we say that the neural network is trained and is capable of mapping a 35-bit representation of A into a five-bit representation of the A.

Now let us consider the arrangement in Figure 7.11, where we have a 35-bit representation of a B as an input to the neural network and a five-bit representation of the B as a desired output, which in this case is (0, 1, 0, 1, 0). If we use the neural network we have just trained for an A and apply the B at the input matrix, we can continue the same procedure used before to calculate the output of the neural network and compare it with the desired output. Although there is a small probability that we might get the right output initially, the most likely outcome is that the actual output and the desired output will be quite different (i.e., some of the outputs will be wrong and others will be between 0.1 and 0.9 and hence indeterminate). This

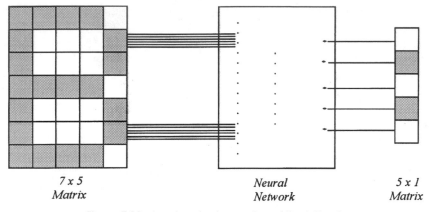

Figure 7.11 Input–output mapping of the letter *B*.

produces another error signal that becomes the basis for adjusting the weights further beyond the training provided for the A input. We continue this training process until every time we apply B at the input matrix on the left, we get the desired $(0, 1, 0, 1, 0)$ output at the five-bit matrix at the right. At this point we have trained the neural network to map a 35-bit representation of a B into a five-bit representation.

Now suppose we again apply an A to this neural network that has been trained for an A and a B. Are we likely to get the desired output? Maybe, maybe not. If not, we can carry out additional training, until we achieve the desired results. Then we can apply the B again. Will we get the desired output? Maybe; maybe not. If not, we can carry out more training. This process of going back and forth between the A and B can be continued until every time we apply an A to the input matrix we get the desired $(1, 0, 1, 0, 1)$ output and every time we apply a B to the input matrix we also get the desired $(0, 1, 0, 1, 0)$ output. Now we have a network that is capable of mapping both an A and a B into the five-bit representations we specified.

Now let us apply a C to the input matrix as shown in Figure 7.12 and specify the desired output matrix as being a $(1, 0, 0, 1, 0)$. When we apply the C, there is a high probability that we will not get the desired output that we have chosen. So we start the training process again and continue it until every time that we apply a C to the left-hand matrix, we get the desired $(1, 0, 0, 1, 0)$ output. We now have a network that is capable of mapping the 35-bit representation of a C into the desired five-bit representation.

If we now apply the A to this trained network, will we get the desired output? Maybe, maybe not. If not, we perform more training until we achieve the desired result. Then we can apply a B. If we don't get the desired output, we carry out additional training. Then we can apply a C. If we don't get the

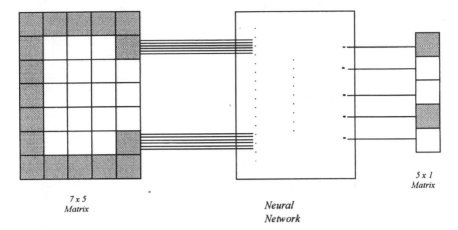

Figure 7.12 Input–output mapping of the letter *C*.

desired output, we carry out more training. We repeat this process over and over until every time we put in an A, a B, or a C, we get the desired output. At this point the neural network has been trained to map an A or B and a C into the desired representation that we have chosen.

At this point we could continue the process with D, E, and F and work our way through the alphabet. Since we have a five-bit binary output, the total number of possible mappings is 2^5, or 32. Hence, we can represent the whole alphabet plus six punctuation symbols. However, this is not a very efficient process to go through the complete training process for one symbol before starting the training process for another symbol. A more realistic and appropriate way would be to choose the 32 training sets (i.e., an A and its five-bit representation, a B and its five-bit representation, etc.) and, after randomizing the weights, to apply all 32 training sets, one after the other until we go all the way through the 32 letters and punctuation symbols once. This set of 32 input and desired output pairs, known as an *epoch*, is applied again and again until all 32 letters or symbols are mapped into the five-bit codes we specified.

Overall error is a better determination of the status of the training than it is of whether all the outputs are correct or not. This is simply the summation of all the errors between outputs of the neural network and the corresponding desired outputs (0s and 1s) for all pairs in the epoch. Ideally, this overall error should approach zero. If it does not, additional training should be carried out. However, if there is any noise in the inputs and/or outputs, an overall error of zero is never attained. Indeed, it is possible to overtrain a neural network until it fits the noise pattern rather than the underlying relationship.   □

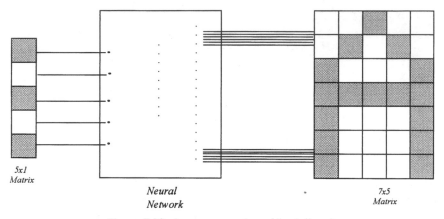

Figure 7.13 Inverse mapping of the letter *A*.

Example 7.2 Data Compression and Expansion. Now let us consider the arrangement shown in Figure 7.13. Here we have an input matrix that is a five-bit representation of an A and an output that is a 35-bit representation of the letter A. Is it possible to train a neural network to map a five-bit representation into a 35-bit representation? Yes, it is. The process is exactly the same as we went through in mapping the reverse arrangement. In this case, we have data expansion instead of data compression.

We can even have an arrangement that combined the networks shown in Figures 7.10 and 7.13—that is, a 35-bit representation that is compressed into a five-bit representation of an A and then is expanded back out to a 35-bit representation of the letter A as shown in Figure 7.14. Why would we want such an arrangement? Suppose we were sending information down a narrow-band data channel. We could compress the data (in this case by a factor of seven), send it down the channel, and then expand it back to the original symbol. This process is used in many practical situations. □

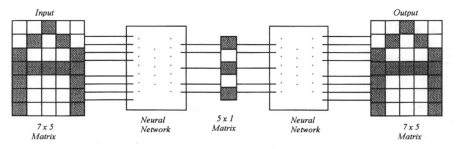

Figure 7.14 Compression and expansion using neural networks.

Example 7.3   Distortion Correction. A variation of the compression–expansion arrangement is shown in Figure 7.14, where the input to the neural network is exactly the same as the output. This arrangement is called an *autoassociative neural network*, which simply means that the input and the output are exactly the same. In this case, we randomly adjust the weights, apply the 35-bit representation of A as the input, apply the same 35-bit representation of the A on the right as the desired output, and start the training the process that we have used in the previous examples until we can consistently get an output that is equal to the desired output. Why would we want to do this? Suppose, after training the network, instead of using the 35-bit representation of an A as the input to the trained ANN, we put in a representation of a distorted A. The output of the neural network would be an undistorted A, because this is the only output pattern the neural network is trained to produce.

Suppose we go further and train this autoassociative network to represent all 26 letters of the alphabet plus the six punctuation symbols that we discussed earlier. It would then be reasonable to expect that every time that you applied a distorted symbol as the input, you would get the correct symbol as the output. In general, this is true, but there are exceptions. Suppose we put in a distorted B with the distortion in the lower right-hand side. This network might produce a B as the output or it might produce an *R*. The choice by the network would depend upon whether the distorted B input was closer, in a least square sense, to the B or the R that was used in the training. The same is true for other, similar combinations of letters—for example, Q & O, R & P, C & G, and perhaps others.   □

7.6   FEATURES OF ARTIFICIAL NEURAL NETWORKS

What makes neural networks different from artificial intelligence or traditional computing? Generally, there are four features that are associated with artificial neural networks:

- They learn by example.
- They constitute a distributed, associative memory.
- They are fault-tolerant.
- They are capable of pattern recognition.

Neural networks are not the only systems capable of *learning by example*, but this feature certainly is an important characteristic of neural networks. Indeed, one of the most important characteristics of artificial neural networks is the ability to utilize examples taken from data and to organize the information into a form that is useful. Typically, this form constitutes a model that represents the relationship between the input and output variables. In essence, this is what we were doing with the mapping exercises that we went through in the last section.

A neural network memory is both *distributed* and *associative*. By distributed, we mean that the information is spread among all of the weights that have been adjusted in the training process. These connection weights are the memory units of neural networks, and the values of the weights represent the current state of the knowledge of the network. Hence, each individual unit of knowledge is distributed across all the memory units in the network. Furthermore, it shares these memory units with all other items of information stored in the network.

The memory in a neural network is also *associative*. This means that if the trained network is presented with a partial input, the network will choose the closest match in the memory to that input and generate an output that corresponds to a full input. This is the process that was discussed with the autoassociative network in Figure 7.15, where the presentation of partial input vectors to the network resulted in their completion.

Neural networks are also fault-tolerant, since the information storage is distributed over all the weights. For instance, in the example in Figure 7.10, the information is distributed over 800 weights. Hence, the destruction or misadjustment of one or a few of these 800 weights does not significantly influence the mapping process between the inputs and outputs. In general, the amount of distortion is approximately equal to the fraction of the weights that have been destroyed.

Furthermore, even when a large number of the weights are destroyed, the performance of the neural network degrades gradually. While the performance suffers, the system does not fail catastrophically because the information is not contained in just one place but is, instead, distributed throughout the network. When neural networks are implemented in hardware, they are

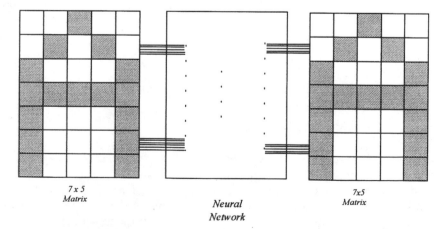

7 x 5
Matrix

Neural
Network

7x5
Matrix

Figure 7.15 Input \output mapping in an autoassociative neural network.

very fault-tolerant, as contrasted to von Neumann-type computers where the failures of a single component can, in theory, lead to catastrophic results. For this reason, neural networks show great promise for use in environments where robust, fault-tolerant pattern recognition is necessary in a real-time mode, and the incoming data may be distorted or noisy. Such applications might include: nuclear power plants, missile guidance systems, space probes, or any system that is inaccessible for repair or where continuous performance is critical.

Pattern recognition requires the neural network to match large amounts of input information simultaneously and generate a categorical or generalized output with a reasonable response to noisy or incomplete data. Neural networks are good pattern recognizers, even when the information comprising the patterns is noisy, sparse, or incomplete. For a complex system with many sensors and possible fault types, real-time response is a difficult challenge to both human operators and expert systems. While the training time for a neural network may be long, once it has been trained to recognize the various conditions or states of a complex system, it only takes one cycle of the neural network to detect or identify a specific condition or state.

Neural computing networks consist of interconnected units that act on data instantly in a massive parallel manner. Indeed, when a neural network is implemented in hardware, such computation occurs virtually instantaneously. Such a neural computer provides an approach that is closer to human perception and recognition than that of conventional computers, and it can produce reasonable results with noisy or incomplete inputs.

7.7   HISTORICAL DEVELOPMENT OF NEURAL NETWORKS[3]

Artificial intelligence had its beginning at the Dartmouth Summer Research Conference in 1956, which was organized by Marvin Minsky (learning machines), John McCarthy (symbolic languages), Nathaniel Rochester (neural systems), and Claude Shannon (information theory). This conference led to the development of computer programs capable of making machines perform human-like or intelligent tasks and to the development of machines that used mechanisms modeled after studies of the brain to become "intelligent." The conference inspired Frank Rosenblatt to develop his concept of the perceptron, a generalization of the 1943 McCulloch–Pitts concept of the functioning of the brain by adding learning. The McCulloch–Pitts abstract model of a

---

[3]The history of the development of neural networks has been well documented by a number of books in the past few years: Caudill and Butler (1989, 1992), DARPA (1988), Hecht-Nielsen (1989), Maren, Pap, and Harston (1990), Miller, Sutton, and Werbos (1990), Nelson and Illingsworth (1990), Pao (1989z0, Simpson (1990), Wasserman (1989, 1993), and White and Sofge (1992). We will limit this review to descriptions of Rosenblatt's Perceptron, Minsky and Papert's review entitled *Perceptrons*; and Widrow's ADALINE because all had a profound influence on the development of neural networks.

brain cell was based on the theory that the probability of a neuron firing depended on the input signals and the voltage thresholds in the soma. It introduced the idea of a step threshold, but it did not have the ability to learn.

The first learning machine was actually built by Minsky and Dean in 1951 (before the Dartmouth conference) at the Massachusetts Institute of Technology. It had 40 processing elements, which, when described in neural network terms, were neurons with synapses that adjusted their weights according to their success in performing a specific task. Each neuron or processing element required six vacuum tubes and a motor/clutch/control system. The machine utilized Hebbian learning and was able to learn enough that it could "run a maze." It worked surprisingly well, considering the state of electronics and the understanding of the learning process at that time.

Rosenblatt's Perceptron

After the Dartmouth conference, Frank Rosenblatt of Cornell Aerolaboratory developed a computational model for the retina of the eye, called the "perceptron." The perceptron (see Figure 7.16) was inspired by the

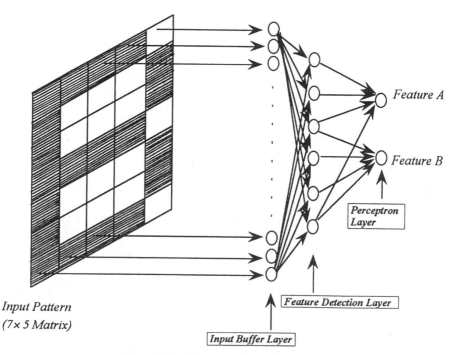

Input Pattern
(7 × 5 Matrix)

Figure 7.16 Diagram of the perceptron.

McCulloch–Pitts model and incorporated Hebbian learning, which he summarized as follows:

> When the synaptic input and the neuron output were both active, the strength of the connection is enhanced.

The perceptron was a pattern classification system that could identify both abstract and geometrical patterns. The first perceptron was primarily an optical system that had a grid of 400 photocells connected to associator units in the input buffer, which collected electrical impulses from the photocells. The photocells were randomly connected to the associators and received optical stimuli. The output of the sensors were connected to a hard-wired genetically predetermined set of processing elements (called demons) that recognized particular types of patterns. The output of each demon was connected to a threshold logic unit which had no output until a certain level and type of input was received. Then the output rose linearly with the input. This concept was inspired by the observation that the neuron does not fire until the balance of input activity exceeds some threshold, and that the firing rate is increased in proportion to certain characteristics of the input. It was quite robust and capable of some learning, it possessed a great deal of plasticity (i.e., information could be retained after some of the cells had been destroyed), and it was capable of making limited generalizations. It could properly categorize patterns despite noise in the input.

Rosenblatt studied both the two-layer and three-layer perceptrons. He was able to prove that the two-layer perceptron could separate inputs into two classes only if the two classes were linearly separable. In some systems, supervised learning was used in which the weights were adjusted in proportion to the error between the desired and actual output. While his attempts to extend the learning procedure to the three-layer perceptrons were encouraging, he could not find a mathematical basis for distributing credit (or blame) for the output errors between the two layers of weights. Hence, there was no mathematical basis for making corrections to the weighting functions. Actually Amari had solved the credit assignment problem in 1967, but it went unnoticed, because it was published in the Japanese literature. Had Amari's work been more widely known, it could have mitigated the impact of the critical book entitled *Perceptrons* by Minsky and Papert discussed later in this section.

The perceptron paradigm was designed to explain and model the pattern recognition capabilities of the visual system. The perceptron was a feedforward network without any feedback, without connections between neurons in the same layer, and without any randomness about the operation of the network. It was basically a three-layer network in which the input layer was a buffer (fanout) layer that mapped a rectangular pixelized sensor pattern to a linear array. The second layer, consisting of a set of feature detectors or feature demons, was either fully or randomly connected to the input layer.

This layer used either linear or nonlinear threshold activation functions to condition the outputs. The output layer contained "pattern recognizers" or "perceptrons." The weights of the inputs to the second layer were randomized and then fixed while the weights of the output layer were "trainable." The artificial neurons in the output or perceptron layer each had an input tied to a bias with a value of $+1$. The activation functions on the neurons in the output layer sometimes were "threshold-linear" functions in which the output signal is zero until the sum of the weighted inputs becomes positive, at which time the output increased to the weighted summation of the inputs. An alternate activation function sometimes used was a threshold function in which the output was zero if the weighted sum was zero or negative and equal to one if the weighted summation of the input was positive.

The basic learning algorithm procedure for training the perceptron is as follows:

- If the output is correct, leave the weights unchanged.
- If the output should be 1 but is instead 0, increment the weights on the active input lines (an active input line is defined as one that has a positive input).
- If the output should be 0 but is instead 1, decrement the weights on the active input lines.

The amount that the weights were changed depended upon the learning scheme that has been chosen. The three basic types of learning used in the perceptrons were as follows:

- A fixed increment or decrement.
- A variable amount of increment or decrement based upon the error (defined as the difference between the weighted sum and the desired output).
- A combination of both a fixed increment and an increment proportional to the error.

To classify a wide variety of shapes, the number of feature neurons must be quite large. By selective use of feedback, it is possible to radically reduce the number of neurons required. Another scheme used with the perceptron involved (a) the segmentation of the image into smaller pieces and (b) the creation of neurons that were specific to particular areas.

Minsky and Papert's Perceptrons

In the mid-1960s, Marvin Minsky began studying the "limitations" of the perceptrons, because of concern that Rosenblatt was making claims that were not being substantiated. (Fierce competition between Minsky and Rosenblatt is alleged to have extended back to the time when both were students at the Bronx High School of Science, which was probably the top technical high school in the United States at that time.) He and Seymour Papert showed that the two-layer perceptron was rather limited because it could only work problems with a linearly separable solution space. The exclusive-or (*XOR*) problem was cited as an elementary system that the perceptron was unable to solve. They emphasized the inability of the perceptron to assign credit for the errors to the different layers of weights. After their book entitled *Perceptrons* was published in the late 1960s, virtually all support for research in the neural networks field was ended by the various U.S. funding agencies.

A quotation from *Perceptrons* is indicative of the nature of the criticism by Minsky and Papert.

> The perceptron has shown itself to be worthy of study despite (and even because of!) its severe limitations. It has many features to attract attention: its linearity; its intriguing learning theorem; its clear paradigmatic simplicity as a kind of parallel computer. There is no reason to suppose that any of these virtues carry over to the many layered version. Nevertheless, we consider it an important research problem to elucidate (or reject) our intuitive judgment that the extension is sterile. Perhaps some power conversion theorem will be discovered, or some profound reason for the failure to produce an interesting 'learning theorem' for the multi-layered machine will be found.

The criticism in *Perceptrons*, while generally fair in the contexts of the state of knowledge at that time, was absolutely wrong in one respect. The virtues cited by Minsky and Papert for the two-layer network indeed did carry over to the many-layered version, and in fact a three-layer perceptron was capable of separating linearly inseparable variables, including the XOR problem. Rosenblatt died in a boat accident shortly after publication of *Perceptrons*, and unfortunately, the criticism of Minsky and Papert was never properly refuted at that time.

Widrow's Adaline

Adaline (*ada*ptive *lin*ear *e*lement) is a neural network that adapts a system to minimize the "error" signal using supervised learning. It acts as a filter to sort input data patterns into two categories. Up until the last decade, it was perhaps the most successful application of neural networks, because it is used in virtually all high-speed modems and telephone switching systems to cancel out the echo of a reflected signal in a transmission line or corridor. It was invented by Bernard Widrow and M. E. (Ted) Hoff of Stanford University in the early 1960s. (Hoff is also generally credited with being the inventor of the microprocessor as we know it today and was the founder of Intel Corporation.)

The basic design of the Adaline is shown in Figure 7.17. This arrangement is substantially the same as that for the perceptron discussed earlier, with the

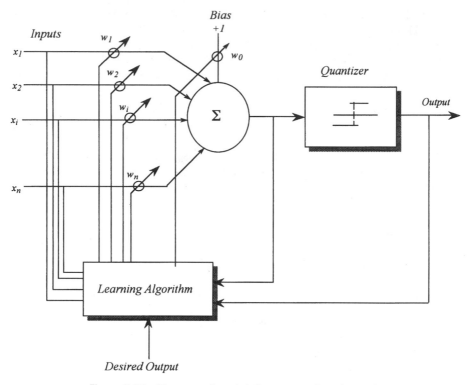

Figure 7.17 Diagram of an Adaline processing element.

expectation that the quantizer is a threshold-type nonlinear function with $+1$ and -1 as the limiting values (i.e., if the summation of the weighted inputs is positive, the output of the system will be $+1$; and if it is zero or negative, the output will be -1). The learning algorithm uses the difference between the desired output and the output of the summation (not the output of the quantizer) to produce the error ϵ function used to adjust the weights. In contrast, Rosenblatt used the difference between the actual output of the system and the desired output as the error functions to adjust the weights in the perceptron.

Prior to the beginning of the training, all weights must be adjusted to random values. With the Adaline, an input pattern is presented to the processing elements that filters it for a specific category. If the input matches the category, the processing element output is $+1$, and if it does not match the category, the processing element output is -1. The learning rule is the "Delta Rule," also known as the "Widrow–Hoff Learning Rule," which is, in fact, a least mean squares minimization of error procedure and is discussed extensively in Section 8.2. It involves the adjustment of weights according to the error in the processing element, computing a "delta" vector

$$\Delta w_i = \frac{\eta \cdot \epsilon \cdot x_i}{|\mathbf{X}|^2} \qquad (7.7\text{-}1)$$

where η is the learning constant, ϵ is the error, x_i is the ith input (-1, or $+1$), and $\mathbf{X}$ is the input vector. This Widrow–Hoff learning rule is discussed in Section 8.2 of the next chapter.

The learning algorithm for Adaline involves the application of the input (which may be noisy) to the single processing element or neuron. The application of the desired output and a computation of the error, defined as the difference between the weighted sum prior to the quantizer and the desired output, provides the input to the learning module. Then each weight is adjusted so that the error is equally distributed among the weights (including the weight for the bias). Equation (7.7-1) becomes

$$\Delta w_i = \frac{\eta \cdot \epsilon \cdot x_i}{|\mathbf{X}|^2} = \frac{\epsilon \cdot x_i}{(N+1)|\mathbf{X}|^2} \qquad (7.7\text{-}2)$$

where $(N+1)$ is the number of inputs plus the bias input and $1/(N+1)$ replaces the learning constant η. This means that the error is uniformly assigned to the $(N+1)$ inputs.

Since all inputs are $+1$ or -1, this adds or subtracts a fixed amount to the weight for each element input, depending on the sign. This process is repeated over and over again for each set of inputs in the epoch, and the epochs are repeated until the error is reduced to the desired value. Since both the desired output and actual inputs are binary, it is possible to have complete agreement between the desired outputs and the quantizer output even though the error (the difference between the desired value and the neuron output before the quantizer) has a substantial value. Further training beyond the time when the quantizer output is equal to the desired output is performed because the minimization of the error makes the system more tolerant of noise fluctuations in the input signals. This algorithm has been shown to guarantee convergence, provided that a set of weights exist that will minimize the error in a least squares sense.

Most of the time, the convergence of the learning process in the Adaline is very fast. However, the nature of the initial randomization can have a major effect on the speed of convergence. In a limited number of cases, convergence will not occur at all. Some of the real-world problems dealing with the Adaline occur where input patterns may not be perfect examples of the categories they represent. For instance, suppose we consider separating "circle" from "noncircle." The pertinent question is, How perfect does the circle have to be before it is considered a circle; or, alternately, How much deviation from a perfect circle is necessary before a figure is considered as a noncircle? Another restriction associated with the Adaline, as it was originally conceived, is that it is capable of classifying only linearly separable patterns. Later versions involving multilayers of Adalines proved more powerful and capable of separating input space even though the variables were not linearly separable.

One of the major applications by Widrow of the Adaline is in adaptive noise reduction. Every telephone has different transfer characteristics which can change during a single transmission. The use of an adaptive network to adjust the input signals spectrum so as to keep the *signal-to-noise-ratio* high for the given state of the line was one of the early applications. Other applications of the Adaline by Widrow and his students at Stanford University include: (1) adaptive antenna arrays, (2) adaptive blood pressure regulation, (3) adaptive filtering, (4) seismic signal pattern recognition, (5) weather forecasting, (6) long-distance and satellite telephone adaptive echo cancellation, (7) cancellation of correlated interference in acoustical and electronic instruments, (8) separation of a fetal heartbeat from its mother's heartbeat, and (9) signal equalization in all high-speed modems in use today.

Widrow's Madaline

A Madaline (which is an acronym for "Many Adalines") involves the use of several Adalines as the middle layer of a three-layer neural network. The input layer, as in the case of the perceptron, is an input buffer to ensure that all inputs go to each of the Adalines, and the output layer is a single unit that combines the outputs of all Adalines in a prescribed way. Sometimes this output unit gives a $+1$ when the majority of the inputs are $+1$, and a -1 when they are not (i.e., voting majority). In other cases, it will give a $+1$ only when all of the output of all Adaline's are $+1$ (an "*AND*" output). In another situation, the output unit will give a $+1$ when any of the outputs of the Adaline are $+1$ (an "*OR*" output).

Since Madaline has a binary output, it can only be used to discriminate between two classes. It is possible of course to use many independent Madalines to discriminate between more than two classes. One Madaline is needed for each pair of classes added. Typically, the final classification is the class that constitutes the most outputs. A typical Madaline network architecture is shown in Figure 7.18. The Madaline combiner unit does not have a bias input, and the weights on the input to the Madaline unit are fixed; that is, they are adjusted initially to represent the importance of the specific Adaline output. Here again the stability of the Madaline relates to the stability of individual Adalines, and convergence is seldom a problem. Adaline elements in a Madaline network evolve as detectors for a specific input features. This is particularly useful when the Madaline is used in control systems.

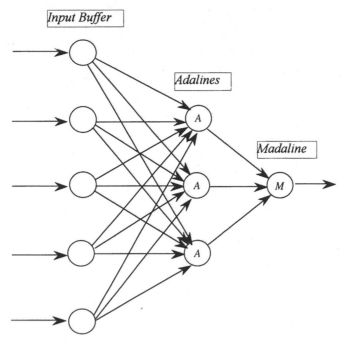

Figure 7.18   Diagram of a Madaline network architecture.

7.8   SEPARATION OF NONLINEARLY SEPARABLE VARIABLES

The ability of an Adaline to separate linearly separable variables can be readily demonstrated with a processing element or neuron that involves two inputs (x and y) and two weights (w_x and w_y). The neuron shown in Figure 7.19a sums the two weighted inputs—that is,

$$I = xw_x + yw_y \qquad (7.8\text{-}1)$$

The output z is equal to 1 if the sum is greater than the threshold value T, and it is equal to 0 (or -1 in an Adaline) if the sum is less than or equal to the threshold value. A special case occurs when the output is equal to the threshold T (i.e., the case that divides the two regions). Equation (7.8-1) becomes

$$xw_x + yw_y = T \qquad (7.8\text{-}2)$$

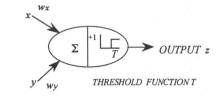

(a)

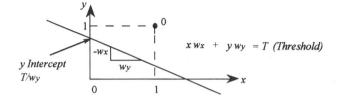

(b)

Figure 7.19 Separation of variables by a neural network: (*a*) Processing element with two inputs. (*b*) Division of *x–y* plane by a processing element.

which can be rearranged to

$$y = \left[-\frac{w_x}{w_y} \right] x + \left[\frac{T}{w_y} \right] \qquad (7.8\text{-}3)$$

This is the equation of a straight line where the slope is equal to $[-w_x/w_y]$ and the y intercept is $[T/w_y]$, which divides the plane into values that are below the threshold and above the threshold as shown in Figure 7.19*b*.

This concept can be extended further with a three-layer network as shown in Figure 7.20, where the first layer is a buffer with two inputs x and y; the middle layer has two neurons fully connected to the buffer layer with weights on each connection. The output layer is a single processing element whose input weights are set at 0.5 and whose threshold is set at 0.75. This configuration represents a logic "and" function, where both processing elements in the middle layer must produce a 1 to give a 1 in the output layer. The two processing elements in the middle layer are threshold functions with

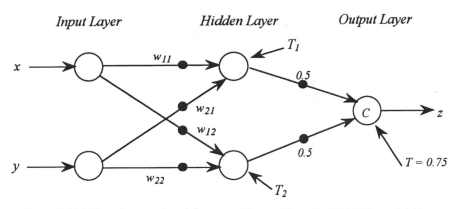

Figure 7.20 Three-layer network for separating nonlinearly separable variables.

thresholds T_1 and T_2. Hence, the outputs are either 0 or 1, depending upon whether the summation of the weighted inputs is less than or greater than the threshold values. As in the previous paragraphs, each of these threshold values effectively allow the plane to be divided.

The equations for the cases where the outputs of the two middle layer neurons are equal to the threshold values T_1 and T_2 are

$$xw_{11} + yw_{21} = T_1 \qquad (7.8\text{-}4)$$

$$xw_{12} + yw_{22} = T_2 \qquad (7.8\text{-}5)$$

which can be rearranged into the classical equation form for a straight line:

$$y = -\frac{w_{11}}{w_{21}}x + \frac{T_1}{w_{21}} \qquad (7.8\text{-}6)$$

$$y = -\frac{w_{12}}{w_{22}}x + \frac{T_2}{w_{22}} \qquad (7.8\text{-}7)$$

It is readily apparent that the use of two processing elements in the hidden layer provides for a double division of the plane as shown in Figure 7.21. The location and orientation of these two lines are determined by six quantities, namely, the values of the four weights and the two thresholds. Since only four parameters are needed to define these two lines unambiguously, there is a wide range of values of weights and threshold values that will define any two particular lines.

Since the outputs of the two neurons in the middle layer are either 0 or 1, it is apparent that both outputs must be 1 if the output layer is to produce a 1. It is readily shown that this corresponds to coordinates x and y being located in only one particular "quadrant" produced by these two intersecting

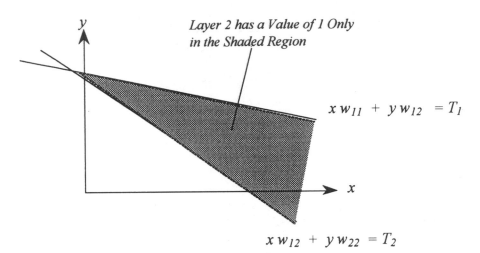

Figure 7.21 Division of x–y plane by two neurons.

lines. Hence, only one "quadrant" of this area will have a value of 1 while the other three "quadrants" will have a value of 0.

The use of three hidden processing elements in the middle layer further subdivides the plane with three lines, producing a triangular closed region. Writing the equations for the outputs of the processing elements in the middle layer of the neural networks shown in Figure 7.22 and setting them equal to the thresholds give the three dividing lines:

$$v_3 = xw_{13} + yw_{23} = T_3 \tag{7.8-8}$$

$$v_4 = xw_{14} + yw_{24} = T_4 \tag{7.8-9}$$

$$v_5 = xw_{15} + yw_{25} = T_5 \tag{7.8-10}$$

where T_3, T_4, and T_5 are the thresholds. Again, these equations can be put in the classical form for a straight line where the coefficient of the x terms are the slopes of the three lines and the three constant terms involving thresholds T_3, T_4, and T_5 are the y intercepts of the three straight lines:

$$y = -\frac{w_{13}}{w_{23}}x + \frac{T_3}{w_{23}} \tag{7.8-11}$$

$$y = -\frac{w_{14}}{w_{24}}x + \frac{T_4}{w_{24}} \tag{7.8-12}$$

$$y = -\frac{w_{15}}{w_{25}}x + \frac{T_5}{w_{25}} \tag{7.8-13}$$

which are shown in Figure 7.23.

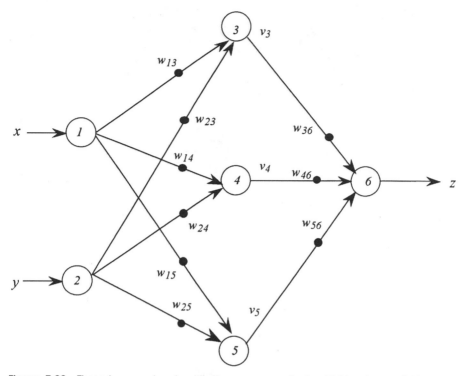

Figure 7.22 Three-layer network with three neurons in the hidden layer divides the plane with three lines.

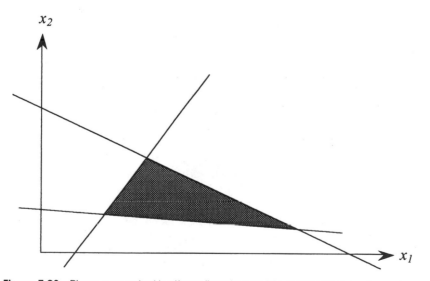

Figure 7.23 Plane separated by three lines to provide a closed triangular region.

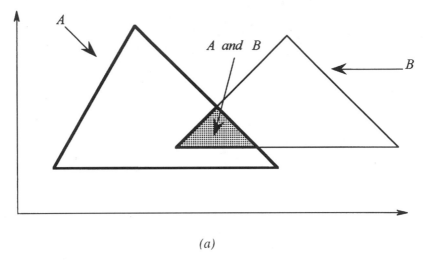

(a)

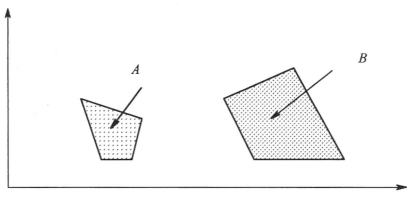

(b)

Figure 7.24 Areas enclosed by use of many neurons: (a) Closed triangular area with re-entrant area created by six lines (six nodes in middle layer of Figure 7.22). (b) Separate closed areas created by eight lines (eight nodes in middle layer of Figure 7.22).

It is readily apparent that the use of additional processing elements in the hidden layer allows us to generate virtually any type of enclosed area desired, ranging from an approximation of a circle to a convex polygon with re-entrant regions as shown in Figure 7.24a. Indeed, not all of the outputs of the middle layer need to overlap, and hence it is possible to use such an artificial neural network to enclose multiple regions as shown in Figure 7.24b.

REFERENCES

Caudill, M., and Butler, C., *Naturally Intelligent Systems*, MIT Press, Cambridge, MA, 1989.

Caudill, M., and Butler, C., *Understanding Neural Networks: Computer Explorations*, Vols. 1 and 2, MIT Press, Cambridge, MA, 1992.

DARPA (Defense Advanced Research Projects Agency), *DARPA Neural Network Study*, AFCEA International Press, Fairfax, VA, 1988.

Hebb, D., *Organization of Behavior*, John Wiley, New York, NY, 1949.

Hecht-Nielsen, R., *Neurocomputing*, Addison-Wesley, Reading, MA, 1989.

Maren, A. J., Harston, C. T., and Pap, R. M., *Handbook of Neural Computing Applications*, Academic Press, New York, 1990.

Miller, W. T., Sutton, R. S., and Werbos, P. J., *Neural Networks for Control*, MIT Press, Cambridge, MA, 1990.

Nelson, M. M., and Illingsworth, W. T., *A Practical Guide to Neural Networks*, Addison-Wesley, Reading, MA, 1990.

Pao, Y. H., *Adaptive Pattern Recognition and Neural Networks*, Addison-Wesley, Reading, PA, 1989.

Simpson, P., *Artificial Neural Systems*, Pergamon, Elmsford, NY, 1990.

Wasserman, P. D., *Neural Computing: Theory and Practice*, Van Nostrand Reinhold, New York, 1989.

Wasserman, P. D., *Advanced Methods in Neural Computing*, Van Nostrand Reinhold, New York, NY, 1993.

White, D. A., and Sofge, D. A., *Handbook of Intelligent Control*, Van Nostrand Reinhold, New York, 1992.

PROBLEMS

1. In the linear associator of Figure 7.8, the input vector x is a three component vector $(0.3, -0.7, 0.2)$ and the output vector is a four component vector $(-0.8, -0.3, 0.6, 0.9)$. Calculate the weights.

2. Discuss the differences between Widrow's Adaline and Madaline networks and Rosenblatt's perceptron. How do they differ as far as error input is concerned?

3. The three-layer network of Figure 7.22 divides the plane with three lines forming a triangle. Calculate the weights that will give a triangle with its vertices at (x, y) coordinates $(0, 0)$, $(1, 3)$, and $(3, 1)$.

4. Design a three-layer network as shown in Figure 7.20 to separate the non-linearly separable variables for the "exclusive-nor" function having the following truth table:

	Input x	
	0	1
Input y 0	1	0
1	0	1

5. The five-bit code for the letter Q is 01011. Develop a storage matrix W and correct it so that chaotic oscillations will not occur. Show that this storage matrix can produce a correct memory state, even when an erroneous code for q is applied. Use the erroneous representation of Q to be 01010. Show all steps involved.

6. A weight matrix M is given by

$$M = \begin{vmatrix} 1 & 2 & 3 & 2 \\ -2 & 1 & 3 & -1 \\ 3 & 1 & 2 & -3 \end{vmatrix}$$

Draw a 2-layer neural network in which the given matrix represents the weights.

8

BACKPROPAGATION AND RELATED TRAINING ALGORITHMS

8.1 BACKPROPAGATION TRAINING

Backpropagation is a systematic method for training multiple (three or more)-layer artificial neural networks. The elucidation of this training algorithm in 1986 by Rumelhart, Hinton, and Williams (1986) was the key step in making neural networks practical in many real-world situations. However, Rumelhart, Hinton, and Williams were not the first to develop the backpropagation algorithm. It was developed independently by Parker (1982) in 1982 and earlier by Werbos (1974) in 1974 as part of his Ph.D. dissertation at Harvard University. Nevertheless, the backpropagation algorithm was critical to the advances in neural networks because of the limitations of the one- and two-layer networks discussed previously. Indeed, backpropagation played a critically important role in the resurgence of the neural network field in the mid-1980s. Today, it is estimated that 80% of all applications utilize this backpropagation algorithm in one form or another. In spite of its limitations, backpropagation has dramatically expanded the range of problems to which neural network can be applied, perhaps because it has a strong mathematical foundation.

Prior to the development of backpropagation, attempts to use perceptrons with more than one layer of weights were frustrated by what was called the "weight assignment problem" [i.e., how do you allocate the error at the output layer between the two (or more) layers of weights when there is no firm mathematical foundation for doing so?]. This problem plagued the neural network field for over two decades and was cited by Minsky and Papert as one of the criticisms of multilayer perceptrons. Ironically, this need not have been the case, because Amari developed a method for allocating

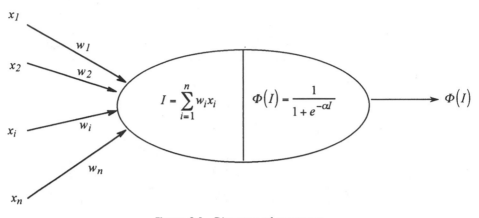

Figure 8.1 Diagram of a neuron.

weights in the 1960s that was not widely disseminated (Amari, 1972). Even more ironic is the fact that Rosenblatt's method of using a random distribution of the weight values in the middle neuron layer and adjusting only the weights for the output neuron layer has been shown to provide adequate training of the network in most cases.

Let us consider a typical neuron as shown in Figure 8.1, with inputs x_i weights w_i, a summation function in the left half of the neuron, and a nonlinear activation function in the right half. The summation of the weighted inputs designated by I is given by

$$I = x_1 w_1 + x_2 w_2 + \cdots + x_n w_n = \sum_{i=1}^{n} x_i w_i \qquad (8.1\text{-}1)$$

The nonlinear activation function used is the typical sigmoidal function and is given by

$$\Phi(I) = \frac{1}{(1 + e^{-\alpha I})} = (1 + e^{-\alpha I})^{-1} \qquad (8.1\text{-}2)$$

This function is, in fact, the logistic function, one of several sigmoidal functions which monotonically increase from a lower limit (0 or -1) to an upper limit ($+1$) as I increases. A plot of a logistic function is shown in Figure 8.2a, in which values vary between 0 and 1, with a value of 0.5 when I is zero. An examination of this figure shows that the derivative (slope) of the curve asymptotically approaches zero as the input I approaches minus infinity and plus infinity, and it reaches a maximum value of $\alpha/4$ when I equals zero as shown in Figure 8.2b. Since this derivative function will be utilized in backpropagation, let us reduce it to its most simple form. If we

$$\Phi(I)$$

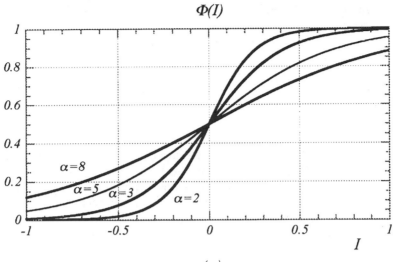

(a)

$$\partial\Phi(I) \, / \, \partial I$$

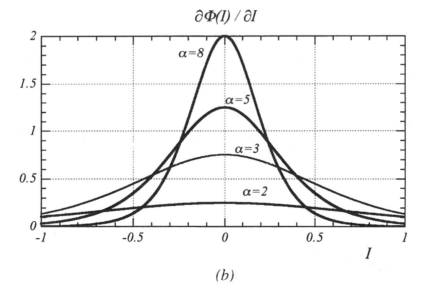

(b)

Figure 8.2 (a) Logistic activation function (α = 2, 3, 5, and 8) and (b) its first derivative (slope).

take a derivative of equation (8.1-2), we get

$$\frac{\partial\Phi(I)}{\partial I} = (-1)(1 + e^{-\alpha I})^{-2}e^{-\alpha I}(-\alpha)$$

$$= \alpha e^{-\alpha I}(1 + e^{-\alpha I})^{-2} = \alpha e^{-\alpha I}\Phi^2(I) \qquad (8.1\text{-}3)$$

If we solve equation (8.1-2) for $e^{-\alpha I}$, substitute it into equation (8.1-3), and simplify, we get

$$\frac{\partial \Phi(I)}{\partial I} = \alpha \frac{1 - \Phi(I)}{\Phi(I)} \Phi^2(I) = \{\alpha[1 - \Phi(I)]\Phi(I)\} = \alpha(1 - \Phi)\Phi$$

$$(8.1\text{-}4)$$

where $\Phi(I)$ has been simplified to Φ by dropping (I).

It is important to point out that multilayer networks have greater representational power than single-layer networks only if nonlinearities are introduced. The logistic function (also called the "squashing" function) provides the needed nonlinearity. However, in the use of the backpropagation algorithm, any nonlinear function can be used if it is everywhere differentiable and monotonically increasing with I. Sigmoidal functions, including logistic, hyperbolic tangent, and arctangent functions, meet these requirements. The arctangent function, denoted as $\tan^{-1}$, has the form

$$\Phi(I) = \frac{2}{\pi} \tan^{-1}(\alpha I) \qquad (8.1\text{-}5)$$

where the factor $2/\pi$ reduces the amplitude of the arctangent function so that it is restricted to the range -1 to $+1$. The constant α determines the rate at which the function changes between the limits of -1 and $+1$ and to the slope of the function at the origin is $2\alpha/\pi$. It influences the shape the arctangent function in the same way that α influences the logistic function in Figure 8.2a. The arctangent function has the same sigmoidal shape as shown in Figure 8.3a. The derivative is

$$\frac{\partial \Phi(I)}{\partial I} = \frac{2}{\pi}\left[\frac{\alpha}{1 + \alpha^2 I^2}\right] \qquad (8.1\text{-}6)$$

which would be used in place of equation (8.1-4) if the arctangent replaced the logistic activation function.

The hyperbolic tangent function has the form

$$\Phi(I) = \tanh(\alpha I) = \frac{e^{\alpha I} - e^{-\alpha I}}{e^{\alpha I} + e^{-\alpha I}} \qquad (8.1\text{-}7)$$

and its shape is shown in Figure 8.3b. Its derivative is

$$\frac{\partial \Phi(I)}{\partial I} = \alpha \operatorname{sech}^2(\alpha I) \qquad (8.1\text{-}8)$$

$$\Phi(I) = \frac{2}{\pi}\arctan(\alpha I)$$

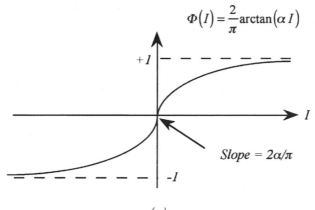

Slope = 2α/π

(a)

$$\Phi(I) = \tanh(\alpha I) = \frac{e^{\alpha I} - e^{-\alpha I}}{e^{\alpha I} + e^{-\alpha I}}$$

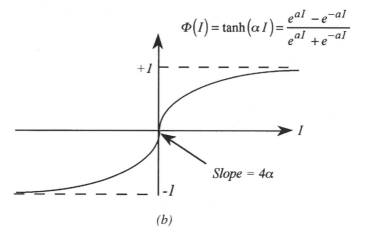

Slope = 4α

(b)

Figure 8.3 Alternate activation functions for backpropagation: (*a*) Arctangent. (*b*) Hyperbolic tangent.

The slope of $\Phi(I)$ at the origin is 4α, and it determines the rate at which the function changes between the limits of -1 and $+1$ in the same general way that α influences the shape of the logistic function in Figure 8.2*a*.

The use of a sigmoidal (squashing) function provides a form of "automatic gain control"; that is, for small values of I near zero, the slope of the input–output curve is steep, producing a high gain, since all sigmoidal activation functions have derivatives with bell shapes of the type shown in Figure 8.2*b*. As the magnitude of I becomes greater in a positive or negative direction, the gain decreases. Hence, large signals can be accommodated without saturation. This is shown in Figure 8.2a.

8.2 WIDROW–HOFF DELTA LEARNING RULE

The Widrow–Hoff delta learning rule can be derived by considering the node of Figure 8.4, where T is the target or desired value vector and I is defined by equation (8.1-1) as the dot product of the weight and input vectors and is given by

$$I = \sum_{i=1}^{n} w_i x_i \qquad (8.2\text{-}1)$$

For this derivation, no quantizer or other nonlinear activation function is included, but the result presented here is equally valid when such nonlinear elements are included.

From Figure 8.4, we see the error function ε as a function of all weights w_i, and we see the squared error ε^2 to be

$$\varepsilon = (T - I) \qquad (8.2\text{-}2)$$

$$\varepsilon^2 = (T - I)^2 \qquad (8.2\text{-}3)$$

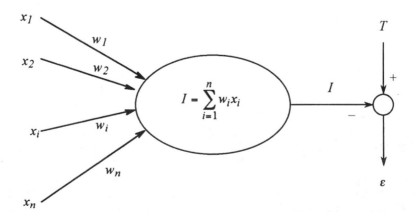

Figure 8.4 Neuron without activation function but with a target value T and an error ε.

The gradient of the square error vector is the partial derivatives with respect to each of these i weights:

$$\frac{\partial \varepsilon^2}{\partial w_i} = -2(T - I)\frac{\partial I}{\partial w_i} = -2(T - I)x_i \qquad (8.2\text{-}4)$$

Since this gradient involves only the ith weight component, the summation of equation (8.2-1) disappears.

For demonstration purposes, let us consider a neuron with only two inputs, x_1 and x_2. The square error is now given by

$$\varepsilon^2 = [T - w_1 x_1 - w_2 x_2]^2$$
$$= T^2 + w_1^2 x_1^2 + w_2^2 x_2^2 - 2Tw_1 x_1 - 2Tw_2 x_2 + 2w_1 x_1 w_2 x_2$$
$$= w_1^2[x_1^2] + w_1[-2x_1(T - w_2 x_2)] + [(T - w_2 x_2)^2]$$
$$= w_2^2[x_2^2] + w_2[-2x_2(T - w_1 x_1)] + [(T - w_1 x_1)^2] \qquad (8.2\text{-}5)$$

The minimum square error occurs when the partial derivatives of square error with respect to the weights w_1 and w_2 are set equal to zero:

$$\frac{\partial \varepsilon^2}{\partial w_1} = -2[T - w_1 x_1 - w_2 x_2]x_1 = 0 \qquad (8.2\text{-}6)$$

$$\frac{\partial \varepsilon^2}{\partial w_2} = -2[T - w_1 x_1 - w_2 x_2]x_2 = 0 \qquad (8.2\text{-}7)$$

Since x_1 and x_2 cannot be zero, the quantities in the brackets, which are identical for both equations, must be zero. This gives

$$T - w_1 x_1 - w_2 x_2 = 0 \qquad (8.2\text{-}8)$$

from which the location of the minimum in the w_1 and w_2 dimensions are

$$w_1 = \frac{T - w_2 x_2}{x_1} \qquad (8.2\text{-}9)$$

$$w_2 = \frac{T - w_1 x_1}{x_2} \qquad (8.2\text{-}10)$$

Substitution of either of these values into equation (8.2-5) gives the minimum square error to be zero. Technically, this is correct, but in the real world the minimum square error is never equal to zero because of nonlinearities, noise, and imperfect data. The presence of noise with a sigmoidal activation function will give a minimum square error that is not zero which we designate as ε^2_{min}.

Examination of equation (8.2-5) shows that plots of ε^2 versus w_1 or w_2 will be parabolic in shape. The parabolic curve of squared error ε^2 versus w_1 is shown in Figure 8.5 for two cases of minimum square error: zero and ε^2_{min}. For both cases, the minimum square error occurred at a value of w_1 given by equation (8.2-9). An identical result can be obtained for squared error versus w_2, where the minimum value occurs at the value of w_2 given by equation (8.2-10). Hence, the minimum square error surface for the two dimensional weight case is a paraboloid of revolution with the ε^2 axis located at (w_1, w_2).

A geometrical interpretation of the delta rule is that it involves a gradient descent algorithm to minimize the square error. When the square error is viewed in three dimensions $(w_1, w_2, \varepsilon^2)$ the square error surface is a paraboloid of revolution with the weight vector descending toward the minimum value along a gradient vector on the surface of the paraboloid. The projection of this gradient vector on the w_1–w_2 plane is the delta vector as shown in Figure 8.6. The delta rule moves the weight vector along the negative gradient of the curved surface toward the ideal weight vector

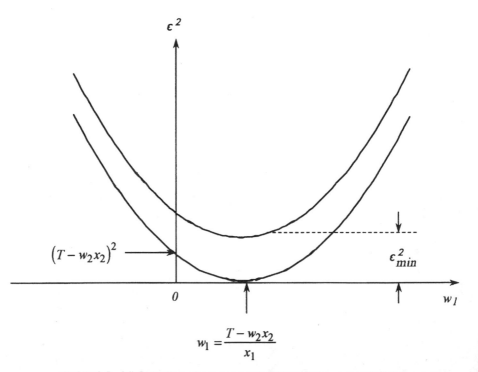

Figure 8.5 Minimization of square error during Widrow–Hoff training.

position. Because it follows the gradient, it is called a *gradient descent* or *steepest descent* algorithm. Since the gradient is the most efficient path to the bottom of the curved surface, the delta rule is the most efficient way to minimize the square error. There is, however, one caveat that must be added here: This statement is true only if the weight vector is descending toward a global minimum. If there are local minima, which are common with multidimensional problems, other techniques must be used to ensure that a solution (i.e., a weight configuration) is not trapped in one of these local minima.

The Widrow–Hoff delta training rule provides that the change in each weight vector component is proportional to the negative of its gradient:

$$\Delta w_i = -K \frac{\partial \varepsilon^2}{\partial w_i} = K \cdot 2(T - I)x_i = 2K\varepsilon x_i \qquad (8.2\text{-}11)$$

where K is a constant of proportionality. The negative sign is introduced because a minimization process is involved. It is common to normalize the

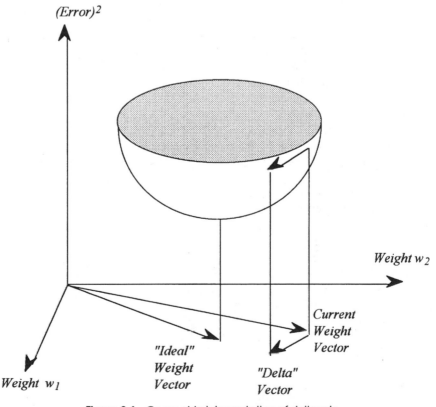

Figure 8.6 Geometric interpretation of delta rule.

input vector component x_i by dividing by $|\mathbf{X}|^2$. Equation (8.2-11) now becomes

$$\Delta w_i = \left[2K|\mathbf{X}|^2\right] \frac{\varepsilon x_i}{|\mathbf{X}|^2} = \frac{\eta \varepsilon x_i}{|\mathbf{X}|^2} \qquad (8.2\text{-}12)$$

which agrees with equation (7.6-1) if we define the learning constant η to be equal to the terms in the brackets:

$$\eta = 2K|\mathbf{X}|^2 \qquad (8.2\text{-}13)$$

8.3 BACKPROPAGATION TRAINING FOR A MULTILAYER NEURAL NETWORK[1]

Before discussing the details of the backpropagation process, let us consider the benefits of the middle layer(s) in an artificial neural network. A network with only two layers (input and output) can only represent the input with whatever representation already exists in the input data. Hence, if the data are discontinuous or nonlinearly separable, the innate representation is inconsistent, and the mapping cannot be learned. Adding a third (middle) layer to the artificial neural network allows it to develop its own internal representation of this mapping. Having this rich and complex internal representation capability allows the hierarchical network to learn any mapping, not just linearly separable ones.

Some guidance to the number of neurons in the hidden layer is given by Kolmogorov's theorem as it is applied to artificial neural networks. In any artificial neural network, the goal is to map any real vector of dimension m into any other real vector of dimension n. Let us assume that the input vectors are scaled to lie in the region from 0 to 1, but there are no constraints on the output vector. Then, Kolmogorov's theorem tells us that a three-layer neural network exists that can perform this mapping exactly (not an approximation) and that the input layer will have m neurons, the output layer will have n neurons, and the middle layer will have $2m + 1$ neurons. Hence, Kolmogorov's theorem guarantees that a three-layer artificial neural network will solve all nonlinearly separable problems. What it does not say is that (1) this network is the most efficient one for this mapping, (2) a smaller network cannot also perform this mapping, or (3) a simpler network cannot perform the mapping just as well. Unfortunately, it does not provide enough detail to find and build a network that efficiently performs the mapping we want. It does, however, guarantee that a method of mapping does exist in the form of an artificial neural network (Poggio and Girosi, 1990).

[1]The analysis presented here is the classical approach in which the hidden and output layer neurons have sigmoidal activation functions. An alternate approach in which the output neurons have linear activation functions is presented in Section 8.7.

Let us consider the three-layer network shown in Figure 8.7, where all activation functions are logistic functions. It is important to note that back-propagation can be applied to an artificial neural network with any number of hidden layers (Werbos, 1994). The training objective is to adjust the weights so that the application of a set of inputs produces the desired outputs. To accomplish this the network is usually trained with a large number of input–output pairs, which we also call examples.

The training procedure is as follows:

1. Randomize the weights to small random values (both positive and negative) to ensure that the network is not saturated by large values of weights. (If all weights start at equal values, and the desired perfor-mance requires unequal weights, the network would not train at all.)
2. Select a training pair from the training set.
3. Apply the input vector to network input.
4. Calculate the network output.
5. Calculate the error, the difference between the network output and the desired output.
6. Adjust the weights of the network in a way that minimizes this error. (This adjustment process is discussed later in this section.)

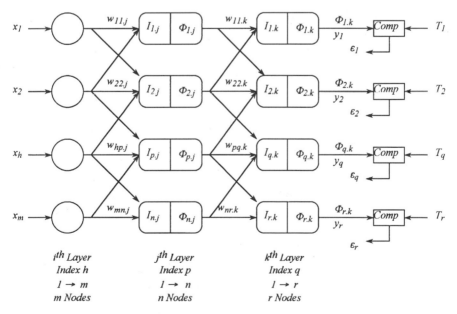

Figure 8.7 Sketch of multilayer neural network showing the symbols and indices used in deriving the backpropagation training algorithm.

7. Repeat steps 2–6 for each pair of input–output vectors in the training set until the error for the entire system is acceptably low.

Training of an artificial neural network involves two passes. In the forward pass the input signals propagate from the network input to the output. In the reverse pass, the calculated error signals propagate backward through the network, where they are used to adjust the weights. The calculation of the output is carried out, layer by layer, in the forward direction. The output of one layer is the input to the next layer. In the reverse pass, the weights of the output neuron layer are adjusted first since the target value of each output neuron is available to guide the adjustment of the associated weights, using the *delta rule*. Next, we adjust the weights of the middle layers. The problem is that the middle-layer neurons have no target values. Hence, the training is more complicated, because the error must be propagated back through the network, including the nonlinear functions, layer by layer.

Calculation of Weights for the Output-Layer Neurons

Let us consider the details of the backpropagation learning process for the weights of the output layer. Figure 8.8 is a representation of a train of neurons leading to the output layer designated by the subscript k with neurons p and q, outputs $\Phi_{p.j}(I)$ and $\Phi_{q.k}(I)$, input weights $w_{hp.j}$ and $w_{pq.k}$, and a target value T_q. The notation (I) in $\Phi_{q.k}(I)$ will be dropped for convenience. The output of the neuron in layer k is subtracted from its target value and squared to produce the square error signal, which for a layer k neuron is

$$\varepsilon = \varepsilon_q = \left[T_q - \Phi_{q.k} \right] \tag{8.3-1}$$

since only one output error is involved. Hence

$$\varepsilon^2 = \varepsilon_q^2 = \left[T_q - \Phi_{q.k} \right]^2 \tag{8.3-2}$$

The delta rule indicates that the change in a weight is proportional to the

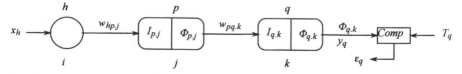

Figure 8.8 Representation of a train of neurons for calculating the change of weight for an output-layer neuron in backpropagation.

rate of change of the square error with respect to that weight—that is,

$$\Delta w_{pq.k} = -\eta_{p.q} \frac{\partial \varepsilon_q^2}{\partial w_{pq.k}} \tag{8.3-3}$$

where $\eta_{p.q}$ is constant of proportionality called *learning rate*. To evaluate this partial derivative, we use the chain rule of differentiation:

$$\frac{\partial \varepsilon_q^2}{\partial w_{pq.k}} = \frac{\partial \varepsilon_q^2}{\partial \Phi_{q.k}} \frac{\partial \Phi_{q.k}}{\partial I_{q.k}} \frac{\partial I_{q.k}}{\partial w_{pq.k}} \tag{8.3-4}$$

Each of these terms are evaluated in turn. The partial derivative of equation (8.3-2) with respect to $\Phi_{q.k}$ gives

$$\frac{\partial \varepsilon_q^2}{\partial \Phi_{q.k}} = -2\left[T_q - \Phi_{q.k}\right] \tag{8.3-5}$$

From equation (8.1-4), we get

$$\frac{\partial \Phi_{q.k}}{\partial I_{q.k}} = \alpha \Phi_{q.k}\left[1 - \Phi_{q.k}\right] \tag{8.3-6}$$

From Figure 8.7 we see that $I_{q.k}$ is the sum of the weighted inputs from the middle layer—that is,

$$I_{q.k} = \sum_{p=1}^{n} w_{pq.k}\Phi_{p.j} \tag{8.3-7}$$

Taking the partial derivative with respect to $w_{pq.k}$ gives

$$\frac{\partial I_{q.k}}{\partial w_{pq.k}} = \Phi_{p.j} \tag{8.3-8}$$

Since we are dealing with one weight, only one term of the summation of equation (8.3-7) survives. Substituting equations (8.3-5), (8.3-6), and (8.3-8) into equation (8.3-4) gives

$$\frac{\partial \varepsilon_q^2}{\partial w_{pq.k}} = -2\alpha\left[T_q - \Phi_{q.k}\right]\Phi_{q.k}\left[1 - \Phi_{q.k}\right]\Phi_{p.j} = -\delta_{pq.k}\Phi_{p.j} \tag{8.3-9}$$

where $\delta_{pq.k}$ is defined as

$$\delta_{pq.k} \equiv 2\alpha\left[T_q - \Phi_{q.k}\right]\Phi_{q.k}\left[1 - \Phi_{q.k}\right]$$

$$= 2\varepsilon_q \frac{\partial \Phi_{q.k}}{\partial I_{q.k}} \qquad (8.3\text{-}10)$$

Substituting equations (8.3-9) into equation (8.3-3) gives

$$\Delta w_{pq.k} = -\eta_{p.q}\frac{\partial \varepsilon_q^2}{\partial w_{pq.k}} = -\eta_{p.q}\delta_{pq.k}\Phi_{p.j} \qquad (8.3\text{-}11)$$

$$w_{pq.k}(N+1) = w_{pq.k}(N) - \eta_{p.q}\delta_{pq.k}\Phi_{p.j} \qquad (8.3\text{-}12)$$

where N is the number of the iteration involved. An identical process is performed for each weight of the output layer to give the adjusted values of the weights. The error term $\delta_{pq.k}$ from equation (8.3-10) is used to adjust the weights of the output layer neurons using equation (8.3-11) and (8.3-12). It is useful to discuss why the derivative of the activation function is involved in this process. In equation (8.3-10) we have calculated an error which must be propagated back through the network. This error exists because the output neurons generate the wrong outputs. The reasons are (1) their own incorrect weights and (2) the middle-layer neurons generate the wrong output. To assign this blame, we backpropagate the errors for each output-layer neuron, using the same interconnections and weights as the middle layer used to transmit its outputs to the output layer.

When a weight between a middle-layer neuron and an output-layer neuron is large and the output layer neuron has a very large error, the weights of the middle layer neurons may be assigned a very large error, even if that neuron has a very small output and thus could not have contributed much to the output error. By applying the derivative of the squashing function, this error is moderated, and only small to moderate changes are made to the middle-layer weights because of the bell-shaped curve of the derivative function shown in Figure 8.2b.

Calculation of Weights for the Hidden Layer Neurons

Since the hidden layers have no target vectors, the problem of adjusting the weights of the hidden layers stymied workers in this field for years until backpropagation was put forth. Backpropagation trains hidden layers by propagating the adjusted error back through the network, layer by layer, adjusting the weight of each layer as it goes. The equations for the hidden layer are the same as for the output layer except that the error term $\delta_{hp.j}$ must be generated without a target vector. We must compute $\delta_{hp.j}$ for each neuron in the middle layer that includes contributions from the errors in each neuron in the output layer to which it is connected. Let us consider a single neuron in the hidden layer just before the output layer, designated with the subscript p (see Figure 8.8). In the forward pass, this neuron propagates its output values to the q neurons in the output layer through the interconnecting weights $w_{pq.k}$. During training, these weights operate in reverse order, passing the value of $\delta_{pq.k}$ from the output layer back to the hidden layer. Each of these weights is multiplied by the value of the neuron through which it connects in the output layer. The value of $\delta_{hp.j}$ needed for the hidden-layer neuron is produced by summing all such products.

The arrangement in Figure 8.9 shows the errors that are backpropagated to produce the change in $w_{hp.j}$. Since all error terms of the output layer are involved, the partial derivative involves a summation over the r outputs. The procedure for calculating $\delta_{hp.j}$ is substantially the same as calculating $\delta_{pq.k}$. Let us start with the derivative of the square error with respect to the weight for the middle layer that is to be adjusted. Then, in a manner analogous to equation (8.3-3), the delta rule training gives

$$\Delta w_{hp.j} = -\eta_{h.p}\frac{\partial \varepsilon^2}{\partial w_{hp.j}} = -\eta_{h.p}\sum_{q=1}^{r}\frac{\partial \varepsilon_q^2}{\partial w_{hp.j}} \tag{8.3-13}$$

where the total mean square ε^2 is now defined by

$$\varepsilon^2 = \sum_{q=1}^{r}\varepsilon_q^2 = \sum_{q=1}^{r}\left[T_q - \Phi_{q.k}\right]^2 \tag{8.3-14}$$

since several output errors may be involved. The learning constant $\eta_{h.p}$ is

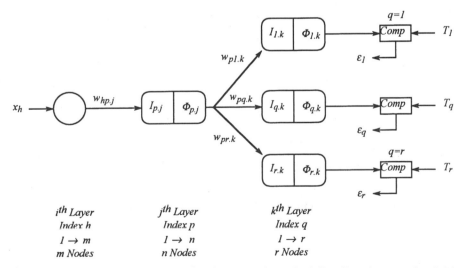

i^{th} Layer	j^{th} Layer	k^{th} Layer
Index h	Index p	Index q
$1 \rightarrow m$	$1 \rightarrow n$	$1 \rightarrow r$
m Nodes	n Nodes	r Nodes

Figure 8.9 Representation of a train of neurons for calculating the change of weight for a middle (hidden) layer neuron in backpropagation.

usually, but not necessarily, equal to $\eta_{p.q}$. Again, we can evaluate the last term of equation (8.3-13) using the chain rule of differentiation, which gives

$$\frac{\partial \varepsilon^2}{\partial w_{hp.j}} = \sum_{q=1}^{r} \frac{\partial \varepsilon_q^2}{\partial \Phi_{q.k}} \frac{\partial \Phi_{q.k}}{\partial I_{q.k}} \frac{\partial I_{q.k}}{\partial \Phi_{p.j}} \frac{\partial \Phi_{p.j}}{\partial I_{p.j}} \frac{\partial I_{p.j}}{\partial w_{hp.j}} \tag{8.3-15}$$

Each of these terms is similar to those in equation (8.3-4) and are evaluated in the same manner.

The first two terms are already given by equations (8.3-5) and (8.3-6), which are

$$\frac{\partial \varepsilon_q^2}{\partial \Phi_{q.k}} = -2(T_q - \Phi_{q.k}) = -2\varepsilon_q \tag{8.3-16}$$

$$\frac{\partial \Phi_{q.k}}{\partial I_{q.k}} = \alpha \Phi_{q.k}(1 - \Phi_{q.k}) \tag{8.3-17}$$

Taking the partial derivative of equation (8.3-7)

$$I_{q.k} = \sum_{p=1}^{n} w_{pq.k}\Phi_{p.j} \tag{8.3-18}$$

with respect to $\Phi_{p.j}$ gives

$$\frac{\partial I_{q.k}}{\partial \Phi_{p.j}} = w_{pq.k} \tag{8.3-19}$$

The summation over p disappears because only one connection is involved. Changing subscripts on equations (8.3-6) to correspond to the middle layer gives

$$\frac{\partial \Phi_{p.j}}{\partial I_{p.j}} = \alpha \Phi_{p.j}[1 - \Phi_{p.j}] \tag{8.3-20}$$

Changing subscripts on equation (8.3-7) and substituting the ith-layer input x_h for the jth-layer input $\Phi_{p.j}$ gives

$$I_{p.j} = \sum_{h=1}^{m} w_{hp.j}x_h \tag{8.3-21}$$

Taking the partial derivative of equation (8.3-21) gives

$$\frac{\partial I_{p.j}}{\partial w_{hp.j}} = x_h \qquad (8.3\text{-}22)$$

Again, the summation over h in equation (8.3-21) disappears because only one connection is involved. Substitution of equations (8.3-17) through (8.3-22) into equation (8.3-15), use of equation (8.3-14) and the definition of $\delta_{pq.k}$ in equation (8.3-10) gives

$$\frac{\partial \varepsilon^2}{\partial w_{hp.j}} = \sum_{q=1}^{r} (-2)\,\alpha\,(T_q - \Phi_{q.k})\big[\Phi_{q.k}(1 - \Phi_{q.k})\big]w_{pq.k}\,\alpha\big[\Phi_{p.j}(1 - \Phi_{p.j})\big]x_h$$

$$= -\sum_{q=1}^{r} \delta_{pq.k}w_{pq.k}\frac{\partial \Phi_{p.j}}{\partial I_{p.j}}x_h \qquad (8.3\text{-}23)$$

If we define $\delta_{hp.j}$ as

$$\delta_{hp.j} \equiv \delta_{pq.k}w_{pq.k}\frac{\partial \Phi_{p.j}}{\partial I_{p.j}} \qquad (8.3\text{-}24)$$

then equation (8.3-23) becomes

$$\frac{\partial \varepsilon^2}{\partial w_{hp.j}} = -\sum_{q=1}^{r} \delta_{hp.j}x_h \qquad (8.3\text{-}25)$$

Since the change in weights as given in equation (8.3-13) is proportional to the negative of the rate of change of the square error with respect to that weight, then, substitution of equation (8.3-23) and (8.3-24) into equation (8.3-13) gives

$$\Delta w_{hp.j} = -\eta_{h.p}\frac{\partial \varepsilon^2}{\partial w_{hp.j}} = \eta_{h.p}\sum_{q=1}^{r} \delta_{pq.k}w_{pq.k}\frac{\partial \Phi_{p.j}}{\partial I_{p.j}}x_h$$

$$= \eta_{h.p}x_h\sum_{q=1}^{r} \delta_{hp.j} \qquad (8.3\text{-}26)$$

and hence

$$w_{hp.j}(N+1) = w_{hp.j}(N) + \eta_{h.p}x_h\sum_{q=1}^{r} \delta_{hp.j} \qquad (8.3\text{-}27)$$

If there are more than one middle layer of neurons, this process moves through the network, layer by layer to the input, adjusting the weights as it goes. When finished, a new training input is applied and the process starts the whole process again. It continues until an acceptable error is reached. At that point the network is trained.

Example 8.1 Updating Weights Through Backpropagation. A simple, fully connected feedforward neural network is shown in Figure 8.10, where bias inputs of $+1$ and adjustable weights w_{1C}, w_{1D} and w_{1E} have been added to neurons C, D, and E, respectively. (See Section 8.4 for a discussion of bias.) All neurons have the same logistic activation function with $\alpha = 1$ and the same learning constants with $\eta = 0.5$.

The desired output of neuron E is 0.1. The weights are randomized to the values shown, and training is started. Then the backpropagation process of learning is applied in the backward direction and the process is carried through one cycle, i.e., the change in each of the six weights and the new values of the weights are calculated.

For the weights between layers j and k, substitution of equation (8.3-9) into equation (8.3-11) gives the changes to be

$$\Delta w_{pq.k} = -\eta_{p.q}\left[-2\alpha\left[T_q - \Phi_{q.k}\right]\Phi_{q.k}\left[1 - \Phi_{q.k}\right]\Phi_{p.j}\right] \quad \text{(E8.1-1)}$$

For the weights between layers i and j, substitution of equation (8.3-23) into

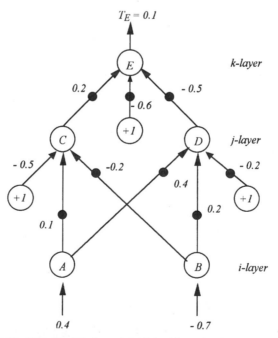

Figure 8.10 Weight adjustment (training) in a simple neural network.

equation (8.3-26) gives the changes to be

$$\Delta w_{hp.j} = -\eta_{h.p}\left[\sum_{q=1}^{r} -2\alpha(T_q - \Phi_{q.k})\Phi_{q.k}(1 - \Phi_{q.k})w_{pq.k}\alpha\Phi_{p.j}(1 - \Phi_{p.j})x_h\right]$$

$$(E8.1\text{-}2)$$

The subscripts h, p, q, and indices m, n, and r are defined in Figure 8.7. Subscript 1 refers to the $+1$ bias terms.

First, calculate the outputs of each neuron using the logistic function with $\alpha = 1$.

$$I_C = 0.4 \times 0.1 + (-0.7) \times (-0.2) + 1 \times (-0.5) = 0.04 + 0.14 - 0.50$$

$$= -0.32 \qquad\qquad \Phi(I_C) = 0.42$$

$$I_D = 0.4 \times 0.4 + (-0.7) \times 0.2 + 1 \times (-0.2) = 0.16 - 0.14 - 0.2$$

$$= -0.18 \qquad\qquad \Phi(I_D) = 0.46$$

$$I_E = 0.42 \times 0.2 + 0.46 \times (-0.5) + 1 \times (-0.6) = 0.08 - 0.23 - 0.60$$

$$= -0.75 \qquad\qquad \Phi(I_E) = 0.32$$

Substitution of these values and other network parameters into equations (E8.1-1) and (E8.2-2) gives

$$\Delta w_{CE} = -0.5 \times (-2) \times 1 \times (0.10 - 0.32) \times 0.32 \times (1.00 - 0.32) \times 0.42$$

$$= -0.020$$

$$\Delta w_{DE} = -0.5 \times (-2) \times 1 \times (0.10 - 0.32) \times 0.32 \times (1.00 - 0.32) \times 0.46$$

$$= -0.022$$

$$\Delta w_{1E} = -0.5 \times (-2) \times 1 \times (0.10 - 0.32) \times 0.32 \times (1.00 - 0.32) \times 1$$

$$= -0.048$$

$$\Delta w_{AC} = -0.5 \times (-2) \times 1 \times (0.10 - 0.32) \times 0.32 \times (1.00 - 0.32) \times 0.20$$

$$\times 1 \times 0.42 \times (1.00 - 0.42) \times 0.4$$

$$= -0.00093 = -0.001$$

$$\Delta w_{AD} = -0.5 \times (-2) \times 1 \times (0.10 - 0.32) \times 0.32 \times (1.00 - 0.32)$$

$$\times (-0.50) \times 1 \times 0.46 \times (1.00 - 0.46) \times 0.4$$

$$= 0.00238 = 0.002$$

$$\Delta w_{BC} = -0.5 \times (-2) \times 1 \times (0.10 - 0.32) \times 0.32 \times (1.00 - 0.32) \times 0.20$$

$$\times 1 \times 0.42 \times (1.00 - 0.42) \times (-0.7)$$

$$= 0.00163 = 0.002$$

$$\Delta w_{BD} = -0.5 \times (-2) \times 1 \times (0.10 - 0.32) \times 0.32 \times (1.00 - 0.32)$$
$$\times (-0.50) \times 1 \times 0.46 \times (1.00 - 0.46) \times -0.7$$
$$= -0.00142 = -0.001$$
$$\Delta w_{1C} = -0.5 \times (-2) \times 1 \times (0.10 - 0.32) \times 0.32 \times (1.00 - 0.32) \times 0.20$$
$$\times 1 \times 0.42(1.00 - 0.42) \times 1$$
$$= -0.0023 = -0.002$$
$$\Delta w_{1D} = -0.5 \times (-2) \times 1 \times (0.10 - 0.32) \times 0.32 \times (1.00 - 0.32)$$
$$\times -(0.50) \times 1 \times 0.46(1.00 - 0.46) \times 1$$
$$= 0.0059 = 0.006$$

Adding these changes to the original weights gives the new weights.

$$w_{CE} = 0.200 - 0.020 = 0.180$$
$$w_{DE} = -0.500 - 0.022 = -0.522$$
$$w_{IE} = -0.600 - 0.048 = -0.648$$
$$w_{AC} = 0.100 - 0.001 = 0.099$$
$$w_{AD} = 0.400 + 0.002 = 0.402$$
$$w_{BC} = -0.200 + 0.002 = -0.198$$
$$w_{BD} = 0.200 - 0.001 = 0.199$$
$$w_{IC} = -0.500 - 0.002 = -0.502$$
$$w_{ID} = -0.200 + 0.006 = -0.194$$

This process is repeated until all sample pairs in the epoch have been utilized. After these weight changes have been calculated, the total square error is then calculated. If it is more than the specified amount, the learning algorithm is again applied to the network using another epoch of training data. A better alternative is to continue the training process until monitoring of the total square error for a test set of data starts to increase, even though the total square error for the training set continues to decrease. □

8.4 FACTORS THAT INFLUENCE BACKPROPAGATION TRAINING

Adding a bias (a +1 input with a training weight, which can be either positive or negative) to each neuron is usually desirable to offset the origin of the activation function. This produces an effect equivalent to adjusting the threshold of the neuron and often permits more rapid training. The weight of the bias is trainable just like any other weight except that the input is always $+1$.

Momentum

Another technique to reduce training time is the use of *momentum*, because it enhances the stability of the training process. Momentum is used to keep the training process going in the same general direction analogous to the way that momentum of a moving object behaves. This involves adding a term to the weight adjustment that is proportional to the amount of the previous weight change. In effect, the previous adjustment is "remembered" and used to modify the next change in weights. Hence, equation (8.3-11) now becomes

$$\Delta w_{pq.k}(N + 1) = -\eta_{pq}\,\delta_{pq.k}\Phi_{p.j} + \mu\,\Delta w_{pq.k}(N) \qquad (8.4\text{-}1)$$

where μ is the *momentum coefficient* (typically about 0.9). This relationship is shown in Figure 8.11. The new value of the weight then becomes equal to the previous value of the weight plus the weight change of equation (8.3-11), which includes the momentum term. Equation (8.2-12) now becomes

$$w_{pq.k}(N + 1) = w_{pq.k}(N) + \Delta w_{pq.k}(N + 1) \qquad (8.4\text{-}2)$$

This process works well in many problems, but not so well in others. Another way of viewing the purpose of momentum is to overcome the effects of local

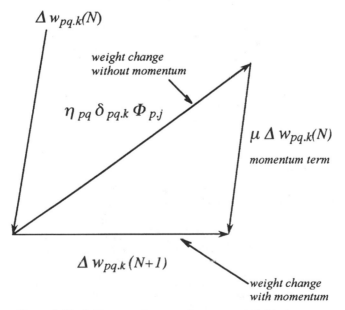

Figure 8.11 Influence of momentum upon weight change.

minima. The use of the momentum term will often carry a weight change process through one or more local minima and get it into a global minima. This is perhaps its most important function.

There is a substantial number of advanced algorithms or other procedures that have been proposed as means of speeding up the training of backpropagation networks. Sejnowski and Rosenberg (1987) proposed a similar momentum method that used exponential smoothing. However, the results were mixed. In some cases it improved the speed of the training, whereas in other cases it did not. Parker (1987) proposed a method called the "second-order" backpropagation that used the second derivative to produce a more accurate estimation of the correct weight change. The computational requirements were greater and were generally viewed as not being cost effective compared to other methods. It was, however, clear that higher-order (greater than 2) backpropagation systems were not effective. Stornetta and Huberman (1987) pointed out that the 0–1 range of sigmoidal function is not optimal for binary inputs. Since the magnitude of a weight adjustment is proportional to the output level of the neuron from which it originates, a level of 0 results in no modification. With binary inputs, half of the outputs (on the average) will be zero, and weights do not train. The proposed solution was to change the input range of the activation function from $-1/2$ to $+1/2$ by adding a bias of $-1/2$. They demonstrated that for binary functions this procedure reduces the training time by 30–50%. Today a more common method of accomplishing this is to use the arctan or hyperbolic tangent activation function.

Despite some spectacular results, it is clear that backpropagation is not a panacea. The main problem is the long and sometimes uncertain training time. Some artificial neural networks have been known to require days or weeks of training, and in some cases the network simply will not train at all. This may be the result of a poor choice of training coefficients or perhaps the initial random distribution of the weights. However, in most cases failure to train is usually due to local minima or *network paralysis*, where training virtually ceases due to operation in the flat region of the sigmoid function.

Stability

The proof of convergence of backpropagation by Rumelhart, Hinton and Williams (1986) used infinitesimal weight adjustments. This is impractical because it requires infinite training time. In the real world, if the step size is too small, the training is too slow; if the step size is too large, instability may result. However, recent efforts that involve the use of large steps initially with automatic reduction as the training proceeds have been quite successful in reducing training time.

Another issue is temporal instability. If a network is to learn the alphabet, it is of no value to learn the letter B if it destroys the learning of letter A. The network must learn the entire training set without disrupting what is already learned. Rumelhart's convergence requires the network to process all training examples before adjusting any weights. Furthermore, backpropagation may not be useful if the network faces a continuously changing environment where the inputs are continuously changing, because the process may never converge. There are alternate networks discussed later that are useful in such situations.

Adjusting α Coefficient in Sigmoidal Term

Sometimes weights become very large in value and force the neurons to operate in a region where the sigmoidal function is very flat—that is, its derivative is very small. Since the error sent back for training in backpropagation is proportional to the derivative of the sigmoid function, very little training takes place. This network paralysis can sometimes by avoided by reducing the training coefficient, which unfortunately results in extending the training time.

A better method of coping with network paralysis if to adjust the α coefficient on the exponential term in the logistic term. By decreasing α, we effectively spread out the sigmoidal function over a wider range. Values of I that gave $\Phi(I)$ of 0.99 now gives smaller values, like 0.75 or perhaps 0.35, depending upon the value of α. The training process is now operating in a range where the derivative of the sigmoidal is much greater, and hence training will proceed much faster.

For large negative values of I, the logistic activation function squeezes the $\Phi(I)$ values close to zero. Use of hyperbolic tangent and arctangent activation functions spreads these values down into the range between 0 and -1, thereby eliminating the network paralysis. The use of a trainable bias term as an input to each neuron, which is standard practice in most commercial neural network software, is also useful in avoiding network paralysis.

Dealing with Local Minima

Perhaps the major problem of backpropagation is local minima. Since backpropagation employs a form of gradient descent, it is assumed that the error surface slope is always negative and hence constantly adjusting weights toward the minimum. However, error surfaces often involve complex, high-dimensional space that is highly convoluted with hills, valleys, folds, and gullies. It is very easy for the training process to get trapped in a local minimum. One of the most practical solutions involves the introduction of a shock—that is, changing all weights by specific or random amounts. If this fails, then the most practical solution is to re-randomize the weights and start the training over.

Another alternative is to utilize *simulated annealing*, a technique that is used to search for global minima in a search surface in which states are updated based on a statistical rule rather than deterministically. This update rule changes to become more deterministic as the search progresses. (*Simulated annealing* is discussed later in Chapter 9.) The procedure is to use backpropagation until the process seems to stall. Then *simulated annealing* is used to continue training until the local minimum has been left behind. Then the simulated annealing is stopped and the backpropagation training continues until a global minimum is reached. In most cases, only a few simulated annealing cycles of this two-stage process are needed. If the mean square error of the outputs stalls in its descent, then the annealing process may have to be used again. The final training step in this process is backpropagation to minimize the overall error of the process.

Learning Constants

Choosing the correct learning constant η is important in backpropagation training. First, η cannot be negative because this would cause the change of the weight vector to move away from the ideal weight vector position. Of course, if η is equal to 0, then no learning takes place. Therefore, η must always be positive. It can be shown both analytically and experimentally that if η is greater than 2, then the network is unstable, and if η is greater than 1, then the weight vector will overshoot the ideal position and oscillate, rather than settle into a solution. Hence, η should be in the range between 0 and 1.

If η is large (0.8 or thereabouts), then the weight vector will take relatively large steps and will find the minimum faster. However, if the input data patterns are not highly compacted around the "ideal" example, this will cause the network to jump wildly each time a new input pattern is presented. If the value of η is small (0.2 or thereabouts), then the weight vector will take small steps toward the "ideal" position, and will not vary wildly if the input data patterns are not very close to an "ideal" example. However, the network will require a longer time to learn the patterns with many iterations of the data needed.

As a compromise, large values of η are used when the input data patterns are close to the ideal whereas small values of η are used when they are not. When the nature of the input data patterns are not known, then it is better to use a moderate value of η. As suggested earlier, an even better method is to change the value of η as the network learns, beginning with a large value initially and reducing it as the learning progresses. Then the leaning process is not distracted by minor variations in the input data. It is important to remember that real data patterns are never perfect examples of a category, and that the separating hyperplanes cannot always separate all the input data into a specific number of categories.

Variations of the Standard Backpropagation Algorithm

In addition to such standard techniques are adding momentum, adjusting learning rate, adjusting the exponential decay constant in the sigmoidal function, and using other activation functions, there are a number of other variations that are often useful in many situations. Most of them modify the standard backpropagation algorithm in one way or another.

Cumulative Update of Weights

A variation of backpropagation training (called "cumulative backpropagation") that seems to be helpful in speeding up training is the cumulative update of weights. In this case, the individual weight changes for each weight are accumulated for an epoch of training, summed, and then the cumulative weights changes are made in the individual weights. This procedure significantly reduces the amount of computation involved, and there usually is no noticeable effect on the final training of the network.

Fast Backpropagation

This variation of backpropagation introduced by Tariq Samad of Honeywell (Samad, 1988) involves the following changes in standard backpropagation. A multiple of the error at layer k is added to the k-layer activation value $\Phi_{q.k}$ prior to doing the weight update for weights on connections between the j and k layers. This can dramatically increase the speed of training, usually by more than an order of magnitude. Furthermore, it has been shown in one case to reach convergence when standard backpropagation training failed to do so after 10,000 iterations.

Quickprop Training

Fahlman's quickprop training algorithm is one of the more effective algorithms in overcoming the step-size problem in backpropagation. The $\partial \varepsilon^2 / \partial w$ values are computed as in standard backpropagation (Fahlman, 1988). However, a second-order method related to Newton's method is used to upgrade the weights in place of simple gradient descent. Fahlman reports that quickprop consistently outperforms backpropagation, sometimes by a wide margin.

Quickprop's weight update procedure depends on two approximations: (1) Small changes in one weight produce relatively little effect on the error gradient observed at other weights, and (2) the error function with respect to each weight is locally quadratic. The slopes and weights for current and previous iterations are used to define a parabola. The algorithm then goes to the minimum point of this parabola as the next weight. This process continues going through all weights for an epoch. If the error is sufficiently small, the training process is terminated; if not, training continues for another epoch.

Use of Different Error Functions

The error function defined in equation (8.3-2) is proportional to the square of the Euclidean distance between the desired output and the actual output of the network for a particular input pattern. As an alternative, we can substitute any other error function whose derivatives exist and can be calculated as the output layer. Errors of third and fourth order have been used to replace the traditional square error criterion. NeuralWorks[2] has suggested cubic and quadratic errors of the forms

$$\varepsilon^2 = \sum_{q=1}^{r} \left[|T_q - \Phi_{q.k}| \right]^3 \tag{8.4-3}$$

$$\varepsilon^2 = \sum_{q=1}^{r} \left[T_q - \Phi_{q.k} \right]^4 \tag{8.4-4}$$

which have local errors, analogous to equation (8.3-10), of

$$\delta_{pq.k} = -3(T_q - \Phi_{q.k})^2 \frac{\partial \Phi_{q.k}}{\partial I_{q.k}} \tag{8.4-5}$$

$$\delta_{pq.k} = -4(T_q - \Phi_{q.k})^3 \frac{\partial \Phi_{q.k}}{\partial I_{q.k}} \tag{8.4-6}$$

It is not clear whether the benefits of cubic and quadratic error functions compensate for the additional complexity introduced.

Delta-Bar-Delta Networks

Most changes of the standard backpropagation algorithm involve one of two methods: (1) Incorporate more analytical information to guide the search such as second-order backpropagation, which has not proven particularly successful, and (2) use heuristics (often intuition) that are reasonably accurate. The delta-bar-delta algorithm is an attempt introduced by R. A. Jacobs (Jacobs, 1988) to improve the speed of convergence using heuristics. Empirical evidence suggests that each dimension of the weight space may be quite different in terms of the overall error surface. By using past values of the gradient, heuristics can be used to imply the curvature of the local error surface, from which intelligent steps can be taken in weight optimization. Since parameters for one weight dimension may not be appropriate for all weight dimensions, each neuron has its own learning rate which is adjusted over time (i.e., reduced as the training progresses). This method has proven to be effective in reducing the time required for training neural networks.

[2]Copyright held by NeuralWare Inc., Pittsburgh, PA.

Extended Delta-Bar-Delta

Minai and Williams (1990) have extended the delta-bar-delta algorithm by (1) adding a time-varying momentum term and (2) adding a time-varying learning rate. The rates of change in momentum and learning rate decrease exponentially with weighted gradient components so that greater increases will be applied in areas of small slope or curvature than in the areas of high curvature. To prevent wild jumps and oscillations in weight space, ceilings are placed on the individual learning and momentum rates.

8.5 SENSITIVITY ANALYSIS IN A BACKPROPAGATION NEURAL NETWORK

Sensitivity analysis is an extremely useful tool in many practical applications. The introduction of a small perturbation in one of the input neurons usually produces perturbations in the outputs of all neurons connected, directly or indirectly, to that input neuron (Guo and Uhrig, 1992; Hashem, 1992). The ratio of the magnitude of the perturbation in the output of a specific output neuron to the perturbation in the input of a specific input node is defined as the *sensitivity*. A perturbation of x_h of the multilayer neuron network in Figure 8.7 produces perturbations in all values of y_q. Hence, the sensitivity $\sigma_{q.h}$ is given by

$$\sigma_{q.h} = \frac{\Delta y_q}{\Delta x_h} = \frac{\partial y_q}{\partial x_h} \tag{8.5-1}$$

where the ratio of the Δ perturbations are replaced with partial derivatives (after taking the appropriate limit). Using the notation of Figure 8.7, equation (8.5-1) becomes

$$\sigma_{q.h} = \frac{\partial y_q}{\partial x_h} = \frac{\partial \Phi_{q.k}}{\partial x_h} = \frac{\partial \Phi_{q.k}}{\partial I_{q.k}} \frac{\partial I_{q.k}}{\partial \Phi_{p.j}} \frac{\partial \Phi_{p.j}}{\partial I_{p.j}} \frac{\partial I_{p.j}}{\partial x_h} \tag{8.5-2}$$

The first term is given by equation (8.3-6) to be

$$\frac{\partial \Phi_{q.k}}{\partial I_{q.k}} = \alpha \Phi_{q.k} \left[1 - \Phi_{q.k}\right] \tag{8.5-3}$$

The second term is found by taking the partial derivative of equation (8.3-7) with respect to $\Phi_{p\,j}$, which gives

$$\frac{\partial I_{q.k}}{\partial \Phi_{p.j}} = \sum_{p=1}^{n} w_{pq.k} \tag{8.5-4}$$

The third term is given by equation (8.3-20) to be

$$\frac{\partial \Phi_{p.j}}{\partial I_{p.j}} = \alpha \Phi_{p.j}\left[1 - \Phi_{p.j}\right] \tag{8.5-5}$$

The fourth term is found by taking the partial derivative with respect to x_h of equation (8.3-21) to be

$$\frac{\partial I_{p.j}}{\partial x_h} = w_{hp.j} \tag{8.5-6}$$

Again, the summation disappears because only one input is involved. Substituting these four terms into equation (8.5-2) gives

$$\sigma_{q.h} = \alpha \Phi_{q.k}\left[1 - \Phi_{q.k}\right] \sum_{p=1}^{n} w_{pq.k} \alpha \Phi_{p.j}\left[1 - \Phi_{p.j}\right] w_{hp.j} \tag{8.5-7}$$

which becomes

$$\sigma_{q.h} = \alpha^2 \Phi_{q.k}\left[1 - \Phi_{q.k}\right] \sum_{p=1}^{n} w_{pq.k} \Phi_{p.j}\left[1 - \Phi_{p.j}\right] w_{hp.j} \tag{8.5-8}$$

Experimental Evaluation of Sensitivity Coefficients

An experimental evaluation of sensitivity coefficients, sometimes called the "dither" method, is possible after a neural network has been trained. It involves the introduction of small perturbations of each input x_i, one at a time, of about 0.5% in both the positive and negative directions. The resultant perturbations of each output y_j in each direction for each input perturbation is measured and averaged. The sensitivity coefficient is then taken as

$$\sigma_{q.h} = \frac{\overline{\Delta y_q}}{\overline{\Delta x_h}} \tag{8.5-9}$$

where the averages are taken over the perturbations in the positive and negative directions. If there are h inputs and q outputs, then there are hq sensitivity coefficients that can be evaluated experimentally.

In general, the usefulness of such measurements are limited. The principal problem is that the values are valid only for the particular location in problem space represented by the values of the inputs and outputs before they are perturbed.

8.6 AUTOASSOCIATIVE NEURAL NETWORKS

Although any layer in an neural network can have any number of neurons, the most common backpropagation network starts with a large number of neurons in the input layer and has relatively few neurons in the output layer. The reason is that many problems involve a complex description of a situation or condition as the input, and a limited number of classes or conditions as the output. There is no rule prescribing the number in neurons of a middle or hidden layer of a three-layer neural network. Kolmogorov's theorem gives us a number of neurons that guarantee the existence of a mapping function between the input and output, but as indicated earlier, this may not be the optimal or most appropriate number of neurons in any given situation. As a rule of thumb, the number of neurons in the middle layer should be less than the number of data sets in an epoch so that the neural network does not *memorize* the various input data sets; that is, a particular neuron in the hidden layer becomes associated with a particular data set of an epoch.

Autoassociative neural networks, in which the output is trained to be identical to the input, have many unusual characteristics that can be exploited in many applications. Autoassociative neural networks as defined here are feedforward, fully connected, multilayer perceptrons usually (but not always) trained using backpropagation (Masters, 1993). The number of neurons in the hidden layer(s) may be greater or smaller than the number in the input and output. All neurons in the middle layer(s) must have nonlinear sigmoidal (logistic, arctangent, or hyperbolic tangent) activation function, but the output layer may have either a linear or nonlinear activation function. Kramer (1991, 1992) has investigated the special case where the middle layer has fewer neurons (which Kramer called a "bottleneck") and reported features that lend themselves to diagnostic and monitoring as well as to the identification of nonlinear principal components. This bottleneck layer prevents a simple one-to-one of "straight through" mapping from developing during the training of the network.

Let us consider the autoassociative backpropagation network shown in Figure 8.12, where there are 100 neurons in the input and output layers and 40 neurons in the hidden (bottleneck) layer. We start with the desired output of the output layer being exactly equal to input to the input layer, and we proceed with the training using backpropagation. The data for the training set must be chosen so that individual input–output values cover the range over which the network will have to accept inputs in the future. Eventually, within some tolerable error, the input and output of the network arc the same. This indicates that the information contained in the input vector which has 100 components is approximately equal to the information contained in the output vector which also has 100 components. Furthermore, the information in the input vector passes through the middle layer, where it is represented by a 40-component vector. This compression of the information into

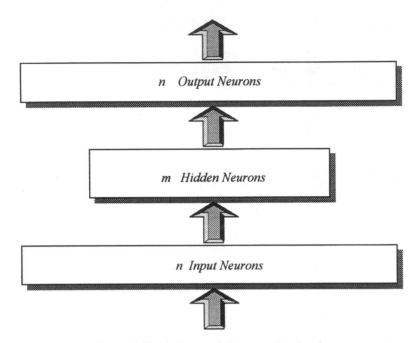

Figure 8.12   Autoassociative neural network.

40 components is a very useful property for certain specific applications. In effect, the 100-dimensional data set has been reduced to a 40-dimensional data set and then reexpanded to a 100-dimensional data set with the error minimized in a least square sense. This means that the 100-dimensional input must be reproduced at the output with only 40 independent variables represented by the outputs of the neurons in the middle layer. In effect, least squares training induces the network to model correlations and redundancies in the input data in order to reproduce the input data at the output with minimal distortion under the dimensional restriction of the "bottleneck" layer.

A quantity that is often of interest is the compressed representation of the input variable consisting of 40 values in the hidden layer. The values from this middle layer are often extracted and utilized as a compressed representation of the input information. Since we are now effectively dealing with a two-layer network (the input layer and the middle layer, which is now an output layer), the validity of this representation is dependent upon the nature of the input data. There is no "hidden" layer to capture the patterns of features of the input data and present it in an organized manner to the middle layer. The same situation exists with respect to the conversion of a

40-dimension representation of the data contained in the middle layer to a 100-dimension representation of the data contained in the output layer. These situations can be overcome by putting in two "intermediate" layers, one between the input and the bottleneck layer and another between the middle layer and the output layer, to give a five-layer neural network. Such an arrangement is shown in Figure 8.13. The second and fourth layers, each having 135 neurons, are effectively feature extraction layers. Kramer calls them "mapping" and "demapping" layers, respectively. These layers have more nodes than the input and output layers so that they are capable of representing nonlinear functions of arbitrary form. However, care must be taken to avoid memorization by the neurons in these layers. With this configuration, the signal now coming from the third (middle) layer of this five-layer neural network now is an appropriate 40-dimensional representation of the input data. In the case of this five-layer autoassociative neural network, the second and fourth layers must have nonlinear activation functions, but the middle and output layers may have either linear or nonlinear activation functions. Kramer indicates that to capture linear functionality efficiently, linear bypasses can be allowed from the input layer to the bottleneck layer and from the bottleneck layer to the output layer, but not across the bottleneck layer.

The functioning of autoassociative neural networks should not be confused with associative memory of the type illustrated in the last chapter, where a neural network that was trained to recognize the letters of the alphabet could recognize distorted versions of the letters. In effect, the distorted letter was closer (in a least squares sense) to the corresponding undistorted letter stored in the associative memory. In contrast, the autoassociative neural network has no stored quantities and no discrete classes, and its outputs are continuous variables.

Once the autoassociative neural network of Figure 8.13 has been trained, it can be split into two separate neural networks. The first consists of the input, mapping, and bottleneck layers as shown in Figure 8.14a, in which the input is separated into a reduced-dimensional representation of the input. If all neurons in this autoassociative neural network had linear activation functions, the output of the middle layer would be the linear principal components. With nonlinear activation functions, these components represent nonlinear principle components of the input, and the number of principal components obtained is equal to the number of neurons in the bottleneck layer.

The second network consists of the bottleneck, demapping, and output layers as shown in Figure 8.14b, in which the reduced-dimensional representation of the middle layer is expanded into a good representation of the original input signal. This separation can occur only after training has taken place; otherwise there are no target values available for the neurons in the middle layer.

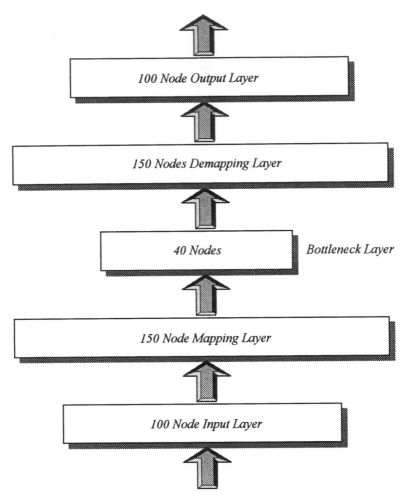

Figure 8.13   Five-layer autoassociative neural network.

Use of an Autoassociate Neural Network for Filtering

The process described above is a form of "filtering" where the amount of information lost is related inversely to the number of neurons in the "bottleneck" layer. If the input is a time series taken from an fluctuating analog signal with a sampling rate of 100 samples per second, then the 100 input components represents 1 sec of data. (Similarity, if the sampling rate is 5 or 1000 samples per second, the input data represents 20 or 0.1 sec of data, respectively.) The training proceeds by using data sets consisting of successive groups of 100 samples, shifted by one sample, which are entered into the neural network softwarc or hardwarc as both the input and desired output. Each successive group of 100 values consists of the next sample value and the

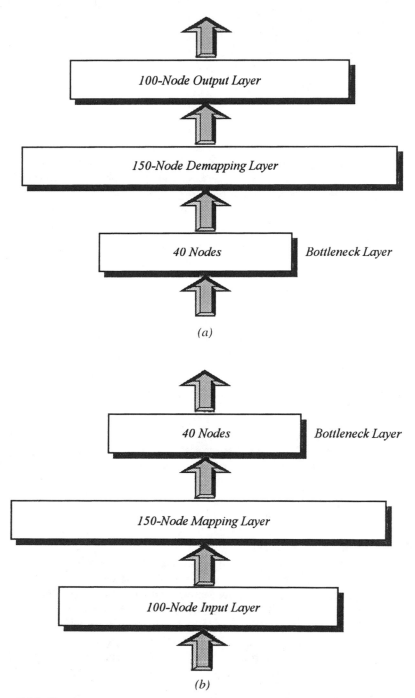

Figure 8.14 Five-layer autoassociative neural network broken into three-layer networks.

preceding 99 sample values. The oldest sample is dropped. The neural network then undergoes another training cycle, and then the data are shifted again to include the next data point, and the training process proceeds. Care must be taken to ensure that the range of variables to which the 100 inputs and outputs are subjected during training covers the range of input data expected in the future. When this training is complete, a time series can be introduced into the neural network by introducing successive 100-sample groups of data to the input layer. Because the bottleneck forces the variable to be represented by fewer dimensions, some information is lost. In this case, it is the high-frequency information that is lost since more dimensions are required to represent a high-frequency variable. In effect, the autoassociative neural network behaves as a low-pass filter.

It is in this filtering application that the true nature of the autoassociative network process is revealed. In effect, the high-frequency components in the signal are eliminated by the middle layer. The smaller the number of neurons in the middle layer, the lower the cutoff frequency and the greater the number of high-frequency components that are eliminated. Hence, it is very clear that the input and output signals can never be identical, because information has been filtered out of the input signal. The output can be only a low-frequency approximation of the original input signal. As the number of neurons in the hidden layer decreases, the cutoff frequency is reduced further, thereby eliminating more of the higher-frequency components.

Use of Autoassociative Neural Networks in Systemwide Monitoring

Neural networks, in combination with other artificial intelligence technologies, offer means of interpreting data and measurements in ways that are not otherwise possible. The unique characteristics of three- and five-layer autoassociative neural networks in which the outputs are trained to emulate the inputs over an appropriate dynamic range have been explored and found to be useful in systemwide monitoring. Many (typically 10–20) variables of complex systems (power plants, chemical or manufacturing processes, social systems, etc.) that have some degree of correlation (typically > 0.3) with each other constitute the inputs. Hence, each output receives some information from almost every input. During training to make each output equal to the corresponding input, the interrelationships between all the input variables and each individual output are embedded in the connection weights of the network. As a result, any specific output, even the corresponding output shows only a small fraction of the input change over a reasonably large range. This characteristic allows the autoassociative neural network to detect drift, deterioration, or failure of a sensor by simply comparing each input with the corresponding output.

Upadhyaya and Eryurek (1992) have demonstrated the feasibility of such an application using data from the EBR-2 (Experimental Breeder Reactor

#2). A necessary prerequisite for such an application is that the various inputs have some degree of correlation. As indicated above, when all of the inputs to an autoassociative neural network are correlated to some degree, then each output is dependent on all the inputs. Hence, the deterioration of one input signal will have only a slight influence on the outputs. The change in the channel that corresponds to that input would be larger than the changes in the other channels, but significantly smaller than the change in the input because of the influence of the correlation with the inputs from the other channels.

In principle, all that is necessary is to detect a deteriorating sensor or instrumentation channel is to compare each input with the corresponding output, calculate the difference, compare it to an allowable difference, and trip an alarm when the difference exceeds the allowable difference. To avoid false alarms, several small deviations beyond the limit may be required in a specified time to trip an alarm. A single large deviation, of course, should trip the alarm. In most practical applications, especially when noise is present, a more sophisticated technique to detect error, such as "sequence probability ratio test," (Wald, 1945) should be employed to minimize the number of missed and false-positive alarms.

An alternative interpretation of the existence of differences between the inputs and the corresponding outputs of the autoassociative neural network might be that the input–output relationship of the system from which the signals come may have changed due to system failure or changes of some sort in the system. All of the results reported in the experimental work discussed in this section are based on the assumption that the underlying system does not change and that only the sensors and related instrumentation channels are being validated. However, in the real world, systems change with time—sometimes slowly, sometimes rapidly—while still behaving normally. Sometimes the changes are anticipated; sometimes these changes come as a surprise. If the changes occur during the training period (or can be artificially introduced), then the relationship between the variables for different conditions can be trained into an autoassociative neural network. This is the case with power ascension from 45% to 100% of full power in the Experimental Breeder Reactor-2 as shown in Example 8.2 (Upadhyaya and Eryurek, 1992).

Corrected Readings from Deteriorating or Failed Sensors

One of the unique advantages of using autoassociative neural networks is the ability to obtain the correct reading for a sensor that has failed. Since the specific sensor that has failed can be identified as discussed above, all that is needed is to carry out an adjustment of the input to the input neuron representing that sensor input to bring the outputs of the autoassociative neural network back to their original values (in a minimization of least squares difference sense). Multiple failures can also be handled in the same way with a multidimensional search for the proper input values, provided that the number of correct sensors is greater than the number of neurons in the bottleneck layer.

Robust Autoassociative Neural Networks

To improve the behavior of the autoassociative neural network, a technique involving the addition of uniform random noise up to 10% to each input, one at a time, while retaining the noise-free values for the desired output, can be employed. This technique is analogous to adding noise to a neural network input to avoid "memorization" and to speed training. Application of this process to all input–output pairs of neurons during training can produce a very robust autoassociative neural network in which the outputs are virtually immune to input change up to 10% of the range of the input (Wrest, 1996).

A critically important issue is how to deal with changing plant configurations and conditions that are not trained into the autoassociative neural network. Fortunately, such changes would be readily detected in most cases by the comparison of outputs with inputs. Differences in more than one input–output pair are almost invariably associated with changes in the system rather than sensor failure, because simultaneous failures of more than one sensor are very rare. Clearly a change in configuration not trained into the autoassociative neural network requires immediate additional training or retraining. Another important issue that needs to be investigated is how the retrained networks relate to the previously trained network. It may be advantageous to retain all consecutive network configuration to have an "audit trail" for the calibration and drift detection. For a slowly changing condition, especially one that is cyclic in nature, is better to train over a whole cycle when possible so that the influence of this quantity is included in the trained network.

Example 8.2   Behavior of an Autoassociative Neural Network as a Plantwide Monitoring System. This autoassociative technique of plant-wide monitoring was applied to data from 18 signals (8 from the primary system and 10 from the secondary system) from the EBR-2 during ascension in power, and the results of these measurements for the primary system are shown in Figure 8.15. Data were collected for the eight primary variables (defined in Table 8.1) as EBR-2 increased in power output from 45% (run #1) to 100% (run #150) of full power. All variables were normalized and scaled into the interval 0.1–0.9. The autoassociative neural network was trained using data collected during the power ascension. Some variables changed rather significantly during the power ascension (between run #1 and run #150) while others changed very little. (The lines connecting the points are used only to indicate that these points belong to the same run.) In all cases, the values predicted by the trained network were within about 0.5% of the measured value. However, when an error was deliberately introduced into variable #1 (as indicated with the point connected with "dashed" lines in Figure 8.15), the corresponding value predicted by the trained network did not change significantly. It is this characteristic behavior of autoassociative neural networks that allows monitoring of many variables to be carried out simultaneously by simply comparing network inputs and outputs. □

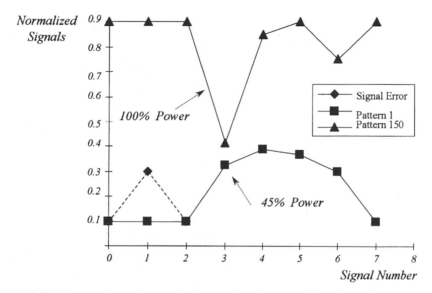

Figure 8.15 Comparison of actual output with output predicted by autoassociative neural network at 45% power and 100% power. (Dashed lines indicate effect of an error deliberately introduced into input signal #1.)

Table 8.1 EBR-2 variables of the primary coolant system monitored by an autoassociative neural network

Variable Number-- Variable

0 -- Power Level (%)
1 -- Core Exit Temperature (°F)
2 -- Control Rod Position (°F)
3-- Primary Pump Flow rate (%)
4-- High Pressure Plenum Sodium Temp. (°F)
5-- Low Pressure Plenum Sodium Temp. (°F)
6-- IHX Primary Outlet Sodium Temp. (°F)
7 --Core Upper Sodium Temp. (°F)

Inferential Measurements Using Neural Networks

Inferential measurement, as the name implies, is the inferring of a measurement value from its relationship, usually its physical correlation, with other variables (Guha, 1992). Reasons for an inferential measurement are that the quantity cannot be measured directly, measurements are difficult or expensive, a sensor is failing or has failed, the measurement process itself is deteriorating, or comparison of an inferred value with an actual value will assist in the identification and diagnoses of problems. The mapping ability of neural networks are ideal for such inferential measurements because they can map plant characteristics to the quantity whose measurement is to be inferred. Typically, the neural network is a simple multilayer perceptron with only a few (three to five) inputs and a single output. The number of neurons in the middle layer is usually not important as long as memorization and overtraining are avoided. The inputs must have some degrees of correlation with the quantity to be inferred, because using an input with no relation to the output would only deteriorate the quality of the measurement.

Examples of where inferential measurements have been used advantageously include the following examples:

1. Inferential measurements of nitrous oxide emissions from a gas-line pumping station have been used to demonstrate compliance with regulatory requirements. This avoided the placement of a chemical analysis unit at each of many pumping stations that often are located at remote sites as well as the need for technical personnel required to carry out the measurements.

2. Inferential measurements of feedwater flow in a nuclear power plant (an important quantity in the thermal power calibration) have been carried out using a neural network to map four related inputs to the flow. The neural network is trained using data gathered immediately after the venturi flowmeter has been cleaned and calibrated. As the venturi fouls due to water chemistry phenomena, a 1–2% difference develops between the inferred (correct) value and the (incorrect) measurement by the fouled flowmeter.

3. Inferential measurements are also being used for sensor validation. Again the process is one of mapping several related inputs to a single measured quantity. If the actual measurement deviates significantly from the value predicted by the trained neural network, then sensor failure or deterioration may be involved.

8.7   AN ALTERNATE APPROACH TO NEURAL NETWORK TRAINING

In Section 8.3, the traditional approach to backpropagation training of a neural network was examined. The neural network was a traditional three-layer network with the input layer as a buffer layer and the hidden and output layers having nonlinear activation functions (a logistic function in this case). Such a network has been shown by Kolmogorov (1965) to be capable of mapping any arbitrary function into any other arbitrary function, i.e., it is a *universal approximator*. This holds true for neural networks with several hidden layers, each having neurons with nonlinear activation functions. While such configurations were shown to be sufficient for the arbitrary mappings of a universal approximator, they were not shown to be necessary.

In the past few years, a simpler network configuration has been shown to be equally effective in performing arbitrary mappings (Cybenko, 1989; Funahashi, 1989; Hornik, 1989). This network has one or more hidden layers with nonlinear activation functions, but the output layer has a linear activation function. Because of these linear activation functions of the neurons in the output (kth) layer, its output vector Φ_k is proportional to the summation vector I_k. Since it is common practice to set the constant of proportionality equal to unity (because this constant would simply scale the weights associated with the output layer), we now have Φ_k equal to I_k, which in turn is equal to the desired output vector T if we set the error vector ε equal to zero.

Such a configuration was explored extensively almost a decade ago by Lapedes and Farber and found to be very useful, especially when the output of the network were analog variables (Lapedes, 1988). This configuration is proposed here as an alternative to conventional backpropagation training that takes advantage of the linearity of the output of the output layer of neurons to speed up training. Generally, this procedure is implemented using a matrix type software program such as MATLAB[©3] so that a set of training vector pairs (X, T), the corresponding layer summation and output vectors I_j, I_k, Φ_j, and Φ_k, and the error vector ε are handled as matrices of vectors for the z pairs of training samples.

Let us apply the methodology and notation of Section 8.3 to such a network with a single hidden layer. (It is equally applicable to the last layer weight matrix in neural networks with more than one hidden layer.) The process starts in the same manner as cumulative backpropagation with the randomization of the weights, the application of the input matrix X of the training set (X, T) containing z patterns to the input layer, and the calculation of the output matrix of the hidden (j-*th*) layer Φ_j.

Let us assume as a starting point that the error matrix ε is zero. Now we have the target matrix T equal to the output Φ_k, which in turn is equal to I_k because of the linear activation function. We now have a deterministic input–output relationship for the output layer weight matrix W_k. The middle layer output matrix Φ_j (which we have just calculated) is the input and T

---

[3]MATLAB is a registered trademark owned by MathWorks Inc., Natick, MA.

(which is obtained from the training set) is the output, both related by

$$\mathbf{T} = \mathbf{\Phi}_k = \mathbf{W}_k^* \, \mathbf{\Phi}_j \tag{8.7-1}$$

which can be solved for the weight matrix $\mathbf{W}_k$ either exactly by matrix inversion or approximately by regression. Because $\mathbf{T}$ and $\mathbf{\Phi}_j$ either represent or were calculated from experimental data, $\mathbf{\Phi}_j$ usually cannot be inverted, and regression is really the only practical method for solving for $\mathbf{W}_k$.

This regression calculation of $\mathbf{W}_k$ is a first approximation of the weight matrix $\mathbf{W}_k$. It is dependent on the weight matrix $\mathbf{W}_j$ that was created by a randomization process. The critical question then is whether a neural network with such a combination of $\mathbf{W}_j$ and $\mathbf{W}_k$ can model adequately the process from which the training data set was obtained. Masters (1993, p. 170) has pointed out that

> The essence of neural networks is that they activate hidden neurons based on patterns in the input data. What we are really interested in is the weights that connect the inputs to the hidden layer (and interconnect hidden layers if more than one is used). Once these weights are determined, computation of the weights that lead to the output layer is almost an afterthought.

This view is further enhanced by recent work of Lo (1996) that indicates that the essence of the representation of the model of a system is contained in the weights associated with the hidden layer(s) (whose neurons have nonlinear activation functions), a view consistent with the concept of the hidden layer being a "feature detection" layer as described in Chapter 7. Hence, we need do more than create the hidden layer weights by a randomization process. Indeed, randomization of weights is only a convenient starting condition that avoids some training problems and is useful only when associated with a long training process (e.g., backpropagation) that allows these randomized hidden layer weights to be adjusted sufficiently that they represent adequately the system model. Hence, we need to proceed with a modified form of backpropagation in which a regression method is used to solve for the weight matrix $\mathbf{W}_k$. The weight matrix $\mathbf{W}_j$ is then iteratively determined using backpropagation.

One reason that conventional backpropagation training is very slow to converge is that the error terms are propagated back through the output layer weights to the hidden layer to provide an error correction. When these error terms are backpropagated through the non-optimal output layer weights, the changes in the hidden layer weights are far from optimal. Hence, the use of a regression method to provide a good first approximation of $\mathbf{W}_k$, albeit based on randomized weights, greatly speeds up the training process.

Use of Regression to Solve for the Weight Matrix $\mathbf{W}_k$

We can solve for the weight matrix $\mathbf{W}_k$ in equation (8.7-1) using a least square error regression method. Given the input matrix $\mathbf{\Phi}_j$, the output matrix

I_k and the weight matrix W_k, a standard regression equation for equation (8.7-1) is

$$\Phi_k = W_k \Phi_j \qquad (8.7\text{-}2)$$

which we can use to solve for the weight matrix using the general least squares procedure

$$W_k = \left(\Phi_j \Phi_j'\right)^{-1} \Phi_j' I_k \qquad (8.7\text{-}3)$$

In this solution, the mean squared error is minimized with extremely high precision. The weight values produced by this algorithm may contain very large values (5 or 6 orders of magnitude) that result in very poor network generalization. Instead of learning the general trend of the data, the network has also learned the noise. The fit to the training data would be excellent, but the generalization would be very poor. There are several methods by which this least-squares operation may be carried out properly; the method suggested as best by Masters (1993), "Singular Value Decomposition," is discussed below.

At this point, we have a neural network whose state of training is much better than the typical neural network after the first iteration of backpropagation training. We now proceed with conventional backpropagation as described in Section 9.3, modified to include the regression method for calculating W_k. This combination of a good starting state and an algorithm for optimally calculating one layer of weights speeds up the training process dramatically. Application of this methodology has sped up the training of neural networks 40-fold (Uhrig, 1996), and resulted in networks with superior generalization capabilities. After this initial pass through the network, we start again with the application of the input matrix X of the training set to the buffer input layer and calculate our way through the network to produce the hidden layer output matrix Φ_j. Then we use the regression method to calculate the changes in W_k and conventional backpropagation to calculate the change in W_j. Then the weight matrices are updated, and the input matrix X is applied to the input layer, starting a new cycle of this hybrid training process. As the training proceeds, the features of model of the system under study is embedded in the weight matrix W_j associated with the hidden layer(s).

Singular Value Decomposition

The hidden layer output matrix Φ_j is broken down into its singular value composition given by

$$\Phi_j = USV^T \qquad (8.7\text{-}4)$$

where

$\mathbf{\Phi}_j = z \times n$ matrix of hidden layer outputs

$\mathbf{U} = z \times n$ matrix of principle components

$\mathbf{S} = n \times n$ diagonal matrix of single values

$\mathbf{V} = n \times n$ matrix of right singular values (orthonormal matrix)

$\mathbf{T} = z \times n$ matrix of target outputs

In this method, only the most relevant information is retained to compute the weights. The least important information is discarded because it is most likely to result from noise. The amount of noise that is removed from the solution to the system is determined relative to the largest weight expected in the network. This is achieved by altering the diagonal matrix of singular values ($\mathbf{S}$). Singular values in the matrix $\mathbf{S}$ that are less than a cutoff value (e), are changed to zero in the inverse matrix. The weights are then calculated by

$$\mathbf{W}_k = \mathbf{V}\mathbf{S}^{-1}\mathbf{U}^T\mathbf{T} \tag{8.7-5}$$

Experimental results show that keeping the magnitude of the weights within ± 10 provides excellent network generalization with no loss of important information.

Adaptation of Models to Changing Conditions

The concentration of the model information in the hidden layer weights pointed out by Lo (1996) as described above also addresses one of the most troublesome problems in using neural networks to model many real-world situations, namely, the nonstationary system with small but often continuous changing of the operating conditions over time. Adapting the neural network model to changing operating conditions without the model representation suffering deterioration can be performed by adjusting the linear output layer weights using the regression method (i.e., singular value decomposition) as described above. This updates effectively the neural network to actual plant conditions (Hines, 1996).

There is, however, risk that using this method for continuous adaptation of the neural network model may mask a continuously deteriorating condition of the system. Hence, it may be useful to have duplicate neural network models, one that is continually adjusted and one that is not adjusted. By monitoring the difference between the outputs of these two models, it is possible to assess whether the changing conditions representation represents deteriorating or normal behavior.

8.8   MODULAR NEURAL NETWORKS[4]

The modular neural network is another rather special extension of the backpropagation network. It is comprised of (a) several independently operating expert networks competing to produce the correct response to individual input vectors and (b) a gating network mediating this competition. Basically, the modular neural network consists of the input layer, a processing "superlayer" comprising the expert networks and the gating network, and the output layer. The input units are fully connected to the input units of both the expert networks and the gating network, but there are no weights in these connections. The expert network output units are fully connected to the modular neural network output units, with connections having weights whose values correspond to the values produced in the output units of the gating network. The output units perform a summation of the weighted incoming signals without applying any activation function to the result. A topology of the modular neural network with two input units, three expert networks, and a single output unit is shown in Figure 8.16.

For the function approximation task the expert networks are typically simple perceptrons (i.e., often backpropagation networks with only two layers) with their output neurons performing only a simple summation without using an activation function. However, the expert networks can also be the classical three-layer backpropagation networks) or even another modular neural network, thus creating a complex hierarchical structure), and their output nodes may use nonlinear activation functions. The structure of the expert networks has to be chosen in such a way that the task of concern cannot be solved by a single expert network, because otherwise the modular neural network would gradually degenerate during training into a network with the structure of this expert network (i.e., all the input vectors would be processed solely by one of the expert networks). In any case, all the expert networks in the modular neural network have to have the same structure with the same number of layers and identical processing units in them.

The gating network is a fully connected feedforward neural network, having typically only two layers. There is one output neuron in the gating network for each expert network in the whole modular neural network (hence, the weights in all connections between the output neurons of each expert network and the output neurons of the whole modular neural network are identical). Similarly to the backpropagation network processing neurons, the gating network output neurons sum the weighted signals received from the input units and filter the result through an activation function. This activation function, however, is the so-called "softmax" function which essen-

---

[4]Part of this section was extracted from a thesis "Modeling a Probabilistic Safety Assessment Using Neural Networks" by Vaclav Hojny, a graduate student at the University of Tennessee, 1993–1995.

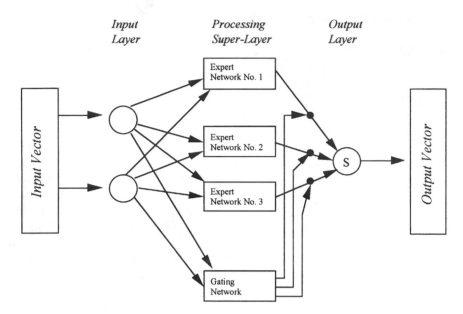

Figure 8.16 Topology of the modular neural network.

tially normalizes activations of output units and amplifies differences between them. The activation function for the softmax function as given in Figure 8.17 is

$$\Phi_N = \frac{e^{I_N}}{\sum_{j=1}^{m} e^{I_j}} \qquad (8.8\text{-}1)$$

Operating of the modular neural network consists of the following: division of a complex task to be solved into several simpler subtasks, finding of separated solutions for these subtasks, and combination of these subsolutions into the desired solution of the original complex task. This approach is sometimes referred to as a principle of "divide and conquer." To achieve this, the modular neural network utilizes a special combination of supervised and unsupervised training based on maximization of the likelihood function, which represents a product of the probabilities of generating the correct output vectors for individual input–output vector pairs. These probabilities are typically modeled using a mixture model (i.e., a linear combination) of multivariate Gaussian distributions, which characterize conditional probabilities of producing the correct output vectors by the individual expert networks for a given input vector from the training set. To maximize the likelihood function, its known parameters have to be optimized—that is, properly adjusted during training. In this modular neural network, these parameters represent (a) the weights in connections of the expert network processing units and (b) the weights in connections of the expert network output units

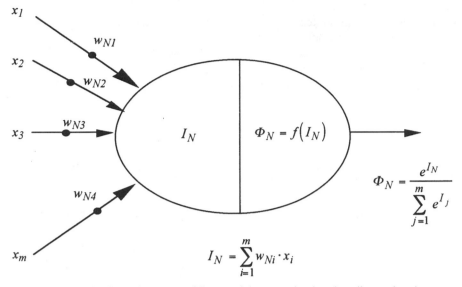

Figure 8.17 Output neuron of the modular neural network gating network.

and the output units of the whole modular neural network. All these weights are trained simultaneously starting from small arbitrary values. During the training, the modular neural network is presented with individual input vectors from the set of training samples to which individual expert networks and gating networks respond with certain output vectors.

The basic training of the expert networks is supervised and utilizes a procedure very similar to that used by the backpropagation neural network. The resulting update values of the weights for each expert network are, however, modified by a probability that the particular expert network is allowed to produce the particular modular neural network output vector. The value of this probability is determined as a product of the distance (typically Euclidean) between the output vector produced by the given expert network and the target output vector multiplied by the value produced by the gating network output corresponding to this expert network. Values of these products, determined for individual expert networks, are then processed through the softmax function with the aim of amplifying the outcome of the competition of the expert networks. The resulting values of the products effectively represent posterior probabilities that the particular expert networks are allowed to produce the particular modular neural network output vector, while the values produced by the corresponding gating network outputs represent prior probabilities of it. The training of the gating network is essentially unsupervised and aimed at minimization of the differences between these prior and posterior possibilities. The training procedure for the gating network is, in fact, the same one used in backpropagation networks. The elements of the target output vector of the gating network are determined by the posterior probabilities of the corresponding expert networks (which are, to a great extent, independent on the modular neural network target output vector). All the input–output vector pairs in the training set are repeatedly presented to the modular neural network until its error decreases to an acceptable level, or until the modual neural network reaches a steady state when the values of its weights stop changing.

8.9   RECIRCULATION NEURAL NETWORKS

Another variation of the basic backpropagation network is the recirculation neural network (RNN) introduced by Geoffrey Hinton and James McClelland (1988) as a neurally plausible alternative to the autoassociative backpropagation network. They considered backpropagation to be neurally implausible and hard to implement in hardware, because it requires that all connections be used backwards, that these connections be symmetrical, and that the units are different input–output functions for the forward and backward passes. In a backpropagation network, errors are passed backwards through the same connections that are used in the forward pass, but they are scaled by the derivative of the feed forward activation function. In a recirculation neural network, data are processed through weights in only one direction.

The recirculation neural network is a four-layer autoassociative type network as shown in Figure 8.18. in which the input and output layers are buffer layers with the same number of neurons. The other two layers are called the "visible" and the "hidden" layers. In a recirculation neural network, the visible and hidden layers are fully connected to each other in both directions by separate links with separate sets of weights. The visible-to-hidden connections involve what is called the *bottom-up weights*, and the hidden-to-visible connections contain the *top-down weights*. Each neuron in the visible and hidden layers is connected to a bias element with a trainable weight.

The learning schedule involves two complete passes between the visible and hidden layers. The learning is carried out using only local knowledge—that is, the state of the processing element and the input values of the particular connection to be adapted. The purpose of the learning rule is to construct in the hidden layer an internal representation of the data presented at the visible layer. Recirculation neural networks use unsupervised learning, in the sense that no desired vector is required to be present at the output layer. A bias term is used for all neurons in the hidden and visible layers. The learning process proceeds in the following manner. Initially, all weights, including the bias weights, are randomly set to small values. The data are first presented at the visible layer (time 0), then filtered through the bottom-up weights to the hidden layer (time 1), and then circulated back to the visible layer through the top-down weights (time 2). Finally, the data are passed for a second time (recirculated) to the hidden layer through the

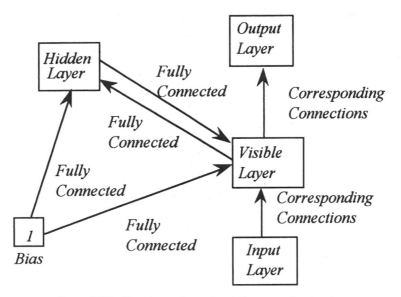

Figure 8.18   Structure of a recirculation neural network.

bottom-up weights (time 3) and back to the visible layer through the topdown weights (time 4) and on to the output buffer layer. If desired, there could be a second output buffer layer connected to the hidden layer where the compressed version of the output could be made available.

Learning occurs only after the second pass through the network. The output of the visible layer at time 2 is the reconstruction of the original input vector from the compressed vector in the hidden layer at time 1. The aim of the learning is to minimize the error between the original input and the reconstructed vector at time 2 by adjusting the top-down weights, as well as to minimize the error between the compressed vectors at times 1 and 3 by adjusting the bottom-up weights. All summations over the hidden layer (times 1 and 3) or visible layer (times 0 and 2) include the bias terms. During training the output of the hidden layer at time 1 is the compressed version of the input data. In Hinton and McClelland's simulations, cumulative learning is used—that is, changes in the weights are accumulated over an epoch—and the actual weights are changed only at the end of an epoch.

The state of the visible layer at time 2 is the top-down response of the network to the initial bottom-up stimulus. Hinton and McClelland used sigmoid functions for the activation functions for both visible and hidden layers, although their analysis assumes that the activation function for the hidden layers is linear and that the activation function for the visible layer is any smooth monotonic nonlinear function with bounded derivatives.

The recirculation neural network has full connectivity between the hidden and visible layers in both directions. Learning for the top-down weights in the connections from the jth hidden layer to the ith visible layer and bottom-up weights in the connections from the ith visible layer to the jth hidden layer are given by

$$\Delta w_{ji} = \varepsilon_{y_j}^{(1)} \left[y_i^{(0)} - y_i^{(2)} \right] \quad \text{(top down weight change)} \quad (8.9\text{-}1)$$

and

$$\Delta w_{ij} = \varepsilon_{y_i}^{(2)} \left[y_j^{(1)} - y_j^{(3)} \right] \quad \text{(bottom up weight change)} \quad (8.9\text{-}2)$$

where $y_i^{(0)}$ is the state of the ith visible neuron at time 0, $y_i^{(2)}$ is the state of the ith visible neuron after the activity has passed around the loop once, $y_j^{(1)}$ is the state of the jth visible neuron at time 1 (first pass around the loop), $y_j^{(3)}$ is the state of the jth neuron at time 3 (second pass around the loop), and $\varepsilon_{y_j}^{(1)}$ and $\varepsilon_{y_i}^{(2)}$ are the errors after recirculation.

This learning process for the recirculation neural network approaches gradient descent under certain specific conditions. The error at each processing element in the visible layer between its state at time 0 and at time 2 is referred to as the "reconstruction error." The learning in the top-down weights seeks to reduce the reconstruction error. Hinton and McCleland have shown that under certain conditions the learning in the bottom-up weights also performs gradient descent learning in the reconstruction error. For the visible-to-hidden connections, the changes are partially related to the gradient descent. Early changes do not necessarily improve the state of the system, but as learning progresses, these changes tend to agree with the gradient descent and total agreement occurs after the hidden-to-visible weights are approximately aligned with the visible-to-hidden weights.

Example 8.3   An Application of Recirculation Neural Networks[5]. One of the applications of the recirculation neural network is to transform narrow peaks in a Fourier transform of undamped vibration data of rotating machinery into a pattern where the information in the peaks is spread over the entire frequency range. An example of the influence of small changes in the amplitude and the frequency for a single narrow peak is shown in the next four figures. In Figure 8.19 we have a single peak at a specific frequency in which the amplitude decreases for the series of six examples. When these six peaks are subject to transformation by the use of the RNN, the results are shown in Figure 8.20, where the individual values in the spectrum have been connected. It is clear that the information contained in the single peak has been spread throughout the frequency range and that the shape changes as the amplitude is decreasing. However, this transformation is even more drastic when there is a small shift in the frequency. In this case, small changes in the frequency in the original spectrum in Figure 8.21 produces drastic changes in the transformed spectrum shown in Figure 8.22. $\square$

---

[5] Results presented in Figures 8.19 through 8.22 were produced by Dr. Israel E. Alguidingue, when he was a graduate student at the University of Tennessee, 1989–1993.

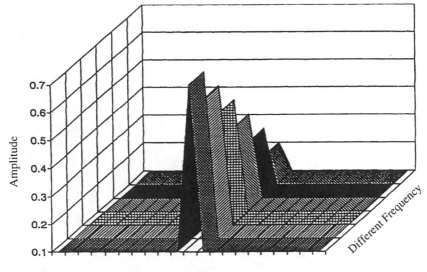

Figure 8.19 Single peaks in a power spectral density plot for several different amplitudes at a specific frequency.

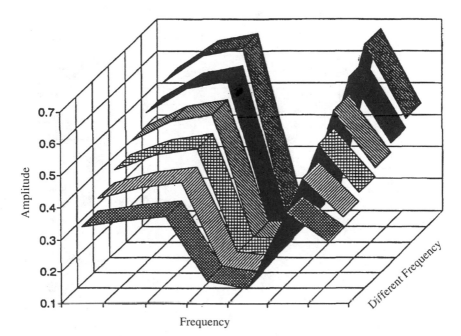

Figure 8.20 Results of processing of peaks in Figure 8.14 with a recirculation neural network.

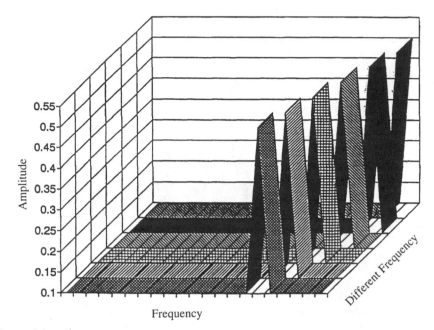

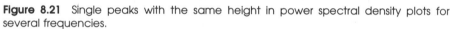

Figure 8.21 Single peaks with the same height in power spectral density plots for several frequencies.

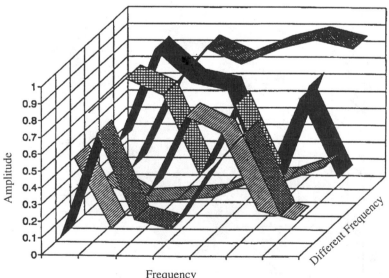

Figure 8.22 Results of processing of peaks in Figure 8.17 with a recirculation neural network.

8.10 FUNCTIONAL LINKS

In conventional backpropagation networks, weights are applied only to connecting links leading to neurons. The hidden layer in backpropagation networks provides the ability to form complex relationships between input pattern elements. However, if the data pattern presented is in a form that already has complex elements in the form of functional links, then the hidden layer may not be necessary. In functional link networks, developed by Yoh-Han Pao (1989) connections (links) provide information to the network by incorporating a representation of the relationships between the input and output patterns. This involves adding inputs that are functions of the normal inputs. While it is possible for backpropagation to learn complex relationships (e.g., $x^2, xy, \cos^2(x), \sin(x)$, etc.), functional link networks establish these relationships directly. The difficulty is knowing which functions to use. Generally, this requires an understanding of the nature of the problem involved. If the problem can be represented by a polynomial, then simple power and cross terms (e.g., $x^2, xy, x^2 y^3$, etc.) may be appropriate as additional inputs to the network. If a problem has cyclic terms for frequencies that are important, then sine and cosine terms may be appropriate. Functional link networks are feedforward networks that use standard backpropagation training. Clearly, the output layer neurons must have nonlinear activation functions if there are only two layers in the network.

There are two kinds of functional links. The first type is the outer product (tensor) model where each component of the input is multiplied by the entire input vector $x_i (1 \le i \le n)$ or $x_i x_j$, where $i \le j \le n$ and $i \le j$. Representation of the input space is enhanced, making it easier for the model to learn.

The second general type of functional link is functional expansion where the input variables are individually acted upon by the appropriate functions —that is, $\sin(x)$, $\cos(x)$, $\sin(2x)$, and so on. The functions selected may be a subset of the orthonormal basic functions. The overall effect is to map the input vector into a larger pattern space, enhancing the representation. Of course, it is possible to combine the tensor and functional expansion types of functional links.

Although functional links offer an attractive analytical alternative to the general problem of specifying the architecture of a network, they have their own limitations. The principal concern with functional links are the following:

1. As the number of inputs increases, the number of connections (with weights) increases.

2. There are indications that a smaller number of training examples relative to the number of connections can influence the ability of the network to generalize.

3. With fewer examples per connection, the model may learn to reproduce the training set errors and not generalize. This problem is especially difficult with noisy data.

8.11 CASCADE-CORRELATION NEURAL NETWORKS

Cascade-correlation is a supervised learning algorithm for neural networks that adjusts the network architecture as well as the weights (Fahlman and Lebiere, 1990). Cascade-correlation starts with a minimal network and adds new hidden neurons one-by-one, creating a multilayer network in the learning process. Once a new hidden neuron has been added to the network, its input-side weights are fixed, and it becomes a permanent part of the network, helping to serve the function of hidden layers (i.e., feature detection). This architecture attempts to overcome the issues which cause backpropagation to be so slow in training a neural network. Two specific issues that are addressed are (1) the step-size problem and (2) the moving target problem.

The "step problem" arises in backpropagation because only infinitesimal changes during the learning process (which implies an infinite training time) can reasonably ensure that a global minimum can be reached. Large changes, which would speed up the training process, tend to cause backpropagation to reach local minima; and various methods, such as the use of momentum of simulated annealing, must be used to reach a global minimum.

The "moving target problem" arises because each neuron in the interior of the network is trying to evolve into a feature detector that will play some useful role in the network's overall computation, but its task is complicated by the fact that it cannot communicate to other neurons to which it is connected (both directly and indirectly), which are changing all the time. One way of decreasing the moving target problem is to allow only a few of the weights in the network to change at once. In a sense, this is reducing the dimensionality of the training process.

There are two related primary features of the cascade-correlation training process: (1) the cascade architecture in which hidden neurons with fixed (nontrainable) inputs are added to the network one at a time and (2) the learning algorithm which creates and installs the new hidden neurons. All neurons have bias inputs with trainable weights and nonlinear activation functions which may be any of the sigmoidal functions. All weights are initially randomized between -1 and $+1$.

The cascade-correlation neural network starts as a two-layer, fully connected perceptron with adjustable weights on every connection which are initially randomized. The direct input–output connections are trained using the Widrow–Hoff delta training rule (or any other training algorithm for two layer networks such as Quickprop). Training is terminated when the weight values approach an asymptotic value, based on a "patience" parameter set by the used. Then the overall error is measured, and a decision is made as to whether to continue training.

Training is continued by adding a single neuron to create a hidden layer. It is connected to all input neurons through connections with fixed weights and to all output neurons through trainable (randomized) weights. The fixed input weighs of the new neuron are set by a "pretraining" process before the outputs are connected to the output layer. In this pretraining process, a number of training sets are applied to this single neuron, and the input weights are adjusted after each pass to maximize the sum (over all outputs) of the magnitude of the correlation between the neurons output and the residual output error of the neuron.

Training proceeds the same as previously because the fixed input weights allow the network to be treated as if it were a two-layer network. When the training stops because the "patience" parameter is reached, the overall error is calculated and a decision is again made whether to proceed. If so, another single neuron is added, in the same manner as the first neuron, fully connected with fixed weights (set by the optimization pretraining process described above) to the inputs, and fully connected with trainable (randomized) weights to the output layer. However, the only connection between the two added neurons is from the output of the first neuron to the input of the second neuron through a fixed weight. This means that we have effectively added a second hidden layer with a single neuron.

The process of adding one neuron at a time continues until the user is satisfied with the overall error of the network for the training data. The multitude of single-neuron hidden layers presents a very powerful feature detector, but it also leads to a large fanout of the input connections and a very "deep" network. Fahlman and Lebiere (1990) indicate that strategies for addressing these issues are being investigated.

8.12 RECURRENT NEURAL NETWORKS

The backpropagation neural networks previously discussed are strictly "feedforward" networks in which there are no feedbacks from the output of one layer to the inputs of the same layer or earlier layers of neurons. However, such networks have no memory since the output at any instant is dependent entirely on the inputs and the weights at that instant.

There are situations (e.g., when dynamic behavior is involved) where it is advantageous to use feedback in neural networks. When the output of a neuron is fed back into a neuron in an earlier layer, the output of that neuron is a function of both the inputs from the previous layer at time t and its own output that existed at an earlier time—that is, at time $(t - \Delta t)$, where Δt is the time for one cycle of calculation. Hence, such networks exhibit characteristics similar to short-term memory, because the output of the network depends on both current and prior inputs.

Neural networks that contain such feedback are called *recurrent neural networks*. Although virtually all neural networks that contain feedback could be considered as recurrent networks, the discussion here will be limited to those that use backpropagation for training (often called "recurrent backprop networks"). Let us consider the elementary feedforward network shown in Figure 8.23*a*, where the input, middle, and output layers each have only one neuron, and where neuron *h* is a buffer neuron that instantaneously send the input *x* to neuron *p*. When the input $x(0)$ (*x* at time 0) is applied to the input, the outputs of neurons *p* and *q* at time (0), $v(0)$, and $y(0)$, respectively, are

$$v(0) = \left\{ \Phi\left[w_{ij} x(0) \right] \right\} \tag{8.12-1}$$

$$y(0) = \Phi\left\{ w_{jk}\left[v(0) \right] \right\} = \Phi\left\{ w_{jk}\left\{ \Phi\left[w_{ij} x(0) \right] \right\} \right\} \tag{8.12-2}$$

where Φ is the activation function operator (usually a sigmoidal function) and (0) indicates the value at time 0.

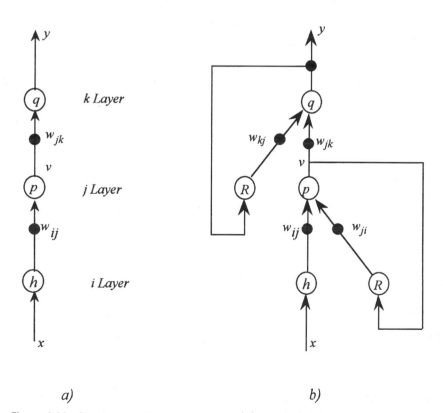

Figure 8.23 Simple neural networks without (*a*) and with (*b*) recurrent feedback.

For nonrecurrent networks (no feedback), this relationship remains valid for times $1, 2, 3, \ldots, n, \ldots, N$, and so on. Hence above equation becomes

$$v(n) = \{\Phi[w_{ij}x(n)]\} \tag{8.12-3}$$

$$y(n) = \Phi\{w_{jk}[v(n)]\} = \Phi\{w_{jk}\{\Phi[w_{ij}x(n)]\}\} \tag{8.12-4}$$

To make this network recurrent, we add feedback from the output neuron to the middle layer and from the middle neuron to the input layer through the recurrent neurons (which are buffer neurons) labeled R and the corresponding weights w_{kj} and w_{ji}, respectively, as shown in Figure 8.23b. The outputs to neurons p and q must exist before there can be any feedback. Hence, as the process proceeds step by step, the feedback term of neurons p and q will not come into play until time 1. Hence, the equation (8.11-2) for $y(0)$ is valid for a recurrent network for time 0, but the feedback terms must be added for all subsequent times. The output of the network for neurons p and q at time 1 are

$$v(1) = \Phi\{w_{ij}[x(1)] + w_{ji}[v(0)]\}$$

$$= \Phi\{w_{ij}x(1) + w_{ji}\{\Phi[w_{ij}x(0)]\}\} \tag{8.12-5}$$

$$y(1) = \Phi\{w_{jk}v(1) + w_{kj}y(0)\}$$

$$= \Phi\{w_{jk}\{\Phi\{w_{ij}x(1) + w_{ji}\{\Phi[w_{ij}x(0)]\}\}$$

$$+ w_{kj}\{\Phi\{w_{jk}\Phi[w_{ij}x(0)]\}\}\}\} \tag{8.12-6}$$

For time 2, the outputs of neurons p and q are

$$v(2) = \Phi\{[w_{ij}x(2)] + [w_{ji}v(1)]\}$$

$$= \Phi\{[w_{ij}x(2)] + w_{ji}\{\Phi[w_{ij}x(1)] + w_{ji}\{\Phi[w_{ij}x(0)]\}\}\} \tag{8.12-7}$$

$$y(2) = \Phi\{[w_{jk}v(2)] + [w_{kj}y(1)]\}$$

$$= \Phi\{w_{jk}\{\Phi[w_{ij}x(2)] + w_{ji}\{\Phi[w_{ij}x(1)] + w_{ji}\{\Phi[w_{ij}x(0)]\}\}\}$$

$$+ w_{kj}\{\Phi\{w_{jk}\{\Phi[w_{ij}x(1)] + w_{ji}\{\Phi[w_{ij}x(0)]\}\}$$

$$+ w_{kj}\{\Phi\{w_{jk}\{\Phi[w_{ij}x(0)]\}\}\}\}\}\} \tag{8.12-8}$$

For time 3, the outputs of neurons p and q are

$$v(3) = \Phi\{[w_{ij}x(3)] + [w_{ji}v(2)]\}$$
$$= \Phi\Big\{[w_{ij}x(3)] + w_{ji}\Big\{\Phi\{[w_{ij}x(2)] + w_{ji}\{\Phi[w_{ij}x(1)]$$
$$+ w_{ji}\{\Phi[w_{ij}x(0)]\}\}\}\Big\}\Big\}$$

$$(8.12\text{-}9)$$

$$y(3) = \Phi\{[w_{jk}v(3)] + [w_{kj}y(2)]\}$$
$$= \Phi\Big\{w_{jk}\Big\{\Phi\{[w_{ij}x(3)] + w_{ji}\{\Phi\{[w_{ij}x(2)] + w_{ji}\{\Phi[w_{ij}x(1)]$$
$$+ w_{ji}\{\Phi[w_{ij}x(0)]\}\}\}\}\}\Big\} + w_{kj}\Big\{\Phi\{w_{jk}\{\Phi[w_{ij}x(2)]$$
$$+ w_{ji}\{\Phi[w_{ij}x(1)] + w_{ji}\{\Phi[w_{ij}x(0)]\}\}\}\}$$
$$+ w_{kj}\{\Phi\{w_{jk}\{\Phi[w_{ij}x(1)] + w_{ji}\{\Phi[w_{ij}x(0)]\}\}$$
$$+ w_{kj}\{\Phi\{w_{jk}\{\Phi[w_{ij}x(0)]\}\}\}\Big\}\Big\}\Big\}$$

$$(8.12\text{-}10)$$

Magnitude of Terms

Note that the equation (8.11-10) for $y(3)$ has $x(0)$, $x(1)$, $x(2)$, and $x(3)$ as inputs. An equation for $y(4)$ would add $x(4)$ to the list of inputs; $y(5)$ would add $x(5)$, and so on. Furthermore, an examination of the terms of the equation for $y(3)$ indicates that the magnitudes of the earlier inputs decrease when later inputs are added. This is seen more readily if we assume that the activation function is linear—that is,

$$\Phi\{x\} = x \qquad (8.12\text{-}11)$$

Then the equations for $y(0)$, $y(1)$, $y(2)$, and $y(3)$ become

$$y(0) = w_{jk}w_{ij}[x(0)] \qquad (8.12\text{-}12)$$

$$y(1) = w_{jk}w_{ij}\{[x(1)] + [w_{ji} + w_{kj}][x(0)]\} \qquad (8.12\text{-}13)$$

$$y(2) = w_{jk}w_{ij}\{[x(2)] + [w_{ji} + w_{kj}][x(1)]$$
$$+ [w_{ji}^2 + w_{ji}w_{kj} + w_{kj}^2][x(0)]\} \qquad (8.12\text{-}14)$$

$$y(3) = w_{jk}w_{ij}\{[x(3)] + [w_{ji} + w_{kj}][x(2)]$$
$$+ [w_{ji}^2 + w_{ji}w_{kj} + w_{kj}^2][x(1)]$$
$$+ [w_{ji}^3 + w_{ji}^2 w_{kj} + w_{ji}w_{kj}^2 + w_{kj}^3][x(0)]\} \qquad (8.12\text{-}15)$$

Since the weights are usually less than 1, the increasing power of the weights for the earlier terms [w^2 for $x(3)$, w^3 for $x(2)$, w^4 for $x(1)$, and w^5 for $x(0)$] causes the coefficients to decrease rapidly. When the sigmoidal (or any nonlinear activation function) is used, the presence of multiple nonlinear activation functions in the above equations for $y(0)$, $y(1)$, $y(2)$, and $y(3)$, each with a maximum value of 1, reduces the early terms much faster than for a linear activation function. As later inputs are introduced, the influence of the earlier terms become negligible. This reduced weighting of the earlier terms is analogous to the decrease of influence of earlier values in a convolution transformation.

The complexity introduced by feedback connections, even for the elementary system of Figure 8.23a, is readily apparent. For networks more complicated than the elementary system in Figure 8.23b, the same principles are applied, but the complexity grows even more rapidly. However, the increase in complexity is often compensated for because the feedback almost always drastically reduces the number of cycles needed to train a neural network significantly. Feedback can often be used advantageously to speed up the training of a neural network and to avoid local minima. Indeed, it is sometimes possible to train a neural network after feedback has been added, whereas it may not have been previously possible to train it to the desired level of error. However, capturing dynamic behavior in a model is the most common justification for the use of feedback, and recurrent neural networks (or neural networks with delayed inputs) are almost always used for dynamic signals.

REFERENCES

Amari, S.-I, Characteristics of Random Nets of Analog Neuron-like Elements, *IEEE Transactions on Systems, Man and Cybernetics*, Vol. SMC-2, pp. 643–647, 1972.

Fahlman, S. E., Faster-Learning Variations of Backpropagation: An Empirical Study, *Proceedings of the 1988 Connectionist Models Summer School*, Morgan Kaufmann, San Mateo, CA, 1988.

Fahlman, S. E., and Lebiere, C., The Cascade-Correlation Learning Architecture, in *Advances in Neural Information Processing Systems* 2, D. Touretzky, ed., Morgan Kaufmann, San Mateo, CA, 1990, pp. 534–532.

Guha, A., Neural Network Based Inferential Sensing and Instrumentation, in *Proceedings of the* 1992 *Summer Workshop on "Neural Network Computing for the Electric Power Industry*, D. J. Sobajic, ed., Stanford, CA, August 17–19, 1992a.

Guo, Z., and Uhrig, R. E., Use of Artificial Neural Networks to Analyze Nuclear Power Plant Performance, *Nuclear Technology*, Vol. 99, pp. 36–42, July 1992.

Hashem, S., Sensitivity Analysis for Feedforward Artificial Neural Networks with Differentiable Activations Functions, in *Proceedings of the IJCNN*, Baltimore, MD, June 1992.

Hines, J. W., Wrest, D. J., and Uhrig, R. E., "Plant-Wide Sensor Calibration Monitoring," Proceedings of the *1996 IEEE International Symposium on Intelligent Control*, Detroit, MI, Sept. 15–18, 1996.

Hinton, G., and McClelland, J., Learning Representations by Recirculating," in *Proceedings of the 1987 IEEE Conference on Neural information Processing Systems—Natural and Synthetic*, American Institute of Physics, New York, 1988, pp. 358–366.

Hojny, V., Modeling a Probabilistic Safety Assessment Assessment Using Neural Networks, M. S. Thesis, University of Tennessee Library, Knoxville, TN, 1995.

Jacobs, R. A., Increased Rates of Convergence Through Learning Rate Adaptation, *Neural Networks*, Vol. 1, pp. 295–307, 1988.

Kramer, M. A. Nonlinear Principal Component Analysis Using Autoassociative Neural Networks, *Journal of the American Institute of Chemical Engineers*, Vol. 37, pp. 233–243. 1991.

Kramer, M. A. Autoassociative Neural Networks, *Computers in Chemical Engineering*, Vol. 16, No. 4, pp. 313–328, 1992.

Masters, T. *Practical Neural Network Recipes in C + +*, Academic Press, San Diego, CA, 1993.

Minai, A. A., and Williams, R. D., Acceleration of Backpropagation Through Learning and Momentum, in *Proceedings of the International Joint Conference on Neural Networks*, Vol. 1, 1990, pp. 676–679.

Pao, Y-H., *Adaptive Pattern Recognition and Neural Networks*, Addison-Wesley, Reading, MA 1989.

Parker, D., Learning Logic, Invention Report, S81-64, File 1, Office of Technology Licensing, Stanford University, Palo Alto, CA, 1972.

Parker, D., Optimal Algorithms for Adaptive Networks: Second Order Back Propagation, Second order Direct Propagation, and Second Order Hebbian Learning, *Proceedings of the IEEE First International Conference on Neural Networks*, Vol. II, San Diego, CA, 1987, pp 593–600.

Poggio, T., and Girosi, F., Regularization Algorithms for Learning that Are Equivalent to Multiple-Layer Networks, *Science*, Vol. 247, pp. 978–982, 1990.

Rogers, S. K., and Kabrisky, M., *An Introduction to Biological and Artificial Neural Networks for Pattern Recognition*, SPIE Optical Engineering Press, Belingham, WA, 1992.

Rumelhart, D. E., Hinton, G. R., and Williams, R. J., Learning Internal Representations by Error Propagation, in *Parallel Distributed Processing*, Vol. 1, D. E. Rumelhart, and J. L. McClelland, eds., MIT Press, Cambridge, MA, 1986.

Samad, T., Backpropagation is Significantly Faster If the Expected Value of the Source Unit Is Used for Update, *The International Network Society Conference Abstracts*, 1988.

Samad, T., Backpropagation Extensions, Honeywell SSDC Technical Report, 1000 Boone Ave. N., Golden Valley, MN 55427.

Sejnowski, T., and Rosenberg, C., Parallel Networks that Learn to Pronounce English Text, *Complex Systems*, Vol. 1, pp. 145–168, 1987.

Stornetta, W., and Huberman, B., An Improved Three-Layer Backpropagation Algorithm, *Proceedings of the IEEE First International Conference on Neural Networks*, Vol. II, San Diego, CA, 1987, pp. 637–644.

Tsoukalas, L. H., Ikonomopoulos, A., and Uhrig, R. E., Generalized Measurements in Neuro-Fuzzy Systems, in *Proceedings of the International Workshop on Neural Fuzzy Control*, Muronan, Japan, March 22–23, 1993.

Uhrig, R. E., Hines, J. W., Black, C., Wrest, D. J., and Xu, X., "Instrumentation Surveillance and Calibration Verification System, Final Report, Sandia National Laboratory contract AQ-6982 (Contractor: University of Tennessee), March 1996.

Upadhyaya, B. R., and E. Eryurek, Application of Neural Networks for Sensor Validation and Plant Monitoring, *Journal of Nuclear Technology*, Vol. 97, pp. 170–176, February 1992.

Wald, A., Sequential Tests on Statistical Hypotheses, *Annals of Mathematical Statistics*, Vol. 16, pp. 117–186, 1945.

Werbos, J. P., *The Roots of Backpropagation*, John Wiley & Sons, New York, 1994.

Werbos, P., Beyond Regression: New Tools for Prediction and Analysis in the Behavioral Sciences, Ph.D. Dissertation, Harvard University, Boston, MA, 1974.

Wrest, D., Hines, J. W., and Uhrig, R. E., Instrument Surveillance and Calibration Verification to Improve Nuclear Power Plant Reliability and Safety Using Autoassociative Neural Networks, *Proceedings of the International Atomic Energy Agency Specialist Meeting on Monitoring and Diagnosis Systems to Improve Nuclear Power Plant Reliability and Safety*, Barnwood, Gloucester, United Kingdom, May 14–17, 1996.

PROBLEMS

1. The backpropagation training algorithm is developed in Section 8.3, using the logistic function as the activation function where the derivative has the convenient form given in Equation 8.1-4. Derive the backpropagation training algorithm for the case where the activation function is an arctan function where the derivative is given by Equation 8.1-6.

2. Derive the backpropagation training algorithm for the case where the neurons in the hidden layer have a logistic function for the activation function and the neurons in the output layer have linear activation functions. Compare the results with those obtained in Section 8.7 where this arrangement of activation functions are used.

3. In the recurrent neural network of Section 8.12, the feedback of both loops comes into action at time 1. If the feedback from the output neuron does not come into action until time 2, how will the Equations 8.11-3 through 8.11-15 change?

4. Discuss the results shown in Figures 8.19 through 8.22 for a recirculation neural network and their implications.

5. In the operation of backpropagation, contrast how the changes of learning constant, changes of momentum coefficient, and changes in the α value in the logistic function influence the time required to train a neural network.

6. Derive the backpropagation training algorithm for the simple recurrent neural network shown in Figure 8.23b using the approach given in Section 8.3.

7. Discuss the role of the "bottleneck" layer in a five-layer autoassociative neural network with respect to the identification of principal components. (*Hint*. see references by Kramer and McAvoy at the end of this chapter.)

8. Compare the reduced representation of an input vector in the hidden layer of a recirculation neural network with the reduced representation of in input vector in the middle layer of three- and five-layer autoassociative neural networks.

9

COMPETITIVE, ASSOCIATIVE, AND OTHER SPECIAL NEURAL NETWORKS

9.1 HEBBIAN LEARNING

Donald Hebb (1949) introduced a nonmathematical statement of biological learning in 1949. The Hebbian system was the first truly self-organizing system developed. Even today, it is very prevalent throughout the neural network field because there are many paradigms based on Hebbian learning. Hebb's law can be summarized as follows:

As A becomes more efficient at stimulating B during training, A sensitizes B to its stimulus, and the weight on the connection from A to B increases during training as B becomes sensitized to A.

One problem with Hebb's law is that it is too vague. Questions such as the following arise: How much should a weight increase? How active does B need to be for training to occur? Furthermore, there is no way for the weights to decrease. In theory, they can increase to infinity. Inhibitory synapses are not allowed, whereas it is well known that real biological systems clearly have inhibitory (negatively weighted) connections.

Corrections to Hebb's law involve normalizing the weights to force them to stay within limited bounds and forcing them to both increase and decrease to retain the normalization. There are many variations of Hebbian learning that are utilized. One of these is the Neo-Hebbian learning put forward by Steven Grossberg, who developed an explicit mathematical statement for the weight change law of the form

$$w_{AB}^{\text{new}} = w_{AB}^{\text{old}}(1 - \alpha) + \beta x_B x_A \qquad (9.1\text{-}1)$$

where w_{AB} is the weight on the synapse connecting neuron A with neuron B, α is the "forgetting" term that accounts for the fact that biological systems forget slowly with time, and β is a "learning" constant that accounts for simultaneous firing of neurons A and B. The right-hand term is called the "Hebbian learning term" because it ties the learning rate to the product of the neuron outputs. Hebbian learning is characterized by the product of two neuron activities. Hence, anytime such a product appears in an equation, Hebbian learning is involved. Generally, both α and β are in the range between 0 and 1.

If we rearrange equation (9.1-1) and put it into the form

$$\frac{w_{AB}^{\text{new}} - w_{AB}^{\text{old}}}{\Delta t} = \frac{\Delta w_{AB}}{\Delta t} = \frac{dw_{AB}}{dt} = -\alpha w_{AB}^{\text{old}} + \beta x_B x_A \qquad (9.1\text{-}2)$$

and consider only the terms involving w_{AB}, it is clear that the forgetting term involves a slow exponential decay with time constant α. Even so, neo-Hebbian learning does not permit the weights to decrease when the neuron outputs decrease.

To overcome this problem, Grossberg introduced differential Hebbian learning. It has the same mathematical form as Hebbian learning of equation (9.1-1) except that it uses the product of rates of change in the outputs for the neurons A and B, as given

$$w_{AB}^{\text{new}} = w_{AB}^{\text{old}}(1 - \alpha) + \beta \frac{dx_B}{dt} \frac{dx_A}{dt} \qquad (9.1\text{-}3)$$

9.2 COHEN–GROSSBERG LEARNING

Pavlov's Experiments

Cohen–Grossberg learning comes from an attempt to mathematically explain the observations from psychological conditioning experiments carried out by Pavlov. Let us consider the various types of conditioning. The first is "observational conditioning," which involves copying the actions of others and is sometimes described as "monkey see, monkey do." The second type of conditioning is "operational conditioning," which involves an action and a response. It is described as "push button, get food." The third type of conditioning is "classical conditioning," and it involves a stimulus and a response. The experiments of Pavlov fall into the classical conditioning category.

The psychological model used in the Cohen–Grossberg learning is Pavlovian learning in which a dog is offered food at the same time a bell rings. Eventually, the dog associates food with the bell ringing and salivates when the bell rings even when no food is presented. This behavior is illustrated in

Table 9.1 Stimuli and corresponding responses in Pavlov's experiment

Stage	*Stimulus*	*Response*
1.	Unconditional Stimulus ====> (Plate of Food)	Unconditioned Response (Dog Salivates)
2.	Unconditional Stimulus (Plate of Food) *plus* ====> Conditioned Stimulus (Bell Rings)	Conditioned Response (Dog Salivates)
3.	Conditioned Stimulus ====> (Bell Rings)	Conditioned Response (Dog Salivates)

Example 9.1 to be Grossberg outstar learning. The three stages of Pavlovian learning are shown diagrammatically in Table 9.1.

Instars and Outstars

Next, we must develop the concept of "instars' and "outstars" to explain Pavlovian conditioning or learning. Every neuron receives hundreds or thousands of inputs through its own synapses from the axon collaterals of other neurons. Schematically, this can be represented as a "star" with radially inward paths called the instar. Indeed, every artificial neuron in a neural network is an instar.

Every neuron also sends out hundreds or thousands of collaterals which branch off from the main axon and go to the synapses of other neurons. This also can be represented by a "star" with radial outward paths. This configuration is called an "outstar," and again every neuron is effectively an outstar. A geometrical interpretation of the instar and outstar configurations is shown in Figure 9.1. The instar has many inputs and a single output whereas the outstar has a single input and many outputs. The fanout of a neuron in the

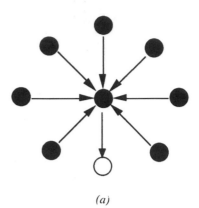

(a)

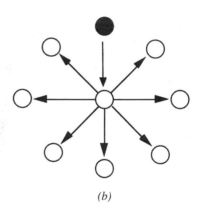

Figure 9.1 Graphical representations of an
"instar" (a) and an "outstar" (b).

(b)

input layer of a neural network can be considered as an outstar, whereas a
neuron in the output layer can be considered as an instar.

Development of Cohen–Grossberg Learning Equations—
Instar Activity

Let us consider Pavlovian learning from the perspective of an instar. The
activity of the instar processing element or neuron has a number of requirements:

1. The activity must grow when there is an external stimulus.
2. It must rapidly decrease if it is no longer stimulated from the outside.
3. It must respond to stimuli from other neurons in the network.

Let us consider the arrangement in Figure 9.2, in which an instar y_j receives

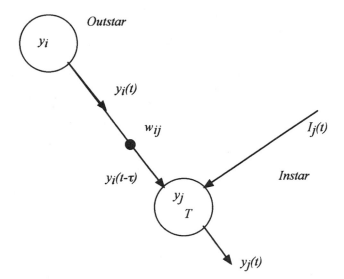

Figure 9.2 An instar y_j receives a signal from the outstar y_i through a weight w_{ij}. (T is the threshold for the incoming signal to node y_j, and τ is the time required for the signal to travel from y_i to y_j.)

signals $y_i(t)$ from outstars y_i through weights w_{ij}. The activity of the instar $y_j(t)$ can be represented by a differential equation

$$\frac{dy_j(t)}{dt} = -Ay_j(t) + I_0 + B \sum_{i=1}^{n} w_{ij} y_i(t) \qquad (9.2\text{-}1)$$

where $y_i(t)$ is the activity of the ith neuron, I_0 is the external stimulus, w_{ij} is the weight between the ith and jth neurons, and A and B are constants. The first term on the right-hand side of equation (9.2-1) allows the activity of the instar to decrease exponentially with a time constant A when it is no longer stimulated by I or inputs from other neurons. The second term I_0 corresponds to an external stimulus, and the third term represents the stimuli from the n neurons in the network. We need to allow for signals received at neuron j that were actually generated in the neuron i at some previous time τ ago and transmitted to neuron j, where τ is the "average" transmission time to from neuron i to neuron j. We also need to put a threshold (T) on the intraneuron inputs so that random noise will not interfere with the network's operation. Hence, we can modify equation (9.2-1) as follows:

$$\frac{dy_j(t)}{dt} = -Ay_j(t) + I_0(t) + B \sum_{i=1}^{n} w_{ij}[y_i(t-\tau) - T]^+ \qquad (9.2\text{-}2)$$

where the superscript $+$ means that only positive values are used.

Instar's Learning Law

The two processes involved in instar's learning law are Hebbian learning and forgetting. We need a process that will explicitly allow forgetting to occur—that is, to allow the weight to slowly decay. Also, we need to put a threshold on the incoming activity term and to account for the transmission time between neurons i and j. Hence, we can write the instar learning law, which controls the adjustment of the weights between neurons i and j in the form given in Equation (9.1-2)

$$\frac{dw_{ij}(t)}{dt} = -Fw_{ij}(t) + Gy_j(t)[y_i(t-\tau) - T]^+ \qquad (9.2\text{-}3)$$

where F is the forgetting time constant that should be much smaller (i.e., a slower decay rate) than the activity decay constant A, and G is the gain or learning constant. The value of F is never greater than about 0.01, and the superscript $+$ means that we should use only the positive values.

Equation (9.2-3) incorporates the "simple" Hebbian learning, and hence it typically does not allow the weights to decrease except for the very slow decay process associated with the forgetting term. A more appropriate Cohen–Grossberg learning law would be one in which the first derivative of the activities with respect to time are substituted for the activities—that is,

$$\frac{dw_{ij}(t)}{dt} = -Fw_{ij}(t) + G\frac{dy_j(t)}{dt}\left[\frac{dy_i(t-\tau)}{dt} - T\right]^+ \qquad (9.2\text{-}4)$$

This is a version of the differential Hebbian learning given in equation (9.1-3).

Grossberg Learning in Outstars

The minimum number of artificial neurons that must be activated to cause recall of a complex spatial pattern is only one, the hub of the outstar. Repeatedly applying a stimulus on the hub neuron and simultaneously putting a pattern stimuli on m neurons on the rim of the outstar (see Figure 9.1b), or to a grid of neurons (see Figure 9.3), each connected to the hub neuron, eventually will cause the weight pattern to reflect the input pattern on the rim due to the Hebbian learning. The mathematical mode of spatial learning in outstars describes the outcome of the standard psychological experiments conducted by Pavlov.

Grossberg outstar learning is based on Hebbian learning; that is, if a stimulus arrives at a receiving neuron at the time when it is active, then the weight associated with that link will be increased. In outstar learning, the weight increase depends on the product of the input and the output signals to the grid neuron; that is, the grid neurons corresponding to bright spots in a

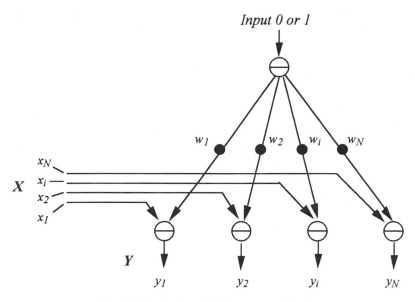

Input 0 or 1

Figure 9.3   An "outstar" learning network.

pattern will have large outputs at the time the stimulus arrives. Hence, the weights will be increased. The grid neurons corresponding to medium spots will have lower outputs, and the weights will change less. After a number of cycles, bright spots will correspond to large weights while medium spots will correspond to medium weights. How about the dark spots? They are a problem. The use of neo-Hebbian learning adds a forgetting term. Hence, weights subject to neo-Hebbian learning that do not increase will slowly decrease.

Example 9.1   Grossberg Outstar Learning. This example illustrates "outstar learning." Consider the neural network shown in Figure 9.3 that has a single outstar neuron and N instar neurons in the second layer. The input to the outstar is a binary signal that switches alternatively between 0 or 1 with a period Δ. The output is a vector **Y** having N components $y_1, y_2, y_i, \ldots, y_N$. The desired output is a vector **X** having components $x_1, x_2, x_i, \ldots, x_N$. The weights $w_1, w_2, w_i, \ldots, w_N$ are to be adjusted using the Hebbian learning algorithm.

Initially, the weights are set to randomly small values. When a 1 (which is considered a "high") is applied to the outstar neuron at the same time the vector **X** (whose components may be "high," "low," or any value in between) is applied to the second layer, Hebbian learning requires that the weight on the connections between two neurons increase in proportion to the product of the magnitudes of the two weights. Hence, those weights on connections leading to neurons in the second layer with large values of x_i increase significantly since the outstar output is unity. The weights on the connections to neurons with medium values of x_i increase somewhat less. The weights on the connections to neurons with low values of x_i increase only slightly or do not increase at all. This process occurs each time the outstar signal switches to 1. The changes in weights become smaller as the weights increase and eventually stop increasing. Now, the magnitudes of the weight vector components w_i mimic the magnitudes of the corresponding values of the input vector components x_i.

At this point, the desired output vector **X** can be removed. When the input is 0, all the components y_i of output vector **Y** are equal to 0. When the input is 1, the components y_i of the output vector **Y** are proportional to components w_i of the trained weight vector **W**; that is, the output vector **Y** mimics the desired output vector **X** even though it is no longer applied. Hence, except for a constant of proportionality, the output vector **Y** is the same as the desired vector **X**. This means that all that is necessary to produce the desired output at the second layer is to apply a 1 to the outstar.

If we return to the analogy with Pavlovian learning, the pattern **X** is analogous to the food, the unconditioned stimulus; the input to the outstar is analogous to ringing the bell, the conditioned stimulus; the stimulus of the grid output is analogous to dog salivating when the food is presented initially, the unconditioned response; and the output **Y** after the pattern **X** is eliminated is analogous to the dog salivating after the food is eliminated, the conditioned response. □

Driver Reinforcement Learning

Driver reinforcement learning is a variation of outstar learning that uses differential Hebbian learning in which the weight increase depends on the product of the change in the output signal of the receiving neuron and a time-weighted sum of the changes of the inputs to that neuron over a period of time. The weight used in weighing the sum of the changes is the weight of the neuron at that time.

During training, we artificially cause grid neurons to display the image we want the network to reproduce. Hence, grid neurons corresponding to bright spots on the pattern have large outputs when the outstar stimulus is received, and weights are increased with each repetition. Eventually, the outstar's stimulus alone is sufficient to cause the neurons to produce the pattern without the input. This process is the essence of driver reinforcement learning.

9.3   ASSOCIATIVE MEMORIES

An associative memory is any memory system that stores information by associating each data item with one or more other stored data items. The characteristics of associative memories are such that they usually store information in a distributed form. They are content addressable memories; that is, data are accessed by its content, not by its address. They are robust and can usually handle garbled or incomplete data inputs and can usually operate with some failed elements. In many ways, they are very similar to human memory.

Data are stored as patterns of activity in an associative memory. As a result, associative memories are insensitive to minor differences in details. This provides the robustness which allows garbled inputs to be still understood, and minor errors or damage to the network do not cause loss of functionality.

It is useful to distinguish between heteroassociative and autoassociative memories. In a heteroassociative memory network, the input X and the output Y are different patterns; that is, the input and output are not the same. In the case of the autoassociative memory networks, the input and the output patterns are the same. At first glance, this may seem like a trivial system, but it is very useful. When the input to an autoassociative memory is somewhat garbled (e.g., it is a distorted version of X), the network will produce a "correct" version of X as the output.

While associative memories do not have to involve neural networks, there are several types of neural networks that constitute associative memories. The most common types are the crossbar associative memory and the adaptive filter (e.g., the Adaline neural network can be used as an associative memory). There are other architectures in neural networks that can be used as associative memories, but they are not common. Generally, associative networks are important because they provide robust and efficient storage of pattern data and are generally considered to be essential to any "intelligent" system.

Another classification of associative memories is that they are accretive or interpolative. This indicates how they interpret data. Suppose we have an associative memory that associates the color red with the numerical value 1, the color blue with 2, and the color green with 3; that is, if we put the color magenta into the associative network, an accretive associative network will return the value of either 1 or 2, depending upon whether the shade of magenta is closer to red or blue. If the magenta corresponded to a value of 1.6, then the accretive network would give an output of 2. On the other hand, an interpolative associative memory would return the actual numerical value (1.6 in this case).

Crossbar Structure

Crossbar networks have the structure of an early twentieth-century telephone exchange system from which they take their name. They typically have one or two layers of artificial neurons, and each neuron or layer is fully interconnected with the other neurons or layers. Crossbar representations that are commonly used are matrix representation, energy surface representation, and a feedback competition representation. The matrix representation is perhaps the most common because the weights are stored as elements of a matrix. The matrix weight representation discussed here is substantially the same as those given in Figures 7.6 and 7.7. This representation is popular because matrix mathematics and operations are well understood by most researchers. They are also mathematically tractable, allowing simple explanations of characteristics. If we consider the fully connected network shown in Figure 9.4, the input is a column vector $\mathbf{X}$ with components x_1, x_2, x_3, and x_4, and the output is a column vector $\mathbf{Y}$ with components y_1, y_2, and y_3; then we can say that

$$\mathbf{Y} = \mathbf{W} \cdot \mathbf{X} \tag{9.3-1}$$

where $\mathbf{W}$ is the weight matrix. If we expand the terms in equation (9.3-1), it becomes

$$\begin{bmatrix} y_1 \\ y_2 \\ y_3 \end{bmatrix} = \begin{bmatrix} w_{11} & w_{21} & w_{31} & w_{41} \\ w_{12} & w_{22} & w_{32} & w_{42} \\ w_{13} & w_{23} & w_{33} & w_{43} \end{bmatrix} \cdot \begin{bmatrix} x_1 \\ x_2 \\ x_3 \\ x_4 \end{bmatrix} \tag{9.3-2}$$

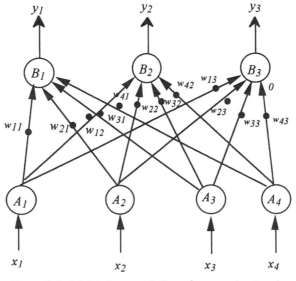

Figure 9.4 Matrix representation of a neural network.

Bidirectional Associative Memory

Mathematically, a bidirectional associative memory (BAM) developed by Kosko (1988) is also a matrix; technically it is a crossbar network with symmetric weights. Each neuron in each layer has one input from the outside and inputs from each of the neurons in the other field of the neurons.

Suppose we construct a BAM to store three pattern pairs: $[\mathbf{X}_1, \mathbf{Y}_1]$, $[\mathbf{X}_2, \mathbf{Y}_2]$, and $[\mathbf{X}_3, \mathbf{Y}_3]$. Since a BAM is bidirectional we can enter any $\mathbf{X}$ and retrieve the corresponding $\mathbf{Y}$, or we can enter any $\mathbf{Y}$ and retrieve the corresponding $\mathbf{X}$.

The process in a BAM is fundamentally different than the operation of other types of neural networks. For instance, in the backpropagation network discussed previously, the weights are trained to provide the desired input–output mapping. In the case of a BAM, the weight matrix is not trained; it is constructed using the input–output pairs. The process involves constructing a matrix for each input–output pair and then combining them into a master matrix. $\mathbf{X}_i$ and $\mathbf{Y}_i$ are treated as column vectors, and then the matrix is produced by taking the product of the $\mathbf{X}_i$ vector and transpose of the $\mathbf{Y}_i$ vector, $\mathbf{Y}_i^T$. Let us consider the three pairs shown below:

$$\mathbf{X}_1: \quad (+1 - 1 - 1 - 1 - 1 + 1) \Leftrightarrow (-1 + 1 - 1) \quad :\mathbf{Y}_1 \quad (9.3\text{-}3)$$

$$\mathbf{X}_2: \quad (-1 + 1 - 1 - 1 + 1 - 1) \Leftrightarrow (+1 - 1 - 1) \quad :\mathbf{Y}_2 \quad (9.3\text{-}4)$$

$$\mathbf{X}_3: \quad (-1 - 1 + 1 - 1 - 1 + 1) \Leftrightarrow (-1 - 1 + 1) \quad :\mathbf{Y}_3 \quad (9.3\text{-}5)$$

Weight Matrix Representation   Since $\mathbf{X}_1$ has 6 elements and $\mathbf{Y}_1$ has 3 elements, the matrix for each set of inputs result in a 6×3 matrix. It is important to note that each of the patterns is made up of $+1$ and -1 values, which means that the components are bipolar. If the patterns values are binary (i.e., made up of 1 and 0 values), they should be converted to bipolar form by substituting -1 for each 0 before they are used in a BAM. The correlation matrices $\mathbf{M}_i$ for equations (9.3-3), (9.3-4), and (9.3-5) are obtained by cross product of $\mathbf{X}_i$ and $\mathbf{Y}_i$—that is,

$$\mathbf{M}_i = \mathbf{X}_i \times \mathbf{Y}_i^T \tag{9.3-6}$$

The three correlation matrices are

$$\mathbf{M}_1 = \mathbf{X}_1 \times \mathbf{Y}_1^T = \begin{bmatrix} +1 \\ -1 \\ -1 \\ -1 \\ -1 \\ +1 \end{bmatrix} \times \begin{bmatrix} -1 & +1 & -1 \end{bmatrix} = \begin{bmatrix} -1 & +1 & -1 \\ +1 & -1 & +1 \\ +1 & -1 & +1 \\ +1 & -1 & +1 \\ +1 & -1 & +1 \\ -1 & +1 & -1 \end{bmatrix}$$

$$(9.3\text{-}7)$$

$$\mathbf{M}_2 = \mathbf{X}_2 \times \mathbf{Y}_2^T = \begin{bmatrix} -1 \\ +1 \\ -1 \\ -1 \\ +1 \\ -1 \end{bmatrix} \times [\,+1 \quad -1 \quad -1\,] = \begin{bmatrix} -1 & +1 & +1 \\ +1 & -1 & -1 \\ -1 & +1 & +1 \\ -1 & +1 & +1 \\ +1 & -1 & -1 \\ -1 & +1 & +1 \end{bmatrix}$$

$$(9.3\text{-}8)$$

$$\mathbf{M}_3 = \mathbf{X}_3 \times \mathbf{Y}_3^T = \begin{bmatrix} -1 \\ -1 \\ +1 \\ -1 \\ -1 \\ +1 \end{bmatrix} \times [\,-1 \quad -1 \quad +1\,] = \begin{bmatrix} +1 & +1 & -1 \\ +1 & +1 & -1 \\ -1 & -1 & +1 \\ +1 & +1 & -1 \\ +1 & +1 & -1 \\ -1 & -1 & +1 \end{bmatrix}$$

$$(9.3\text{-}9)$$

Note that each value in the above matrices is a product of two quantities, one component of $\mathbf{X}$ and one component of $\mathbf{Y}$. This product $x_i \cdot y_j$ is a classical indication that Hebbian learning is involved.

In order to obtain an associative weight memory (called the master weight matrix) capable of storing the three pairs in equations (9.3-3), (9.3-4), and (9.3-5), we simply add the three correlation matrix equations (9.3-7), (9.3-8), and (9.3-9). The result is

$$\mathbf{M} = \mathbf{M}_1 + \mathbf{M}_2 + \mathbf{M}_3 \qquad\qquad (9.3\text{-}10)$$

$$\mathbf{M} = \begin{bmatrix} -1 & +3 & -1 \\ +3 & -1 & -1 \\ -1 & -1 & +3 \\ +1 & +1 & +1 \\ +3 & -1 & -1 \\ -3 & +1 & +1 \end{bmatrix} \qquad\qquad (9.3\text{-}11)$$

Matrices can be added only if they are the same size. Hence, this means that all of the $\mathbf{X}_i$ vector patterns must have the same number of components, and all of the $\mathbf{Y}_i$ vector patterns must have the same number of components. However, the number of components in the $\mathbf{X}_i$ pattern can be different from the number of components in the $\mathbf{Y}_i$ patterns (as is the case in this example).

In order to put in any $\mathbf{X}_i$ and get back any $\mathbf{Y}_i$ (or put in any $\mathbf{Y}_i$ and get back any $\mathbf{X}_i$), we have to take the product of the input vector and the matrix. This is equivalent to taking the dot product of the vectors and the master matrix. The result is

$$\mathbf{X}_i = \mathbf{M} \cdot \mathbf{Y}_i \qquad\qquad (9.3\text{-}12)$$

and

$$\mathbf{Y}_i = \mathbf{M}^T \cdot \mathbf{X}_i \qquad (9.3\text{-}13)$$

where the **M**'s are 6×3 weight matrices, $\mathbf{X}_i$ is a 6×1 column vector, and $\mathbf{Y}_i$ is a 3×1 column vector. Note that we must use the transpose of the master matrix to get $\mathbf{Y}_i$; that is,

$$\mathbf{M}^T = \begin{bmatrix} -1 & +3 & -1 & +1 & +3 & -3 \\ +3 & -1 & -1 & +1 & -1 & +1 \\ -1 & -1 & +3 & +1 & -1 & +1 \end{bmatrix} \qquad (9.3\text{-}14)$$

Example 9.2   Using a Bidirectional Associative Memory. Let us use equation (9.3-12) to obtain $\mathbf{X}_2$ from $\mathbf{Y}_2$:

$$\mathbf{X}_2 = \begin{bmatrix} -1 & +3 & -1 \\ +3 & -1 & -1 \\ -1 & -1 & +3 \\ +1 & +1 & +1 \\ +3 & -1 & -1 \\ -3 & +1 & +1 \end{bmatrix} \cdot [\,+1 \quad -1 \quad -1\,] = \begin{bmatrix} -3 \\ +5 \\ -3 \\ -1 \\ +5 \\ -5 \end{bmatrix} \Rightarrow \begin{bmatrix} -1 \\ +1 \\ -1 \\ -1 \\ +1 \\ -1 \end{bmatrix} \qquad (9.3\text{-}15)$$

which is the correct pattern. The last step in equation (9.3-15) is accomplished through the use of the threshold rule; that is, the component is -1 if the original value is < 0, and it is $+1$ if the original value is ≥ 0.

Operation of a BAM. The master matrix now has three pairs stored in it. If any of the **X**s or **Y**s are introduced to this matrix in the proper way, the corresponding response is given immediately. The problem comes when the input is a distorted version of **X** (or **Y**) which we will call **X*** (or **Y***) is introduced, especially if **X*** has some similarity to more than one of the **X**s. The initial response obtained as the dot product of **X*** and **M** may not be any of the **Y**s stored in the matrix, but may be some combination of two or more of the **B**s which we will call **Y'**. In turn, **Y'** is sent back through the BAM to give **X'** as the dot product of **Y'** and **M**^{T}. **X'** moves back across the BAM to give **Y"** as the dot product of **X'** and **M**. **Y"** then moves back across the BAM to give **X"** as the product of **Y"** and **M**^{T}. This process continues until an equilibrium condition is attained when successive values of $\mathbf{X}^i$ and $\mathbf{Y}^i$ do not change.

The sequence of events are as follows:

1. An **X** input pattern is presented to the BAM.
2. The neurons in field **X** generate an activity pattern that is passed to field **Y** through the weight matrix **M**
3. Field **Y** accepts input from field **X** and then generates a response back to field **X** through the transpose weight matrix **M**^{T}

4. Field **X** accepts the return response from **Y**, and then it generates a response back to field **Y** through the weight matrix **M**.
5. The activity bounces back and forth until a "resonance" is achieved, which means that no further changes in the patterns occur (i.e., successive values of **X**s and **Y**s are the same). At this point, the output **Y** is one of the **Y** values stored in the master matrix, and it is the correct response for the distorted **X** input.

In summary, we constructed a master matrix with three pairs of input patterns $[X_1, Y_1]$, $[X_2, Y_2]$, and $[X_3, Y_3]$. We transposed this matrix (if necessary, depending upon which half of the pair we used as input) and then applied it to the input patterns. The result following thresholding was the other half of the pattern pair. This apparently arbitrary methodology always generates a memory matrix from which we can recall the input patterns used to produce it. Where are the patterns actually stored in the BAM matrix? They are not stored in any individual element of the weight matrix, because if we were to change one of the three patterns and reconstruct the matrix, we would get an entirely different weight matrix with virtually every element changed. Changing an individual pattern doesn't just change one row or one column of the matrix; it changes every element. We must therefore conclude that the information is stored not in an individual elements but in the matrix as a whole, and each pattern is distributed over the entire matrix.

Adding and Deleting Pattern Pairs to the Master Matrix. We can add another pattern pair $[X_4, Y_4]$ to our matrix by adding its matrix M_4 to get to the memory matrix **M**:

$$\text{New } M = M_1 + M_2 + M_3 + M_4 \qquad (9.3\text{-}16)$$

Alternately, we can "forget" or erase a pattern pair by subtracting the matrix for that pattern pair from the memory matrix. For instance, if we wanted to remove the pair $[X_2, Y_2]$ from the memory, we could do it by subtracting the matrix M_2 from the memory matrix:

$$\text{New } M = M - M_2 \qquad (9.3\text{-}17)$$

This system has all the requisite features of a memory system. It can store into memory, it can recall from memory, it can write new information, and it can erase old information. □

Capacity and Efficiency of a Crossbar Network

The capacity of a crossbar network of size $N \times N$ neurons is theoretically limited to approximately N patterns. In reality, the actual capacity of the crossbar networks is more on the order of 10–15% of N. In the matrix example cited earlier, 288 bits were needed (18 elements of 16 bits per element) to store three patterns of 9 (total) bits each, or 27 bits total. The storage and recall operation required several major matrix operations (multiplication, transposition, and addition). With current technology this is much less efficient than simply storing the data conventionally.

Disadvantages of Crossbars

There are many disadvantages of crossbars that need to be considered. These are as follows:

1. *The Number of Connections.* A 100-node network has 100×100, or 10,000, total connections.
2. *Binary-Only Input.* To implement an analog problem, some suitable transformation must be used to convert analog quantities to binary signals.
3. *Capacity.* The theoretical storage is low for the number of connections, and the real storage capacity is even much lower.
4. *Orthogonality.* For best results, the stored data patterns should be as orthogonal as possible to minimize the overlap.
5. *Spurious Results.* In energy surface representation, spurious minima or "energy wells" that have nothing to do with the problem are sometimes produced. These are the so-called "localized minima."

There are, however, some mitigating circumstances. Near-orthogonality is usually adequate because the capacity is so low. There are few spurious minima because low capacity implies sparse coding (lots of zeros) in the data. Finally, the efficiency could be dramatically improved when practical optical systems become a reality.

Hopfield Networks

Dr. John Hopfield is the person perhaps most responsible for the rejuvenation of the neural network field after publication of *Perceptrons*. His contributions include work conceptualizing neural networks in terms of an energy model (based on spin glass physics). He showed that an energy function exists for the network and that processing elements with bistable outputs are guaranteed to converge to a stable local energy minimum. His presentation at the National Academy of Science meeting in 1982 triggered the subsequent large-scale interest in neural networks. A crossbar associative network is called the Hopfield network in his honor (Hecht-Nielsen 1990).

A typical Hopfield network is shown in Figure 9.5. It has only one computing layer, called the Hopfield layer, and the other two layers are the input and output buffers. In contrast to the backpropagation network discussed earlier, the Hopfield network has feedback from each neuron to each

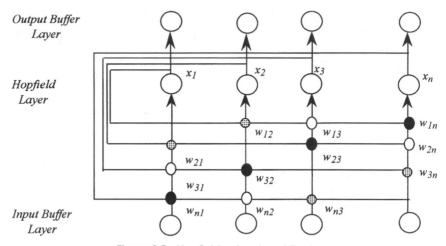

Figure 9.5   Hopfield network architecture.

of the other neurons, but not to itself. The Hopfield layer of neurons computes the weighted sum of the inputs, and it quantizes the output to 0 or 1. (This restriction was later relaxed.) The activation function used was a sigmoid with a reactive (resistor–capacitor) delay. An examination of this network shows that the weights are symmetrical; that is,

$$w_{ij} = w_{ji} \tag{9.3-18}$$

The basic Hopfield learning rule is

$$\Delta w_{ij} = (2x_i - 1)(2x_j - 1) \tag{9.3-19}$$

where x_i and x_j have values of 0 and 1, x_j is the current neuron, x_i is the input to the neuron, and w_{ji} is the connection between the jth neuron and the ith neuron. Furthermore, symmetry dictates that weight changes are symmetrical; that is,

$$\Delta w_{ji} = \Delta w_{ij} \tag{9.3-20}$$

Examination of equation (9.3-19) shows that the learning is Hebbian; that is, the change in weight is the product of two activities, and the change in weight is proportional to this product. Since x_i and x_j can only have values of 0 and 1, then the $(2x - 1)$ terms effectively convert the binary inputs and outputs into bipolar inputs and outputs; that is, the 1 remains a 1 and the 0 becomes a -1. The connections are strengthened (i.e., made more excitatory) when the output of a neuron is the same as the input (i.e., both are 0 or both are 1). Connections are weakened (i.e., made more inhibitory) when the input differs from the output of the neuron (when one input or output is a 0 and the other is a 1). Hopfield has shown that for a large number of neurons the upper bound for the memory capacity is $0.5 \times (N/\log N)$, where N is the number of processing elements.

Energy Surface Representation

The key to the popularity of crossbar networks is that the "state" of the network can be represented by an "energy surface" in which data storage corresponds to "sculpting" energy minima in the energy surface. This view is mathematically equivalent to well-understood physical systems known as "spin glasses." Each energy well in the energy surface has a corresponding area within which all states will move to the bottom of that well or "basin of attraction."

A crossbar associative (Hopfield) memory operates by attracting the network state to an energy minimum. If we consider the energy function of the crossbar associative memory as a foam sheet with dents of different depths, the bottoms of the dents are the energy minima. They correspond to the data stored in the crossbar network. It is sometimes said that the network state is falling down into the nearest energy well, which may or may not be the global minimum.

Simulated Annealing

Simulated annealing is a process used in neural networks to reach a global minimization of an error function. It is analogous to the annealing process in metallurgy in which a metal is heated beyond a transition temperature, allowing the preexisting structure to change physically (relieving residual stresses, changing the metallographic structure, eliminating dislocations and disruptions in the crystal lattice, etc.) due to thermal agitation. Then the temperature is lowered slowly to room temperature, allowing the metal structure to slowly go through a transformation and grow structures by which it attempts to attain a global "minimum energy" configuration. In practice, the annealing process does not take place suddenly, but instead the transition starts almost simultaneously at many locations, creating many homogeneous regions that are usually separated by dislocations. Hence, there is no guarantee that the final energy level will be lower, but it usually is lower.

When a minimization process is trapped in a spurious local minimum, one of the few ways to get out of this trap is to add noise to the function until it literally escapes the minimum. This is equivalent to raising the temperature in the annealing process. When the high noise level has driven the function away from the spurious local minimum, the noise level (temperature) can be gradually lowered, allowing the function to gradually approach a global minimum. The success of this process is dependent upon the temperature used and the programmed cooling rate. If the global minimum is not reached, the process can be repeated as many times as necessarity using other temperatures and cooling rate curves. The details of this process are discussed in detail in several references (Hecht-Nielsen, 1990; Korn, 1992; Aarts and Korst, 1989).

It is common to combine simulated annealing with other minimization or training processes. An example of this was discussed in Section 8.4 of Chapter 8, in which simulated annealing was used after the backpropagation training process had become stuck in a local minimum. After simulated annealing has moved the process away from the local minimum, backpropagation was resumed to complete the training of the neural network.

Stochastic Neural Networks

Stochastic neural networks use noise processes in their operation in an effort to reach a global minimum of an error function. The process involved in virtually all statistical networks is simulated annealing. Examples of statistical neural networks are the Boltzmann machine and the Cauchy machine. The Boltzmann machine is a discrete-time Hopfield net in which the processing element transfer function is modified to use the annealing process. The Cauchy machine is similar to the Boltzmann machine, in which different temperatures, cooling rate patterns, and procedures are used. Both allow the error to increase under some conditions in order to move out of a local minimum.

9.4 COMPETITIVE LEARNING: KOHONEN SELF-ORGANIZING SYSTEMS

"Self-organization" refers to the ability of some networks to learn without being given the corresponding output for an input pattern. Self-organizing networks modify their connection strengths based only on the characteristics of the input patterns. The Kohonen feature map, perhaps the simplest self-organization system, consists of a single layer of neurons (called the Kohonen layer) which are highly interconnected (lateral connections) within the Kohonen layer as well as to the outside world through an input buffer layer that is fully connected to the neurons in the Kohonen layer through adjustable weights.

Lateral Inhibition

Kohonen networks utilize lateral inhibition (i.e., connections between neurons within a layer) to provide (a) positive or excitatory connections to neurons in the immediate vicinity and (b) negative or inhibitory connections to neurons that are further away. The strengths of the connections vary inversely with distance between the neurons, that is, the strengths are stronger when neurons are close, but they are weaker when the neurons are distant. The inputs from lower levels and outputs to higher levels (if any) are the same as for other networks. Generally, there is no feedback from higher to lower layers.

Lateral connections moderate competition between neurons in the Kohonen layer. When an input pattern is presented to the Kohonen layer, each neuron receives a complete copy of the input pattern modified by the connecting weights, and the varying responses establish a competition that flows over the intralayer connections. The purpose of the competition is to determine which neuron has the strongest response to the input. Each neuron in the layer tries to enhance its output and the output of its immediate neighbors and inhibit the output of the remaining neurons that are further away. Lateral connections can cause oscillations in networks, but the output eventually stabilizes with the output of the neuron, with the strongest response being declared the winner and being transmitted to the next layer if there is one. The activity of all other neurons is squashed, as the network determines for itself which neuron has the greatest response to the input pattern. The relative impact of a neuron's interlayer inhibition is also permitted to decrease with training. Initially, it starts fairly large and is slowly reduced to include only the winner and possibly its immediate neighbors. It has been shown that similar systems exist in the brain with regard to vision.

The complexity of intralayer connections makes lateral inhibition and excitation hard to implement. An alternative which is much easier to implement is to use a "max" function to determine the neuron with the greatest response to the input and then assign this neuron a +1 value to the output while assigning a zero to all other neurons in that layer. The winning neuron represents the category to which the input pattern belongs. This is not a true implementation of lateral inhibition, but it generally gives the same result as a true implementation, and it is far more efficient when implemented using serial computers. An even simpler alternative is to merely compute the dot product of each of the weight vectors with the input and then select the winner from this list.

In training, the Kohonen network classifies the input vector components into groups that are similar. This is accomplished by adjusting the Kohonen layer weights, so that similar inputs activate the same Kohonen neurons. Preprocessing the input vectors is very helpful. This involves normalizing all inputs before applying them to the network—that is, divide each component of the input vector by the vector's length:

$$x_i' = \frac{x_i}{\left[x_1^2 + x_2^2 + x_3^2 + \cdots + x_N^2\right]^{1/2}} \tag{9.4-1}$$

When building a Kohonen layer, two new things are required:

1. Weight vectors must be properly initialized. Generally, this means that the weight vectors point in random directions.

2. Weight vectors and input vectors must be normalized to a constant fixed length, usually "unity." Such normalization can cause loss of information in some situations, and there are methods of dealing with it if it occurs.

Let us assume that the weight vectors are randomly distributed and then determine how close each neuron's weight vector is to the input vector. The neurons then compete for the privilege of learning. In essence, the neuron with the largest dot product of the input vector and a weight component is declared the winner. This neuron is the only neuron that will be allowed to generate an output signal; all other neuron outputs will be set to zero. Furthermore, this neuron and its immediate neighbors are the only ones permitted to learn in this presentation. Only the winner is permitted to have an output (i.e., winner takes all).

Kohonen Learning Rule

Determining the winner is the key to training a Kohonen network. Only the winner and its immediate neighbors modify the weights on their connections. The remaining neurons experience no training. The training law used is

$$\Delta w_i = \eta \left[x_i - w_i^{\text{old}} \right] \qquad (9.4\text{-}2)$$

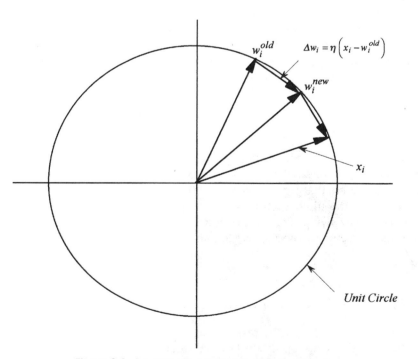

Figure 9.6 Learning in a Kohonen neural network.

where η is the learning constant whose value may vary between 0 and 1, with a typical value of about 0.2, and x_i is the input along the ith connection. It can be shown that this learning rule is a variation of the Woodrow–Hoff learning rule. This is shown graphically in Figure 9.6 for the two-dimensional case. Learning is illustrated for the case of an input x_i and a weight w_i^{old}. The difference between these two unit vectors is a vector from the tip of w_i^{old} to the tip of x_i. In Figure 9.6, this vector is broken into two parts (cords of the unit circle) so that the w_i^{new} will have a unit length. The vector from the tip of w_i^{old} to w_i^{new} represents the change in the weight vector due to learning and is equal to $\eta(x_i - w_i^{old})$.

If we consider the collection of weights for a given neuron as the components of an n-dimensional weight vector **W**, and we consider the corresponding inputs as the components of an m-dimensional input vector **I**, then Kohonen learning merely moves the weight vector so that it is more nearly aligned with the input vector. Since both input vectors and weight vectors are generally normalized to a unit magnitude, each vector points to a position on the unit circle. The winning neuron is the one with the weight vector closest to the input vector. Each training pass nudges the weight vector closer to the input vector. The winner's neighbors also adjust weights using the same learning equation, and their weight vectors move closer to the input vector. Training a Kohonen layer begins with a fairly large neighborhood size that is slowly decreased as training proceeds. The learning constant also starts with a large value and decreases as training progresses.

Let us consider the unsupervised training process for three input vectors, each with eight components and hence eight weights (a small training set solely for illustrative purposes) at three different stages of training: initial random weight distribution, partially trained weights, and fully trained weights. These three conditions are shown in two dimensions in Figure 9.7. Initially, the weight vectors are randomly scattered around the unit circle. As

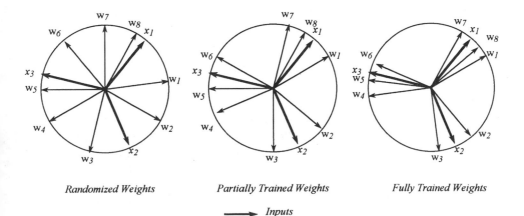

Randomized Weights   Partially Trained Weights   Fully Trained Weights

⟶ Inputs
⟶ Weights

Figure 9.7 Training of weights in a Kohonen neural network.

training proceeds, the weight vectors move toward the nearest input. When the system is fully trained, the weights cluster around the three inputs so that the centroids of the three local weight clusters are on the three input vectors.

Training of a Kohonen Neural Network.

Let us consider a Kohonen neural network [sometimes called a self-organizing map (SOM)] with an input buffer layer (typically a linear array) and a Kohonen layer (typically a rectangular array or grid) that are fully connected. An input vector is applied to the buffer layer, and its component vectors are transmitted to each neuron in the Kohonen layer through randomized connecting weights. The neuron in the Kohonen layer with the strongest response (let's call it neuron q) is declared the winner and its value is set equal to 1. Then the weights connecting all component vectors from the buffer layer to the winning neuron undergo training in accordance with the process shown graphically in Figure 9.6. Neurons immediately adjacent to the winner are also allowed to undergo training. Then a second input vector is applied to the buffer layer, another neuron in the Kohonen layer is declared the winner, its value is set equal to unity, and it and its neighbors are allowed to undergo training. This process continues until all the inputs in the epoch of data have been applied to the buffer input layer. In the training process, the weights tend to cluster around the input vectors as indicated in Figure 9.7. Training stops when a criterion relating the nearness of the weights in the clusters to the relevant input vector is satisfied. Kohonen neural networks train relatively rapidly compared to backpropagation neural networks. Often a single cycle through an epoch of data, especially if the data set is large, constitutes adequate training.

It is important to note that just because the input vector selected neuron q as the most active in the first cycle of training does not mean that it will select neuron q in the second or subsequent cycles of training, because the weights on connections to neuron q change during training and perhaps training caused by adjacent neurons being winners in the first cycle. Furthermore, it is common for a particular Kohonen layer neuron to be the winner for many inputs. During the training process, input vectors that have similar characteristics move into a cluster of neurons in a particular area of the Kohonen layer, often to a single neuron. Other input vectors that have similar characteristics, which are different than those of the first cluster, move toward another area of the Kohonen layer. There will be as many clusters as there are types of inputs if an appropriate-sized rectangular array size is chosen for the Kohonen layer. More clusters require larger rectangular arrays of neurons and, the larger the number of neurons in the Kohonen layer, the longer the training process.

A Kohonen network models the probability distribution function of the input vectors used during training. Many weight vectors cluster in portions of the hypersphere that have relatively many inputs, and few weight vectors cluster in portions of the hypersphere that have relatively few inputs. Kohonen networks perform this statistical modeling, even in cases where no closed-form analytical expression can describe the distribution. The Kohonen network can achieve this modeling spontaneously, with no outside tutor.

Kohonen networks work best when the networks are very large. The smaller the network, the less accurate the statistical model will be. Kohonen neural networks are very fast, even while training. Activation of the network is a single-pass, feedforward flow. Thus, Kohonen networks have the potential for real-time application learning. Kohonen neural networks can literally learn continuously. Hence, if the statistical distribution of the input data changes over time, it can automatically adapt to those changes and continually model the current distribution of the input pattern. The statistical modeling capabilities of the Kohonen network are unmatched by any other neural network.

The learning rate coefficient η is always less than 1; it usually starts at about 0.7 and is gradually reduced during training. If only one input vector were to be associated with each Kohonen neuron, the Kohonen layer could be trained in one calculation per weight. The weights of a winning neuron would be adjusted to the components of the training vector (with $\eta < 1$). Usually, a training set has many input vectors that are similar, and the network should be trained to activate the same Kohonen neuron for each of them.

The weight vector must be set before training begins. It is common practice to randomize these weights to small values. For the Kohonen network training, the randomized weights must be normalized. After training, the weights must end up equal to the normalized input vectors. Prenormalization to unit vectors will start weight vectors closer to their final states, thereby shortening the training process.

The most desirable arrangement is to distribute the weight vectors according to the density of the input vectors that must be separated. This places more weight vectors in the vicinity of the input vectors. Although this is impractical to implement directly, there are several techniques that approximate this ideal arrangement (Masters, 1993).

Example 9.3   Valve Status Classification Using Kohonen Neural Networks.
Uhrig et al. (1994) have reported a method of utilizing an accelerometer spectrum to determine the status of check valves under full flow conditions. The same data have also been applied to a Kohonen SOM to illustrate its ability to classify check valves by type and condition. The procedure involved acquiring an analog time record from an accelerator mounted on a check valve, digitizing the time record, performing a fast Fourier transform of the data to produce many spectra, and then introducing these spectra as input vectors to a Kohonen SOM. Typically, four seconds of data, sampled at 25,000 samples per second and filtered through a band pass filter with 50- and 10,000-Hz cutoff frequencies, were fast Fourier transformed to produce

390 spectra, each with 128 values. Hence the input buffer layer had 128 neurons. The size of the Kohonen rectangular array was varied, depending upon the number, type, and condition of the valves that were being investigated. In several cases, spectral measurements were taken on the same valve when it was faulty or broken and again after it had been repaired. The results of several runs are shown in Figure 9.8 through 9.10.

Figure 9.8 shows a plot of a Kohonen SOM for five 18-inch-diameter valves, three of them (V63, V124, and V251) being identical single-disk swing check valves and two of them (V148 and V150) being identical "duo-check" valves that have two moving vanes, each covering half of the valve opening. Clearly, the spectra for the swing check valves clustered in the lower

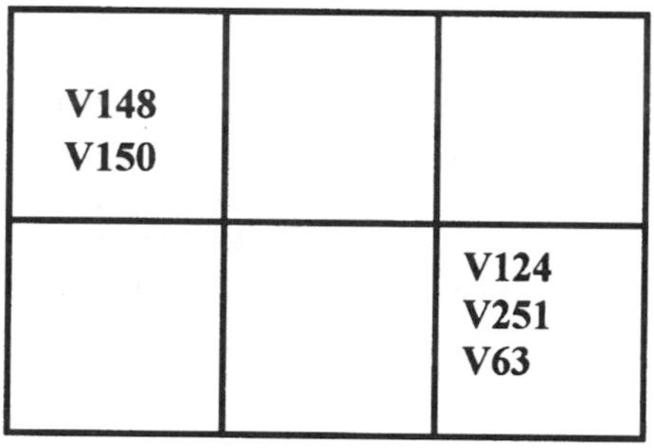

V148 V150		
		V124 V251 V63

V148 and V150 are Identical 18-Inch "Duo-Check" Valves
V63, V124, and V251 are Identical 18-Inch Swing Check Valves

Figure 9.8 SOM plot for 18-inch check valves.

V34 V64 V65		Vo65 Broke Disk
		Vo34 Loose Bolt

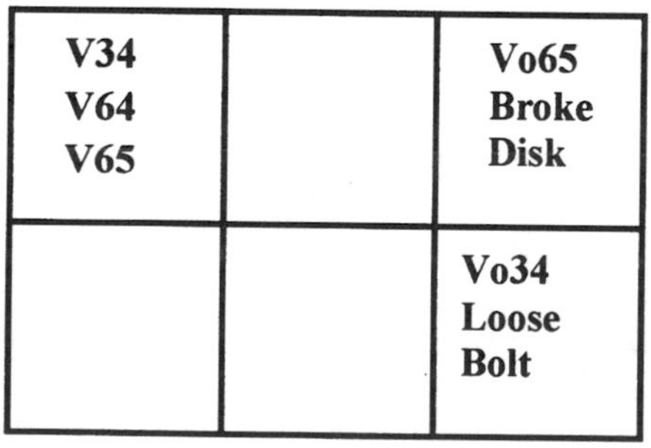

V34, V64, and V65 are Identical 14-Inch Swing Check Valves
Vo34 is V34 Operating with a Loose Bolt in the Disk Structure
Vo65 is V65 Operating with a Broken Disk

Figure 9.9 SOM plot for 14-inch swing check valves.

3-Inch 30 % SP Degrad.		3-Inch 1st Good Valve	3-Inch 2nd Good Valve	3-Inch 30 % HP Degrad.	
					3-Inch Stuck Disk
		Both 4-Inch Good Valves			4-Inch 30 % SP Degrad.
4-Inch 30 % HP Degrad.					4-Inch Stuck Disk

Figure 9.10 SOM plot for similar 3-inch check valves operating under normal and degraded conditions.

right-hand neuron, whereas the spectra for the duo-check valves clustered in the upper left-hand neuron. Identical results were obtained for spectra from accelerometer measurements on the upstream side near the hinge pin and on the downstream side near the backstop.

Figure 9.9 shows a plot of a Kohonen SOM for three identical 14-inch swing check valves, two of them having measurements taken both in a degraded or broken condition and after they were repaired. All the spectra for the good valves clustered in the upper left-hand neuron, whereas the spectra for the degraded and broken valves clustered in the two right-hand neurons. Again, identical results were obtained from accelerometer measurements taken on the upstream and downstream sides of the valve.

Figure 9.10 shows a plot of a Kohonen SOM for similar 3-inch and 4-inch swing check valves that were tested in a test flow loop facility for both normal and degraded conditions. Both valves were tested under normal conditions twice and deliberately subjected to 30% degradation of the hinge pin (a less serious problem), 30% degradation of the stud pin (a more serious problem), and a stuck disk condition (a very serious problem). Spectra for the 3-inch valve tended to cluster in the top rows of neurons, whereas spectra for the

4-inch valve tended to cluster in the bottom rows of neurons. Spectra for normal conditions (with one exception) clustered in the third column of neurons, whereas spectra for degraded conditions (with one exception) clustered in the leftmost or rightmost columns of neurons. In all cases, the spectra for the hinge pin (HP) degraded condition (a lesser problem) clustered closer to the neurons for normal (good) conditions than the spectra for the stud pin (SP) degraded conditions and the stuck disk conditions (both more serious problems).

The choice in the size and arrangement of the Kohonen layers were arrived at by trial and error, although similar results usually were found over a range of sizes. One of the interesting aspects of this work is that there was no way of controlling where the clustering occurred within the Kohonen network. This is to be expected since self-organizing is involved here, which means that the results are dependent only on the data that is introduced into the Kohonen SOM. □

Learning Vector Quantization

A variation of the learning scheme and the addition of an output layer of neurons can make the Kohonen network into a classification network called learning vector quantizer (LVQ). The modification involves changing the training scheme from an unsupervised system to a supervised procedure. This requires a collection of training examples, each assigned to one of a set of known categories or classes. The number of neurons in the output layer is equal to the number of classes in the training data set. The Kohonen layer is trained first using a modified training procedure. LVQ training proceeds in a manner similar to that of Kohonen feature map training. An input pattern is presented to the network, and the winning node is determined by selecting the neuron with weight vector closest to the input vector. This neuron responds with its assigned category and is allowed to update its weights. The Kohonen training law is modified as shown here:

$$\Delta w_i = \eta\left(x_i - w_i^{\text{old}}\right) \qquad \text{(if answer is correct)} \qquad (9.4\text{-}3)$$

$$\Delta w_i = -\eta\left(x_i - w_i^{\text{old}}\right) \qquad \text{(if answer is not correct)} \qquad (9.4\text{-}4)$$

That is, if the winning weight vector (the one closest to the input vector) is the correct category for the input pattern, the weight vector is nudged closer to that input pattern. If, however, the winning vector is the wrong one, the weight change repels the weight vector from the input pattern vector. This should allow another weight vector to win the next time that input pattern, or one similar to it, is presented to the network. LVQ systems, including some more elaborate variations on the basic idea presented here, can achieve performance that is nearly as good as an optimal Bayesian decision system. The system is mathematically simple to implement and does not require knowledge of the probabilities involved (Haykin, 1994). After the training is complete, the activated neurons in the Kohonen layer for input vectors associated with each class are connected directly to the output neuron for that class. Kohonen layer neurons not activated by any input vectors are not connected to the output layer.

9.5   COUNTERPROPAGATION NETWORKS

The counterpropagation network was developed by Robert Hecht-Nielsen in 1987 (Hecht-Nielsen, 1990) as an alternative to the back propagation network. It can reduce training time by a factor of 100, but it is not as general in its application. The counterpropagation network is a combination of two networks, a self-organizing Kohonen network and a Grossberg outstar network. This combination yields properties not available in either alone. In many respects this network can function as a "look-up" table that is capable of generalization. It has a supervised learning process, because the training associates input vectors with the corresponding output vectors (which may be binary or continuous). Once the network is trained, applying an input produces the desired output, even with partially incomplete input. It is useful for pattern recognition, pattern completion, and signal enhancement. The counterpropagation network combines the categorization capability of the Kohonen self-organizing network with the conditioning capabilities of the outstar network.

Robert Hecht-Nielsen, the inventor of counterpropagation, realized its limitations, indicating that counterpropagation is obviously inferior to backpropagation for most applications. Its advantages are its simplicity, the fact that it forms a good statistical model of the input vector environment, its ability to train rapidly, and its ability to save large amounts of computing time. It can be useful for rapid prototyping of systems where great accuracy is not required or a quick approximation is adequate. Furthermore, the ability to generate a function and its inverse is often useful.

Unidirectional Counterpropagation Network

Figure 9.11 shows the connection scheme of a unidirectional counterpropagation network. For clarity, only a few of the input neurons' connections to the middle layer are shown as well as only a few of the middle layers connections to the output layer. At first glance, this appears to be very similar to a fully connected backpropagation network, with connections between the input and output layers that bypass the Kohonen middle layer, but it is very different.

Consider a mapping of pattern A of size n elements to pattern B of size m elements. The objective is to introduce the A pattern to the network and get back to corresponding B pattern. The input layer receives both the A and B patterns, and thus it must be of size $m + n$. The output layer must be able to

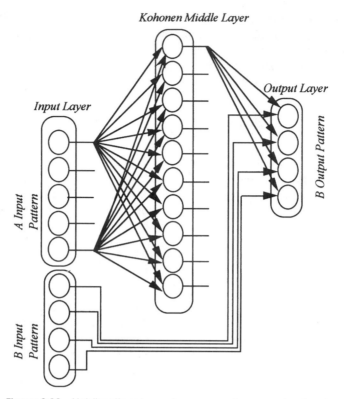

Figure 9.11 Unidirectional counterpropagation neural network.

reproduce only the B patterns so it must be a size m. There is also a direct connection between the B input and the output neurons. These connections are shown in Figure 9.11. The size and geometry of the middle layer also has to be determined.

The input layer is split into two subsections, one which receives the incoming A pattern and the other that receives the incoming B pattern. The middle layer is a competitive layer in which only a single neuron generates an output signal for each input. This output is normally set to $+1$ as it is with the Kohonen network. As a result, each neuron in the output layer merely receives a single signal, $a + 1$ representing the input patterns category, from the middle layer and an output from the B section of the input layer. The connections between the middle layer and the output layer obey the neo-Hebbian (outstar) learning law as demonstrated in Example 9.1, thereby producing the B output pattern when the A pattern is applied to the input layer.

The output layer's main function is to associate the correct output pattern for each category generated by the middle Kohonen layer. Because the outstar uses a supervised learning procedure corresponding to classical conditioning, the direct connections from the input layer's B subsection to the output layer are used to provide each neuron with an "external" input or unconditioned stimulus that defines the correct output for each of middle layer's categories. The single $+1$ signal that arrives from the middle (Kohonen) layer neurons acts as the condition stimulus during the training of the output layer. Hence, the weight on the connection between the Kohonen layer and an output neuron is trained using neo-Hebbian learning.

The operation of a trained unidirectional counterpropagation network can be summarized as follows: An input pattern is presented to the A subsection of the input layer and is categorized by the middle (Kohonen) layer. The output layer treats the category generated by the middle layer as an outstar stimulus, because the output layer itself corresponds to the grid of an outstar network. After training is completed, an input of a particular category presented to the A input section of the network causes the output layer to generate the correct output pattern for that category without any input from the B input section.

Although the operation is simple, training the counterpropagation network is not simple, because this network involves two very different learning methods, Kohonen and outstar. Kohonen uses unsupervised training, whereas outstar requires supervised training. Training such a hybrid network normally involves a two-step procedure. In the counterpropagation network the middle Kohonen layer and the output outstar layer are separately trained. First the Kohonen layer is trained on input patterns and develops a valid feature map for the input data. Generally, the Kohonen layer is trained until it adequately recognizes the input patterns and categorizes them into the correct number of categories. During this training period the output layers and learning constants are set to very low values (or even zero) because the output of the network does not matter at this time. Once the middle layer is adequately trained, the weights between the input and middle layer are frozen. The learning constants for the middle layer are set to zero to ensure that no further changes occur, because the middle layer has learned the correct category for each input. Now it is up to the outstar to reproduce the correct output for each category. The learning constants for the outstar layer are increased so that learning occurs and continues until the output layer is appropriately trained.

Bidirectional Counterpropagational Network

The unidirectional counterpropagation network really has little advantage over other networks and systems that perform a mapping function between the input pattern and output pattern. The only advantages over backpropagation is that the middle layer does give a probability distribution mapping of the input data and the training may be faster.

The bidirectional counterpropagation network is shown in Figure 9.12 with only a few of the connections. Both the A and B input layer subsections

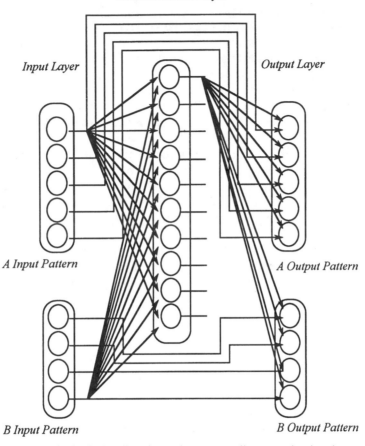

Figure 9.12   Bidirectional counterpropagation neural network.

connect to the Kohonen middle layer, and each also has a one-to-one connection to the corresponding subsection of the output layers. This effectively constitutes two counterpropagation networks, one to map A patterns to B patterns and one to map B patterns to A patterns, operating with a single Kohonen layer. A bidirectional network can accept either kind of pattern as its input and respond with a corresponding pattern with the opposite type. An A input yields a B pattern output, and a B input pattern yields an A output pattern.

In bidirectional counterpropagation networks, the input and output layers are the same size, both have A and B input sections with full connections to the middle Kohonen layer, and the middle layer is fully connected to both sections of the output layer. Each input neuron in each section has direct connections to the corresponding output neuron. The middle layer receives input from all elements of the input layer and transmits its output to the entire output layer. Training is the same two-step process as for the unidirectional network. First the middle layer is trained using Kohonen learning to categorize the inputs correctly. Then the output layer is trained to produce the correct output for each category of input using outstar learning. The major difference is that bidirectional network can only learn one-to-one mappings. If several A patterns generate the same B pattern, then when that B pattern becomes an input, it cannot determine which A pattern to produce. Hence, one-to-many or many-to-one mappings are not possible with a bidirectional counterpropagation neural network.

Characteristics of Counterpropagation Neural Networks

Counterpropagation networks have the same disadvantages as do both the Kohonen and the outstar networks. The problem encountered most frequency is getting a variety of winners in the Kohonen layer so that the input patterns are categorized correctly. It is not unusual to find that the Kohonen layer has only a few distinct clusters during the early part of the training section, particularly if the weight vectors are randomly distributed through *n*-dimensional space.

Counterpropagation networks tend to be larger than corresponding backpropagation networks. If a certain number of mapping categories are to be learned, the middle layer must have at least that number of neurons. Training is usually faster if the number of neurons in the middle layer is substantially larger than the number of mappings. However, the counterpropagation network can do inverse mappings. It can provide "ungarbled" versions of A and B when supplied with garbled versions. Very few networks have a bidirectional ability; most require two networks to achieve the same result. Furthermore, the self-organization of the features of the Kohonen layer is lost, and the outputs must be supplied for supervised training.

9.6 PROBABILISTIC NEURAL NETWORKS

The probabilistic neural network (PNN) developed by Donald Specht provides a general technique for solving pattern classification problems. In mathematical terms, an input vector (called a feature vector) is used to determine a category (e.g., the spectral energy values from a sensor system can be represented as a feature vector), and the network classifiers are trained by being shown data of known classifications. The PNN uses the training data to develop distribution functions that are in turn used to estimate the likelihood of a feature vector being within several given categories. Ideally, this can be combined with *a priori* probability (relative frequency) of each category to determine the most likely category for a given feature vector.

Bayesian Probability

The PNN is a neural network implementation of Bayesian classifiers. There-fore, let us look at how Bayesian probability works. Bayes inversion formula gives

$$P(X|Y) = \frac{P(Y|X) \cdot P(X)}{P(Y)} \tag{9.6-1}$$

This equation indicates that for an event X with a certain known probability $P(X)$, the probability of event X given event Y has occurred $[P(X|Y)]$ can be computed from the probability of Y occurring given that X has occurred $[P(Y|X)]$ and the overall probability that Y will occur at all $[P(Y)]$. The relationship $P(Y|X)$ is called the *a posteriori* (the posterior) probability indicating that the probability is known only before after the event X itself has occurred.

The Bayesian formula also provides a method for categorizing patterns. In this formulation, Y is interpreted as a possible category into which a pattern might be placed and X is interpreted as the pattern itself. The decision function can be associated with each possible category (all values of Y). Bayesian decision theory tries to place a pattern in the category that has the greatest value of its decision function. However, in real-world problems, we rarely have known probabilities and must estimate or approximate such Bayesian probabilities. A probabilistic neural network has this capability. Bayesian classifiers require probability density functions that can be con-structed using Parzen estimators which are used to obtain the probability density function over the feature space for each category. This allows the computation of the chance a given vector lies within a given category. Then, combining this information with the relative frequency of each category, the PNN selects the most likely category for a given feature vector. The PNN is a simple network that categorizes by estimating the probability distribution function. Like the Kohonen feature map, input data to the PNN is often normalized to a standard value, usually one.

Structure of Probabilistic Neural Networks

The probabilistic neural network consists of four layers as shown in Figure 9.13. The first layer is the input layer, which is a "fanout" or buffer layer. The second or pattern layer is fully connected to the input layer, with one neuron for each pattern in the training set. Each of the neurons in the pattern layer performs a weighted sum of its incoming signals from the input layer and then applies a nonlinear activation function to give that neuron's output.

The third layer is the summation layer to which each pattern layer neuron transmits its output to a single summation layer neuron. The weights on the

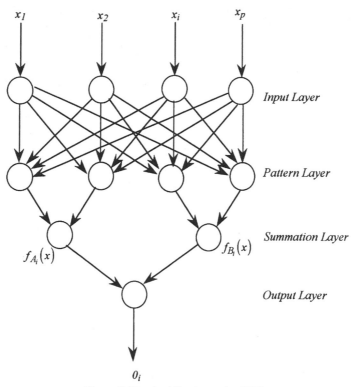

Figure 9.13 Architecture of a PNN.

connections to the summation layer are fixed at 1.0 so that the summation layer merely adds the outputs from the pattern layer neurons, which generates the networks category choice. There is one summation layer neuron per category.

The nonlinear activation function used by pattern layer neutrons is not a sigmoidal function but instead is an exponential function as shown in Figure 9.14. This activation function is given by

$$\Phi(I_i) = \exp\left[(I_i - 1)/\sigma^2\right] \tag{9.6-2}$$

where I is the weighted input to the neuron and the σ is the smoothing parameter that determines how smooth the surface separating categories will be. A reasonable range of values for σ is 0.1 to 10. The reason for the exponential activation function is that it is a simplification of the Parzan estimator of a Bayesian surface. Using a Bayesian estimating function in the pattern layer neurons allows the PNN to approximate Bayesian probabilities in categorizing patterns.

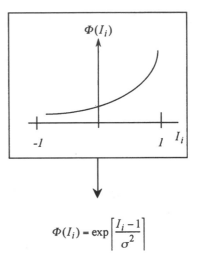

Figure 9.14 Exponential activation function for pattern layer neurons in a probabilistic neural network.

$$\Phi(I_i) = \exp\left[\frac{I_i - 1}{\sigma^2}\right]$$

The pattern layer has one neuron for each pattern in the training set. if there are 20 patterns in the training set, 12 in category A, and 8 in category B, then there are 20 neurons in the pattern layer. Each of these neurons has a set of weighted connections between it and the input layer. Each pattern layer neuron is assigned to one of the 20 training patterns, which connects to the summation layer neuron that represents its patterns category. Since in this case the summation layer has two neurons, the category A neuron receives inputs only from the 12-pattern layer neurons that represent category A, and the category B neuron receives inputs only from the eight-pattern layer neurons that represent categories B patterns. The weights on the connections from the pattern layer to the summation layer are fixed at unity.

Each neuron in the output layer received only two inputs, one from each of two summation units. One weight is fixed with a strength of unity; the other weight has a variable strength equal to

$$w^t = -[h_B/h_A][l_B/l_A][n_A/n_B] \tag{9.6-3}$$

where h refers to *a priori* probability of patterns being in category A or B, l is the loss associated with identifying a pattern as being in one category when it is in reality in the other category, and n is the number of A or B patterns in the training set. The values for h_A, h_B, n_A, and n_B are determined by the data pattern themselves, but the losses must be based on knowledge of the application. In many real-world cases there is no difference in loss if the categorization is wrong in one direction or the other. If so and if the training samples are present at approximately the ratio of their overall likelihood of occurrence, this weight reduces to unity.

The PNN is trained by setting the weights of one neuron in the pattern layer to the magnitude of each training pattern's elements. That neuron is then connected to the summation unit corresponding to that pattern's category. With a single pass through the training set the network is trained.

The Smoothing Parameter

A smoothing parameter which affects the generality of decision boundaries can be modified without retraining. The PNN usually needs a reasonable number of training samples for good generalization, but it can give good results with a small number of training samples. Since each training sample is represented by a neuron in the pattern layer, this serial implementation of a PNN will typically take longer in the recall mode than a backpropagation model. Inputs need not be normalized, which in some cases may distort the inputs space in an undesirable way. However, some implementations of PNN do normalize inputs for convenience.

The smoothing parameter σ varies between zero and infinity, but neither limit provides an optimal separation. A degree of averaging of nearest neighbors provides better generalization where the degree of averaging is dictated by the density of the training samples. Figure 9.15 shows the smoothing parameter σ as it varies between 0.1 and 1.0. For the lowest value the estimated probability density function has five distinct neurons, whereas for the largest variable there is a very severe flattening of the probability density function between -3 and $+3$.

Advantages and Disadvantages of the PNN

The advantage of probabilistic neural networks is that the shape of the decision surface can be made as complex as necessary using the smoothing parameter. The decision surface can approach Bayes optimal solutions, and the neural network tolerates erroneous samples and works reasonably well with sparse data. For time-varying statistics, old patterns can be overwritten with new patterns.

The PNN operates in parallel without feedback, and training is almost instantaneous. As soon as one pattern per category has been observed, the network can begin to generalize. As new patterns are included, the decision boundary becomes more complex and better defined, and the entire training set must be stored. Testing (recalling) requires that the entire data set be used. The amount of computation required for PNN to classify any unknown pattern is proportional to the size of the training set. Unfortunately, the PNN is not as general as other neural network algorithms.

One of the serious drawbacks of the PNN is that it cannot deal with extremely large training sets. Since there must be one neuron in the pattern layer for each example in the training set, the network memory requirements can increase very rapidly with the size of training sets. In effect, the entire

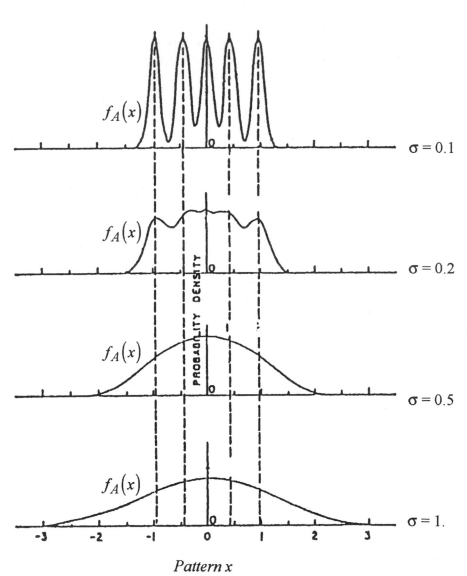

Figure 9.15 Influence of smoothing parameter σ on probability density function $f_A(x)$, the output of the summation neurons in the PNN.

training set is stored continually and retained during the classification of all feature patterns. On the other hand, the PNN signal training technique provides an extremely fast training time, particularly in comparison with an iterative network such as the backpropagation network. Furthermore, the network can deal with problems that have only a few samples of some of the categories.

9.7 RADIAL BASIS FUNCTION NETWORK

The radial basis function network (RBFN) (NeuralWare, 1993; Hush and Horne, 1993; Moody and Darken, 1989; Wasserman, 1993) always consists of three layers: the input layer, the pattern (or hidden) layer, and the output layer (i.e., the topology of the RBFN is thus identical to the backpropagation neural network). It is a fully connected and feedforward network with all connections between its processing units provided with weights. The individual pattern units compute their activation using a radial basis function; typically the Gaussian kernel function as shown in Figure 9.16 is used where σ is the width of the radial function. The activations of pattern units essentially characterize the distances of centers of radial basis functions of the pattern units from a given input vector. The radial basis functions thus produce localized, bounded, and radially symmetric activations—that is, activations rapidly decreasing with the distance from the function's centers (in contrast, the backpropagation network sigmoidal activation functions produce global and unbounded activations). Use of the radial basis activation functions requires a careful choice of the number of the pattern units to be used for a specific application, especially when a good generalization is needed; the areas of significant activation have to cover all the input space while overlapping in just the right way. For function approximation applications, this means that the samples included in the training set have to evenly represent all possible input vectors. The output units of the RBFN simply sum the weighted activations of individual pattern units without using any activation function. To speed up the training, the pattern layer neurons are augmented with bias units which have their activation values fixed to one.

The training of the RBFN differs substantially from the training used for the backpropagation network. It consists of two separate stages. During the

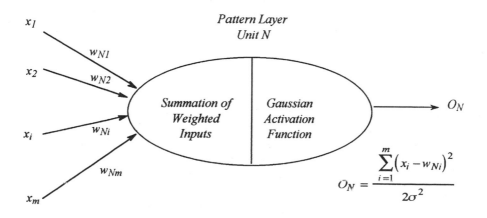

Figure 9.16 Pattern unit of the radial basis function network.

first stage, parameters of the radial basis functions (i.e., their centers and widths) for individual pattern units are set using the unsupervised training. The centers of radial basis functions of the individual pattern units define input vectors causing maximal activation of these units. Location of these centers is the first step of the training and is conducted with the help of some clustering algorithm (typically, the k-means algorithm is used). The clustering algorithms usually operate iteratively, and the clustering process is finished when locations of the centers for individual pattern units stabilize. The resulting values of individual elements of the center vectors are then directly used as values of the weights in connections between the input units and the corresponding pattern units. The widths of radial basis functions of the individual pattern units (denoted σ in Figure 9.16) determine the radii of the areas of the input space around the centers where activations of these units are significant. Their determination is the next step of the training and is performed using the "nearest-neighbor" heuristic.

In the second training stage, the weights in connections between the pattern units and the output units are determined using the supervised training based (as when training the backpropagation network) on minimization of a sum of squared errors of RBFN output values over the set of training input–output vector pairs. Before the training starts, these weights are randomized to small arbitrary values. At that stage, the weights in connections between the input units and the pattern units and the parameters of the radial basis functions of the pattern units are already set as determined in the first training stage and are not subject to any further changes. During this training, the RBFN is presented with individual input vectors from the set of training samples and responds with certain output vectors. These output vectors are compared with the target output vectors also given in the training set, and the individual weights are updated in a way ensuring a decrease of the difference between the actual and target output vectors (typically, the steepest descent optimization algorithm is used). The individual input–output training pairs are presented to the RBFN repeatedly until the error decreases to an acceptable level.

9.8 GENERALIZED REGRESSION NEURAL NETWORK

The generalized regression neural network (GRNN) (NeuralWare, 1993; Wasserman, 1993; (Specht, 1991; Caudill, 1993) is a special extension of the RBFN. It is a feedforward neural network based on nonlinear regression theory consisting of four layers: the input layer, the pattern layer, the summation layer, and the output layer (see Figure 9.17). It can approximate any arbitrary mapping between input and output vectors. While the neurons in the first three layers are fully connected, each output neuron is connected only to some processing units in the summation layer. The function of the input and pattern layers of the GRNN is exactly the same as it is in the

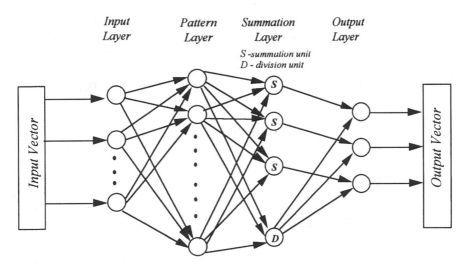

Figure 9.17 Topology of the generalized regression neural network.

RBFN. The summation layer has two different types of processing units: the summation units and a single division unit. The number of the summation units is always the same as the number of the GRNN output units; their function is essentially the same as the function of the output units in the RBFN. The division unit only sums the weighted activations of the pattern units without using any activation function. Each of the GRNN output units is connected only to its corresponding summation unit and to the division unit; there are no weights in these connections. The function of the output units consists in a simple division of the signal coming from the summation unit by the signal coming from the division unit. The summation and output layers together basically perform a normalization of the output vector, thus making the GRNN much less sensitive to the proper choice of the number of pattern units than the RBFN. The overlapping of radial basis functions of individual pattern units is not a problem for the GRNN; in fact, it turns out to be an important parameter allowing the user to influence generalization capabilities of the GRNN. In general, larger values of the width of radial basis functions of the pattern units results in a smoother interpolation of the output vectors values among the values corresponding to the centers of radial basis functions of the individual pattern units.

The training of the GRNN is quite different from the training used for the RBFN. It is completed after presentation of each input–output vector pair from the training set to the GRNN input layer only once; that is, both the centers of the radial basis functions of the pattern units and the weights in connections of the pattern units and the processing units in the summation layer are assigned simultaneously. The training of the pattern units is unsupervised, as in the case of the RBFN, but employs a special clustering algorithm which makes it unnecessary to define the number of pattern units in advance. Instead, it is the radius of the clusters that needs to be specified before the training starts. The first input vector in the training set becomes the center of the radial basis function of the first pattern unit. The next input vector is then compared with this center of the first pattern unit and is assigned to the same pattern unit (cluster) if its distance from this center is less than the prespecified radius; otherwise it becomes the center of the radial basis function of the next pattern unit. In the same manner, all the other input vectors are compared one-by-one with all the pattern units already set, and the whole pattern layer is thus gradually built. During this training, the determined values of individual elements of the center vectors are directly assigned to the weights in connections between the input units and the corresponding pattern units. Owing to the much lower sensitivity of the GRNN to the overlapping of the radial basis functions of the pattern units, the widths of radial basis functions of the individual pattern units need not be set according to the resulting structure of the pattern layer. Instead, their setting typically becomes the subject of experimentation as their values determine generalization properties of the GRNN.

Simultaneously with building the pattern layer, the values of the weights in connections between the neurons in the pattern layer and the summation layer are also set using the supervised training. The weights in connection between each pattern unit and the individual summation units are directly assigned with values identical to the elements of the output vector corresponding in the training set to the input vector which formed the center of the radial basis function of that particular pattern unit. In case that some additional input vectors in the training set are assigned to the same pattern unit, values of the elements of their corresponding output vectors are simply added to the previous values of these weights. At the same time, the weight in the connection of each pattern unit and the division unit, which was originally set to zero, is increased by one for each input vector from the training set which is assigned to this pattern unit.

9.9   ADAPTIVE RESONANCE THEORY (ART-1) NEURAL NETWORKS

Adaptive resonance neural networks are among the more complex neural networks in use today. They are based on adaptive resonance theory (ART) developed by Carpenter and Grossberg (1986). Three general types of ART networks are used: (a) ART-1, which can handle only binary inputs and was developed in 1986; (b) ART-2, which can handle gray-scale inputs and was developed in 1987; and (c) ART-3, which can handle analog inputs better, is more complex, and was developed in 1989 to overcome some limitations of ART-2. We will discuss ART neural networks as a general system, because the principles and characteristics of adaptive resonance in all three versions are the same. However, the implementation becomes increasingly complicated for gray-scale and analog inputs.

General Operations of an ART Neural Network

ART neural networks are two-layer neural networks fully connected with inputs going to the bottom layer, from which they are transmitted through adjustable weights to the top or storage layer. This is a bottom-up or "trial" pattern that is presented to stored patterns of the upper storage layer. The input pattern is modified during its transmission through the "bottom-up" weights to the upper layer, where it tries to stimulate a response pattern in the storage layer that contains several possible responses. Training takes place after every pass of the pattern, up or down. Since the training rule does not matter, it is common to use Hebbian learning for convenience. If this "bottom-up" pattern is selected, then "resonance" occurs, and the input is put into the matching pattern category. If it is not selected, the resulting activity in the "top-down" layer (called the "expectation" pattern or "first guess" pattern) is usually different from the bottom-up pattern because the top-down pattern is presented through the top-down weights to the bottom layer. Then the weights are adjusted, and the process is repeated. After a number of trials, the process is stopped, and a new category of pattern is created in the storage layer. This ability of ART to create new categories is its most important characteristic.

When a pattern fails to produce a match, a new pattern of nodes (from the storage layer) is now free to attempt to reach resonance with the input layer's pattern. In effect, when the trial patterns do not match, a reset subsystem signals the storage layer that a particular guess was wrong. Then that guess is "turned off," allowing another pattern from storage to take its place. This cycle repeats as many times as necessary. When resonance is reached and the guess is deemed acceptable, the search automatically terminates. This is not the only way a search can terminate; the system can terminate its search by learning the unfamiliar pattern being presented. As each trial of the search occurs, small weight changes occur in the weights of both the bottom-up and top-down pathways. These weight changes mean that the next time the trial pattern is passed up to the storage layer, a slightly different activity pattern is received, providing a mechanism for the storage layer to change its guess. If the system cannot find a match and if the input pattern persists long enough, the weights eventually are modified enough that an uncommitted node in the storage layer learns to respond to the new pattern. These changes in weights also explain why the storage layer's second or third guess may prove to be a better choice than the original one. The small weight changes ensure that the activity generated by the bottom-up pattern in the second pass is somewhat different from the activity generated in the first pass. If the input is a slightly noisy version of a stored pattern, it may require a few weight changes before the truly best guess can be matched.

Alternate View of Adaptive Resonance Operation.

The basic mode of operation in adaptive resonance is hypothesis testing. An input pattern is passed to the upper storage layer which attempts to recognize it by making a guess about the category to which the input layer belongs. It is then sent in the form of a top-down pattern to the lower layer. The result is then compared to the original pattern. If the guess is correct (or close enough), the two patterns reinforce each other and all is well. If the guess is incorrect, the upper layer tries again. Eventually, either the pattern is placed into an existing category or it is learned as the first example of a new category. Thus the upper layer forms a hypothesis of the correct category for each input pattern, which is tested by sending it back down to the lower layer. If a good match is made, the hypothesis is validated. However, a bad match results in a new hypothesis. If the pattern excited in the input layer nodes by the top down input is a close match to the pattern excited in the input layer by the external input, then the system is said to be in *adaptive resonance*, because each layer's activity mutually reinforces and strengthens the other layer's activity. It is adaptive because both sets of weights on the interconnections between the layers are continually modified to strengthen the recognition of the input pattern while the patterns resonate. Complexities must be added to carry out all the comparisons and decisions. The implementation of the acceptance/rejection process and storage of patterns are straightforward but complex and based on logical operations.

Vigilance

ART-1 also has the property of *vigilance* by which the accuracy with which the network guesses the correct match can be varied. By setting a new value for vigilance, the user can control whether the network deals with small differences or concerns itself only with global features. A low reset threshold implies high vigilance and close attention to detail. A high threshold implies low vigilance and a more global view of the pattern in the matching process. By controlling the vigilance, the user can differentiate "insignificant noise" and a "significant new pattern." Hence, the coarseness of the categories into which the system sorts patterns can be chosen. High vigilance forces the system to separate patterns into a large number of fine categories, while low vigilance causes the same set of patterns to be lumped into a small number of coarse categories.

Properties of ART-1

ART-1 possesses several of the characteristics needed in a system capable of autonomous learning. The more important characteristics are listed below:

1. It learns constantly but learns only significant information and does not have to be told what information is significant.

2. New knowledge does not destroy information already learned.
3. It rapidly recalls an input pattern it has already learned.
4. It functions as an autonomous associate memory.
5. It can (with a change in the vigilance parameter) learn more detail if that becomes necessary.
6. It recognizes its associative categories as needed.
7. Theoretically, it can even be made to have an unrestricted storage capacity by moving away from single-node patterns in the storage layer.
8. However, it can handle only binary patterns.
9. Its ability to create new categories is its most important attribute.

REFERENCES

Aarts E., and Korst J., *Simulated Annealing and Boltzmann Machines*, John Wiley & Sons, New York, 1989.

Carpenter, G. A., and Grossberg, S., Associative Learning, Adaptive Pattern Recognition, and Cooperative-Competitive Decision Making By Neural Networks, in *Optical and Hybrid Computing*, *SPIE Proceedings*, Vol. 634, H. Szu, ed., Bellingham, WA, 1986, pp. 218–247.

Caudill, M., GRNN and Bear It, *AI Expert*, Vol. 8, No. 5, pp. 28–33, 1993.

Haykin, S., *Neural Networks—A Comprehensive Foundation*, Macmillan, New York, 1994.

Hebb, D., *Organization of Behavior*, John Wiley & Sons, New York, NY, 1949.

Hecht-Nielsen, R., *Neurocomputing*, Addison-Wesley, Reading, MA, 1990.

Hush, D. R., and Horne B. G., Progress in Supervised Neural Networks, *IEEE Signal Processing Magazine*, Vol. 10, No. 1, pp. 8–39, 1993.

Korn, G. A., *Neural Network Experiments on Personal Computers and Workstations*, MIT Press, Cambridge, MA, 1992.

Kosko, B., Bidirectional Associate Memories, *IEEE Trans. Systems, Man and Cybernetics*, Vol. 18, (1), pp. 49–60, 1988.

Masters, T., *Practical Neural Network Recipes in C++*, Academic Press, San Diego, CA, 1993.

Moody, J., and Darken, C. J., Fast Learning in Networks of Locally Tuned Processing Units, *Neural Computation*, Vol. 1, pp. 281–294, 1989.

Neural Ware, *Neural Computing—A Technology Handbook for Professional II/Plus and NeuralWorks Explorer*, NeuralWare, Pittsburgh, PA, 1993.

Specht, D. F., A General Regression Neural Network, *IEEE Transactions on Neural Networks*, Vol. 2, pp. 568–576, 1991.

Uhrig, R. E., Tsoukalas, I. H., Ikonomopoulos, A., Essawy, M., Yancey, S., Travis, M., and Black, C., Applications of Neural Networks, EPRI Report TR-103443-P1-P2, Project 8010-12, Final Report, January 1994.

Wasserman, P. D., *Advanced Methods in Neural Computing*, Van Nostrand Reinhold, New York, 1993.

PROBLEMS

1. Construct a bidirectional associative memory (BAM) that will map the gray code (see Table 17.1) for the ten digits (1, 2, 3, 4, 5, 6, 7, 8, 9, and 0) into the corresponding 4-bit binary code. Introduce a distorted (one bit wrong) gray scale representation for 7 and see if you get the correct binary code (0111). If so, why; if not, why not?

2. A Kohonen network has inputs at 45° and 170° on the unit circle as shown in Figures 9.6 and 9.7. Randomized weights are located at 270° and 90°. Use a learning constant of 0.5. Calculate the new positions of the weight vectors after one upgrade cycle.

3. Discuss the relative benefits and modes of operation for probabilistic neural network, radial basis function network, and the generalized regression neural network. Give examples where each can be used advantageously.

4. The Kohonen network part of the counter propagation neural network uses Hebbian learning. Derive the equation for the training algorithm for this network if backpropagation is used.

10

DYNAMIC SYSTEMS
AND NEURAL CONTROL

10.1 INTRODUCTION

The ability of an artificial neural network to model a system or phenomenon allows it to be used in a variety of ways. Even elementary linear neural systems with a single neuron such as the Adaline (adaptive linear neuron) network introduced by Widrow have proven to be extremely useful. Indeed, much of what is called "adaptive linear systems theory" is directly applicable to artificial neural networks. The ability of neural networks to develop nonlinear models of a system offers an additional advantage that can be useful in many cases. In this chapter, we will explore the use of both simple and complex artificial neural networks to accomplish a variety of dynamic tasks, including control of complex systems.

10.2 LINEAR SYSTEMS THEORY

Linear systems theory is a well-developed field that is extremely important to the processing of data and the application of technologies such as neural networks to practical problems. When combined with random noise theory, the resultant technology becomes a powerful tool for investigating complex systems. It is assumed that the reader is generally familiar with the concepts of both linear systems theory and random noise theory. For those who want a review, there are a number of textbooks, including one written by one of the authors (Uhrig, 1970), that can provide the necessary background. Only the concepts needed for a general understanding of the applications will be presented here.

333

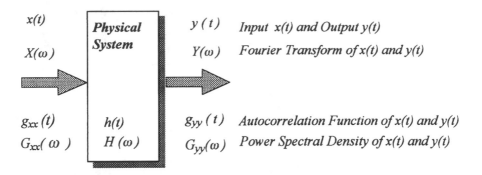

$h(t)$ *Impulse Response Function*
$H(\omega)$ *System Response Function*

Figure 10.1   Simple physical system with input and output.

Autocorrelation and Power Spectral Density Relationships

Let us consider a physical system with an input $x(t)$, an output $y(t)$, an impulse response function $h(t)$, and a system response function $H(\omega)$, as shown in Figure 10.1. $H(\omega)$ is a complex quantity with both amplitude and phase (or real and imaginary components).[1] The system response function is the Fourier transformation of the impulse response function; that is,

$$H(\omega) = \int_{-\infty}^{\infty} h(\tau)e^{-j\omega\tau}\, d\tau \tag{10.2-1}$$

where τ is the variable of integration. The output $y(t)$ of the physical system in the time domain is the convolution of the input $x(t)$ and the impulse response function $h(t)$:

$$y(t) = \int_{-\infty}^{\infty} h(\lambda)x(t-\lambda)\, d\lambda \tag{10.2-2}$$

where λ is the dummy variable of integration. The corresponding relationship in the frequency domain is

$$Y(\omega) = H(\omega)X(\omega) \tag{10.2-3}$$

---

[1]The term *system response function* as used here is the classical meaning of the term; it is the Fourier transformation of the impulse response function of the physical system. It is sometimes erroneously called *transfer function*, which is the Laplace transformation of the impulse response function. The use of the term "transfer function" in neural networks to mean the activation function or a nonlinear filtering element on the output of a neuron is an unfortunate situation that sometimes occurs when two fields are merged.

where $Y(\omega)$ and $X(\omega)$ are Fourier transformations of the output $y(t)$ and input $x(t)$, respectively; that is,

$$X(\omega) = \int_{-\infty}^{\infty} x(\tau)e^{-j\omega\tau}\,d\tau \qquad (10.2\text{-}4)$$

$$Y(\omega) = \int_{-\infty}^{\infty} y(\tau)e^{-j\omega\tau}\,d\tau \qquad (10.2\text{-}5)$$

For a time stationary process (i.e., a process whose characteristics remain constant with time), the autocorrelation function $g_{xx}(\tau)$ of the fluctuating variable input $x(t)$ is defined by

$$g_{xx}(\tau) = \lim_{T\to\infty} [1/2T]\int_{-T}^{T} x(t)x(t+\tau)\,dt = E[x(t)x(t+\tau)] \quad (10.2\text{-}6)$$

and the corresponding power spectral density $G_{xx}(w)$ is given by the Fourier transformation of $g_{xx}(\tau)$ to be

$$G_{xx}(\omega) = \int_{-\infty}^{\infty} g_{xx}(\tau)e^{-j\omega\tau}\,d\tau \qquad (10.2\text{-}7)$$

Relationships identical to equations (10.2-6) and (10.2-7) also apply to the output variable $y(t)$.

The relationship between the autocorrelation functions and the power spectral densities of the input and output variables and the physical system characteristics has been shown to be (Uhrig, 1970)

$$g_{yy}(\tau) = \int_{-\infty}^{\infty}\int_{-\infty}^{\infty} h(\lambda)h(\xi)g_{xx}(\tau - \xi + \lambda)\,d\xi\,d\lambda \qquad (10.2\text{-}8)$$

$$G_{yy}(\omega) = H^*(\omega)H(\omega)G_{xx}(\omega) = |H(\omega)|^2 G_{xx}(\omega) \qquad (10.2\text{-}9)$$

where $H^*(\omega)$ is the conjugate value of $H(\omega)$ (i.e., the sign of the imaginary part of $H(\omega)$ is reversed). Note that $G_{yy}(\omega)$ and $G_{xx}(\omega)$ are real quantities, and hence only the amplitude or modulus of the system response function $|H(\omega)|$ is involved in this relationship. The phase angle of the system response function is not involved.

For the special case where the input $x(t)$ is a white noise—that is, the power spectral density $G(\omega)$ is a constant K over all frequencies, and the autocorrelation function is a Dirac delta function, $2\pi K\delta(t)$—equations (10.2-8) and (10.2-9) become

$$g_{yy}(\tau) = 2\pi K\int_{-\infty}^{\infty}\int_{-\infty}^{\infty} h(\lambda)h(\xi)\delta(\tau - \xi + \lambda)\,d\xi\,d\lambda \quad (10.2\text{-}10)$$

$$G_{yy}(\omega) = K|H(\omega)|^2 \qquad (10.2\text{-}11)$$

While equation (10.2-11) is simple and easily implemented, equation (10.2-10) is complex. However, if we have a simple first-order "lag" system with an exponential impulse response function of the form

$$h_{xx}(\tau) = Ae^{-\alpha\tau} \tag{10.2-12}$$

a situation that often occurs in practical situations, then equation (10.2-10) can be reduced to

$$g_{yy}(\tau) = \left[\pi A^2 K/\alpha\right]e^{-\alpha\tau} = K'e^{-\alpha\tau} \tag{10.2-13}$$

where K' is a constant of proportionality and α is a decay constant. Similar simplifications are possible for other impulse response functions that are more complex than equation (10.2-12).

The autocorrelation relationship of equation (10.2-8) has been developed into a more useful form (Lee, 1960) by introducing the concept of the autocorrelation function of the impulse response function $g_{hh}(\tau)$ which is defined as in equation (10.2-6) to be

$$g_{hh}(\tau) = [1/2T]\lim_{T\to\infty}\int_{-T}^{T}h(t)h(t+\tau)\,dt = E[h(t)h(t+\tau)] \tag{10.2-14}$$

Equation (10.2-8) then becomes

$$g_{yy}(\tau) = \int_{-\infty}^{\infty}g_{hh}(t)g_{xx}(\tau-t)\,dt \tag{10.2-15}$$

Cross-Correlation and Cross-Spectral Density Relationships

In a manner similar to equation (10.2-6), the cross-correlation function $g_{xy}(\tau)$ between two variables $x(t)$ and $y(t)$ is defined by

$$g_{xy}(\tau) = \frac{1}{2T}\lim_{T\to\infty}\int_{-T}^{T}x(t)y(t+\tau)\,dt = E[x(t)y(t+\tau)] \tag{10.2-16}$$

and the corresponding cross-spectral density $G_{xy}(\omega)$ is given by the Fourier transformation of $g_{xy}(\tau)$

$$G_{xy}(\omega) = \int_{-\infty}^{\infty}g_{xy}(\tau)e^{-j\omega\tau}\,d\tau \tag{10.2-17}$$

Note that the cross-spectral density $G_{xy}(\omega)$ is a complex quantity with amplitude and phase, whereas the power spectral density $G_{xx}(\omega)$ is a real quantity with magnitude only.

The time- and frequency-domain input–output relationships for the cross-correlation functions and cross-spectral densities, when applied to the physical system of Figure 10.1, are

$$g_{xy}(\tau) = \int_{-\infty}^{\infty} h(\lambda) g_{xx}(t - \lambda) \, d\lambda \qquad (10.2\text{-}18)$$

$$G_{xy}(\omega) = H_{xy}(\omega) G_{xx}(\omega) \qquad (10.2\text{-}19)$$

For the special case where the input variable $x(t)$ is white noise, these relationships become

$$g_{xy}(\tau) = 2\pi K h(\tau) \qquad (10.2\text{-}20)$$

$$G_{xy}(\omega) = K H_{xy}(\omega) \qquad (10.2\text{-}21)$$

where K is a constant of proportionality equal to the power spectral density of the "white noise."

Influence of Noise on Measurements

The relationships of equations (10.2-8), (10.2-9), (10.2-18), and (10.2-19) are valid as long as the signals are uncorrupted with noise or other extraneous signals. When noise is present, these equations can be modified to include the influence of the noise (which is dependent upon where the noise is introduced), but it is necessary to know the characteristics of the noise to obtain meaningful measurements. Sometimes, the noise is an approximation of "white" noise (i.e., the power spectral density is constant over the frequency range of interest), and this fact may allow the input–output relationships derived here, modified for the noise inputs, to be used advantageously.

For instance, let us consider the arrangement shown in Figure 10.2, which consists of a physical system with an input $x(t)$ and an instrumentation system with an input that is the sum of the output of the physical system and external or detector noise as its input. Application of equation (10.2-9) gives

$$G_{zz}(\omega) = |H_1(\omega)|^2 \big[G_{yy}(\omega) + G_{nn}(\omega) \big]$$

$$= |H_1(\omega)|^2 \big[|H(\omega)|^2 G_{xx}(\omega) + G_{nn}(\omega) \big] \qquad (10.2\text{-}22)$$

If the instrumentation system has a flat response beyond the range of interest and the detector noise G_{nn} is "white" (i.e., the power spectral density of the noise is constant), then equation (10.2-22) becomes

$$G_{zz}(\omega) = K_1 \big[|H(\omega)|^2 G_{xx}(\omega) + K_2 \big] \qquad (10.2\text{-}23)$$

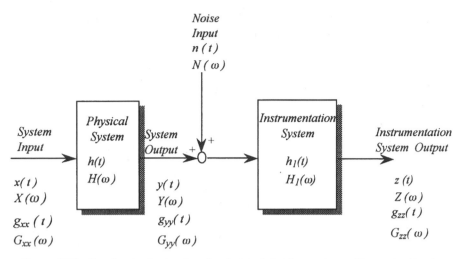

Figure 10.2 Simple physical system having a detection system with a noise input.

where K_1 and K_2 are constants. For the special case where the input $x(t)$ is "white" and $G_{xx}(\omega)$ is constant, equation (10.2-23) can be rearranged to give

$$|H(\omega)|^2 = K_3 G_{zz}(\omega) + K_4 \qquad (10.2\text{-}24)$$

The presence of K_2 and K_4 in equations (10.2-23) and (10.2-24) is due to the detection noise. If K_2 and K_4 are large compared to K_1 and K_3, respectively, then the presence of the noise may seriously deteriorate the quality of the measurement and the ability to evaluate the system response function and the parameters of the system being studied.

Cross-spectral density measurements offer a means of overcoming some of these problems under certain circumstances. Consider the case of the system in Figure 10.2 with two inputs: $x(t)$ that goes through both the physical system and the instrumentation system, and $n(t)$ that goes only through the instrumentation has been analyzed (Uhrig, 1970). The result is

$$G_{xz}(\omega) = H(\omega)G_{xx}(\omega) + H_1(\omega)G_{xn}(\omega) \qquad (10.2\text{-}25)$$

If the noise $n(t)$ is completely uncorrelated with the input $x(t)$, which is almost always the case, then $G_{xn}(\omega)$ is zero, and equation (10.2-25) reduces to

$$G_{xz}(\omega) = H(\omega)G_{xx}(\omega) \qquad (10.2\text{-}26)$$

which is identical to equation (10.2-9) for the noiseless case. This demonstrates the ability of cross-spectral density measurements to eliminate the influence of noise in many practical cases. Equation (10.2-26) can be further reduced for a white noise input to

$$G_{xz}(\omega) = KH(\omega) \qquad (10.2\text{-}27)$$

That is, the amplitude of the transfer function $|H(\omega)|$ is proportional to the amplitude of the cross-spectral density between the input and the output, and the phase angle of the system response function $\Theta(\omega)$ is equal to the phase angle of the cross-spectral density.

Coherence Function as an Index of the Quality of Measurement

As indicated above, the presence of the constants K_2 and K_4 in equations (10.2-23) and (10.2-24) can seriously degrade the measurement. We can get an index of the influence of the presence of noise on measurements by defining the coherence function $\gamma^2(\omega)$ to be the ratio of $|H(\omega)|^2$ as determined by the power spectral density method equation (10.2-9) to $|H(\omega)|^2$ as determined by the cross-spectral density method equation (10.2-19). Since equation (10.2-19) gives a result that is independent of the influence of noise whereas the validity of equation (10.2-9) deteriorates as the noise increases, this ratio is a valid reflection of the adverse influence of noise. Hence

$$\gamma^2(\omega) = \frac{|G_{xy}(\omega)|^2/[G_{xx}(\omega)]^2}{G_{yy}(\omega)/G_{xx}(\omega)} = \frac{|G_{xy}(\omega)|^2}{G_{yy}(\omega)G_{xx}(\omega)} \qquad (10.2\text{-}28)$$

For the case of no input noise, the coherence function is unity. As the noise increases, the quality of the measurement using equation (10.2-9) deteriorates, and the coherence as given by equation (10.2-28) decreases. It is intuitive and readily demonstrated that the coherence function is always less than or equal to unity; that is,

$$\gamma^2(\omega) \le 1 \qquad (10.2\text{-}29)$$

Correlation and Spectral Measurements Using Pseudorandom Binary Variables

In modeling a dynamic system with an artificial neural network, one signal from the system is the input and another is the desired output. Generally, the fluctuations of the input is adequate to ensure training of the network over the desired dynamic range. However, it is sometimes necessary to introduce a small perturbation of the input to make sure that the input signal contains the desired frequency content. Generally this perturbation is either a multiple-frequency signal (sum of sinusoidal signals with frequencies spread evenly on a logarithmic or a linear scale) or a random signal that approximates a "white" noise over the frequency range of interest.

One signal that is particularly useful and easy to implement is the "pseudorandom binary maximum-length shift register sequence" signal (Uhrig, 1970). It is a binary signal that instantaneously shifts between 0 and 1 (or between -1 and $+1$), with the shifts occurring at integral numbers of the time interval Δ. It is easily produced with software or a hardware shift register. It is a periodic signal that has (a) a narrow triangular spike as an autocorrelation function and (b) a power spectral density whose discrete values have an envelope of the general form of $\sin(x)/x$. The period of the signal ($N\Delta$) can be controlled easily by (a) the number of shifts N in one period of a shift register generating the signal and (b) the time interval Δ. By increasing N and decreasing Δ while keeping the period $N\Delta$ constant, the autocorrelation function of the signal becomes narrower, more nearly approximating a Dirac δ function, and the frequency range over which the power spectral density envelope $\sin(x)/x$ remains almost constant increases. Under these conditions, we approach the characteristics of a "white noise," and equations (10.2-20) and (10.2-21) apply. Then, the cross-correlation function between the pseudorandom input signal and the output gives a quantity proportional to the impulse function, and the cross-spectral density between the pseudorandom signal and the output gives a quantity proportional to the system response function. Since all other noise sources are uncorrelated with the pseudorandom input, they have no influence on these measurements. Indeed, two pseudorandom signals of different lengths (i.e., generated with shift registers having different time intervals and number of shifts per cycle) are independent of each other, and the cross-correlation functions with system signals are also independent of each other. Hence pseudorandom signals can be introduced at different locations in the system in order to model system response characteristics of individual components of the system. The power of cross-correlation and cross-spectral density measurements becomes apparent when it is realized that multiple sources of such "white" noise (that are, in fact, deterministic periodic variables) are independent of each other and can be injected into a system without unduly influencing the behavior of the system or the other measurements taking place.

The fact that the pseudorandom signal is periodic usually simplifies the processing of the data to obtain the cross-correlation and cross-spectral density. Furthermore, the shift register configuration for generating the pseudorandom signal can be implemented in software, and its output can be introduced directly into the input of an artificial neural network or mixed with other input signals in any desired manner. There are several variations of this pseudorandom signal and the associated shift register systems (Uhrig, 1970). A three-level pseudorandom signal that deals separately with the linear and nonlinear portions of the system response function has also been demonstrated (Gyftopoulos and Hooker, 1964).

10.3   ADAPTIVE SIGNAL PROCESSING

Much of the early work in the artificial neural network field was carried out by Bernard Widrow and his associates at Stanford University within the framework of "adaptive linear systems and adaptive signal processing" (Widrow and Hoff, 1960). Virtually all of the applications developed in this work can advantageously utilize an artificial neural network with its non-linear capability in the place of the "adaptive linear combiner." Indeed, Widrow's Adaline is an adaptive linear combiner that utilizes a bipolar output element as its nonlinear output device in order to accomplish its tasks.[2]

Adaptive Linear Combiner

The adaptive linear combiner, a nonrecursive adaptive filter, is fundamental to adaptive signal processing, and it is an integral part of an artificial neural network. It is used, in one form or another, in most adaptive filters and control systems. It is the most important element in "learning" systems. It is essentially a time-varying, nonrecursive digital filter that is implemented in many forms, and its behavior and its means of adaptation are well under-stood and readily analyzed.

Multiple-Input Adaptive Linear Combiner

The general form of an adaptive linear combiner is shown in Figure 10.3. The kth input is a vector $\mathbf{X}_k$ with components $x_1, x_2, \ldots, x_n, \ldots, x_N$; that is,

$$\mathbf{X}_k = [x_{0k} x_{1k} \cdots x_{nk} \cdots x_{Nk}]^T \qquad (10.3\text{-}1)$$

where the superscript T indicates the transposed vector (i.e., it is a column vector).

A weight vector $\mathbf{W}_k$ with a constituent set of adjustable weights $w_{0k}, w_{1k}, w_{2k}, \ldots, w_{Nk}$, and a summing unit produces a single output I_k. Most systems also include a bias with an amplitude x_0 equal to unity and an adjustable weight w_0. The configuration shown in Figure 10.3, known as a "multiple input adaptive linear combiner," accepts all of the components of the vector $\mathbf{X}_k$ simultaneously and produces a single output. Then the weights are adjusted, the next input vector is applied, another output is produced, the weights are adjusted again, and so on. This is the form of the classical implementation of almost all neural networks used today.

---

[2]There is nothing inherent in the definition of neural networks that requires that the output of the neurons be nonlinear. However, as pointed out in Chapters 7 and 8, the use of linear activation functions in the hidden layers significantly limits the usefulness and capabilities of a neural network.

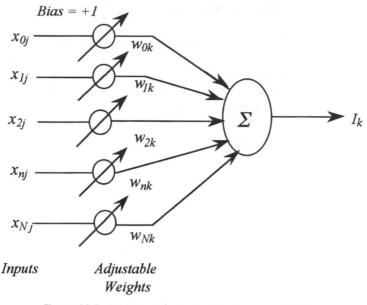

Figure 10.3 Multiple-input adaptive linear combiner.

Single-Input Adaptive Transverse Filter

An alternate form of the adaptive linear combiner (which is equally applicable to artificial neural networks in general) is shown in Figure 10.4, in which the input is applied sequentially to the input layer through a series of time delays, moving down the input layer until it reaches the end. The lines between each pair of delay units are tapped, and the signals (in addition to being sent to the next delay unit) are sent through adjustable weights to a

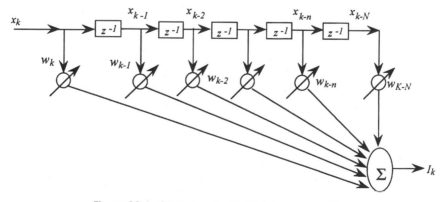

Figure 10.4 Single-input adaptive transverse filter.

summing node. Such a system has historically been called a "tapped delay line," which is quite descriptive of the process when implemented in analog hardware. In a digital computer, the process is implemented by manipulating sequential values of a sampled time series from a file. If a bias is used, it is applied directly to the summing junction through an adjustable weight. However, it is not normally required for single-input systems if the mean value has been removed from the time-varying input signal. The length of the time delay (designated z^{-1} from the "z" transform as used in digital control theory) is normally equal to the sampling interval, or some multiple of it, used in digitizing the input analog variable. The length of the input signal presented to the summing junction is the product of the number of delays in a cycle and the length of the time delay. For instance, a network having 100 input neurons (and 99 time delays) and a sampling rate of 1000 samples per second ($z^{-1} = 0.001$ sec) has an input signal that spans (100×0.001) or 0.1 sec. These parameters are related to the frequency content of the signal as well as the characteristic time constant of the system modeled in the artificial neural network. Such a system is known as a "single-input adaptive transversal filter." Its input vector is given by

$$\mathbf{X}_k = [x_k x_{k-1} \cdots x_{x-N}]^T \qquad (10.3\text{-}2)$$

Input–Output Relationships

The output for the multiple-input adaptive linear combiner of Figure 10.3 is

$$I_k = \sum_{n=0}^{N} w_{nk} x_{nk} \qquad (10.3\text{-}3)$$

while the output for the single-input adaptive transversal filter of Figure 10.4 is

$$I_k = \sum_{n=0}^{N} w_{nk} x_{(k-n)} \qquad (10.3\text{-}4)$$

For both types of systems, we have a weight vector

$$\mathbf{W}_k = [w_{0k} w_{1k} \cdots w_{nk}]^T \qquad (10.3\text{-}5)$$

With the definitions of equations (10.3-1), (10.3-2), and (10.3-5), we can express the outputs of both types of systems with a single relationship:

$$I_k = \mathbf{X}_k^T \mathbf{W}_k = \mathbf{W}_k^T \mathbf{X}_k \qquad (10.3\text{-}6)$$

Desired Response and Error

Although the adaptive linear combiner can be used in both open- and closed-loop adaptive systems, the primary interest here is in closed-loop operation. Hence, the weight vector adjustment depends primarily on the output and its deviation from the desired output. The weight vector $\mathbf{W}_k$ is adjusted or optimized so as to minimize the difference between the actual output I_k and the desired (or target) output T_k—that is, to minimize the square error defined by

$$\varepsilon_k^2 = [T_k - I_k]^2 \qquad (10.3\text{-}7)$$

The arrangement for dealing with this error and target output for the single-input adaptive transversal filter is shown in Figure 10.5. The minimization of square error follows the procedure described in Section 8.2 and the Widrow–Hoff Delta learning rule (1985).

Linear Control Theory

Linear control theory is well documented in the literature and is taught in most undergraduate engineering curricula. The proportional-integral-differential (PID) type of control, the most common linear control system, is discussed briefly in Section 6.1 and Example 6.1 in Chapter 6 ("Fuzzy Control") and in Section 10.5 ("Neural Network Control"). We will presume that the reader has a basic understanding of the PID control system and such concepts as stability, feedback, gain, and so on. Adaptive control and model-reference adaptive control will be explained briefly in the section of this chapter where it is introduced.

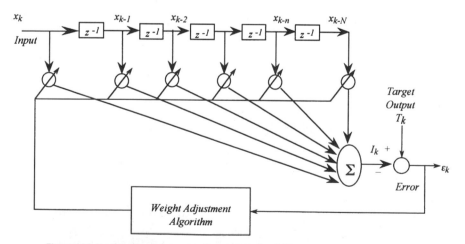

Figure 10.5   Single-input adaptive transversal filter with a target output.

The use of a single-input adaptive transverse filter (shown in Figure 10.5) to model a physical system whose characteristics are not known gives the impulse response function. After the weights have been adjusted to minimize the least squares error between the desired output and the actual output, the values of the weights sequentially from left to right give the values of the impulse response function at the corresponding time. Of course, since the number of weights is finite and an impulse response function $h(t)$ as defined in Section 10.1 extends to infinity, these weights are only an approximation of $h(t)$. As a result, the single-input adaptive transfer filter is often called a "finite impulse response" (FIR) system.

10.4   ADAPTIVE PROCESSORS AND NEURAL NETWORKS

The term "adaptive neural networks" refers to a neural network that adapts its weights to accomplish a mapping of the input to the desired output. Most neural networks that employ supervised learning belong to this category. Least squares adaptation algorithms are the basic learning systems for both adaptive signal processing systems and adaptive neural networks. Least squares minimization was discussed in Chapter 8 in conjunction with the Widrow–Hoff delta learning rule used in backpropagation training.

Linear Versus Nonlinear Systems

Although certain types of adaptive systems called "linear adaptive systems" can become linear when their adjustments are held constant after adaptation, most adaptive systems, by their very nature, have time-varying parameters and are nonlinear. Their characteristics depend on the input and the structure of the adaptive process. Adaptive systems are adjustable and depend on finite-time average signal characteristics rather than on instantaneous values of signals or instantaneous values of the internal system states. The adjustment of adaptive systems are made with the goal of optimizing specific performance measures. Hence, we can take advantage of the neural network's ability to utilize the nonlinear activation function to deal with nonlinearities in the modeling process.

Applications of Adaptive Neural Networks

Widrow and Stearns (1985) have discussed a wide variety of applications of the adaptive signal processing systems. We shall discuss many of these applications in the context of a neural network used to perform the same task. Although some adaptive systems operate in the open-loop mode, we shall deal primarily with closed-loop operation, a mode that provides "performance monitoring." Closed-loop operation, as we normally encounter it, involves measurement of the input and output signals of a system, utili-

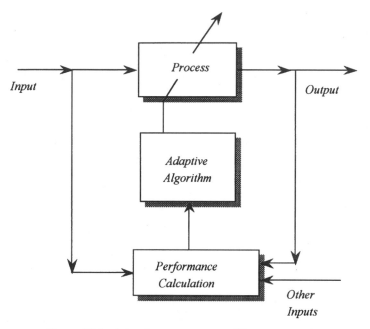

Figure 10.6 Adaptive processor and its components.

zation of a performance index (which can be as simple as a comparison of two values), and automatic adjustment of one or more input parameters of the system. These steps can be sequential discrete operations, but more commonly, all steps take place concurrently (at least in analog systems). Figure 10.6 shows the basic elements of a closed-loop adaptation system that in subsequent discussions in this section will be called an "adaptive processor." The performance calculator can be simple [i.e., the calculation of error or square error as in equation (10.3-7)] or sophisticated (i.e., the calculation of an energy or cost function using many quantities). The adaptation algorithm typically is minimization of least squares error, but it can utilize any optimization process desired. Clearly, a backpropagation neural network can be considered to be an adaptive processor.

In this chapter, the adaptive processor will be implemented as a neural network. This could be viewed as somewhat of a restriction because the type of neural network often determines the type of performance monitoring. For instance, use of a backpropagation neural network inherently involves minimization of least squares error and gradient descent optimization. However, training of neural networks can involve any type of optimization that can be used anywhere. Closed-loop operation can be used in situations where characteristic parameters of the physical system being operated are poorly known or are changing with time. Closed-loop adaptive operation can also compensate for some degree of deterioration of components in the physical system. Hence, performance monitoring can result in a more robust and/or a more reliable system. However, the closed-loop adaptation process is not without difficulties. In some cases, the performance indices to be optimized have more than one minimum that, under some circumstances, can result in the adaptation reaching a "false optimum." Feedback can also lead to instabilities that could degrade the performance or endanger the operation of the physical systems. Even "limit cycle" instabilities (where the effect of instabilities is self-limiting due to saturation) can cause serious difficulties. Feedback systems with adaptation based on performance monitoring are subject to the same stability criteria that apply to other feedback control systems. Nevertheless, performance monitoring with adaptation of the physical system is widely used in complex systems that are difficult to analyze or model analytically.

Configurations of Adaptive Neural Network Systems

Four basic configurations for adaptive systems (Widrow and Stearns, 1985) are as follows:

1. System identification or modeling (used in adaptive control)
2. Inverse modeling (used in equalization and deconvolution)
3. Adaptive interference canceling
4. Adaptive prediction

Each of these systems uses a single adaptive processor or neural network in different configurations to carry out the adapting task in order to accomplish its desired function listed above. Each of these simple systems is subject to the limitations and problems of adaptive systems discussed above. Nevertheless, each system has been extremely successful and is capable of performing its specific task with a minimum of difficulty.

System Identification or Modeling

One of the most common processes for which adaptive neural networks are used is system identification or modeling. This involves placing the neural network in parallel with the physical system, applying the system input to the input of the network, using the system output as the desired output for the neural network, and training the neural network until the error between the system output and the network output reaches an acceptable level. This configuration is shown in Figure 10.7a. The single-input adaptive transversal filter discussed in Section 10.3 is usually implemented by introducing a sampled time-varying signal (a time series) into the input of a neural network where the sampled input values advance laterally along the input layer. (Note that this is not a form of lateral feedback between neurons in the input layer, but instead it is simply a means of introducing the appropriate input values to the network to represent a time series or sampled variable.) At the same

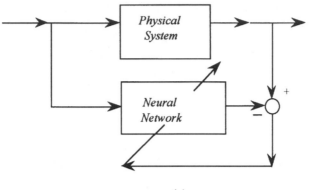

(a)

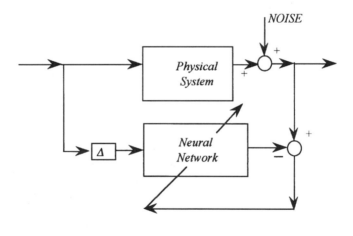

(b)

Figure 10.7 Configurations of adaptive modeling systems: (*a*) Simple modeling system. (*b*) Modeling system with delay Δ and input noise.

time, the corresponding sampled output value advances as the desired output for the single output neuron at the same rate. For multiple input–output systems, individual inputs and desired outputs may be assigned to certain parts of the input and output layers, respectively. Generally, it is necessary to specially design an artificial neural network for multiple-input, multiple-output modeling. One common arrangement is to use individual neural networks for each input and then use another neural network to combine the individual outputs.

After the network is trained, it is expected that the relationship of its input and output is the same as for the input and output of the physical system being modeled. This will be true if the variables have been trained over the appropriate dynamic range for the particular application. For instance, if a neural network is expected to respond over a specific dynamic range (e.g., from 0.1 to 10 Hz), the input signals to the neural network during the training should cover this dynamic range. Therefore, a neural network trained over the above range should not be expected to give proper results for input signals below 0.1 Hz or above 10 Hz. Furthermore, if a periodic signal (other than a single-frequency harmonic) is used in the training, the length of the cycle should be longer than the settling time of the physical system (the time required for the impulse response to approach zero) to ensure proper modeling.

Sometimes it is necessary to introduce a time delay Δ (which has no relation to the sampling time interval z^{-1}) into the configuration in order to model the finite time that is required for a signal to move through a physical process. Indeed, the length of this delay can be a parameter that is adjusted to minimize the residual error in the neural network model. It is often necessary to introduce a noise source into the configuration if such a noise source is inherent in the processes itself. For instance, noise sometimes arises from a random process internal to the physical system (e.g., the measurement of the intensity of a radiation source involves individual events of absorptions or collisions, which are unrelated to the source emissions) and should be included in the model. Even the detection process itself may be an independent source of noise that must be included in any realistic modeling of the process and its measuring system. Figure 10.7b shows a modeling configuration with both a time delay and a noise source.

Inverse Modeling

In a sense, inverse modeling is the same modeling discussed above with the input and desired output signals reversed. (From the standpoint of a neural network, it really does not make any difference where the input and desired output signals originate.) Inverse models are very important in control systems where they are often put in series with the controller or the process. Figure 10.8a gives the configuration normally used for inverse modeling. As in direct modeling, the time for a signal to propagate through a physical system should be included in the modeling process in order to reduce the mean square error to a minimum. Noise sources, if they exist in the actual system, should also be included in the neural network model. Figure 10.8b shows the inverse modeling configuration with a time delay Δ and a noise source.

An inverse model will have frequency response characteristics that are opposite to those of the original system: If the amplitude of the frequency response function at a particular frequency is decreasing in the system, it will be increasing in the inverse model. As a result, the product of the amplitudes of a model and its inverse model is equal to unity when the system and inverse model are connected in series.

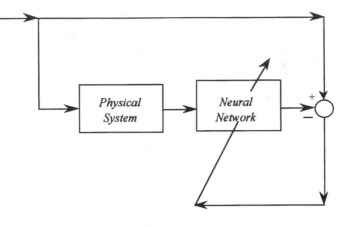

(a)

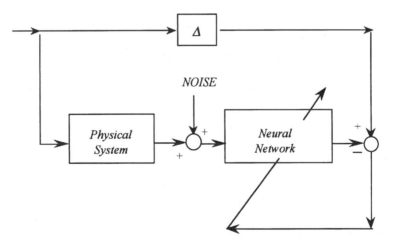

(b)

Figure 10.8 Configurations of adaptive inverse modeling systems: (*a*) Inverse modeling system. (*b*) Inverse modeling system with delay Δ and input noise.

This feature is often used in practical applications to negate the degrading influence of an instrumentation system. For instance, if the frequency response function $H_1(\omega)$ of the instrumentation system in Figure 10.2 falls off and significantly influences the measurement (we had previously assumed that its frequency response was constant over the frequency range of interest), then we can introduce an inverse model of this instrumentation system in series with the instrumentation to negate this adverse influence.

Equalization and Deconvolution

In communications, telephone and radio channels are dispersive; that is, high-frequency signals travel faster and are attenuated more rapidly than low-frequency signals. In dispersive channels, an inverse adaptive filter can be placed at the receiving end to "equalize" the channel—that is, provide a frequency and phase response in the receiver that is the inverse or reciprocal of that of the channel itself. In effect, it "deconvolves" the dissipative influence of the channel characteristic and restores the original signal characteristics. It avoids destructive interference that would virtually destroy the ability of communications channels to transmit information. High-speed transmission of digital signals, which is particularly susceptible to this problem, would not be possible without equalization and deconvolution.

The physical arrangement used for equalization and deconvolution is the same as the one used for inverse modeling in Figure 10.8b. The neural network attempts to recover a delayed version of the signal which may have been altered by the slowing varying system characteristics and which contains noise. The delay is to allow for the propagation time through the system and the neural network. In effect, this system attempts to deconvolve (undo) the effects of the communications channel. It also has applications in control systems.

Interference Canceling

Another very useful application of the adaptive processor is interference canceling. One of the most obvious applications is to cancel out 60-Hz interference from ordinary a.c. electrical power supplies. In this case, the interference frequency is known (60 Hz) and constant, and cancellation is relatively easy. The challenge is to cancel noise when the nature of the interference is unknown and changing. The advantage of noise cancellation as compared to noise filtering is that noise cancellation does not attenuate the signal and hence gives a much higher signal-to-noise ratio under virtually all conditions.

One of the most successful applications of this technology has been the cancellation of background noise from voice communication in small aircraft. The system used is shown in Figure 10.9, where the desired signal S (e.g., voice) is corrupted with background noise N (e.g., aircraft engine noise) when the microphone picks up both signals. The secret to success is to find a source of background noise N' that does not contain the desired signal S but is reasonably correlated with the interference noise N (i.e., a second microphone located away from the speaker's mouth). In this case, the adaptive processor is configured to produce an output Y that closely resembles N which is then subtracted from the speaker's microphone output so that the overall output S' closely resembles the original input signal S.

Another application of interference cancellation is the separation of the weak heartbeat of a fetus from the strong heartbeat of the mother. A

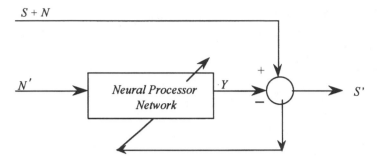

Figure 10.9 Noise cancellation system.

microphonic sensor is attached near the fetus and picks up the heartbeat of both the mother and the fetus, thus providing the $S + N$ signal. Another microphonic sensor is attached to the mother far from the fetus, where only her heartbeat is picked up, thus providing the N' signal. The system of Figure 10.9 then adapts to produce the fetal heartbeat signal S'.

Prediction

Another configuration of the adaptive processor is to predict the future of a signal from its behavior in the past. Figure 10.10 shows the system configuration used. The input signal $S(t)$ is delayed by an amount Δ before it is presented to the neural network. Then the neural network is adjusted so that

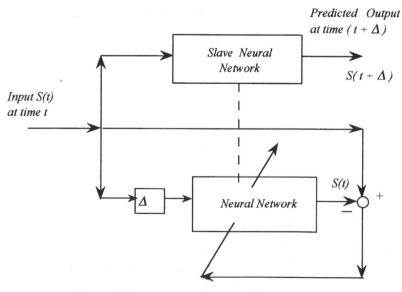

Figure 10.10 Configuration of prediction system.

the error between the delayed input and the undelayed signal is minimized. In effect, the neural network has been trained to produce an output signal $S(t)$ from an earlier input signal $S(t - \Delta)$. What is needed for prediction is a neural network that uses an input $S(t)$ and produces a future value $S(t + \Delta)$. This can be accomplished by using a "slave" neural network which has the same structure as the original neural network and whose weights are updated in real time to be identical to the weights in the original neural network. Such an arrangement is shown in Figure 10.10, where the "slave" neural network has an input signal $S(t)$ and an output $S(t + \Delta)$.

10.5 NEURAL NETWORK CONTROL

The field of control theory and systems is treated exhaustively in literally thousands of books and publications. The purpose of this treatment is to introduce only those concepts that are important and useful in dealing with neural network control systems. Control, by definition, is action taken to achieve a desired result or goal. For instance, the temperature of most modern homes is controlled by a simple, but effective, on–off action activated by a thermostat that turns the furnace on when the temperature falls below a specified temperature and turns it off when room reaches a slightly higher (typically 2–3° higher) temperature. The room temperature is compared with the desired temperature (the setpoint), and the difference constitutes an error signal that activates a simple control system to turn the furnace on and off.

Werbos (1992) has divided neural control into five categories: (1) supervised control, (2) inverse control, (3) adaptive control, (4) backpropagation through time, and (5) adaptive critic methods. Each of these methods will be discussed briefly here, and some of them will be covered in more depth in later sections.

There are many neural-network-based approaches to control problems, but most of them consist in using neural networks for two basic processes that are duplicated and combined as appropriate, along with time lags, to achieve the specific objective of the control system. These two basic processes are system or process modeling (system identification) and some means of control (often based on an inverse model of the system or process). A simple open-loop control system with a single input and output is shown in Figure 10.11a. Neural control systems are sometimes used to complement PID control systems by adaptively tuning the parameters of the controller to match the changing system characteristics. "Black box" identification and model-based controller optimization are commonly used in current neural control systems. For simplicity, single input–output systems are discussed here. However, multiple input–output systems are treated in the same manner; indeed, most neural control systems are multivariable systems.

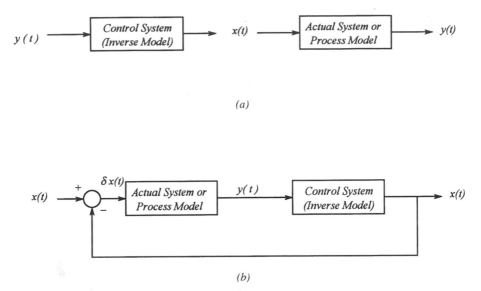

(a)

(b)

Figure 10.11   Simple open-loop (a) and closed-loop (b) control systems. $x(t)$ and $y(t)$ are system input and output, respectively, and $\delta x(t)$ is the change in $x(t)$.

For systems that are linear or can be linearized over the required range of operation, conventional linear control theory is adequate for most applications. However, most complex systems are nonlinear and require either a very sophisticated mathematical treatment or a simulation. Since the needed parameters for mathematical modeling and simulation models of most complex nonlinear systems are usually not available, an experimental determination of the system characteristics often becomes necessary.

More challenging is the mathematical derivation of an inverse model from a system response function (i.e., taking the inverse Fourier transformation if indeed it actually exists). Even for relatively simple linear systems, such an inverse operation can be quite difficult and sometimes impossible. For nonlinear systems, it is almost always impossible, thereby requiring approximations and simplifications or, more commonly, modeling of the system or process and its inverse configuration based on experimental data. This is where neural networks become very useful. Data from tests carried out on the system can be used to train a neural network to emulate the system behavior, thereby providing a neural network model, of the system or process. Because of the unique characteristics of neural networks, it is usually possible to reverse the input and output data provided to another neural network and train an inverse model of the system or process. When connected as shown in Figure 10.11a, these two neural networks become an open-loop control system.

Regulatory Control

Regulatory or closed-loop control is the most common control method; this is the essence of the room temperature control discussed earlier, where the control action is based on an error signal representing the difference between the desired temperature and the actual room temperature. (See Figure 10.12.) Indeed, error feedback is the essence of the training process in most supervised neural networks. For closed-loop control systems, the configuration of Figure 10.11*a* is often reversed so that the feedback signal is the input to the actual system or process model, which is often compared to a desired input $x(t)$. In this case, $x(t)$ is the input for the comparator and $\delta x(t)$ to be the input to the actual system or process model as shown in Figure 10.11*b*.

More common is the PID controller, a second-order control system where the feedback is a weighted sum of three quantities: The deviation of the output variable from its desired value (i.e., the error), the current derivative of the error, and the integral of the error over some time period are fed back in an effort to control the output in the face of fluctuating parameters and/or conditions. This arrangement is shown in Figure 10.13. In recent years, multiple input–output and multiple process systems have become

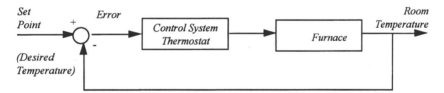

Figure 10.12 Closed-loop control system for a home heating system.

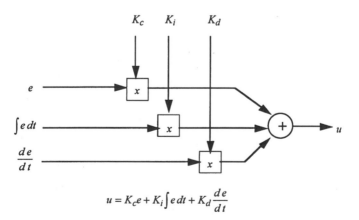

$$u = K_c e + K_i \int e\,dt + K_d \frac{de}{dt}$$

Figure 10.13 The PID controller structure. The symbol *e* is the instantaneous error.

increasingly prevalent, but virtually all of them are second-order linear systems.

Controllers also use control closed-loop operation through adjustment of parameters (e.g., the "tuning" of the gains of the three feedback components of a PID control system). Autotuning (on-line adjusting the parameters of loop controllers, or adjusting their "setpoints" for optimal performance) is a form of adaptive control. The use of a feedforward controller to provide a steady-state process input signal provides faster response and enables a feedback controller to reject noise and improves disturbance handling. Here again, these control systems fall within the linear systems domain.

The assumption of linearity is common, but it does not represent many real-world systems and processes. Most nonlinear systems are linearized over a limited range of operation. In many cases, the assumption of linearity is reasonable and produces good results, especially when a small range of adjustment is involved. Indeed, neural control is often used to complement linear control systems. Nevertheless, neural networks with their inherent ability to model nonlinear behavior show advantages that are important in many cases, particularly with complex systems.

Both the system identification and control portions of neural control can be viewed as nonlinear optimization processes; they seek to find neural network parameters (i.e., the weight matrix) for which some cost function is minimized. The type of cost function differentiates the different types of neural control concepts discussed below.

Example 10.1 Multi I/O System. An example of a multiple-input/multiple-output, open-loop inverse control system is a neural network to control the gas–tungsten arc welding process developed by Mid-South Engineering and Vanderbilt University for the National Aeronautics and Space Administration (Andersen et al., 1991). This approach is necessary because the complexity of the physics of the arc, the molten pool, and the surrounding heat-affected zone were virtually impossible to model using first principles. The lack of reliable, general, and yet computationally fast physical models of such a multivariable system makes the design of a real-time conventional controller a difficult task. Relationships between the various process inputs and outputs are nonlinear and not well-defined, and the process variables are coupled (i.e., a change in any given input parameter affects more than one output parameter). The arc welding process is controlled by a number of parameters, and the objectives of the welding process are specified in terms of the parameters shown in Figure 10.14.

An inverse open-loop system is used here because of the difficulty of providing feedback for closed-loop control purposes since the outputs of the welding process are very difficult to measure in real time. The weld must cool before physical measurements can be made. However, such output parameters as welding bead width and bead penetration can be measured off-line for purposes of modeling a neural network model from which they can be

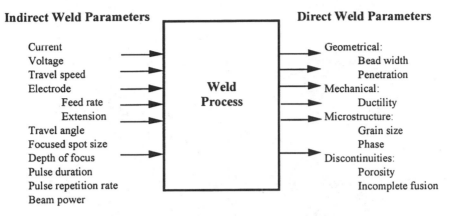

Figure 10.14 Welding illustrated as a multivariable process.

inferred on-line. The input parameters that are controlled (e.g., current, voltage, travel speed, etc.) can be measured on-line. Table 10.1 gives data obtained from test welds conducted over the expected dynamic range of operation for NASA (Andersen et al., 1991). The boldface data were used for testing the validity of the neural network given in Figure 10.15 after the rest of the data was used for training with the backpropagation algorithm. When

Table 10.1 Data used to train and test backpropagation networks for weld modeling and equipment parameter selection

Weld #	Workpiece Thickness [in]	Travel Speed [in/min]	Arc Current [A]	Arc Length[in]	Bead Width [in]	Bead Penetr. [in.]
1	0.125	6.0	80	0.100	0.118	0.024
2	0.125	6.0	100	0.100	0.165	0.051
3	**0.125**	**6.0**	**120**	**0.100**	**0.213**	**0.087**
4	0.125	6.0	140	0.100	0.256	0.126
5	0.125	4.0	100	0.100	0.216	0.071
6	**0.125**	**5.0**	**100**	**0.100**	**0.181**	**0.063**
7	0.125	7.0	100	0.100	0.153	0.043
8	0.125	6.0	100	0.090	0.161	0.063
9	0.125	6.0	100	0.080	0.161	0.075
10	**0.125**	**6.0**	**100**	**0.070**	**0.157**	**0.083**
11	0.250	6.0	80	0.100	0.110	0.028
12	**0.250**	**6.0**	**100**	**0.100**	**0.142**	**0.047**
13	0.250	6.0	120	0.100	0.181	0.063
14	0.250	6.0	140	0.100	0.205	0.075
15	**0.250**	**4.0**	**100**	**0.100**	**0.157**	**0.071**
16	0.250	5.0	100	0.100	0.150	0.059
17	0.250	7.0	100	0.100	0.134	0.043
18	0.250	6.0	100	0.090	0.142	0.051
19	**0.250**	**6.0**	**100**	**0.080**	**0.142**	**0.055**
20	0.250	6.0	100	0.070	0.142	0.059

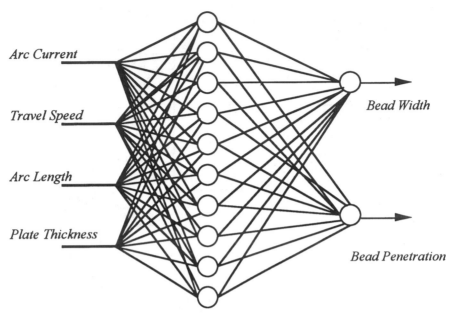

Figure 10.15 The neural network used for weld modeling.

the test data (not used in training) were introduced into the neural network, the predicted bead width and depth of penetration agreed with the actual measurements to within about $\pm 5\%$. This trained neural network now constitutes a (direct) model of the welding process.

The data in Table 10.1 were also used to train an inverse model of the welding process to provide a control neural network for parameter selection for the welder. This inverse model has the desired weld parameters as inputs and the control parameters for the welder as outputs. It can provide the welder controls with the information needed to produce a proper weld with the desired dimensions.

To simulate the performances of such a welding process, the direct and inverse neural network models were coupled together as shown in Figure 10.16 to form a cascade model of the control system (parameter selector) with the welding process. This arrangement is substantially the same as the open-loop control configuration shown in Figure 10.11a. A comparison of the outputs of the two neural network models with the data in Table 10.1 showed the errors of the inverse model output (first neural network) and the process model output (second neural network) to be about 10% and 2%, respectively. The reason for the low error in the process model output is that the errors in neural networks (which effectively are the inverse of each other) when trained on the same data tend to cancel each other out. □

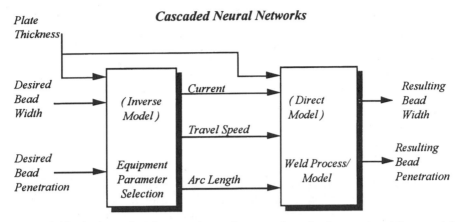

Figure 10.16 A cascade of an equipment parameter selection and welding model networks.

Neural Adaptive Control

Linear adaptive control has been a standard topic in control theory for at least three decades, and there appears to be little advantage of using neural networks in this domain except as a basis for departure into nonlinear neural adaptive control. Narendra et al. utilized neural networks to carry out linear adaptive control as a basis for later work in nonlinear control (Narendra and Parthasarathy, 1990). The goal of adaptive control is to maintain optimal performance as measured by some index of performance (e.g., efficiency, minimum emissions, etc.) under changing plant parameters or operating conditions. Often, the index of performance and optimization algorithms are included in a reference model. The difference (error) between the actual output of the system and the desired output as provided by the optimal reference model serves as the basis for adjusting parameters to improve performance. Because of the important role adaptive control plays in real-world situations, it is discussed extensively in the next section.

Supervised Control

Supervised control involves using a neural network to mimic the behavior of a conventional (i.e., PID) controller or even the behavior of a human being controlling a process or system. The neural network receives the same input and (desired) output as the PID or human controller, and training (typically backpropagation) proceeds in the conventional manner. When training is completed over the appropriate range of variables, the trained neural network replaces the PID or human controller. The major concern here is that the performance of the neural network control system can be no better than the PID or human controller. Even so, PID and human control have often proven to be remarkably effective. It is a well-known fact that a human being's ability to recognize almost imperceptible trends and behavior has resulted in adequate performance for many systems and that the PID controller has been the "workhorse" of the control industry for over half a century. More importantly, supervised control provides a starting point from which more sophisticated control systems can be used to improve performance of a system.

Inverse Control

Inverse modeling as discussed in the previous section can readily be adapted to neural control. Inverse modeling involves training a neural network arranged in accordance with the configuration shown in Figure 10.8b over the appropriate range of variables. Such systems are typically used in an open-loop mode as shown in Figure 10.11a. The operator simply provides an input equal to the output that is desired.

The major concern is that the inverse configuration actually exists and is physically realizable. For instance, if several different inputs produce the same output, then the inverse function does not exist. Another example is a system model where the gain goes to zero under some conditions. Then the inverse model would need infinite gain.

Backpropagation Through Time

In backpropagation through time (BTT), the user specifies a model of the external environment as well as an index of performance (utility function) to be maximized. Backpropagation is used to predict the derivative of this index of performance, summed over all future times with respect to current actions. These derivatives are then used to adapt the artificial neural network that provides the output actions. This approach is used because the designer can select any index of performance to be optimized, and the method accounts precisely for the impact of present actions on future values of the index of performance. BTT is basically equivalent to the calculus of variations as used in control theory. The only essential difference is that BTT includes a way of calculating the derivatives of the utility function. Its disadvantages are that it requires a model of the external environment which should be noise-free and exact and that it requires calculations backwards through time, which is not consistent with real-time learning. However, real-time learning has been implemented by dividing experience into distinct "experiments" and updating weights after each experiment is analyzed (Werbos, 1992).

Adaptive Critics Method of Neural Control

The adaptive critics method of neural control is a form of reinforcement control in which an index of performance to be optimized is supplied by the

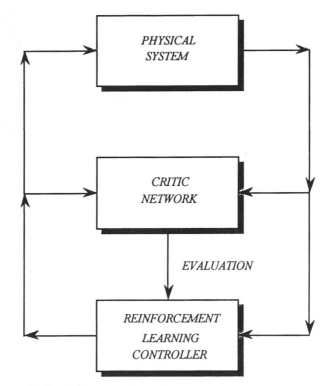

Figure 10.17 Adaptive critic reinforcement learning control system.

user. The long-term optimization problem is solved by using an additional artificial neural network (called the critic network) that evaluates the progress that the system is making and provides input to the reinforcement learning controller. This arrangement, shown in Figure 10.17, is particularly useful for situations where the model of the physical system is vague and ill-defined (e.g., the overall performance of a plant as measured by its total emissions to the atmosphere is such a system).

Example 10.2 Monitoring and Improving Heat Rate of a Power Plant. In the past few years, several systems for monitoring the heat rate (proportional to the reciprocal of efficiency) of power plants have become available. Generally, these systems are based on a first-principles model involving mass flow and energy balance equations applied to the many subsystems of a power plant. Typically, the model involves assumptions of idealized conditions, linearizations, and use of experimental correlation coefficients that are valid over limited ranges. The alternative proposed here is to take advantage of a neural network's ability to model nonlinearities and nonideal conditions inherent in any complex system. The thesis here is that a neural network model is more realistic under "real-world" conditions and that any subsequent analysis (e.g., optimization) is more effective than similar analyses performed on first-principles models. Because of the sensitivity of cost savings to heat rate, even a small improvement in heat rate can have a large financial impact (e.g., an improvement of only 0.1% in efficiency in a 1000-MWe power plant can result in about $500,000 per year additional revenue at current prices).

Guo and Uhrig (1992) carried out the modeling of the thermodynamics of TVA's Sequoyah Nuclear Power Plant, Unit 1, using data taken weekly over about a year. Because of the complexity of the plant, use of ordinary multilayer perceptron networks trained using backpropagation was not adequate as indicated by the fact that the training error was large and the network output was not equal to the desired output used in either the training or testing sets. Rather, it was necessary to utilize a hybrid network (N-Net 210) developed by Pao (1989) in which a Kohonen neural network is used to cluster the data and then the backpropagation neural network is trained to a very small system error using the centroids of the clustered data as the inputs and the corresponding heat rates as desired outputs for training. When the original data (not the centroids) were presented to the trained network, the average error was about 0.06%.

Sensitivity Analysis. The sensitivity analysis procedure discussed in Section 8.3 was applied to the trained network. This process gave sensitivity coefficients that were ranked in order based on absolute value, since the sign only indicates the direction that heat rate moved in response to a positive change in the input perturbation. The three most important variables for heat rate based on this sensitivity analysis were (1) unexpected power deviations, (2) measured condenser backpressure, and (3) condenser circulating water inlet temperature.

Item 3 is a function of the environment and cannot be controlled. Item 2 can be controlled only to the extent that it does not go below the saturation temperature for the condenser circulating water. Item 1 is a calculated value that cannot be controlled. At this point, it appears that little has been gained from the modeling and sensitivity analysis since the three most important inputs to the network model, as far as heat rate is concerned, cannot be controlled. However, a second modeling sensitivity and analysis can be performed to determine which inputs are important to any of these three "most important" variables. This was done for item 1, unexpected power deviations.

This second modeling was carried out by using the same hybrid neural network to cluster the data and predict a new output, item 1, unexpected power deviations. Again a sensitivity analysis was performed indicating that the three most important variables, as far as unexpected power deviations is concerned, are (1) total unaccounted electrical losses, (2) condenser thermal losses, and (3) auxiliary steam loads. This tells us that unexpected power deviations, and hence the plant efficiency, can be improved by reducing electrical losses, condenser thermal losses, and auxiliary steam loads. Each of these items can be investigated individually.

A further study of the original sensitivity analysis shows that the three variables that have the least influence on efficiency are (1) feedwater flow, (2) reactor power (within limits), and (3) impulse pressure at the turbine inlet. Since one's intuition would indicate that all three of these items might be important, this sensitive analysis has directed the plant personnel away from these items and toward those items listed in the previous paragraph that are important.

Iterative Procedure for Improving Plant Performance. Since plant conditions are always changing, keeping the plant at peak efficiency can be carried out using the following iterative process: (1) Train a neural network model of the plant using on-line data, (2) carry out a sensitivity analysis (or other analysis) to determine which two or three variables are the most important to heat rate, (3) adjust one or more of these variables in the direction indicated by the sensitivity analysis, and (4) wait for the plant to reach thermodynamic equilibrium. This four-step process could be carried out every few minutes with the interval being determined by the time necessary to reach thermodynamic equilibrium. This procedure ensures that the plant is always moving toward peak efficiency for the configuration and operating conditions that exist at the time of the analysis. If the plant condition changes (e.g., one train of feedwater heaters is taken out of service for maintenance) or the operating conditions change (e.g., the plant load changes or the condenser water temperature changes due to environmental conditions), it is still possible to move toward a more efficient (but different) configuration under the existing circumstances. □

10.6 SYSTEM IDENTIFICATION

Securing knowledge of the dynamics of a process being controlled is the first step in control. Sometimes *a priori* knowledge about the process is available in the form of a parameterized model where the parameters can be estimated from process input–output data. First-principles models are typically nonlinear, which typically must be linearized; they are usually valid only over a limited range of performance. Often these process models are relatively simple; for example, several second-order systems with lags are often adequate to represent a chemical process. In systems for where first-principles models are not available, the modeling procedure (nonparametric identification) discussed below (sometimes called "black box" modeling) is useful.

Nonparametric Identification

Nonparametric identification develops "black box" models of the input–output relationship as discussed in Section 10.3. Neural network nonparametric process models can be seen as a nonlinear extension of the system identification problem (Tsung, 1991). For instance, the adaptive transverse filter shown earlier in Figure 10.4, when implemented by a neural network, becomes a *finite impulse response* (FIR) network, which is a nonparametric identification system. The network is provided with an input vector of weighted past samples of the variable as a means of modeling dynamic phenomena. The number of samples must be sufficient to provide an interval of time long enough that no input signal prior to x_{k-N} will have any significant effect on the response at time k.

Parametric Identification

Parametric identification identifies structural features and parameter values for models of real-world physical systems. This includes identification of the

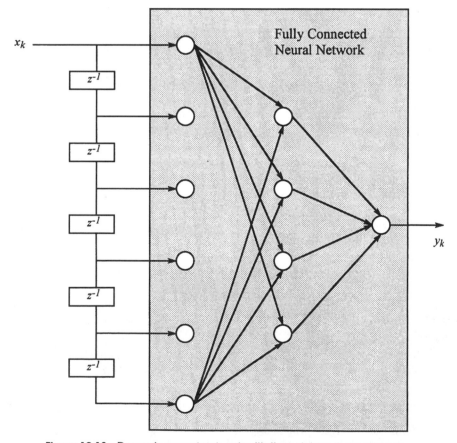

Figure 10.18   Dynamic neural network with time-delayed direct inputs.

model structure (i.e., the form of linear or nonlinear differential or difference equations as well as parameter estimation where the model structure is known). Neural networks trained through supervised learning can be used for both structure identification and parameter estimation. As structure identifiers, they can be trained to select elements of a model structure from a predetermined set. Structure identification with neural networks requires that the space of likely model structures be known in advance. Neural network parameter estimators, however, generate parameter values for a given structure or set of structures (Piovoso et al., 1991; Foslien et al., 1992). Neural network structure identifiers and parameter estimators are both trained off-line with a generalized simulation system. In structure identification, for example, the network output is a vector of structural features and network weights that are optimized to minimize the sum square error between the actual feature and its computer representation.

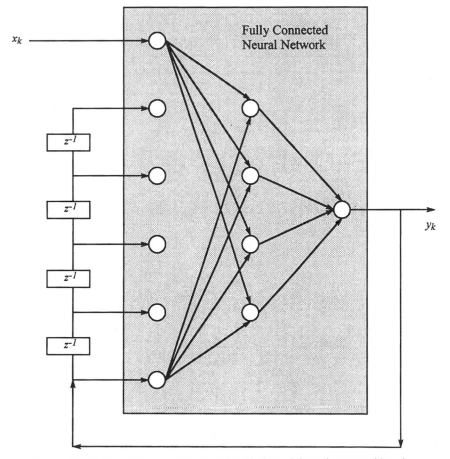

Figure 10.19 Dynamic neural network with time-delayed recurrent inputs.

Models of Dynamic Systems

Since most dynamical systems have temporal behavior, time-delayed versions of the input and/or the output signal are needed to properly model the system. Figure 10.18 shows a neural network with time-delayed versions of the input signal. When a dynamic system's current output usually depends on its previous outputs, recurrent connection with time-delayed versions of the output fed back to the inputs as shown in Figure 10.19 are needed. If the output of the system is not independent of previous inputs, then a time-delayed versions of both the input and output as shown in Figure 10.20 are needed.

In system identification, the neural network is connected in parallel with the system being modeled. Again, dynamic systems require time-delayed direct and recurrent inputs of the type shown in Figures 10.18 to 10.20.

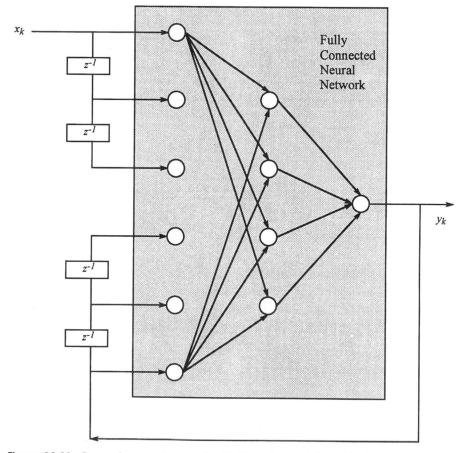

Figure 10.20 Dynamic neural network with time-delayed direct and recurrent inputs.

Figure 10.21 shows the nonrecurrent parallel identification model with time-delayed direct inputs. Figure 10.22 shows the recurrent parallel identification model with time-delayed recurrent (feedback) neurons as inputs. Unfortunately, a recurrent network can become unstable due to the feedback loop between its output and input, and there is no guarantee that the output will converge to a stable configuration. This can be solved by feeding back the signal from the system, not the neural network model output, in the recurrent series–parallel identification model shown in Figure 10.23. Finally, when there is need for both the input and output to be delayed, the general series–parallel identification model shown in Figure 10.24 is used.

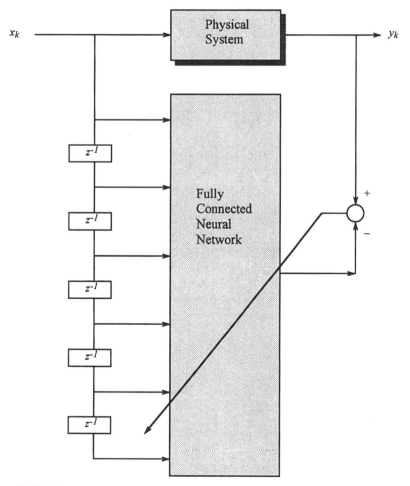

Figure 10.21 Nonrecurrent parallel identification model with time-delayed direct inputs.

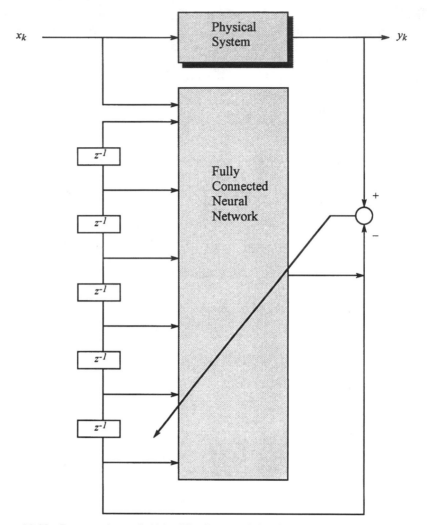

Figure 10.22 Recurrent parallel identification model with time-delayed recurrent outputs from the neural network.

10.7 IMPLEMENTATION OF NEURAL CONTROL SYSTEMS

The status of neural control is well-defined in two recent books based on symposia: *Neural Networks for Control* (Miller et al., 1990) and *Handbook of Intelligent Control—Neural, Fuzzy, and Adaptive Approaches* (White and Sofge, 1992). Particularly valuable contributions in the neural control field include Narendra and Parthasarathy (1990); Narendra (1992), Werbos (1990), Swiniarski (1990), Nguyen and Widrow (1990), Tsund (1991), and Dong and McAvoy (1994). Samad (1993) has explored neural network-based approaches

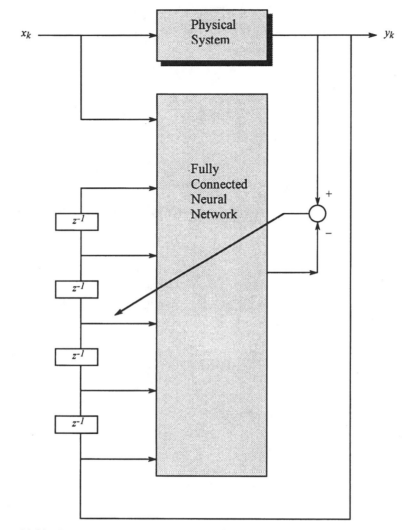

Figure 10.23 Recurrent series–parallel identification model with time-delayed recurrent outputs from the physical system.

to solving control problems, and some of his concepts are incorporated into this chapter.

Inverse Modeling

Inverse modeling develops models that predict corresponding process inputs from process outputs. Inverse models are typically developed with steady-state data and used for supervisory control as feedforward controllers. The appro-

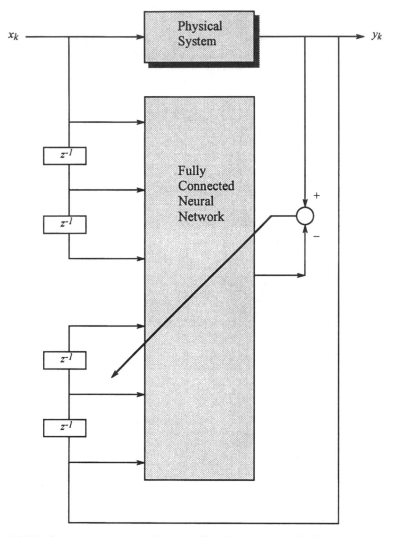

Figure 10.24 General series–parallel identification model with time-delayed direct inputs and time-delayed recurrent outputs from the physical system.

priate steady-state control signal for some setpoint can be determined immediately without the delay associated with the incremental error-correcting operation of feedback control. Neural network inverse models can capture a characteristic source of nonlinearity in many industrial processes (e.g., the variation of process gain with the operating point). Training is readily accomplished since the existing controller output is available.

There are two problems associated with inverse modeling. Many processes have transport delays that imply that any change in the input to a process will only affect the process response after a "dead time." With the introduction of a transport delay into the inverse modeling process and some experimentation with the length of delay, this problem can be overcome in some cases. The second problem is that mapping from steady-state process output to steady-state process input may be one-to-many. Experience indicates that least-mean-squares averaging behavior of many inverse network function approximation models will in such cases lead to control actions that will likely not be effective.

Controller Autotuning

Controller autotuning estimates appropriate values for controller parameters such as PID gains as shown in Figure 10.13. Although there are many traditional methods used to determine these gain constants, neural networks offer a convenient way of dealing with the nonlinearities involved. A nonparametric neural network process model, once trained, can simulate the closed-loop process and serve as the process simulation. An optimization algorithm can then be used to adjust PID gains, in simulation, until some prespecified cost function or evaluation criterion is minimized. The disadvantage of this approach is the computational complexity, since an iterative algorithm is required and each iteration involves a closed-loop simulation using the neural network process model. However, since the algorithm is not being used for closed-loop control, real-time response is not required.

An alternative is to use a neural network as an autotuner in which the output of the neural network is the PID gains. In training the neural network, the three output gains are compared to precomputed optimal PID gains for a set of training examples. An advantage of this approach is that the network can be trained in simulation (i.e., training on actual process data is not necessary) (Swiniarski, 1990; Ruano et al., 1992).

Adaptive Control Systems

The control of a physical system is usually accomplished by a controller that takes information from the outside world and, in the case of a closed-loop system (a system with feedback), from the physical system itself. An industrial drying oven is an example of a system with a closed-loop control system where the difference between the oven temperature and the setpoint is the error signal that provides a basis for changing the power to the oven. If everything remains constant, such a closed-loop system can produce a very uniform product. On the other hand, if the electrical voltage varies from 210 to 250 volts and/or the thermal capacity of the products moving through the oven changes, the end product may not be very uniform. In other situations, some sort of unforeseen (usually nonlinear) temperature dependence may exist or develop with the aging of equipment. Adjustment is needed to take into account the unforeseen variations in the physical system being controlled that take place, based on quality of the product produced. A system capable of adapting to unforeseen changes is an adaptive system. The adaptive processor shown in Figure 10.6 is itself a form of adaptive system. It has a performance calculator capable of providing a quality index based on the input, output, and any other variables available. It has an algorithm by which adjustments can be made, and it has a means of making those adjustments. As utilized in the applications described above, it operates as an autonomous unit to make the desired adjustments. Its functions can also be integrated into a more sophisticated control system to carry out adaptive control of a complex physical system.

The most logical approach to using adaptive control is to place an adaptive controller ahead of the physical system and utilize feedback and other inputs to evaluate the performance. One problem with this "direct adaptive" approach is that the system needs operating data to establish a basis for adaption. Suddenly introducing such a control system without adequate historical data to reach a near normal set of conditions could be disruptive. One approach is to use a conventional PID controller and connect the adaptive system in parallel until it learns to match the behavior of the PID system. A similar approach is to manually control the physical system with feedback through human observation while the adaptive controller accumulates the historical information needed to perform its tasks of performance evaluation.

Adaptive Model Control

In adaptive model control, an external model of the physical system that is not part of the control system is obtained using system identification as discussed in Section 10.5. Then the model is used to determine control inputs to the plant, which will produce the desired system outputs. (Manual control can also be used as a model in this approach.) When we apply these same control inputs to the actual physical system, the system output closely matches the desired output. While this appears to be open-loop control, the loop is actually closed through the adaptive process.

Adaptive Inverse Model Control

In adaptive inverse control, the unknown physical system can be made to track an input command signal when it is applied to a controller whose system response function approximates the inverse of its system response function. (All the following arrangements can be implemented using artificial neural networks as the adaptive inverse models.) This adaptive inverse model becomes the controller whose output is the input to the physical system. Its weight adjustment is slaved to a second adaptive inverse model of the type shown in Figure 10.8b, that is used to minimize the error between the plant

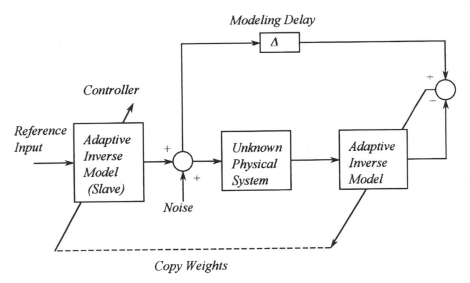

Figure 10.25 Adaptive inverse model control system.

output and the input setpoint. This arrangement is shown in Figure 10.25. An input perturbation or noise is deliberately introduced to ensure that the adaptive process occurs continuously. The introduction of the pseudorandom binary-type noise discussed in Section 10.4 can accomplish this task while at the same time offering the opportunity for cross-correlation and cross-spectral density measurements that could be useful either in the control process, in the identification of process parameters, or the reference model discussed in the next section.

Model Reference Adaptive Control

The configuration of Figure 10.25 can be modified to implement model reference adaptive control. In this concept, a physical system is adapted in such a way that its overall input–output response characteristics best match a reference model response. In this case, the reference model replaces the modeling delay. As a result, the unknown physical system and the adaptive inverse model will be matched to the reference model rather than that of a simple delay.

In some applications, the reference model is a performance model; that is, the performance (efficiency, emissions, etc.) of the system is matched with that of the model rather than matching input–output behavior. The resultant system is usually more complicated than matching input–output response as described above, and it is usually implemented as a form of reinforcement learning as described in the next section.

Reinforcement Learning

Reinforcement learning addresses the problem of improving performance as evaluated by any index of performance the user chooses. The basis for this approach is that desired control signals exist that lead to optimization, but the learning system is not told what they are because there is no system knowledgeable enough to identify them. In reinforcement, the object is to determine desired changes in the controller output that will increase the index of performance, which is not necessarily defined in terms of the desired outputs of the system.

Reinforcement learning involves two issues: (1) how to construct a critic network capable of evaluating physical system performance consistent with the chosen index of performance and (2) how to alter the controller outputs to improve performance as measured by the critic network. These issues are discussed extensively by Barto (1990). Clearly, there is not one unique approach to these issues. Some approaches attempt to introduce knowledge known about the system to bias the learning process in a favorable direction. Others rely on statistical considerations. For instance, a class of reinforcement-learning algorithms known as *stochastic learning automata* probabilistically select actions from a finite set of possible actions and update action probabilities on the basis of evaluative feedback. It is also possible to combine stochastic learning automata with parameter estimation by mapping pattern inputs to action probabilities. As these parameters are adjusted under the influence of evaluative feedback, action probabilities are adjusted to increase the expected evaluation of the index of performance. Research is continuing in these areas.

Reinforcement learning is a very general approach to learning that can be applied when the knowledge required to apply supervised learning is not available. If sufficient information is available, reinforcement learning can readily handle a specific problem. However, it is usually better to use other methods discussed earlier in this section, because they are more direct and their underlying analytical basis is usually well understood.

10.8   APPLICATIONS OF NEURAL NETWORKS IN NOISE ANALYSIS

The integrated use of neural network and noise analysis technologies offers advantages not available by the use of either technology alone. The application of neural network technology to noise analysis offers an opportunity to expand the scope of problems where noise analysis can be used productively. The two-sensor technique, in which the related responses of two sensors on a system whose characteristics are unknown responding to an unknown driving source, is used to illustrate such integration.

In the last three decades, vibration analysis has become a separately identified field of noise analysis that is used as a means of detecting faults in dynamic systems, estimating parameters for models of complex systems, and detecting and identifying loose parts in fluid flow systems. Commercial instruments that quantitatively evaluate the power spectra of signals from accelerometers mounted on rotating machinery and interpret the results automatically (based on a model of the system being tested) are readily available. In other cases, spectra must be interpreted by experts because of the complexity of the system and the complex vibration spectra it produces. In these cases, neural networks with their ability to learn characteristics associated with different types and sources of vibration can enhance our ability to interpret the measurements.

Two-Sensor Technique

A specific example will serve to illustrate the symbiotic relationship between vibration analysis and neural networks. The technique described here involves training a neural network to model the internal behavior of a component or system using vibration data taken from two sensors (accelerometers) located at different positions or mounted in different directions on the component or system. The power spectral density (PSD) (typically 128 values) of a sampled time series (typically 100,000 samples which produces 390 spectra) from one accelerometer is used as the input to the neural network, and the PSD of the sampled time series from the other accelerometer taken at the same time is the desired output of the neural network. The network is trained using the 390 pairs of spectra when the component or system is known to be operating properly. The trained neural network is then put into a monitoring mode to predict the output (second) sensor PSD from the input (first) sensor PSD, and a comparison is made between the predicted and actual output signal PSDs using the method described in Figure 10.26. The mean square difference Δ was used as an index of whether operation was normal or deteriorated. Significant deviations indicate that the interrelationship between the input and output signals has been modified due to a change (failure) in the component or system. The usefulness of this methodology has been demonstrated in the monitoring of the operability of check valves (Ikonomopoulos et al., 1992; Uhrig, 1993) and a pump-motor bearing (Loskiewicz-Buczak et al., 1992). These applications are described in Examples 10.3 and 10.4 later in this chapter.

Noise Analysis Considerations

In almost every situation, both sensors measure output vibrations induced by a driving function (e.g., imbalance in a rotating system or turbulence of the

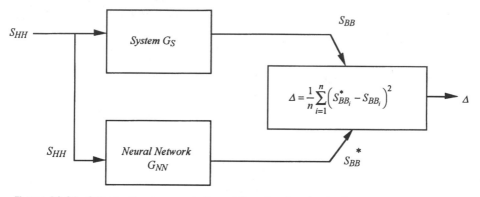

Figure 10.26 Schematic representation of the check-valve testing procedure. $\Delta \approx 0$ if the tested valve is in good condition. $\Delta > 0$ if the tested valve has some degradation. $\Delta \gg 0$ if the tested valve has significant degradation.

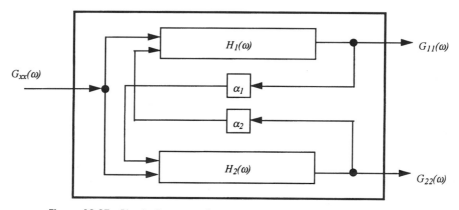

Figure 10.27 Block diagram of a one-signal-input–two-output system.

water as it flows through a pipe or valve) (Uhrig, 1995). This arrangement can be represented in the frequency domain as shown in Figure 10.27, where the PSD of the driving turbulence function is represented by $G_{xx}(\omega)$, the PSDs of the resultant vibration are represented by the output PSDs of the accelerometers $G_{11}(\omega)$ and $G_{22}(\omega)$, α_1 and α_2 are coupling coefficients for the attenuated vibration transmitted between the two sensors, and $H_1(\omega)$ and $H_2(\omega)$ are the system response functions relating the driving turbulence to the resultant outputs of the two accelerometers. Application of the input–output relationships for PSDs (Uhrig, 1970) as given in equation (10.2-9),

$$G_{ii}(\omega) = |H(\omega)|^2 G_{xx}(\omega), \qquad \text{where } i = 1 \text{ or } 2 \qquad (10.8\text{-}1)$$

when applied to the system of Figure 10.27, gives the relationship between $G_{11}(\omega)$ and $G_{22}(\omega)$; after mathematically eliminating $G_{xx}(\omega)$, this becomes

$$G_{22}(\omega) = G_{11}(\omega) \left\{ \frac{\left[1 + \alpha_1 |H_1(\omega)|^2\right] |H_2(\omega)|^2}{\left[1 + \alpha_2 |H_2(\omega)|^2\right] |H_1(\omega)|^2} \right\} \qquad (10.8\text{-}2)$$

Let us postulate a simple model of this phenomena in which these system response functions $H_i(\omega)$ are assumed to be underdamped, second-order systems (i.e., each system response function has a single peak which may be located at any frequency). The frequency response functions and their square moduli can be represented by

$$H_i(\omega) = \frac{K_i}{\left[1 - \tau_i^2 \omega^2\right] + j\lambda_i \omega} \qquad (10.8\text{-}3)$$

$$|H_i(\omega)|^2 = \frac{K_i^2}{\left[1 - \tau_i^2 \omega^2\right]^2 + \left[\lambda_i \omega\right]^2} \qquad (10.8\text{-}4)$$

where λ_i represents damping constants and τ_i represents time constants associated with the natural frequencies of the systems. Substitution of equation (10.8-4) into equation (10.8-2) gives

$$G_{22}(\omega) = G_{11}(\omega) \left\{ \frac{K_2^2 \left[\left[1 - (\tau_1 \omega)^2\right]^2 + \left[\lambda_1 \omega\right]^2 + \alpha_1 K_1^2\right]}{K_1^2 \left[\left[1 - (\tau_2 \omega)^2\right]^2 + \left[\lambda_2 \omega\right]^2 + \alpha_2 K_2^2\right]} \right\} \qquad (10.8\text{-}5)$$

We can obtain the general shape of the curve by considering very high and very low frequencies. For very high frequencies, the term in braces in equation (10.8-5) approaches a constant value of $[K_2^2 \tau_1^4 / K_1^2 \tau_2^4]$. For very low frequencies, the term in the braces approach a constant value of $\{K_2^2[1 + \alpha_1 K_1^2]/K_1^2[1 + \alpha_2 K_2^2]\}$. For mid-range frequencies, the term in the braces is greater than for high or low frequencies. In both $H_1(\omega)$ and $H_2(\omega)$, the amplitudes of their peaks and the frequencies at which the peaks occur are dependent upon λ_i and τ_i, respectively. Hence the peak value of the term in braces and its location are dependent on the values of λ_i and τ_i.

For a more complex model in which the system response functions $H_1(\omega)$ and $H_2(\omega)$ have several peaks, many second-order systems having m and n peaks designated by subscripts i and k, respectively, can be superimposed. Equation (10.8-5) now becomes

$$G_{22}(\omega) = G_{11}(\omega) \left\{ \sum_{i=1}^{m} \sum_{k=1}^{n} \frac{K_{2k}^2 \left[\left[1 - (\tau_{1i} \omega)^2\right]^2 + \left[\lambda_{1i} \omega\right]^2 + \alpha_{1i} K_{1i}^2\right]}{K_{1i}^2 \left[\left[1 - (\tau_{2k} \omega)^2\right]^2 + \left[\lambda_{2k} \omega\right]^2 + \alpha_{2k} K_{2k}^2\right]} \right\}$$

$$(10.8\text{-}6)$$

Again, the values of terms in the braces, which represent the square modulus of the overall system response function for the interior behavior of a complex system, approach constant, but different, values for very low and high frequencies, and there are many peaks over the intermediate frequencies for both $H_1(\omega)$ and $H_2(\omega)$.

This model is consistent with the experimentally measured spectra $G_{11}(\omega)$ and $G_{22}(\omega)$, indicating that the multipeaked representation of the overall system response function of the interior behavior is consistent with the neural network model used in the two-sensor neural network technique for fault identification. Furthermore, this model gives insight into the phenomena modeled by the neural network.

Alternately, the accelerometers can be near each other but mounted so that they measure acceleration in different (usually perpendicular) directions. Under perfect balance conditions for both forces and moments (a rare condition), spectra in different directions at one position would be exactly the same. In almost all real-world conditions, the spectra from sensors located in perpendicular directions are different. Furthermore, the relative shapes of the two spectra change as the systems change or deteriorate.

The diagnosis of faults usually involves measurement and analysis of small fluctuating signals that represent the dynamic behavior of a system or component. Usually this fluctuation is the output signal of an accelerometer measuring vibration or acoustics, but it can be the small random-like fluctuations of a *steady-state* variable (pressure, temperature, etc.). The objectives of such diagnostic work are to identify the existence of abnormalities/deviations and interpreting the results of the monitoring in an intelligent way to identify the fault so that noise specialists and experts are not required for interpretation. While relatively little attention to date has been given to automating these procedures, it is clear that such automation is possible and necessary when this technology is implemented in actual plants and complex industrial/scientific systems. Furthermore, the neural networks can be implemented in microchips to give almost instantaneous outputs.

Typically, measured variables from components or systems are analog variables that must be sampled and normalized to expected peak values before they can be utilized. All the normal precautions associated with digitizing analog data must be exercised to avoid the adverse effects of aliasing and nonstationarity. Often, data must be processed to put them into an acceptable form (e.g., a fast Fourier transformation of the time series to produce a spectrum). In most cases, comparison of predicted results (based on the output(s) of neural network models developed from data taken when the system was working properly) or patterns (learned by neural network models from data presented to it along with actual results or patterns involved) is utilized for fault detection.

Example 10.3 Check-Valve Monitoring. Although there are many possible failure mechanisms for check valves, the most common problems associated with check-valve failures are due to flow induced system disturbances or system piping vibrations. These vibrations and disturbances induce measurable accelerations that produce check-valve component wear and sometimes component failure. Analysis of time records from piezoelectric accelerometers attached to check valves on a large nuclear power plant has been used to demonstrate this process. The procedure uses an autoassociative-like neural network, in which the inputs and desired outputs are values of the PSDs of two related time series representing vibration at two different positions on the valve. It was trained to produce a neural network model of the interrelationship when the valve is operating properly as described in Section 10.8. During monitoring, the output PSD of one accelerometer is used to predict the output PSD of the other accelerometer, which is then compared with the actual PSD. A significant deviation indicates failure of the check valve. The difference in the two spectra are evaluated numerically using the procedure indicated in Figure 10.26 to evaluate the mean square difference Δ.

Comparison of PSD spectra between identical 30-inch check valves (one broken and one normal), operating under identical conditions, demonstrated that this technique can identify the failed valve. Subsequent measurements taken on the broken valve after it was repaired further confirmed the validity of this technique. Tests on three 6-inch check valves (one normal and two that failed for different reasons) operating under identical test conditions has indicated that different kinds of failures give different values for the mean square difference Δ (Ikonomopoulos et al., 1992). Larger values of Δ indicate more serious problems. $\square$

Example 10.4   Large Motor Pump Bearing Failure. The two-sensor technique was also used to analyze the progressive failure of a large (950 HP) motor pump bearing in a nuclear power plant (Loskiewicz-Buczak et al., 1992). A series of measurements of horizontal and vertical components of acceleration for a large motor-pump bearing were taken periodically at intervals of about 6 weeks throughout the operating lifetime of the bearing and as it began to fail. The power spectra of the horizontal and vertical components of acceleration on the bearing during the first four sets of measurements (when the bearing was known to be operating properly) were the input and desired output, respectively, of a neural network while it was being trained. The bearing operated normally for the next three months and then began to fail. For the next five sets of measurements, while the bearing progressed toward failure, the predicted value of the vertical component of acceleration (obtained from the neural network with the horizontal component as the input) was compared with the actual value of the vertical component. The mean square difference Δ (described in Section 10.7) between the predicted and actual vertical spectra grew as the bearing progressed toward failure.

It can be speculated that it might be possible to predict remaining life in bearings. The integrated use of neural networks and noise (vibration) analysis has been shown to perform satisfactorily in monitoring the operability of check valves in power plants and in a large motor-pump bearing. There are many other applications in neural networks to the noise analysis field where the two technologies can be advantageously used together. □

10.9 TIME-SERIES PREDICTION

Neural networks can be used to predict future values in a time series based on current and historical values. Such predictions are, in a sense, a form of inference measurements discussed earlier. They are particularly useful to economists, meteorologists, and planners. Recently, there has been an extraordinary amount of interest in the use of neural networks to predict the stock market behavior, including the publication of at least two commercial periodicals specializing in financial investing (*Artificial Intelligence in Finance*, published by Miller-Freeman Publishing Co. since 1995, and *NeuroVe$t*, published by R. B. Caldwell since 1993). This popularity continues in spite of the fact that the predictions by neural networks cannot be explained. Perhaps the main reasons for the continuing popularity in the field are that neural networks do not require a system model and that they are relatively insensitive to unusual data patterns.

Although backpropagation neural networks are usually used for time-series prediction, it is possible to use any neural network capable of mapping an input vector into an output vector. Typically, the input of a single time series into a neural network is made as shown in Figure 10.28. The fluctuating variable is sampled at an appropriate rate to avoid aliasing, and sequential samples are introduced into the input layer in a manner similar to that used in a transverse filter. At every time increment, a new sample value is introduced into the rightmost input neuron, and a sample value in the leftmost input neuron is discarded. The main difference compared to the transverse filter is that the sample preceding those going to the input is introduced into the single output neuron as the desired output. In this way, the network will be trained to predict the value of the time series one time increment ahead based on the previously sampled values. The network can be trained to predict more than one time increment ahead, but the accuracy of the prediction decreases when predictions are further into the future. Since such systems are often used in real time, or to secure data from historic records, the amount of training data is usually very large. Even so, it is important to periodically check the training to ensure that overtraining does not occur.

Although it is possible to predict multiple outputs, it is best to predict only one value because the network minimizes the square error with respect to all neurons in the output layer. Minimizing square error with respect to a single output gives a more precise result. If multiple time predictions are needed, individual networks should be used for each prediction.

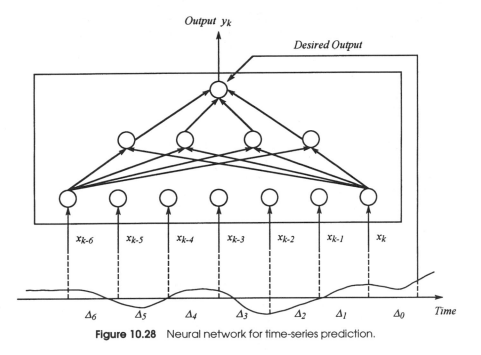

Figure 10.28 Neural network for time-series prediction.

Generally, large-scale deterministic components, such as trends and seasonal variations, should be eliminated from inputs. The reason is that the network will attempt to learn the trend and use it in the prediction. This may be appropriate if the number of input neurons is sufficient for input data to span a complete cycle (e.g., an annual cycle). If trends are important, they can be removed and then added back in later. This allows the network to concentrate on the important details necessary for an accurate prediction.

The standard method of removing a trend is to use a least-squares fit of the data to a straight line, although nonlinear fitting may be appropriate in some cases (e.g., cyclic fluctuations). An alternate method of removing trends and seasonal variations is to pass the data through a high-pass filter with a low cutoff frequency. There are alternative techniques in which a low-pass filter is used to leave only the slowly varying trend which then is subtracted from the original signal, with the difference being the value sent to the neural network input layer.

One of the interesting variations of the above technique for prediction is to use differences between successive sample values as inputs to the neural network. This effectively eliminates constant trends and slowly changing trends by converting them to a constant offset. Even seasonal trends are usually removed. Using differences in predicting is generally useful in such fields as stock price predictions, especially if the difference is scaled relative to the total price of the stock, which is effectively using the percent price change.

REFERENCES

Andersen, K., Barnett, R. J., Springfield, J. F., Cook, G. E., Bjorgvinsson, J. B., and Strass, A. M., A Neural Network Model for the Gas Tungsten Arc Welding Process, *NASA Tech Brief MFS-26209*, George C. Marshall Space Flight Center, AL, 1991.

Barto, A. G., Connectionist Learning for Control, in *Neural Networks for Control*, W. T. Miller III, R. S. Sutton, and P. J. Werbos, eds. MIT Press, Cambridge, MA, 1990, Chapter 1.

Caldwell, R. B., Publisher, *NeuroVe$t*, Haymarket, VA, published since 1993.

Dong, D., and McAvoy, T. J., Sensor Data Analysis Using Autoassociative Neural Networks, *Proceedings of the World Congress on Neural Networks*, Vol, 1, San Diego, CA, 1994, pp. 161–166.

Foslien, W., Konar, A. F., and Samad, T., Optimization with Neural Memory for System Identification, in *Applications of Artificial Neural Networks III*, *Proceedings of the SPIE*, Vol. 1709, S. K. Rogers, ed., Bellingham, WA, 1992.

Guo, Z., and Uhrig, R. E., Use of Genetic Algorithms to Select Inputs for Neural Networks, in *Proceedings of Special Workshop on COGANN* (*Combination of Genetic Algorithms and Neural Networks*), International Joint IEEE-INNS Conference on Neural Networks (IJCNN), Baltimore, MD, June 6, 1992.

Gyftopoulos, E. P., and Hooker, R. J., Signals for Transfer Measurements in Nonlinear Systems, in *Noise Analysis in Nuclear Reactor Systems*, R. E. Uhrig, ed., AEC Symposium Series #4 (TID-7679), June 1964, pp. 335–345.

Ikonomopoulos, A., Tsoukalas, L. H., and Uhrig, R. E., Use of Neural Networks to Monitor Power Plant Components, in *Proceedings of the American Power Conference*, Chicago, IL, April 13–15, 1992.

Lee, Y. W., *Statistical Theory of Communication*, John Wiley & Sons, New York, 1960.

Loskiewicz-Buczak, A., Alguindigue, I. E., and Uhrig, R. E., Vibration Monitoring Using Sensor Readings Predicted by a Neural Network, in *Proceedings of EPRI's 5th Predictive Maintenance Conference*, Knoxville, TN, September 21–23, 1992.

Miller, W. T., III, Sutton, R. S., and Werbos, P. J., eds., *Neural Networks for Control*, MIT Press, Cambridge, MA, 1990.

Miller-Freeman Publishing Co., *Artificial Intelligence in Finance*, San Francisco, CA, published since 1995.

Narendra, K. S., Adaptive Control of Dynamical Systems Using Neural Networks, in *Handbook of Intelligent Control*, D. A. White and D. A. Sofge, eds., Van Nostrand Reinhold, 1992.

Narendra, K. S., and Parthasarathy, K., Identification and Control of Dynamical Systems Using Neural Networks, *IEEE Transactions on Neural Networks*, Vol. 1, pp. 4–27, 1990.

Nguyen, D. H., and Widrow, B., Neural Network for Self-Learning Control Systems, *IEEE Control Systems Magazine*, pp. 18–23, April 1990.

Pao, Y.-H., *Adaptive Pattern Recognition and Neural Networks*, Addison-Wesley, Reading, MA, 1989.

Piovoso, M. J., et al., Neural Network Process Control, in *Proceedings of Analysis of Artificial Neural Networks Applications Conference*, 1991.

Ruano, A. E. B., Fleming, P. J., and Jones, D. I., Connectionist Approach to PID Tuning, *IEEE Proceedings-D*, Vol. 139, pp. 29–285, 1992.

Samad, T., Neurocontrol: Concepts and Practical Applications, in *Intelligent Control Systems: Theory and Practice*, M. M. Gupta and N. K. Sinha, eds., IEEE Press, New York, 1993.

Swiniarski, R. W., Novel Neural Network Based Self-Tuning PID Controller Which Uses Pattern Recognition Technique, in *Proceedings of the American Controls Conference*, 1990, pp. 3023–3024.

Tsung, F. S., Learning in Recurrent Finite Difference Networks, in *Connectionist Models: Proceeding of the 1990 Summer School*, D. E. Touretzky et al., eds., Morgan Kaufmann, San Mateo, CA, 1991.

Uhrig, R. E., Tsoukalas, L. H., Ikonomopoulos, A., Essawy, M., Black, C., and Yancey, S., Using Neural Networks to Monitor the Operability of Check Valves, in *Proceedings of the Conference on Expert System Applications for the Electric Power Industry*, Phoenix, AZ, December 8–10, 1993.

Uhrig, R. E., Integrating Neural Network Technology and Noise Analysis, special edition of *Progress in Nuclear Energy* on IMORN-25 Meeting on Reactor Noise, June 13–15, 1994, published May 1995.

Uhrig, R. E., *Random Noise Techniques in Nuclear Reactor Systems*, Ronald Press (now part of Prentice-Hall), New York, 1970.

Werbos, P. J., Overview of Design and Capabilities, in *Neural Networks for Control*, W. T. Miller III, R. S. Sutton, and P. J. Werbos, eds., MIT Press, Cambridge, MA, 1990, Chapter 2.

Werbos, P. J., Neurocontrol and Supervised Learning; An Overview and Evaluation, in *Handbook of Intelligent Control*, D. A. White and D. A. Sofge, eds., Van Nostrand Reinhold, New York, 1992, Chapter 3.

White, D. A. and Sofge, D. A., eds., *Handbook of Intelligent Control*, Van Nostrand Reinhold, New York, 1992.

Widrow, B., and Hoff, M. E., Adaptive Switching Circuits, 1960 IRE WESCON Convention Record, 96-104, New York, 1960.

Widrow, B., and Stearns, S. D., *Adaptive Signal Processing*, Prentice-Hall, Englewood Cliffs, NJ, 1985.

PROBLEMS

1. Show how Equations 10.1-10 and 10.1-11 are obtained from Equations 10.1-8 and 10.1-9, respectively, for white noise inputs.

2. For the system in Figure 10.2, add a white noise $q(t)$ to the input. Derive the equations for the power spectral density of the output and the

cross-spectral density between the input and the output comparable to Equations 10.1-23 and 10.1-26.

3. In Figure 10.11*b* the control system is located after the actual system and controls by feedback whereas the reverse arrangement is given in Figure 10.12. Discuss the merits of the two arrangements.

4. The difference between the systems in Figures 10.22 and 10.23 is the source of the signal for the recurrent feedback (i.e., from the neural network output in Figure 10.22 and from the system output in Figure 10.23). If the neural network were perfectly trained, these two signals would be identical. Discuss the difference that the source of this signal makes and explain why there is a difference.

5. In time-series prediction, the quality of the predicted signal deteriorates as the time increment into the future increases. Discuss how the quality (by whatever criterion you choose) decreases with future time. How do you determine a practical limit? (*Note*: This problem is also discussed in the context of fuzzy control in Section 15.4.)

11

PRACTICAL ASPECTS OF USING NEURAL NETWORKS

11.1 SELECTION OF NEURAL NETWORKS FOR SOLUTION TO A PROBLEM

Perhaps the best approach to determine whether an application of neural networks is appropriate is to compare its characteristics to those that have been successful in other application. Bailey and Thompson (1990a) have cited a survey of successful neural-network applications developers and gave the following heuristics for successful applications:

- Conventional computer technology is inadequate.
- Problem requires qualitative or complex quantitative reasoning.
- Solution is derived from highly interdependent parameters that have no precise quantification.
- The phenomena involved depend upon multiple-interacting parameters.
- Data are readily available but are multivariate and intrinsically noisy or error-prone.
- There is a great deal of data from specific examples available for modeling the system.
- Some of the data may be erroneous or missing.
- The phenomena involved are so complex that other approaches are not useful, too complicated, or too expensive.
- Project development time is short, but sufficient network training time is available.

Most successful applications of neural networks involve pattern recognition, statistical mapping, or modeling. Successful applications can include signal validation, process monitoring, diagnostics, signal and information processing, and control of complex (often nonlinear) systems. However, problems that can be solved using conventional computer methodologies, especially those that require high precision or involve mathematical rigor, are usually not appropriate for an artificial neural network approach.

Choice of Neural Network Type

The appropriate choice of the type of neural network (supervised, unsupervised, or reinforced) depends on data available. Supervised learning requires pairs of data consisting of input patterns and the correct outputs, which are sometimes difficult to obtain. Unsupervised training classifies input patterns internally and does not need expected results. The data requirements for unsupervised training are thus much easier and less costly to meet, but the capability of the network is significantly less than for supervised learning. A compromise between supervised and unsupervised training is reinforcement learning, which requires an input and only a grade or reward signal as the desired output.

Time required for both the training and recall are also important in the development of neural networks. Most neural networks have relatively long training times, but the recall involves only a single pass through the network. When the neural network is implemented in hardware with the neurons operating in parallel, the recall time is virtually instantaneous. On the other hand, certain paradigms, such as the probabilistic neural network, radial basis function, and general regression neural network, train in a single pass through the network, but the execution time is essentially the same as the training time. Hence, the need to meet on-line requirements (e.g., in an active control system) may dictate the type of neural network used, or it may require that the network be implemented in hardware.

11.2 DESIGN OF THE NEURAL NETWORK

Size of Neural Networks

Neural network size is sometimes related to the experience of the user as much as the nature of the problem. Beginners tend to stick with small networks and reduce the size of the application accordingly. Those with considerable experience with neural networks are usually willing to let the nature of the problem decide the size of the network. With the neural network simulation software available for personal computers and workstations today, a neural network with a thousand neurons and perhaps a hundred thousand connections may no longer be a practical upper limit for nonstatistical paradigms such as backpropagation or counterpropagation.

Choice of Output

The type of output is usually determined by the nature of the application. The activation of the output neurons may be either binary or gray scale (many individual values). Real-number outputs translate into values such as dollars, time units, or distances and may be given in binary form or gray scale. Each of the four common interpretations of neural network outputs—classifications, patterns, real numbers, and optimal choice—has its own specific requirements. For example, since classifications statistically map input patterns into discrete categories, there will usually be two or more output neurons with only one having an output for a given input. In contrast, neural networks that identify patterns such as spectra often have multiple output neurons, all usually active at the same time, which form a pattern in response to the input. Optimization problems usually yield a special pattern that can be interpreted as a set of decisions (Bailey and Thompson, 1990b).

Neuron Activation function

Typically, the activation function is a continuous function that increases monotonically between a lower limit and an upper limit (0 and 1 or -1 and $+1$) as the weighted summation increases in magnitude. Since one of the primary purposes of the activation function is to keep the outputs of the neurons within reasonable limits, it is sometimes called a "squashing" function. By far the most common activation function is the logistic function discussed in Chapter 7, but virtually any function meeting the sigmoidal requirements stated in Chapter 8 will work satisfactorily. Step or signum (threshold) functions are often used for the activation function when the inputs and outputs are binary (0 and 1) or bipolar (-1 and $+1$).

Activation functions that have been used include linear, clamped linear, step, signum, sigmoid, arctangent, and hyperbolic tangent functions. The choice is usually based on both the types of input and output and the learning algorithm to be used. Certain paradigms such as backpropagation require that the derivative of the activation function be continuous, which eliminates step, signum, and clamped linear functions. Many binary (0 and 1) and bipolar (-1 to $+1$) input–output pairs use networks with step and signum functions, respectively, for the activation function. Continuous valued outputs use linear or sigmoidal (or other S-shaped) activation functions.

Number of Layers

Backpropagation networks typically have three layers, but more may be advantageous under some circumstances. It is sometimes better to use two smaller hidden layers rather than one bigger layer. Some neural network paradigms commonly used have a predetermined number of layers. Adalines,

Madalines, Hopfield networks, ART-1, Kohonen self-organizing feature maps, and bidirectional associative memories all require either one or two layers.

Hidden layers act as layers of synthesis, extracting features from inputs. Usually a larger number of hidden layers increases the processing power of the neural network but requires significantly more time for training and a larger number of training examples to train the network properly. As indicated in Chapter 7, one hidden layer (i.e., a three-layer network) with sufficient neurons is capable (theoretically) of representing any mapping. Additional hidden layers should be added only when a single hidden layer has been found to be inadequate. Cascade correlation neural networks start with two layers and add as many one-neuron layers as necessary to satisfy its convergence criterion.

Number of Neurons in Each Layer

The number of neurons in the input and output layers are determined by the nature of the problem. For instance, a problem that utilizes a 128-point power spectral density function as an input and classifies the inputs into 10 categories requires 128 neurons in the input layer and 10 neurons in the output layer. Determining the proper number of neurons for the hidden layer is often accomplished through experimentation. Too few neurons in the hidden layer prevent it from correctly mapping inputs to outputs, while too many impede generalization and increase training time. Too many neurons may allow the network to "memorize" the patterns presented to it without extracting the pertinent features for generalization. Thus, when presented with new patterns, the network is unable to process them properly, because it has not discovered the underlying principles of the system.

For a network with a single hidden layer, it is common practice to initially make the number of neurons equal to about two-thirds of the number in the input layer (Bailey and Thompson, 1990a). When there is more than one hidden layer, the number of neurons is significantly smaller in each hidden layer. Experimentation with greater and smaller numbers of neurons in the hidden layer(s) may change the training time as well as the ability of the neural network to generalize. Often there is a wide range in the number of neurons in the hidden layer that can be used successfully. Harp, Samad and Guha (1989) has utilized an optimization methodology for determining the optimal number of neurons in a single hidden layer of a neural network based on a genetic algorithm optimization process.

Multiple Parallel Slabs

Another method of increasing a neural network's processing power is to add multiple slabs within a single hidden layer. A multiple parallel slab arrangement may use different types of activation functions and different numbers of neurons, because this architecture is attempting to force each slab to extract different features simultaneously.

11.3   DATA SOURCES AND PROCESSING FOR NEURAL NETWORKS

A successful neural network requires that the training data set and training procedure be appropriate to the problem. This includes making the training data set representative of the kinds of patterns the operational network will have to recognize. Furthermore, the training set must span the total range of input patterns sufficiently well so that the trained network can generalize about the data. In order to have extrapolation and interpolation capabilities, neural networks must be trained on a wide enough set of input data to generalize from their training sets. Although most of what is presented here deals with neural networks using the backpropagation training paradigm, much of what is said applies to other, less common neural network paradigms as well.

All data that in any way relate to the application should be reviewed and purged of any data that are considered to be unreliable or impractical for technical or economic reasons. Combining and/or preprocessing data to make it more meaningful can be extremely beneficial. For example, power spectral density functions are much more useful than a time series from sampled time records as inputs to neural networks.

Errors in Databases

Databases are rarely perfect. Hence, a database for 100,000 homes may contain a few with entries such as "200 people in a home or a child aged 1975" Protecting neural networks from such gross errors is essential, because it doesn't take many ridiculous values to distort a neural network's training, especially when the importance of errors is increased by squaring those errors. The use of elementary statistical analysis and time plots of data can help detect such errors.

Subtle data errors that don't involve grossly out-of-range values also occur. Checking the consistency of units will eliminate such errors as one office reporting production in units and another reporting it in dozens. The only way to find errors like this is to remain alert to the possibility of data errors and investigate any suspicions that develop about the sensibleness of the data. Clustering data can often help identify discrepancies. If erratic data cannot be fixed, the impact of discarding the data should be investigated. For instance, discarding sales for the month of December when Christmas sales are very large, even when it is necessary because of erroneous data, can distort the results of an analysis.

Incomplete Data Sets

In spite of the rhetoric that neural networks can work with incomplete data sets, missing data can create serious problems. If the data cannot be found (missing data often has merely been placed in the wrong field or misnamed), the common sense (and technically correct) thing to do is to replace every missing value with the best estimate of what it would have been were it not missing. (Crooks 1992) suggests several ways in which this can be done. The simplest method is to replace missing values with the expected value of the variable. If all other variations in the example are ignored, the expected value of a real-valued variable is the mean of the variable across the sample of cases. If the variable is an arbitrary categorization, the most common value or mode is appropriate. For ordinal values, the median value for the population is the expected value. More sophisticated methods are available, but they invariably involve the assumption that the process is time stationary and that the underlying conditions do not change during the time the missing data are important. Whether this is true is dependent upon the individual situation. The pragmatic approach of using whatever technique seems to make the model train and predict better is usually best, but there is need to ensure that future predictions have some general and reliable basis by using an appropriate test data set.

Time Variations in Data

A neural network can detect trends in time-oriented data such as sales data. Although recurrent neural network models have some sustaining memory of previous data, most networks consider only one example at a time. Since there is no explicit memory of the example from an earlier time, it is not possible to simply present data in a sequential order (i.e., first Monday's data, then Tuesday's data, and so on) and expect it to find the trend. Time-oriented problems such as predicting tomorrow's sales requires that the sales for the last week or two (or some appropriate time period) be utilized. A productive approach that is often used in training is to present input training that spans several time periods (days, weeks, hours, months, etc.) and use data for the next time period as the desired output.

Pictorial Data

When the information is pictorial, the data for the neural network are best suited to a nondistributed representation. For example, in a black-and-white picture, each input neuron receives a number representing the intensity of one pixel (picture element) of the visual field for every point in the picture, as well as parameters indicating the location of the pixel. The biggest problem with pictures is that too many pixels are needed to train a neural network in a reasonable amount of time. If a camera image is 1024 pixels on a side, the number of necessary input neurons is more than a million, and the training time would probably be prohibitive. One approach for feature extraction uses Fourier descriptors of the items to be recognized and feeds them into a neural network for recognition and translation into meaningful results. Characters and graphics have frequency magnitude and phase "signatures" that can be recognized by a neural network. The neural network's output must be formatted into an appropriate form for training and recall.

Data Acquisition

Applications requiring sensory data input to the neural network are impeded when information transfer from the equipment is disrupted. Sensors and data acquisition facilities must be thoroughly tested before being used to provide data to a neural network. Indeed, the influence of sensors and data acquisition facilities on the overall information processing systems needs to be evaluated prior to including them in the system. Situations sometimes arise in which one missing piece of information disrupts the flow (timing) of the system and causes undesirable, sometimes unpredictable, and almost always erroneous results.

11.4   DATA REPRESENTATION

Data may have to be converted into another form to be meaningful to a neural network. How data are represented and/or translated also plays an important role in the network's ability to grasp the problem, that is, a network can learn more easily from some representations than from others. Certain kinds of data (e.g., the time-oriented data used in such problems as forecasting) are especially difficult to handle.

Continuous Valued Versus Binary Representations of Data

Data may be continuous-valued or binary. Sometimes data can be represented either as a single continuous value or as a set of ranges that are assigned binary representations. For instance, temperature could be represented by the actual temperature values or as one of five possible values: frozen, chilled, mild, warm, or hot. When there are naturally occurring groups, the binary categories are often the best method for making correlations. When the values are very continuous, artificially breaking them up into groups can be a mistake, because it is often difficult for the network to learn examples that have values on or near the border between two groups.

Arbitrary Numerical Codes

Using continuous-valued inputs to represent unique concepts can cause problems. Although it may seem perfectly reasonable to represent the months of the year as numbers from 1 to 12, the neural network will presume such data to be continuous-valued and as having "more or less" or "better or worse" qualities. Since the month July (represented by 7) is not more or better than June (represented by 6), individual inputs are required for each month. Discontinuities such as going from 12 for December to 1 for January are also troublesome. Zip codes, bar codes, and marital status are examples of data that require more than one input (Lawrence, 1991).

Variable values represented with numbers don't always behave like numbers, because they sometimes don't reflect any specific sequence or order. For example, although there is some obscure plan behind postal zip codes, it is not possible to add, subtract, or compare zip codes and infer meaningful results. (Generally, a larger zip code indicates the post office is further west in the United States, but there is certainly no sensible interpretation of the sum of two zip codes.) Arbitrary numerical codes should be treated in the same way as mutually exclusive nonnumeric codes (like male/female or apple/orange/pear/banana), that is, assign one input neuron for every possible value. In any single case, only one of the neurons should be set to one, and all of those representing other values should be set to zero. If a categorization has too many possible values, like the states of the United States, it may be necessary to combine some of the categories to produce a taxonomy of fewer values (i.e., combining states into Northwest, Southwest, Northeast, Southeast, and Midwest categories).

"Fairly Continuous" Data

Lawrence (1991) points out that the choice between binary and continuous data representations may not be simple. If the data are fairly continuous but not evenly distributed over the entire range, even the best representation can be tricky. For example, a network that predicts the income level of individuals based upon demographics and personal history might have inputs for the person's education level with values from 0 to 20 years. Alternately, natural groupings occur around traditionally recognized levels of achievement (i.e., high school, baccalaureate, masters, and doctoral graduations), and the data could be grouped into ranges such as less than 13, 13–16, 17–18, and 19–20 years. However, if significant differences occur within a group (e.g., "less than 13" could mean either high-school graduate or grade-school dropout), the representation may not be valid. On the other hand, if one continuous-valued input representing the actual number of years of schooling is used, the neural network might have trouble. For instance, if there is a significant difference in the effect on monetary savings between having a high-school diploma and not having one, the network may not pick it up, because 12 and 11 look very similar in the range 0–20. The best representation may be a combination of the two approaches (e.g., several groups, each continuous-valued). Experimentation with several representations may be necessary to determine the best representation.

Ordering of Variables

At the opposite extreme from nominal numeric values are real or continuous values. Most measurements of natural quantities result in real numbers. Two real values for the same variable can be compared, and the difference will be meaningful. In this context, it is not continuity that matters as much as the orderings possible for the set of numbers used.

Falling between nominal and continuous are variables whose numeric values imply a real ordering, but with undefined intervals between the values. For example, if all of the soldiers in a platoon were lined up and ranked in order from the shortest to the tallest, you could say that soldier #10 was taller than soldier #5, but you couldn't say by how much, and you certainly couldn't say soldier #10 is twice as tall as soldier #5. Crooks (1992) points out that rankings like this are especially troublesome because the rank value depends not only on the height of a particular soldier, but also on how many and which soldiers are compared.

Rank orderings are often handled by converting a ranking into a percentile (more precisely, a percentile divided by 100 to keep the value under 1.0) to make the values independent of the number of cases in the sample taken. Alternately, all observations can be divided into quintiles representing values from the highest quintile as 0.9 with second quintile values at 0.7, and so on, down to 0.1 for the lowest quintile. The approach is the same, but the number of categories is reduced from 100 to five by the use of quintiles. It is essential that boundaries for percentiles and quintiles be based on a good sample of the population of cases to be modeled. It is more important that percentile and quintile values be reliable and repeatable throughout the training and use of a neural network than that they be accurate.

Changes in Values Versus Absolute Values

Another important factor in representing continuous-valued data is whether to use actual values or changes in values. One reason for using changes in values is that the smaller the range, the more meaningful small-value differences are to the network. However, the range of some data, such as the Dow Jones industrial average (DJIA), will probably change over time. The day-to-day change in the DJIA over a month (with rare exceptions) is not likely to exceed ± 200 points. On the other hand, the change over a year might be 1000 points and maybe 3000 points over a decade. The decision whether to use the absolute value or the change depends upon the nature of the problem. If small changes in the day-to-day values of the DJIA are an important consideration in the problem being investigated, then the change in DJIA should be used. If the trend over a decade or over a few years is important, then the actual values should be used, scaled to spread the values between the expected maximum and minimum values over the operating range of the neural network simulator (see Section 11.5).

Distributed Versus Nondistributed Information

A decision whether to describe the information as unique items (i.e., gender, minority status, etc.) or as a set of descriptive qualities (such as height, age, weight, etc.) is necessary. Information that exclusively categorizes a thing or person into one of several possible categories is called a nondistributed representation. Only one neuron is needed when the choice is between two categories (e.g., male or female), but one neuron is needed for each category when there are more than two alternatives (e.g., minority status; Black, Hispanic, Native American, etc.). Using nondistributed information may increase the size of a neural network with resultant training and generalizing problems.

Distributed information involves using a few pieces of information to define a unique pattern. For example, by using three primary color inputs (red, blue, and yellow), many possible color combinations can be represented without adding neurons. Such a distributed input scheme reduces the number of neurons needed to represent a large number of patterns that share common qualities and enhance the generalization ability as well, but there has to be a means of interpreting the results. However, there are potential problems with using a distributed approach for the output. A network with a distributed output layer also has less learning capacity because it has fewer weights. Such a network output sometimes must be decoded twice: first, from neuron activations to the distributed qualities and then to the nondistributed output. For example, if color were expressed as a distributed output pattern such as 0.2 blue, 0.7 yellow, and 0.4 red, this result would have to be decoded again by some external observer or program to designate the color "brown."

Advantages of a distributed output network are that it uses fewer neurons in the output and the hidden layers, has fewer connections, does less computation, and runs faster. Generally speaking, neural networks with a greater number of inputs than outputs perform better. More outputs make it harder to train the neural network to be accurate. Overall error, rather than the error in individual outputs, is minimized.

Encoding Data

An encoding algorithm's function is to take input data and convert it into a form suitable for presenting to the network. A decoding algorithm takes the values of the output layer neurons and converts them into a meaningful answer. Encoding and decoding algorithms are neural-network-specific, but some guiding principles can be applied. Neurons operate with numeric inputs and outputs that correspond to the activation values of the neurons—that is within the range neurons understand (usually 0 to 1, or -1 to $+1$). The input encoding must interpret the raw data—that is, turn it into a sequence of numeric values that the network can understand. The output decoding must take a sequence of numbers that corresponds to the output neurons' values and turn them into the form required for the final output.

As an example, consider a three-neuron output with a binary (0,1) output. This neuron output can represent eight categories of output (i.e., 101 represents the fifth category which can be arbitrarily defined). Since outputs are not likely to be exactly 0 or 1, an output in the range 0.8 to 1 could be interpreted as 1 and an output in the range 0 to 0.2 could be interpreted as 0. Values between 0.2 and 0.8 would then represent ambiguous results. Some investigators arbitrarily split the outputs between 0 and 1 at some arbitrary threshold (not necessarily 0.5).

Fourier analysis of waveforms can also be used for the analysis of acoustical waves, vibration, motion, or electrocardiograph records. The frequency content of the digitally recorded waveform is obtained using the fast Fourier transform technique. The value presented to each input neuron represents the amplitude of the signal at a particular frequency range.

11.5   SCALING, NORMALIZATION, AND THE ABSOLUTE MAGNITUDE OF DATA

Neural networks are very sensitive to absolute magnitudes. If one input ranges from 1000 to 1,000,000 and a second one ranges from 0 to 1, fluctuations in the first input will tend to swamp any importance given to the second, even if the second input is much more important in predicting the desired output. To minimize the influence of absolute scale, all inputs to a neural network should be scaled and normalized so that they correspond to roughly the same range of values. Commonly chosen ranges are 0 to 1 or -1 to $+1$.

Even though one of the great strength of neural networks is that they work well in nonlinear situations, linear relationships are the easiest for neural networks to learn and emulate. Therefore, minimizing the effects of nonlinearity of a problem pays off in terms of faster training, a less complicated network, and better overall performance. Hence, one goal of data preparation is to reduce nonlinearity when its character is known and let the network resolve the hidden nonlinearities that are not understood.

Data Normalization

Numeric data must be normalized or scaled if it has a natural range that is different than the network's operating range. Normalization is simply dividing all values of a set by an arbitrary reference value, usually the maximum value. Use of the maximum value will limit the maximum value to unity. This process, although very commonly used, carries with it the potential for loss of information. It can also distort the data if one or a few values are much larger than the rest of the data (e.g., anomalous spikes) or when all the data are within a narrow band. Scaling, on the other hand, is establishing a linear relationship between two variables over the desired range of each. Normalization is a special case of scaling where the minimum value of both variables is zero.

Data Scaling

Scaling has the advantage of mapping the desired range of a variable (with a range between the minimum and maximum values) to the full "working" range of the network input. For example, let us assume that the values between the minimum and maximum (called the range Δ) must be scaled into the range 0.1 to 0.9 for the neural network input. This linear scaling is shown in Figure 11.1, where the straight line has the form

$$y = mx + b \qquad (11.5\text{-}1)$$

where m is the slope and b is the y intercept. If we substitute the values

$$y = 0.1 \qquad \text{when } x = x_{\min} \qquad (11.5\text{-}2)$$

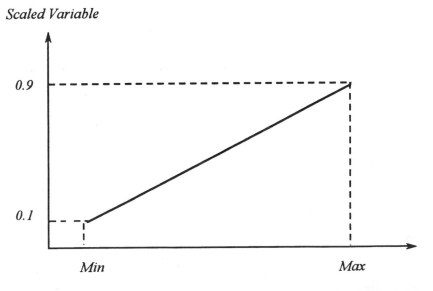

Figure 11.1   Scaling of input variables for artificial neural networks.

and

$$y = 0.9 \quad \text{when } x = x_{\max} \tag{11.5-3}$$

we can solve for the constants m and b to be

$$m = 0.8/(x_{\max} - x_{\min}) = 0.8/\Delta, \tag{11.5-4}$$

and

$$b = [0.9 - 0.8x_{\max}/\Delta] \tag{11.5-5}$$

where

$$\Delta = x_{\max} - x_{\min} \tag{11.5-6}$$

Equation (11.5-1) then becomes

$$y = (0.8/\Delta)x + (0.9 - 0.8x_{\max}/\Delta) \tag{11.5-7}$$

Scaling of the variable between 0.1 and 0.9 is often used to limit the amount of the sigmoid activation function used in the representation of the variables in order to avoid "network paralysis" in the training process. Many neural network simulation software systems perform such scaling automatically. Even so, it is necessary to understand what is occurring so that unforeseen scaling factors are not inadvertently introduced into the process. For instance, if the input is scaled to its maximum and minimum values and the desired output is scaled to its maximum and minimum values which are different, then the recall output has a scale factor that is the ratio of the two input scale factors. This can be avoided by using a single-scale factor for both input and desired output that is based on the maximum and minimum values that occur in both the input and desired output variables. Most commercial software packages automatically use a single-scale factor unless directed to do otherwise.

Crooks (1992) points out that scaling to similar magnitudes is not always adequate. For instance, if one input variable to a neural network fluctuates from 50 to 1000 and a second input variable changes only from 950 to 1000 (even though it may have been very low in the past), it is clear that the region of typical variation is much different for these two variables, even though they have similar magnitudes and historical ranges. Since networks pay attention not only to the magnitude of inputs but to their variability as well, the greater variability of the first variable would tend to distract the network relative to the smaller, but perhaps more important, variation in the second variable.

Z-Scores

An appropriate approach for some problems is to compensate for variability in the scaling of variables. The common way to do this is to scale inputs to their "Z scores" (the number of standard deviations above or below the mean). To perform Z-score scaling on one variable, first calculate the mean and standard deviation for the variable across all of the examples in the data set. Then convert each example value to a Z score by subtracting the mean and dividing the difference by the standard deviation. This procedure partially compensates for both different magnitudes and variabilities. Z-score scaling does not take away some useful information, but instead, it makes the information independent of units of measure.

Crooks (1992) gives two precautions that must be observed when working with Z scoring. First, the calculated mean and standard deviations for an input variable are merely estimates of the mean and standard deviations for the entire population sampled by the data at hand. Hence, if more than one set of training data is used, the scaling for the two sets may be different because the estimated mean and standard deviations will be different for the different samples. This often presents a practical problem when comparing results from different training sets. The solution is to select one estimate for the population mean and one estimate for the population standard deviation and then use them uniformly to scale all data sets in the same way for the selected variable. Second, if one output value is scaled using the Z-score method, the output neurons must represent values throughout the range of about -3.0 to $+3.0$. Often, output neuron sigmoids prevent outputs greater than $+1.0$ or less than 0.0 (or -1.0), which would make it impossible for the network's output to reach 2.0 (or -2.0). Clearly Z scores are not appropriate for cases where this problem arises and is not addressed properly.

Input Transformations

Some fairly simple input transformations, such as ratios, can save a network a good deal of work. While the network can learn to do the division by itself, networks normally perform division by effectively converting the numerator and denominator each to logs, subtracting them, and taking the antilog. Although these three nonlinear operations are often performed by networks, it is more productive to carry out the division in data preprocessing and let the neural network concentrate on establishing relationships from the data.

Besides ratios, nonlinearity in a problem may be reduced by the use of logarithmic scaling for inputs of an exponential or compounding character or the use of exponential scaling for inputs with a logarithmic pattern. Nonlinear scaling is also used to emphasize a particular range of variables. For instance, logarithmic scaling is often used to compress the scale for larger values whereas exponential scaling is used to expand the scale for smaller values.

When a problem may have a geometrical aspect, precalculating relevant distances, areas, and volumes is helpful. In short, when an input is known to have a specific nonlinear tendency, try to counteract the tendency with scaling that will yield a more linear input to the neural network and simplify its operation.

Redundancy in Input Data from Monitored Variables

Experience has shown that the existence of a high degree of redundancy in the data from the monitored variables of a complex process or system can and usually does have an adverse influence of the results of neural network modeling. Decorrelation of the input variables using ordinary statistical methods (Jurik, 1993) can be quite effective in improving the validity of the model. In effect, the methodology of Jurik typically identifies a few special variables in which a high fraction (typically 95%) of the information is contained. These few special variables y_i, (which are, in fact, principal components) have the form

$$y_i = \sum_{j=1}^{N} \alpha_i x_i \qquad (11.5\text{-}8)$$

where i is the index for the number of special variables (principal components) used, and j is the index for the N input variables being monitored in the complex system or process. Once the coefficients a_i for these principal components have been determined using the decorrelation code, they can be implemented by an additional input network ahead of the usual input network with the connecting weights set to equal the values of a_i.

These principal components can also be obtained using autoassociative neural networks as discussed in Section 8.4, where the data are extracted from the "bottleneck" layer. A network of the type shown in Figure 8.14a can be used to provide the principal components as inputs, after the complete network has been trained as an autoassociative neural network.

Genetic algorithms have also been used to select the most important variables for a neural network by Guo and Uhrig (1992) (see Example 17.2 in Chapter 17) and later by Harp, Samad and Guha (1989, 1990).

11.6   DATA SELECTION FOR TRAINING AND TESTING[1]

Kinds of Data

All that is needed to train a neural network is an adequate amount of the kind of information that is important in solving a problem. If there is

---

[1] Many of the suggestions in this section were given by Lawrence (1991).

uncertainty whether specific data are important, it is usually best to include it because a neural network can learn to ignore inputs that have little or nothing to do with the problem, provided that enough examples are provided. Using too much or too many kinds of data is seldom a problem if there is adequate data. If inadequate data are used, correlations become difficult to find. Training time may become excessive when not enough kinds of data exist to make proper associations. This is often the case with backpropagation networks with a very large number of hidden neurons. The end result is memorization of the individual values, and the network trains well but tests poorly on new data.

Difference in Data Requirements in Supervised and Unsupervised Learning

Lawrence (1991) points out that there is a big difference in how data get organized between supervised networks and unsupervised networks. Supervised neural networks are generally used for prediction, evaluation, or generalization. They basically learn to associate one set of input data with the corresponding set of output data, For example, a neural network can associate an increase in agricultural crop yield with certain types of weather patterns; to predict the crop yield, the weather pattern (rainfall, temperature, humidity, cloud cover, etc.), including historical patterns, would be specified as inputs to the neural network. Unsupervised networks, such as Kohonen networks, are best applied to classification or recognition types of problems (e.g., descriptions of diseases can be stored; when a new medical case comes in with a partial description of the symptoms, the Kohonen neural network would look at the description and provide as an output the stored diagnosis that most closely matches one of the descriptions stored in the network).

Generally, the more example sets that are presented to a network for training and testing, the better the training will be. However, there must be enough examples of a sufficient variety for training that the network will be able to make valid correlations and generalizations for unfamiliar cases. The variety must include a good distribution of possible inputs and outputs. Lawrence (1991) cites the following example: If a network is to perform an evaluation such as the operational readiness of an aircraft, examples of good and bad situations should be used in fairly even proportions. However, if 1000 examples of the aircraft being ready and 10 of it not being ready are provided for training, the network will probably not be able to learn those 10 cases. Even if it does learn them, the network may predict that the airplane is ready more often than it should be.

Cases Where Inadequate Data Are Available

There should be sufficient training sets so that a random sampling of data examples can be set aside for testing the neural network. If an inadequate number of training examples are available, creating a data set from simulator runs or using expert evaluations of situations may be necessary and acceptable. Several experts can rate examples, and a single network might be trained on the sum total of the expert's views. Alternately, a network might be trained for each expert's opinion to see which network gives the best results after training.

For fabricated examples, use of "border" patterns (examples in which the output just begins to be different) can be very effective. Research has shown that the success rate of a trained neural network increases rapidly as the number of border training patterns used increases. A manufactured training set using both border patterns and diverse-valued training patterns is substantially better.

If there are only a few data examples, a technique called "leave one out" can be used to train several networks, each with a different subset of most of the examples. Then each network is tested with a different subset. Leaving a different set of examples out of the training set and subsequently testing on a different set will greatly improve assessing the network's effectiveness and may show where more examples are needed. It should also indicate whether a network trained with all of the examples can generalize well.

Randomly chosen training patterns, although often used, may inadvertently emphasize the wrong conceptual points. Since the most easily identifiable patterns must be included for the network to learn the basics, a randomly chosen training set may not include these basic patterns in the proper proportions (Lawrence, 1991).

Data that cover too long a time span can include changes of equipment or other events that make the process nonstationary or even discontinuous. When the behavior has changed over time, data collection should be limited to a time period of similar behavior. For example, the strongest influences on the value of gold today may not be the same as those before the breakup of the Soviet Union. Adding a neuron indication as to whether the data examples were before or after breakup will solve this problem. When the changes are long term rather than associated with specific events, throwing out the oldest data and adding newly collected examples to a training set can be very helpful.

11.7 TRAINING NEURAL NETWORKS

Backpropagation Training

Backpropagation is a gradient descent system that tries to minimize the mean squared error of the system by moving down the gradient of the error curve. In a real situation, however, the network is not a simple two-dimensional system, and the error curve is not a smooth bowl-shaped figure. Instead, it is usually a highly complex, multidimensional, and more-or-less bowl-shaped curve that can have all kinds of bumps, valleys, and hills that the network must negotiate before finding its lowest point (the minimum mean-squared error position). The number of iterations through the data set required to achieve a given level of training will generally increase as the size of the training set increases, as the number of layers increases, and as the size of the middle layer increases. In general, the bigger the network, the slower each pass through the training data, when the network is simulated on a serial-type digital computer.

Dealing with Local Minima

Despite backpropagation's widespread use, it is sometimes difficult to use, and training times are often excessive. Caudill (1991a) has offered some tips on training techniques that have been found to be useful. It is important to note that these suggestions are specific to backpropagation networks and may be unsuitable for other paradigms.

Perhaps the "easiest" way to deal with a neural network that is stuck in local minima created by the hills and valleys and will not train is to start over by reinitializing the weights to some new set of small random values. Geometrically, this changes the starting position of the network so that it has a new set of obstacles and traps to negotiate to get to the bottom of the error surface. It is expected (but certainly not guaranteed) that as a result of starting from a new position there will be fewer obstacles in reaching the global minimum of the error surface. The difficulty is that the user must be willing to forego any progress in training and start over on a path that may be no better, or even worse, than the first path.

A less drastic approach is to "shock" the neural network by modifying the weights in some small random or systematic way. Again, it is expected (but not guaranteed) that a small move in the error surface will provide a path to the global minimum. A good rule of thumb is to vary each weight by adding a random number of as much as 10% of the original weight range (e.g., if the weights range from -1 to $+1$, add random values to each weight in the range -0.1 to $+0.1$). Generally, this technique is used when the network has learned most of the patterns before stalling, whereas starting over is used when the network has been unable to learn very few of the patterns. Such changes should be made only after an integral number of epochs have been presented to the network.

Applying the proper amount of momentum to a backpropagation network is probably the single easiest thing to do to make the network train faster. The momentum term helps a backpropagation network keep moving down the error surface, even when it meets a temporary upward surface. In effect, momentum ensures that if the weights were changing so the error decreased last time, there will be a "force" to make the next weight change reduce the error further.

Another effective way to reduce a network's training time is to use slightly noisy data. Oddly enough, networks actually train faster with noisy data. For example, an input that is a matrix-binary representation (0s and 1s) of an alphabetical letter that is being mapped into an ASCII code for the letter will train faster if the "pure" binary representation is corrupted by the addition of 10–15% random noise. The network never sees two images of the letter that are exactly alike. Hence, this technique forces the network to generalize, a key goal in the training of neural networks.

Monitoring the Training Process

Monitoring the training process includes looking for local minima, overtraining, and network paralysis. Eliminating local minima or overtraining may involve introducing specific or random changes into the weights and often adjusting training parameters (e.g., increasing the momentum or changing the learning and/or activation function constants). If these techniques do not produce results in a reasonable time, it may be necessary to reinitialize the weights and start the training over.

The method of presenting the training set to the network can affect the training results in certain learning algorithms. To mitigate these effects, neural network simulation software often change the order in which the training cases are presented to the network (e.g., present the test cases randomly or in some predetermined order) or delay the adjustment of the weights until an integral number of epochs of training data (or a specific number of data sets) are presented to the network.

After the network has been trained, it is important to test it against both the training set and examples that the network has never encountered before. Increasing the size of the hidden layer usually improves the network's accuracy on the training set, but decreasing the size of the hidden layer generally improves generalization, and hence the performance on new cases. An optimal size can be attained by a balance between the objectives of accuracy and generalization for each particular application. Creating a functioning neural network that provides the most accurate, consistent, and robust model possible requires iterative building, training, and testing to refine the neural network.

Overtraining is probably the most common error in training neural networks. The best method of ensuring that overtraining does not occur is to monitor periodically the sum square error for both the training data and the test data. It is normal for the sum square error for the training data to continue to decrease with training. However, this may be forcing the neural network to fit the noise in the training data. To avoid this problem, stop periodically the training, substitute the test data for one epoch, and record the sum square error. When the sum square error of the test data begins to increase, the training should be stopped. Indeed, if the weights at the previous monitoring are available, they should be used.

Another form of testing uses special inputs to study the neural network's responses (similar to the use of impulse or step functions in testing electrical circuits under specific conditions). Activating (either positively or negatively) an input node and then examining the input–output relationship (e.g., the ratio of the change in a specific output for a given change in a specific input) can give the sensitivity of input–output relationships. If significant problems are found, the neural network or the training process must be debugged. Every aspect of the network and the training process must be examined, including the quality, representativeness, and accuracy of the training data sets, the constants within the learning algorithms, and especially the normalization and scaling (including denormalization and descaling) processes.

Role of the Hidden Layer in Training

The traditional explanation of the functions of the hidden layer of a well-trained three-layer network is that it views the input pattern to determine which features are present in the pattern, and the output layer considers what output should be generated for the particular combination of features identified by the hidden layer. Unfortunately, the hidden layer may memorize the input patterns rather than learning the features, especially if the number of neurons in the hidden layer exceed the number of training cases. If the network memorizes its response instead of generalizing features, it may give the perfect answer for the training input and have no idea at all what to generate for a test-input pattern. If a single neuron responds to a particular input pattern, it is called a "grandmother" cell. Unfortunately, this concentration of information into a single neuron makes the network, which is usually robust, very vulnerable to the failure of a single neuron. Memorization can be prevented by ensuring that the network never sees exactly the same input pattern more than once. This can be done by adding random noise to the each input pattern.

Setting the number on neurons in the middle layer equal to the number of patterns in the training set can encourage the network to assign one neuron in the middle layer to each training pattern, which obviously does not encourage general feature detection and generalization. Solutions include adding noise to the input or reducing the number of neurons in the hidden layer.

Applying a little noise to the training set will generally produce a network that is robust to noisy inputs. Although a network trained with no noise may still do well with noisy inputs in the real world, one trained with an appropriate level of noise will do much better. The exact type and amount of noise depends on the data, but a general rule is that 10–15% perturbation of the signal is a good starting point.

In general, the exact size of the middle layer isn't a critical parameter, and training times don't vary significantly for similar-sized middle layers. Sometimes increasing the size of the middle layer will provide more feature detectors. When the middle layer is just too small, increasing it by 10% or 20% may make a huge difference. However, too large a middle layer will lengthen the training process, and extra degrees of freedom may allow the neural network to "overtrain" (i.e., the neural network will be trained to the point that it fits the noisy fluctuations in the mapping relationship).

REFERENCES

Anderson, J. A., Data Representation in Neural Networks, *AI Expert*, Vol. 5, No. 6, pp. 30–37, 1990.

Bailey, D., and Thompson, D., How to Develop Neural Networks, *AI Expert*, Vol. 5, No. 6, pp. 38–47, 1990a.

Bailey, D., and Thompson, D., Developing Neural Network Applications, *AI Expert*, Vol. 5, No. 9, pp. 34–41, 1990b.

Caudill, M., Using Neural Nets: Diagnostic Expert Nets, *AI Expert*, Vol. 5, No. 9, pp. 43–47, 1990.

Caudill, M., Neural Network Training Tips and Techniques, *AI Expert*, Vol. pp. 56–61, 1991a.

Caudill, M., Evolutionary Neural Networks, *AI Expert*, pp. 28–33, March 1991.

Coleman K. G., and Watenpool, S., Neural Networks in Knowledge Acquisition, *AI Expert*, Vol. 7, No. 1, pp. 36–39, 1992.

Crooks, T., Care and Feeding of Neural Networks, *AI Expert*, Vol. 7, No. 7, pp. 36–41, 1992.

Guo, Z., and Uhrig, R. E., Using Modular Neural Networks to Monitor Accident Conditions in Nuclear Power Plants, *Proceedings of the S.P.I.E. Technical Symposium on "Intelligent Information Systems," Applications of Artificial Networks III*, Orlando, FL, April 20–24, 1996.

Harp, S., Samad, T., and Guha, A., Towards the Genetic Synthesis of Neural Networks, in *Proceedings of the Third International Conference on Genetic Algorithms*, Morgan Kaufmann, San Mateo, CA, 1989.

Harp S., Samad T., and Guha A., Designing Application Specific Neural Networks Using the Genetic Algorithm, in *Advances in Neural Information Processing Systems 2*, D. S. Touretzky, ed., Morgan Kaufmann, San Mateo, CA, 1990.

Jurik, M., *User's Manual for Decorrelator Computer Software*, Jurik Associates, Los Angeles, CA, 1993.

Klimasauskas, C. C., Applying Neural Networks, Part 2: A Walk Through the Application Process, *PC AI*, pp. 27–34, March/April, 1991.

Lawrence, J., Data Preparation for a Neural Network, *AI Expert*, Vol. 6, No. 11, pp. 34–41, 1991.

Pao, Y., Functional Link Nets: Removing Hidden Layers, *AI Expert*, Vol. 4, No. 4, pp, 60–68, 1989.

Tveter, D. R., Getting a Fast Break with Backprop, *AI Expert*, Vol. 6, No. 7, pp. 36–43, 1991.

III

INTEGRATED
NEURAL–FUZZY
TECHNOLOGY

12

FUZZY METHODS
IN NEURAL NETWORKS

12.1 INTRODUCTION

Although fuzzy and neural systems are structurally different, they share a rather complementary nature as far as strengths and weaknesses are concerned. In this chapter we will examine the possibilities of introducing fuzzy operations within individual neurons and networks. Improving the overall expressiveness and flexibility of neural networks is what is sought. In the next chapter we will bring neuronal learning capabilities into fuzzy systems. Making fuzzy systems capable of on-line adaptation would be the desirable objective there. Neuronal enhancements of fuzzy systems as well as the fuzzification of neural systems aim at exploiting the complementary nature of the two approaches through their integration into a soft computing paradigm that permits a certain tolerance for imprecision and uncertainty.

Applying fuzzy methods into the workings of neural networks constitutes a major thrust of neurofuzzy computing (Gupta and Rao, 1994; Gupta, 1994; Pedrycz, 1993; Hirota and Pedrycz, 1993b). Although the field is an active area of research undergoing major changes, in this chapter, at the risk of omitting important research findings and developments, we attempt to introduce some fundamental notions and applications. To begin with, we briefly review the basic model of the artificial neuron we presented in Chapter 7 and then proceed with the "fuzzification" of its workings. Mathematical models of fuzzy neurons employ adaptive fuzzy relations and operators at the synapses in order to convert the external inputs into the synaptic output. Fuzzy logic operators such as min ($\wedge$) and max ($\vee$), and more generally T norms and S norms, are used to perform the confluence and aggregation of dendritic inputs to a neuron's main body, or *soma*. Although fuzziness may be introduced at all aspects of the workings of an artificial neuron (i.e., inputs, weights, aggregation operation, transfer function, and output), the main thrust of fuzzy neural nets has focused on (a) the fuzzification of the dendritic inputs and (b) the aggregation operation of a conventional neuron. The result is a variety of fuzzy neurons differing in properties according to whether, for example, instead of summation we aggregate the inputs through max, min, or some other T-norm and S-norm operation. At the end of the chapter we present a set of applications and a summary of recent developments. It should be stressed, however, that since there is a rapidly growing volume of research dealing with fuzzified neural networks, our survey is partial and unfortunately incomplete.

12.2 FROM CRISP TO FUZZY NEURONS

As we have seen in Chapter 7, a neural network consists of densely interconnected information processing units called *artificial neurons*. The structure of an artificial neuron is schematically reviewed in Figure 12.1. It consists of *external inputs, synapses, dendrites*, a *soma*, and an *axon* through which individual neural output is transmitted to other neurons. Let us call this the jth neuron of the network. We recall that a vector of external inputs $[x_1, x_2, \ldots, x_n]^T$ enters the jth neuron and gets modified by weights $w_{1j}, w_{2j}, \ldots, w_{nj}$ representing the synaptic junctions of the neuron. In earlier chapters, we considered these weights as simple gains—that is, scalars modifying via multiplication the external input vector $[x_1, x_2, \ldots, x_n]^T$. In general, however, the synaptic weights may be functions of the external

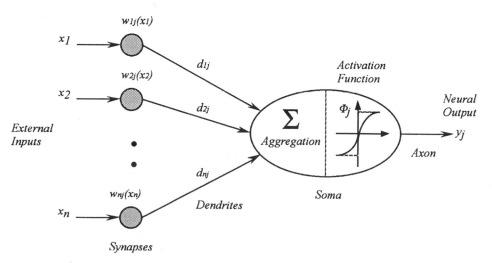

Figure 12.1 Simplified model of a neuron as an information processor.

inputs—that is, $w_{1j}(x_1), w_{2j}(x_2), \ldots, w_{nj}(x_n)$. Each synaptic output constitutes an input to the soma, called the *dendritic input*. Thus, the input to the jth neuron's soma is the vector of dendritic inputs $[d_{1j}, d_{2j}, \ldots, d_{nj}]^T$, where each dendritic input is a transformed version of an external input x_i; that is,

$$d_{ij} = w_{ij}(x_i) \tag{12.1-1}$$

The weighting function $w(\cdot)$ that models the synaptic junction between the axon of the transmitting neuron and the dendrite of the receiving neuron is thought of as a memory of the neuron's past experience, capable of adapting to new experiences through learning.

The neuron produces an output response when the aggregate activity of all dendritic inputs exceeds some threshold level T_j. Computing this aggregate input activity is an essential somatic operation as seen in Figure 12.1. Mathematically, this is usually expressed as

$$I_j = \sum_{i=1}^{n} d_{ij} \tag{12.2-2}$$

where n is the number of dendritic inputs to the neuron. It should be mentioned, however, that there is nothing sacred about summation as the aggregation operator in equation (12.2-2). We could, and indeed we will, use other aggregation operators—for example, min, max, and more generally T norms and S norms—in place of summation.[1]

Finally, the output y_j of the jth neuron is produced by the other essential operation within a neuron's soma, which is that performed by the activation (or *transfer*) *function* Φ_j (really a *decision function*). The neural output y_j is mathematically expressed as

$$y_j = \Phi_j[I_j, T_j] \tag{12.2-3}$$

where Φ_j is the activation function that describes the degree to which the jth neuron is active, I_j is the total aggregate input activity incident on the *soma* of the neuron, and T_j is the inherent threshold level for this neuron. The *perceptron*, for example, is an artificial neuron with a neural output given by

$$y_j = \text{sign}\left[\sum_{i=1}^{N} w_{ij} x_i + T_j \right] \tag{12.2-4}$$

where the activation function is assumed to be a binary "on–off" function given by sign [·], the aggregation operator is the summation of weighted inputs, and the inherent threshold T_j is a negative bias value. For the

---

[1]See Appendix for a discussion of T norms and their co-norms, called S norms.

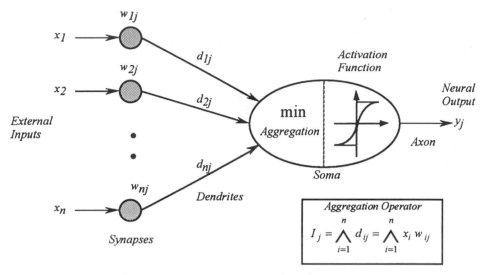

Figure 12.2   An example of a min (*AND*) fuzzy neuron.

perceptron, all external inputs and the resultant neural response are assumed to be binary (± 1). The synaptic weights, w_{ij}, may be either positive (*excitatory*) or negative (*inhibitory*) real numbers. Both the synaptic weights and the threshold level are assigned to the neuron during *training*.

Fuzzy Neurons and Fuzzy Neural Networks

A fuzzy neuron has the same basic structure as the artificial neuron shown in Figure 12.1, except that some or all of its components and parameters may be described through the mathematics of fuzzy logic. There are many possibilities for fuzzification of an artificial neuron and hence one encounters a variety of fuzzy neurons in the literature, all possessing interesting logic-oriented information processing properties. Figure 12.2 shows a fuzzy neuron where the external input vector $\mathbf{x} = [x_1, x_2, \ldots, x_n]^T \in R^n$ is defined over the unit hypercube $[0, 1]^n$ and is comprised of fuzzy signals bounded by graded membership over the unit interval $[0, 1]$.[2] The external inputs, after being modified by synaptic weights w_{ij} (also defined over the unit interval), become dendritic inputs d_{ij} to the soma. Input modification may be done through straightforward multiplication $d_{ij} = x_i w_{ij}$ or taking the maximum between input and weight $d_{ij} = x_i \vee w_{ij}$ (i.e., like an *OR-gate*).

The dendritic inputs are processed by an aggregation operator I_j that selects the minimum ($\wedge$) of the product (or max) modifications; for example,

$$I_j = \bigwedge_{i=1}^{n} d_{ij} = \bigwedge_{i=1}^{n} x_i w_{ij} \qquad (12.2\text{-}5)$$

[2]For simplicity we will use x_i instead of μ_i, and so on, when referring to fuzzy signals.

This type of fuzzy neuron may be thought of as the implementation of fuzzy conjunction (*AND-gate*). Generally, fuzzy neurons use aggregation operators such as min and max and more generally T norms and S norms instead of summation as in equation (12.2-2).

As far as the meaning and purpose of neuronal fuzzification goes, we can say that each fuzzy neuron may be thought as the representation of a linguistic value such as *LOW, MEDIUM*, and so on.[3] Hence the output of the neuron y_j in Figure 12.2 could be associated with membership to some linguistic value; that is, y_j expresses the degree to which the input pattern $[x_1, x_2, \ldots, x_n]^T$ belongs to a given linguistic category. In other words, the output y_j is a real value in the interval $[0, 1]$ indicating the degree to which the applied external inputs are able to generate the given linguistic value. The jth neuron after receiving n inputs $[x_1, x_2, \ldots, x_n]^T$ and producing an output y_j can subsequently convey this degree to the $m - 1$ other fuzzy neurons in a network consisting of m neurons.

The synaptic operations, but most importantly the aggregation operator, and the activation function determine the character of a fuzzy neuron. Using different aggregation operators and activation functions results in fuzzy neurons with different properties. Thus, many different types of fuzzy neurons can be defined. Consider, for example, the following neurons (Kwan and Cai, 1994).

Max (*OR*) Fuzzy Neuron

A *max fuzzy neuron* is a neuron that uses an aggregation function that selects the maximum ($\vee$) of the dendritic inputs to the soma; that is,

$$I_j = \bigvee_{i=1}^{n} x_i w_{ij} \tag{12.2-6}$$

(A max fuzzy neuron is an implementation of a *logical OR*; hence we can also call this an *OR fuzzy neuron*.[4])

Min (*AND*) Fuzzy Neuron

A *min fuzzy neuron* is a neuron that uses an aggregation function that selects the minimum ($\wedge$) of the dendritic inputs; that is,

$$I_j = \bigwedge_{i=1}^{n} x_i w_{ij} \tag{12.2-7}$$

---

[3]Actually fuzzy neurons may model *if/then* rules also, as we shall see later on.
[4]A special class of *OR* and *AND* fuzzy neurons that has been defined by Pedrycz in terms of T norms will be examined later in the chapter.

(A min fuzzy neuron is an implementation of a logical *AND*; hence it can also be called an *AND fuzzy neuron.*)

In addition, one can define *input fuzzy* neurons such as the *fan-in neurons* that we have seen in Chapter 7 whose purpose is simply to distribute input signals to other neurons. An *input fuzzy neuron* is an element used in the input layer of a fuzzy neural network, and it has only one input x such that

$$y = x \qquad (12.2\text{-}8)$$

In general, the weights, the activating threshold, and the output functions which describe the interaction between fuzzy neurons could be adjusted via a learning procedure resulting in neurons that are adaptive. The aim is, of course, to synthesize fuzzy neural networks capable of learning from experience.

12.3   GENERALIZED FUZZY NEURON AND NETWORKS

Let us consider a neural network consisting of m fuzzy neurons, each admitting n inputs. As noted in the previous section, fuzziness may be introduced at the synaptic inputs (weights), the aggregation operation, and the transfer function of individual neurons. Thus, fuzzy sets can be used to describe various aspects of neuronal processing (Gupta and Knopf, 1992). The following are conventions frequently encountered in fuzzy neural networks.

Synaptic Inputs. The input vector $\mathbf{x} = [x_1, x_2, \ldots, x_n]^T \in R$ to a fuzzy neuron may be thought of as grades of membership to a fuzzy set. For simplicity we do not employ the usual symbol for membership (μ_i). Rather the individual inputs $x_i \in [0,1]$ are taken to represent fuzzy signals bounded by a graded membership over the unit interval.

Dendritic Inputs. For each jth neuron in the network ($j = 1, 2, \ldots, m$) the dendritic inputs are also bounded by a graded membership over the unit interval. Thus, if we let $u \in [0,1]$ designate an element of a generic universe of discourse $[0,1]$, we would use for defining fuzzy quantities the dendritic inputs are fuzzy sets

$$d_{ij} = \sum_{j=1}^{m} \mu_{d_{ij}}(u)/u, \qquad u \in [0,1] \qquad (12.3\text{-}1)$$

Aggregated Values. The output of the aggregation operator in each of the m fuzzy neurons of the network can also be thought of as graded membership over the unit interval. Thus we have

$$I_j = \sum_{j=1}^{m} \mu_{I_j}(u)/u, \qquad u \in [0,1] \qquad (12.3\text{-}2)$$

Neuron Output. Finally, the output y_j of each of the m fuzzy neurons may also be thought of as a grade of membership to a fuzzy set; that is,

$$y_j = \sum_{j=1}^{m} \mu_{y_j}(u)/u, \qquad u \in [0,1] \qquad (12.3\text{-}3)$$

The weighting function w_{ij} transforming an external input x_i into dendritic signal d_{ij} for the jth fuzzy neuron does not have to be just a simple gain. It can, in general, be a *fuzzy relation* defined over the Cartesian product $w_{ij} = x_i \times d_{ij}$. Such a synaptic junction fuzzy relation between the external inputs x_i and dendritic inputs d_{ij} may assume many forms, with the simplest and actually the most common being $d_{ij} = x_i w_{ij}$. More generally, however, the dendritic inputs d_{ij} may be given by the composition $x_i \circ w_{ij}$ of fuzzy input signals and the weight relation; that is,

$$d_{ij} = x_i \circ w_{ij} \qquad (12.3\text{-}4)$$

The concept of fuzzy negation is used in order to produce both *excitatory* and *inhibitory* inputs to a fuzzy neuron.[5] Consider a jth fuzzy neuron such as the one shown in Figure 12.3. The synaptic outputs, d_{ij}, may be modified to produce *excitatory* or *inhibitory* effects by defining a new variable δ_{ij} to denote both excitatory and inhibitory inputs received by the soma, and a negation operation that modifies d_{ij} as follows:

$$\delta_{ij} = \begin{cases} d_{ij} & \text{(for excitatory inputs)} \\ \bar{d}_{ij} & \text{(for inhibitory inputs)} \end{cases} \qquad (12.3\text{-}5)$$

where the inhibitory inputs are fuzzy complements of the excitatory inputs

$$\bar{d}_{ij} = 1 - d_{ij} \qquad (12.3\text{-}6)$$

Consider, for example, the fuzzy neuron shown in Figure 12.3. This neuron receives four dendritic inputs: $d_{1j}, d_{2j}, d_{3j}, d_{4j}$. The first two are *excitatory* inputs sent to the aggregation operator just as they are, while the second two are *inhibitory* inputs, which are complemented in a fuzzy sense according to equation (12.3-6). We graphically indicate the inhibitory signals by small (white) circles at the end of the corresponding arrows as shown in Figure

---

[5]Since we use the [0, 1] range, it is not possible to use negative values for inhibitory inputs.

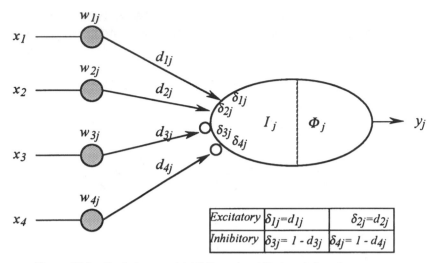

Figure 12.3 *Excitatory* and *inhibitory* dendritic inputs to a fuzzy neuron.

12.3. Thus this neuron's aggregation operator I_i will aggregate the following signals:

$$\delta_{1j} = d_{1j}$$
$$\delta_{2j} = d_{2j}$$
$$\delta_{3j} = 1 - d_{3j}$$
$$\delta_{4j} = 1 - d_{4j}$$

(12.3-7)

The result of the aggregation will be subsequently modified by the function Φ_j to produce the neuron's output y_j.

12.4 AGGREGATION AND ACTIVATION FUNCTIONS IN FUZZY NEURONS

In a fuzzy neuron the aggregation operator I_j may be a T norm (see Appendix) mathematically expressed as

$$I_j = \mathop{T}_{i=1}^{n} \delta_{ij}$$

(12.4-1)

Often, but not always, fuzzy neurons do not explicitly use a threshold; thresholding may instead be contained within the choice of the activation function. The activation function Φ_j is a mapping operator that transforms the membership of the aggregate fuzzy set I_j into the fuzzy set of the neuronal response y_j. In a sense, this mapping operation corresponds to a linguistic modifier such as *VERY* and *MORE-OR-LESS* (see Chapter 2). The role of this modification is to enhance or diminish the degree to which the external inputs give rise to the fuzzy value represented by the jth fuzzy neuron, before becoming an external input to neighboring neurons. Thus, a general expression of the response of the jth fuzzy neuron may be written as

$$y_j = \Phi_j[I_j] = \Phi_j\left[\mathop{T}_{i=1}^{n} \delta_{ij}\right] \tag{12.4-2}$$

where each dendritic input δ_{ij} is given by equations (12.3-5) and (12.3-6).

If the activation function is assumed to be a linear relationship with unit slope (i.e., $y_j = I_j$), we have an interesting special case (we will see more of it in the following sections), a simplified fuzzy neuron whose response can be written as

$$y_j = \mathop{T}_{i=1}^{n} \delta_{ij} \tag{12.4-3}$$

The concepts of T norm and S norm (or T conorm), originally used in the field of probability theory, provide a means for generalizing and parametrizing fuzzy set operations such as *union* and *intersection* as well as implication operators, fuzzy inferencing, and fuzzy neurons (Dubois and Prade, 1980; Gupta and Qi, 1991; Terano et al., 1992; Terano et al., 1994).

A T norm can be thought of as a circuit (gate) with two inputs (x_1, x_2) and one output $T(x_1, x_2)$, also written as $x_1 T x_2$. The most widely used T norm is the min; that is, $T(x_1, x_2) = x_1 \wedge x_2$ (but also *algebraic product, bounded product, and drastic product* are all T norms; see Appendix).

An S norm can be thought of as a circuit with two inputs (x_1, x_2) and one output $S(x_1, x_2)$, also written as $x_1 S x_2$. A very common S norm is the *logical sum* or max; that is, $S(x_1, x_2) = x_1 \vee x_2$ (but also the *algebraic sum, bounded sum*, and *drastic sum* are some other S norms).

The relationship between T norms and S norms is given by *fuzzy De Morgan's laws*, which may be written as

$$T(x_1, x_2) = \bar{S}(\bar{x}_1, \bar{x}_2)$$
$$S(x_1, x_2) = \bar{T}(\bar{x}_1, \bar{x}_2) \tag{12.4-4}$$

where T is the T norm and S is the S norm and the bar over the symbols indicates negation.

Let us consider the simplified fuzzy neuron, shown in Figure 12.4a, using a linear transfer function and output given by equation (12.4-3). For simplicity we assume that the dendritic inputs are directly received from the external inputs ignoring any weight function modifications. This neuron can be

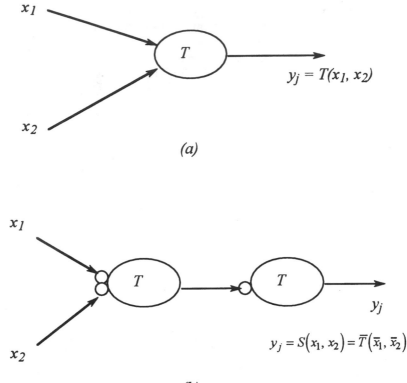

$$y_j = T(x_1, x_2)$$

(a)

$$y_j = S(x_1, x_2) = \overline{T}(\overline{x}_1, \overline{x}_2)$$

(b)

Figure 12.4 A simplified fuzzy neuron using *T*-norm aggregation and a linear transfer function can perform both (*a*) *T* norm and (*b*) *S* norm operations on the signals (x_1, x_2).

thought of as the realization of a T norm operation. With the aid of equations (12.4-4), this simplified fuzzy neuron can be used to construct a network of neurons, such as the one shown in Figure 12.4*b*, that realizes an S norm. As indicated by the small circles in the left neuron of Figure 12.4*b* the inputs are first negated in a fuzzy sense and then aggregated by a T-norm aggregation. The output of this neuron is complemented again by the second neuron, in accordance with equation (12.4-4). Thus the network of neurons in Figure 12.4*b* is a realization of an S norm, made out of cascaded neurons that individually use T norm for the aggregation operation.

12.5   *AND* AND *OR* FUZZY NEURONS

A special class of fuzzy neurons are the *AND* and *OR* neurons shown in Figure 12.5. These neurons employ the T norm and S norm operations for

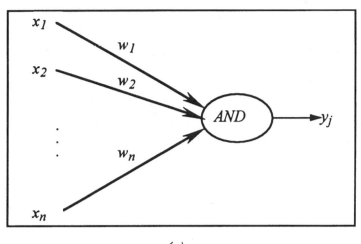

(a)

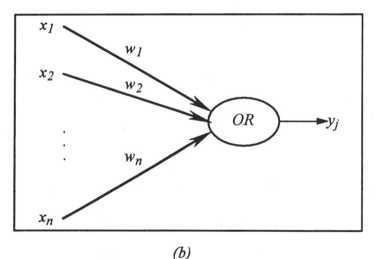

(b)

Figure 12.5 (*a*) An *AND* fuzzy neuron and (*b*) an *OR* fuzzy neuron.

forming dendritic inputs to the soma and for aggregating them (Pedrycz, 1993, Rueda and Pedrycz, 1994; Pedrycz and Rocha, 1993).

The *AND* neuron first uses equation (12.3-4) to perform an *S* norm or *OR* operation between external input x_i and corresponding weight w_{ij}; that is,

$$d_{ij} = x_i \ OR \ w_{ij} \qquad (12.5\text{-}1)$$

Subsequently, it uses a *T* norm or *AND* operation to carry the following aggregation of its dendritic inputs (assuming a linear activation function):

$$y_j = (x_1 \; OR \; w_{1j}) \; AND \; (x_2 \; OR \; w_{2j}) \; AND \; \cdots \; AND \; (x_n \; OR \; w_{nj}) \quad (12.5\text{-}2)$$

It should be noted that *AND* and *OR* are generally realized by taking any T norm and S norm—for example, logical product (min), logical sum (max), algebraic product, algebraic sum, and so on. In practice, however, the min and max interpretations are most commonly used.

The output of an *AND* neuron can succinctly be written using T norms and S norms as

$$y_j = \underset{i=1}{\overset{n}{T}} \; (x_i \; S \; w_{ij}) \quad (12.5\text{-}3)$$

The *OR* neuron, on the other hand, performs a complementary computation:

$$y_j = \underset{i=1}{\overset{n}{S}} \; (x_i \; T \; w_{ij}) \quad (12.5\text{-}4)$$

Both the *AND* and *OR* neurons given above are intrinsically excitatory in their behavior; that is, higher values for the x_i's imply higher values for y_j. To allow for inhibitory behaviors of such *AND* or *OR* fuzzy neurons (and still maintain the standard $[0, 1]$ range of the grades of membership) we include negated values of x_{ij}—that is, $1 - x_{ij}$ [as we have seen before in equation (12.3-6)]—thus potentially doubling the size of the input vector. The *AND* and *OR* neuron can now handle both inhibitory and excitatory behaviors, depending on the numerical values of the connections.

Now let us look at some interesting boundary cases, say in the *AND* neuron (Pedrycz, 1993). First, suppose that all the weights of a neuron equal zero—that is, $w_{ij} = 0$. Then we should have $x_i \; S \; w_{ij} = x_i$ (e.g., $x_i \; S \; 0 = x_i$). Second, if all the weights are unity, $w_{ij} = 1$, we have $x_i \; S \; 1 = 1$; that is, the input does not have any influence on the output. To deal with such extremes, a bias term may be added as an additional term in (12.5-3) driven by a constant input signal always equal to 0, say $0 \; S \; w_{0j}$, where w_{0j} denotes the connections associated with this input. The *AND* neuron incorporating such a bias is given by

$$y_j = \underset{i=0}{\overset{n}{T}} \; (x_i \; S \; w_{ij}) \quad (12.5\text{-}5)$$

where, by convention, we put $x_0 = 0$. A similar bias term may be added to an

OR neuron, making equation (12.5-4) look like

$$y_j = \mathop{S}_{i=0}^{n} \left(x_i \, T \, w_{ij} \right) \tag{12.5-6}$$

In equations (12.5-5) and (12.5-6) we have assumed that the neuronal output is produced immediately after aggregation; in other words, the activation function used is linear. However, a nonlinear activation function such as a sigmoidal function may also be used.

12.6 MULTILAYER FUZZY NEURAL NETWORKS

The fuzzy neurons discussed in the previous section can be put together to construct more general computational structures with enhanced representational capabilities. While in Part II of the book we used networks composed of identical neurons, the networks built out of fuzzy neurons are often heterogeneous; that is, they are composed of neurons with different computational characteristics—for example, *AND* or *OR* fuzzy neurons—organized into several layers (most commonly three).

Let us look at a three-layer neural network built out of *AND* and *OR* neurons [originally proposed by Pedrycz (1993)]. Each layer in the network is constructed out of neurons of the same type (i.e., *AND* or *OR* only). A hidden layer is used to enhance the representational capabilities of the entire structure. In Figure 12.6*a* the hidden layer is made out of *AND* neurodes, while in Figure 12.6*b* the hidden layer has only *OR* neurons. These are actually two different types of networks: One uses *AND* neurons in the hidden layer, with the output layer consisting of a single *OR* neuron, whereas the other uses *OR* neurons in its hidden layer and a single *AND* neuron in the output layer. As seen in Figure 12.6*a*, the first network has an input layer consisting of 2*n* input neurodes; both networks use direct signals and their complements, namely, $x_1, x_2, \ldots, x_n, \bar{x}_1, \bar{x}_2, \ldots, \bar{x}_n$. Because the neurons of the first layer are fan-in neurodes [see equation (12.2-8)], (they simply distribute the input signals to all the nodes of the hidden layer.

The hidden layer itself is composed of *p AND* neurons, each one of them sending to the output layer signal

$$z_l = AND(w_l, x), \qquad l = 1, 2, \ldots, p \tag{12.6-1}$$

The weight vector of connections, w_l, captures information about the connections between the *l*th node of the hidden layer and the input nodes; that is,

$$z_l = \left[\mathop{T}_{i=1}^{n} \left(x_i \, S \, w_{li} \right) \right] T \left[\mathop{T}_{i=1}^{n} \left(\bar{x}_i \, S \, w_{l(n+i)} \right) \right] \tag{12.6-2}$$

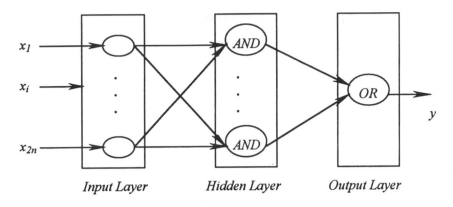

Input Layer   Hidden Layer   Output Layer

(a)

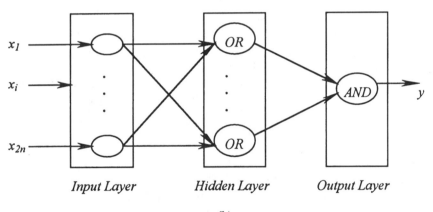

Input Layer   Hidden Layer   Output Layer

(b)

Figure 12.6 Three-layer networks with fuzzy (*a*) *AND* and (*b*) *OR* neurons in the hidden layer.

where $l = 1, 2, \ldots, p$. The output layer consists of a single *OR* performing an aggregation of z's:

$$y = \overset{p}{\underset{l=1}{S}} (z_l \, T \, v_l) \qquad (12.6\text{-}3)$$

If we put *OR* neurons in the hidden layer and an *AND* neuron at the output layer (see Figure 12.6*b*, we perform a similar sequence of computations, except we interchange T norm and S norm operations in equations (12.6-2) and (12.6-3).

Other network architectures are also possible. Consider, for example, the homogeneous network shown in Figure 12.7. Here only *T*-norm aggregating (*AND*) neurons are used to realize a network structure of three layers emulating a system of *m* fuzzy *if/then* rules, that is a fuzzy rule base, receiving *n* inputs and producing one output. The first layer consists of *m* fuzzy neurons, with each neuron being a representation of an *if/then* rule. As seen in Figure 12.7, the output of each fuzzy neuron in layer 1 becomes an external input to a single *OR* neuron realized as cascaded *AND* neurons (see Figure 12.4) comprising layers 2 and 3. This three-layered neural network architecture can be used to simulate a situation when *n* fuzzy inputs are applied to *m* fuzzy inference rules (Gupta and Knopf, 1992).

12.7 LEARNING AND ADAPTATION IN FUZZY NEURAL NETWORKS

The process of learning in fuzzy neural networks consists of modifying their parameters by presenting them with examples of their past experience. How can this be done in practice? Typically by adjusting the weights of the networks so that a certain performance index is optimized (maximized or minimized). This requires that a collection of input–output pairs be specified and also requires a performance index that expresses how well the network maps inputs x_k into the corresponding target values of the output t_k.

Let us recall that an important difference between a crisp (nonfuzzy) neuron and a fuzzy neuron lies in the model of the synaptic connection.

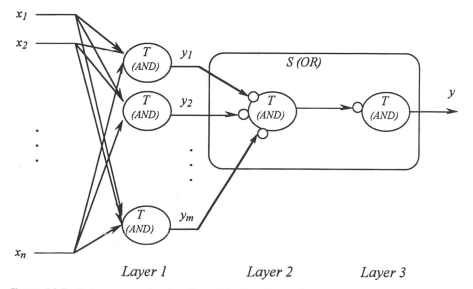

Figure 12.7 A fuzzy neural network architecture for *m* fuzzy rules accepting *n* inputs.

Synaptic connections in a crisp neuron are linear gains multiplying inputs x_i. Any adaptation or learning occurring within an individual neuron involves modifying the values of these gains by adjusting w_{ij}. For a fuzzy neuron, synaptic connections are represented as a two-dimensional fuzzy relation between synaptic inputs and outputs. Hence, learning in fuzzy neurons, in the most general case, involves changing a two-dimensional relation surface at each synapse.

Consider one synaptic connection to the jth neuron as shown in Figure 12.8. For a given external input to this synapse at time k, $x_i(k)$, we want to determine the corresponding fuzzy relation, $w_{ij}(k)$, such that we have minimum error $e_j(k)$ between the fuzzy neuron response and the desired target response $t_j(k)$. In order to achieve this, we can employ the following adaptation rule to modify the fuzzy relation surface:

$$w_{ij}(k + 1) = w_{ij}(k) + \Delta w_{ij}(k) \tag{12.7-1}$$

The term $\Delta w_{ij}(k)$ is the change in the fuzzy relation surface given as a function $F[\cdot]$ of the error $e_j(k)$; that is,

$$\Delta w_{ij}(k) = F[e_j(k)] = F[t_j(k) - y_j(k)] \tag{12.7-2}$$

In multilayer networks, learning involves matching t_k (up to some error) with the output of the entire network y. For this purpose, a distance function—for example, Euclidean distance between y and t_k—may be used. Then a performance index Q (a global error term to be minimized) may be defined as follows

$$Q = \sum_{k=1}^{N} [y(\mathbf{x}_k) - t_k]^2 \tag{12.7-3}$$

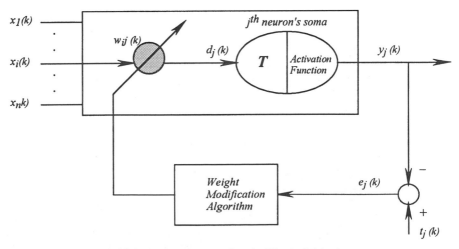

Figure 12.8   Learning at the level of the individual neuron.

Q in Equation (12.7-3) reflects quantitatively the state of the networks learning process. Optimization includes all the weights of the network between the input layer and the hidden layer as well as between the hidden layer and the output layer. The simplest update scheme is that in which the modifications are driven by a gradient of the performance index taken with respect to the connections themselves (see Chapter 8). The learning formula can be expressed as

$$\Delta(\text{connections}) = -\eta\frac{\partial Q}{\partial(\text{connections})} \tag{12.7-4}$$

where η denotes a *learning factor*, $\eta \in (0, 1)$. Detailed computations can be performed once the performance index Q and a parametric description of the network have been defined, as is done in the example that follows.

Example 12.1 Learning and Adaptation in *AND / OR* Neurons. Given a three-layer network having a hidden layer of *AND* neurons and output with an *OR* fuzzy neuron (as shown in Figure 12.6a) and an error based on sum of squared errors, we want to derive an on-line learning algorithm for modifying its weights. The network's neurons use *algebraic sum* for the S norm and use *product* for the T norm [see Appendix and Pedrycz (1993)].

A single pair of input–output data involves $\mathbf{x}$ and t.[6] For the hidden layer (see Figure 12.6a) we have the following intermediate outputs based on equation (12.6-2):

$$z_h = \left[\mathop{T}_{i=1}^{n}(x_i \; S \; w_{hi})\right]T\left[\mathop{T}_{i=1}^{n}(\bar{x}_i \; S \; w_{h(n+i)})\right] \tag{E12.1-1}$$

While the output layer gives [according to equation (12.6-3)]

$$y = \mathop{S}_{h=1}^{p}(z_h \; T \; v_h) \tag{E12.1-2}$$

where p denotes the dimension of the hidden layer. In order to adjust the weights, we differentiate the performance (error) function of (12.7-3) with respect to hidden layer weights, that is,

$$\frac{\partial Q}{\partial w_{hj}} = \frac{\partial Q}{\partial y}\cdot\frac{\partial y}{\partial w_{hj}} = \frac{2[y(x) - t]\partial y}{\partial w_{hj}}, \quad h = 1, 2, \ldots, p, \quad j = 1, 2, \ldots, 2n$$

$$\tag{E12.1-3}$$

and then we differentiate the error function with respect to the intermediate

---

[6]Since we develop an on-line learning version [where each pair $(\mathbf{x}_k, t_k)$ immediately affects the connections of the network], the index k denoting the element in the training set can be dropped.

weights (we use the letter v for those);

$$\frac{\partial Q}{\partial v_h} = \frac{2[y(x) - t]\partial y}{\partial v_h}, \qquad h = 1, 2, \ldots, p \qquad \text{(E12.1-4)}$$

Now we have that

$$\frac{\partial y}{\partial v_h} = \frac{\partial}{\partial v_h}\left[\underset{h=1}{\overset{p}{T}}(z_h \, T \, v_h)\right] = \frac{\partial}{\partial v_l}[A \, S \, (z_l \, T \, v_l)] \qquad \text{(E12.1-5)}$$

where A is a shorthand notation defined as $A \equiv \underset{h \neq 1}{\overset{p}{S}}(z_h \, S \, v_h)$.

Since we use algebraic sum and product for the T norm and S norm, we have

$$\frac{\partial}{\partial v_l} = [A + v_l z_l - A v_l z_l] = z_l(1 - A) \qquad \text{(E12.1-6)}$$

$$\left[A \, S \, (z_{h_i} \, T \, v_{h_i})\right] = \left[A \, S \, z_{h_i} v_{h_i}\right] = A + z_{h_i} v_{h_i} - A z_{h_i} v_{h_i} \qquad \text{(E12.1-7)}$$

Hence we can compute that the output change with respect to input weights is given by the following expression:

$$\frac{\partial y}{\partial w_{hj}} = \sum_{h=1}^{p} \frac{\partial y}{\partial z_h} \frac{\partial z_h}{\partial w_{hj}} \qquad \text{(E12.1-8)}$$

The above sum reduces to single component, since only one term contributes; that is,

$$\frac{\partial z_h}{\partial w_{h_1 j}} = 0, \qquad \forall h \neq h_l \qquad \text{(E12.1-9)}$$

Hence, we obtain

$$\frac{\partial y}{\partial z_{hl}} = \frac{\partial}{\partial z_{hl}}\left[\underset{i=1}{\overset{p}{S}}(z_i \, T \, v_i)\right] = \frac{\partial}{\partial z_{hl}}[B \, S \, (z_{hl} \, T \, v_{hl})] = \frac{\partial}{\partial z_{hl}}[B \, S \, z_{hl} v_{hl}]$$

$$= \frac{\partial}{\partial z_{hl}}[B + z_{hl} v_{hl} - B z_{hl} v_{hl}] = v_{hl} - B v_{hl}$$

$$\frac{\partial y}{\partial w_{h_1 j}} = \frac{\partial y}{\partial z_{h_l}} \frac{\partial z_{h_1}}{\partial w_{h_1 j}}$$

$$\frac{\partial y}{\partial z_{h_l}} = v_1(1 - B), \qquad B = \underset{h \neq h_1}{\overset{p}{S}}(v_h \, S \, z_h)$$

$$\frac{\partial z_{h_l}}{\partial w_{h_1 j}} = \frac{\partial}{\partial w_{h_1 j}}\left[\prod_{i=1}^{n}(w_{h_l i} + x_i - w_{h_l i} x_i)\prod_{i=1}^{n}(w_{h_l(n+i)} + \bar{x}_i - w_{h_l(n+i)}\bar{x}_i)\right]$$

$$\text{(E12.1-10)}$$

Thus we can write

$$
\frac{\partial h_l}{\partial w_{h_{lj}}} = \begin{cases} C_1(1 - x_j), & \text{if } j \leq n \\ C_2(1 - \bar{x}_j), & \text{if } j > n \end{cases} \tag{E12.1-11}
$$

where C_1 and C_2 stand for the product terms in (E12.1-10), not including x_j or its complement.

If we use nondifferentiable T norm and S norm such as minimum and maximum, the derivatives must be judiciously defined since they can severely affect the learning algorithm as we saw in Chapter 8. For example, the derivative of $(x \wedge w)$ with respect to w is

$$
\frac{\partial (x \wedge w)}{\partial w} = \begin{cases} 1, & \text{if } x \geq w \\ 0, & \text{if } x < w \end{cases} \tag{E12.1-12}
$$

This type of "on–off" weight updates can easily be affected, however, by peculiarities in the connections and the data encountered during learning. One possibility for ameliorating this problem is to replace the above two-valued situation, that is, 0 or 1 in equation (E12.1-12), by some smooth, although very similar, function; for example, (Pedrycz, 1993)

$$
\frac{1}{2}\left[(x + w) - \sqrt{(x - w)^2 + \delta^2} + \delta\right] \qquad \text{(for minimum)}
$$

$$
\frac{1}{2}\left[(x + w) + \sqrt{(x - w)^2 + \delta^2} - \delta\right] \qquad \text{(for maximum)}
$$

where the parameter δ is typically a small positive constant (about 0.05). □

Example 12.2   Steering Control for an Automobile. Let us look at an example of an automatic steering control mechanism (Maeda and Murakani, 1989; Sugeno and Nishida, 1985) and its equivalent fuzzy neural network architecture (Gupta, 1994) (Gupta and Knopf, 1992)

The controller is based on a driver's ability to manipulate both the *position* and *direction* of a moving automobile on a straight highway (assumed for simplicity to travel down the middle of the road). The approximate position and direction for the vehicle with respect to the road edge is used, and hence *position-from-left-side position-from-right-side, direction-angle and change-in-direction-angle* will be fuzzy variables. The output of the controller is another fuzzy variable, namely, the steering-*angle* by which the steering wheel should be turned (Gupta and Knopf, 1992). Figure 12.9 shows the position and direction variables employed in modeling the situation.

The steering control rules consist of two rule bases. The first rule base, called the positioning algorithm, involves a linguistic description that positions the vehicle in the middle of the road, and the second, called the

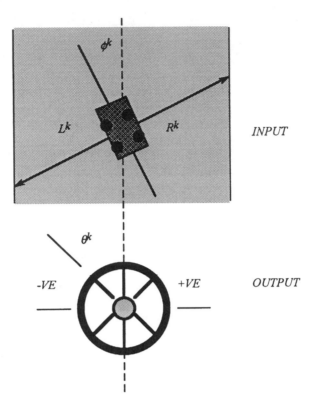

Figure 12.9 The position and direction fuzzy variables in the automobile steering control problem are left and right distances from road side, position, and steering wheel angles.

direction algorithm, involves rules to ensure that the vehicle is parallel to the edge of the road.

We can write the 16 rules of the positioning algorithm succinctly as

$$
\begin{cases}
\text{if } \textit{position-from-left-side} \text{ is } L_j^k \text{ AND } \textit{position-from-right-side} \text{ is } R_j^k \\
\text{then } \textit{steering-angle} \text{ is } \theta_j^k, \forall \, j = 1, \ldots, 16
\end{cases}
$$

$$(E12.2\text{-}1)$$

The direction algorithm has nine rules that can be succinctly written as

if *direction-angle* is ϕ_j^k AND *change-in-direction-angle* is $\Delta \phi_j^k$
then *steering angle* is $\theta_j^k, \forall \, j = 1, \ldots, 9$

$$(E12.2\text{-}2)$$

where L_j^k, R_j^k, ϕ_j^k, $\Delta\phi_j^k$, and θ_j^k are linguistic values for the jth rule and k is the time index. These fuzzy sets are defined as follows:

L^k, R^k	Fuzzy values describing the approximate distances between the road edge and the vehicle (R, right; L, left)
ϕ^k, $\Delta\phi^k$	Fuzzy values describing the angle and change in angle for the direction of the vehicle with respect to centerline
θ^k	Fuzzy value for the output steering angle at time k

The linguistic labels of these values are as follows:

ZE	Approximately zero
S	Small
M	Medium
L	Large
P	Positive
N	Negative
PS	Positive small
PM	Positive medium
PL	Positive large
NS	Negative small
NM	Negative medium
NL	Negative large

The membership functions for the input and output fuzzy sets are shown in Figure 12.10.

Each rule in the fuzzy positioning and direction algorithms above may be represented by a single fuzzy neuron, and the collection of rules in its entirety by a neural network. Hence, for the automatic steering control mechanism the control rules are represented as the network of fuzzy neurons shown in Figure 12.11.

The outputs from the neurons of the first layer in Figure 12.11 become the inputs to one of the two neurons located in the second layer. To obtain the collective decision from either the position or direction control rules we require each neuron in the second layer to perform an S-norm operation. This is achieved by defining the inputs to both neurons as inhibitory. The expression for the position control neuron is

$$y_1 = \mathop{T}_{j=1}^{16} \left[N\left(\theta_j^k \right) \right] \tag{E12.2-3}$$

and the expression for the direction control neuron is

$$y_2 = \mathop{T}_{j=1}^{9} \left[N\left(\theta_j^k \right) \right] \tag{E12.2-4}$$

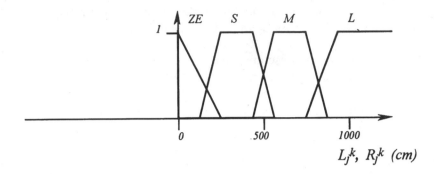

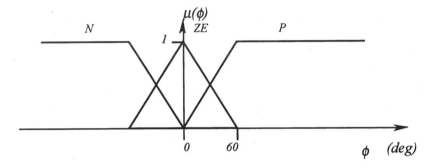

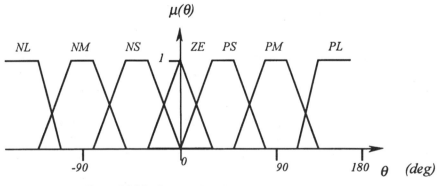

Figure 12.10   Fuzzy values for input–output variables.

The outputs from both fuzzy neurons are then transmitted to a single neuron located in a third layer as shown in Figure 12.11, producing the following output:

$$y_3 = T(y_1, y_2) \tag{E12.2-5}$$

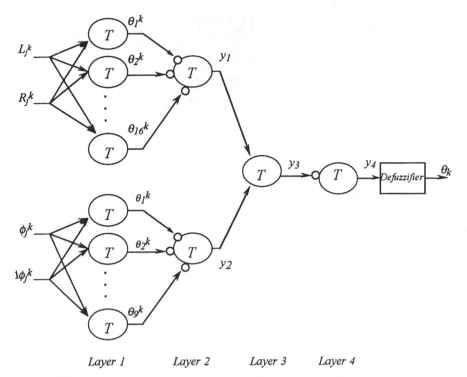

Layer 1   *Layer 2*   *Layer 3*   *Layer 4*

Figure 12.11   Fuzzy-neural network representation of automobile steering controller.

Finally the response of this neuron in the third layer becomes an inhibitory input of a neuron situated in the fourth layer, giving

$$y_4 = T(N(y_3)) \qquad \text{(E12.2-6)}$$

y_4 is generally a fuzzy set, hence the final decision is reached through a defuzzified version of the fuzzy membership function representing y_4. $\square$

12.8   FUZZY ARTMAP

In Chapter 9, we briefly discussed the features of adaptive resonance theory neural networks (ART) with emphasis on its unique ability to create new categories of arbitrary accuracy to accommodate inputs that did not fit into the existing categories. With the introduction of fuzzy concepts, Fuzzy ARTMAP (the MAP refers to mapping inputs to outputs), a synthesis of ART-1 and fuzzy logic, capable of accepting either analog or binary inputs, was developed by Carpenter and Grossberg (1994). Furthermore, it is able to deal with nonstationary time series as inputs.

Fuzzy ARTMAP is a self-organizing architecture that is capable of rapidly learning to recognize, test hypotheses about, and predict the consequences of virtually any input. (There is no ambiguity about the initial configuration since the network literally grows from scratch.) It involves a combination of neural and fuzzy operations that together give these useful capabilities. Like other versions of ART, its use is almost exclusively for classification, and it has only one user-selectable parameter (*vigilance*) which determines the fineness or coarseness of the patterns into which the inputs are fitted. It can learn virtually every training pattern in a few training iterations in an unsupervised mode. Yet, it can use predictive disconfirmations to supervise learning of categories that fit the statistics of patterns being categorized.

Fuzzy ARTMAP operates by autonomously determining how much compression or generalization is needed for each input category to fit the categories of choice. The more general categories have more fuzziness in the feature values that are accepted by the specific category. The acceptable range (or fuzziness) of a particular category is learned through a series of iterations that involve the use of fuzzy logic operators. The fuzzy *AND* (min) and *OR* (max) operators are used to define the range of values that are tolerated by a category for each linguistic variable or feature. The membership functions over the range from 0 to 1 (discussed in Chapters 2 through 5) directly relevant to this determination of the acceptability of an input pattern in a particular category. The min operator helps define features that are "critically present," whereas the max operator helps define features that are "critically absent." The min operator can be realized by nodes that are turned on by an external input, whereas the max operator is realized by nodes that are turned off by an external input. Thus the min and max operators can be introduced at appropriate positions in the neural network by externally controlled on–off switches. The category that best matches an input pattern is chosen by the operation of fuzzy subsethood. Fuzzy logic provides a method by which fuzzy ARTMAP adaptively categorizes analog, as well as binary, input patterns. Hence fuzzy ARTMAP can autonomously learn, recognize, and make rare events, large nonstationary databases, morphological variable types of events, and many-to-one and one-to-many relationships. These features and many other details of fuzzy ARTMAP are discussed extensively by Carpenter and Grossberg (1994).

Although fuzzy ARTMAP has proven itself as a supervised incremental learning system in pattern recognition and M- to N-dimensional mappings by comparison with other techniques, a simplified fuzzy ARTMAP (Kasuba, 1993) has been introduced. It reduces the computational overhead and architectural redundancy of fuzzy ARTMAP with no loss of pattern-recognizing capability. This description follows that of Kasuba (1993).

Normally, when backpropagation neural networks are used for pattern classification, a single output node is assigned to each category of objects that the network is expected to recognize. The creation of these categories are left up to network in both fuzzy ARTMAP and its simplified derivative. Figure

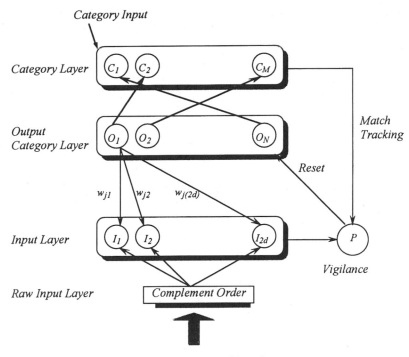

Raw Input Pattern of Size d

Figure 12.12 Simplified fuzzy ARTMAP structure.

12.12 shows the structure of the simplified fuzzy **ARTMAP** to be a two-layer network (input and output category layers) with connection weights, shows a category layer to interpret the results of output layer, and shows a "complement coder" to preprocess the raw input data. This "complement coder" normalizes the input and stretches it to twice its original size to help the network form its decision regions. The vigilance feature (0 to 1) determines the fineness of the categories and thus determines the number of categories to be created.

The expanded input (**I**) from the "complement coder" then flows to the input layer. Weights (**w**) from each of the output category nodes hold the names of the M number of categories that the network has to learn. Since a single output node can only encode a single category, it can only point to a single position in the category layer. Category input is only supplied to the category layer during the supervised training. The "match tracker" portion of the network lets it self-adjust its vigilance during learning from the level set by the user in response to errors in classification during training, thereby controlling the creation of new categories.

Complement coding is an input normalization process that represents the presence or absence of a particular feature vector **a** with d components in the input. Its complement $\bar{\mathbf{a}} = 1 - \mathbf{a}$ is valid since $\mathbf{a}$ has a value between 0 and 1. Therefore the complement-coded input vector $\mathbf{I}$ is given by the $2d$-dimensional vector

$$\mathbf{I} = [\mathbf{a}, \bar{\mathbf{a}}] = \left[a_1, a_2, \ldots, a_d, \bar{a}_1, \bar{a}_2, \ldots, \bar{a}_d \right] \qquad (12.8\text{-}1)$$

For instance, the three-dimensional vector $(0.2, 0.8, 0.4)$ is transformed into the six-dimensional vector $(0.2, 0.8, 0.4, 0.8, 0.2, 0.6)$ through complement coding. This process automatically normalized the input vectors, indicating that the norm of any vector is just the sum of all elements in the vector. Hence, the sum of the elements of a complement-coded vector is equal to the dimensionality of the original non-complement-coded input vector.

When this network is presented with an input pattern, all output nodes become active to some degree. This output activation is denoted by $\mathbf{T}_j$ for the jth output node and its weights $\mathbf{w}_j$. The function to produce this activation is given by

$$\mathbf{T}_j(\mathbf{I}) = \frac{|\mathbf{I} \wedge \mathbf{w}_j|}{\mathbf{a} + |\mathbf{w}_j|} \qquad (12.8\text{-}2)$$

where $\mathbf{a}$ is a small value near zero, usually about 0.0000001. The winning output node is the node with the highest activation; that is, the winner is $\max \mathbf{T}_j$. Hence, the category associated with the winning output node is the network's classification of the current input pattern.

The match function is used to compare the complement-coded input features and a particular output node's weight to help determine if learning has occurred. It calculates the degree to which $\mathbf{I}$ is a fuzzy subset of $\mathbf{w}_j$—that is, whether the match function value indicates that the current input is a good enough match or whether a new output category should be generated. If this match function is greater than the vigilance function, the network is said to be in a state of *resonance*. A mismatch occurs if the match function value is less than the vigilance, indicating that the current output node does not meet the encoding granularity of the vigilance. Once a winning output node $\mathbf{j}$ has been selected to learn a particular input pattern $\mathbf{I}$, the top-down vector $\mathbf{w}_j$ from the output node is updated. The simplified fuzzy ARTMAP neural network is a general-purpose classifier with top-down weight's decision-making facilities so transparent that its classification rules can literally be read out of the network. It can be compared to a self-learning expert system in that it learns from example.

12.9 FUZZY-NEURAL HYBRID DATA REPRESENTATION

During the last few years there has been a large and energetic upswing in research efforts aimed at synthesizing fuzzy logic with neural networks.

Neural networks possess advantages in the areas of learning, classification, and optimization, whereas fuzzy logic has advantages in areas such as reasoning on a high (semantic or linguistic) level. The two technologies nicely complement each other, and a number of synergisms have been proposed [see Bezdek (1995) and Saleem (1994)]. In addition to the fuzzy neurons and networks we have seen, several applications have focused on utilizing and processing fuzzy inputs and outputs in conjunction with conventional networks (Travis and Tsoukalas, 1994; Werbos, 1992). An additional variation is using fuzzy logic to control crisp neural network processes. Let us take a look at some of these.

Fuzzy Representations of Variables that are Inputs and Outputs of Neural Networks

Sometimes dealing with all possible outputs of a neural network requires a large number of neurons, thereby increasing the complexity and training time. For instance, if we consider the temperatures between freezing boiling of water, even on the centigrade scale, there would be 100 integral values. The number can be reduced by grouping these 100 values into groups of 10 successive values and representing each group of 10 values with a single value (e.g., the 10 values in the range 21° to 30° could be represented by 15°). Hence, the scale would become 5°, 15°, 25°, 35°, ..., 95°. Such groupings lead us to considering fuzzy or linguistic representation of the variable, where 0° to 10° might be "extremely cold," 10° to 20° might be "very cold," 20° to 30° might be "cold," 30° to 40° might be "slightly cold," and so on. If one views these temperatures from the standpoint of human comfort, as opposed to the distance along a scale between the freezing and boiling point of water, a nonuniform distribution with fewer values might be more appropriate—that is, 0° to 15°, 16° to 20°, 21° to 23°, 24° to 30°, and 31° to 100°. In linguistic terms, these ranges might be designated *too cold*, *cold*, *comfortable*, *hot*, and *too hot*. Generally, the sequence of events that are involved in utilizing fuzzy data in neural networks is as follows:

1. Crisp (or fuzzy) data are converted into membership functions or sets.
2. These memberships or sets are then subject to fuzzy logic operations.
3. The resultant sets are then defuzzified into crisp data that are presented to the neural network.
4. The neural network may also have their direct inputs that are crisp and do not need the fuzzy processing.
5. The output of the neural network is a crisp set that utilizes a membership function to convert it into a fuzzy variable.
6. This fuzzy output is then operated on fuzzy logic.
7. The fuzzy logic output is then defuzzified to produce a crisp output.

Fuzzy "One-of-n" Coding of Neural Network Inputs

Typically, an input variable is represented by a single input node in the neural network. When an input variable has a special relationship with other variables over only a small portion of its range, the training process of the neural network is made especially difficult. Sometimes a nonlinear transformation is used to emphasize the particular region, but this is usually not a satisfactory process. The difficulty can be overcome by providing the neural network with neurons that focus on one region of the variables' domain. The domain is divided into n regions (where n is typically 3, 5, or 7), and each is assigned a fuzzy set having a triangular membership function. (Of course, the lowest and highest sets have horizontal extensions starting at the minimum and maximum expected values, respectively.) The membership value in each fuzzy set determines the activation level of its associated input neuron. This "one-of-n" coding expands the range of the variable into n network inputs, each covering a fraction of the domain. While the resulting specialization often facilitates learning, the increase in the number of neurons tends to slow down learning. This technique is advantageous only when the importance of the variable changes significantly across its domain.

There is a tendency to want more measurements of imprecise (or linguistic) data in order to compensate for lack of precision. Let us consider the case of two time signals that are to be sampled, digitized, and fast-Fourier-transformed so that one fast Fourier transform (FFT) is the input to a neural network and the other one is the desired output. If we have 100,000 simultaneously sampled data points for each variable and are dealing with spectra that have 128 points each (and another 128 points in the negative frequency range), dividing 100,000 points by 256 points per spectrum gives 390 complete spectra for each evaluation. The traditional approach with such FFTs is to average the 390 spectra to obtain an average spectrum with a high degree of confidence for each variable and to apply these two spectra to the neural network for training. A much better alternative would be to train the neural network using each of the 390 individual spectra, even though each of them is much less precise and would be considered "noisy" or perhaps "fuzzy." Subjecting the 128 components of the input and desired output vectors to "one-of-n" coding in the manner described above is another alternative that should be considered.

Fuzzy Postprocessing of Neural Network Outputs

A neural network can be trained to produce the desired final product, but there are often advantages to training the network to present intermediate values with postprocessing to obtain the desired results. The advantages are that the neural network may be easier to train and that the necessity for retraining if other outputs are desired can be avoided. An example of such a postprocessor might be control of the electrical output of a gas-fired plant when there is competition for the gas with residential users and industrial users, both of which have a higher priority for the gas. A neural network with such inputs as current air temperature and at several earlier times and at several locations, overall demand for industrial products now and at several earlier times, the competitiveness of the products, plant efficiencies as a function of power output, and so on, could be trained to predict the available gas. However, intermediate values such as future temperatures at several locations and future industrial output may be more appropriate since they can reasonably be obtained using an ordinary neural network. However, the relationship between the availability of gas and the intermediate network outputs are fuzzy and should be treated as such.

Fuzzy Control of Backpropagation Learning

Numerous methods of speeding up the learning in backpropagation neural networks have been attempted with varying degrees of success. One of the most common methods have been to adjust the learning rate during the training using an adaptive method that satisfies some index of performance. (The "delta-bar-delta" training procedure is such a method.) Wang and Mendel (Wang, 1994) (Wang and Mendel, 1992) have shown that fuzzy systems may be viewed as a layered feedforward network and have developed a backpropagation algorithm for training this form of fuzzy system to match the input and desired output pairs of patterns or variables. Haykin (1994) has described a method in which an on-line fuzzy logic controller is used to adapt the learning parameters of a multilayer perceptron with backpropagation learning. The system uses the classical four-step fuzzy control process of (1) scaling and fuzzification of the crisp input, (2) development of a fuzzy rule base, (3) fuzzy inference using the fuzzy rule base, and (4) rescaling and defuzzification to give a crisp result or recommended action. The idea is to implement heuristics in the form of fuzzy *if/then* rules that are used for the purpose of achieving a faster rate of convergence. The heuristics (as is the case of almost all supervised training) are based on the behavior of the instantaneous sum of squared errors.

12.10 SURVEY OF ENGINEERING APPLICATIONS

Fuzzy neural networks aspiring to integrate neural learning with the knowledge representation capabilities of fuzzy systems have been actively investigated in recent years. A growing number of researchers in a number of fields have proposed and tested several types of fuzzy neurons. By far the greatest number has turned to the rather simple *AND* and *OR* neurons of Section 12.5 in building fuzzy neural networks. The networks are typically heterogeneous in order to best reflect the logic of a given problem. The layers and nodes of such fuzzy neural networks can be interpreted as a realization of fuzzy *if/then* rules.

Pedrycz and Rocha (1993) introduced a number of neurofuzzy models, using logic operators (*AND, OR, NOT*) encountered in the theory of fuzzy sets superimposed in neural structures. *Aggregation neurons* (*AND* and *OR* neurons) and *referential neurons* (for *matching, dominance, inclusion*) were designed using *T* norms and *S* norms and inhibitory and excitatory characteristics captured by embodying direct and complemented (negated) input signals. The researchers have proposed a number of topologies of neural networks put together with the use of these neurons and demonstrated straightforward relationships between the problem specificity and the resulting architecture of the network (Pedrycz, 1993).

Hirota and Pedrycz (1993a) have also proposed a distributed computational structure called *knowledge-based network* that allows for an explicit representation of domain classification knowledge. The knowledge-based network is composed of basic *AND* and *OR* neurons and has been used in pattern classification problems. Logic-based neurons have also been investigated in conjunction with new architectural aspects of fuzzy neural networks, including those aimed at representing and processing uncertainty associated with the input data (Pedrycz, 1993). Hybrids such as, for example, a multi-variable hierarchical controller for an N-degrees-of-freedom robot manipulator for control tracking problems implemented as a fuzzy-neural network, whose purpose is to select activation levels for local regulators implemented as PD controllers, have also been developed and analyzed (Rueda and Pedrycz, 1994). Lin and Song (1994) have proposed a similar three-layer fuzzy neural network with different types of fuzzy neurons.

The terms *fuzzy-neural* or *neurofuzzy networks* very often in the literature refer to hybrid combinations of fuzzy logic and neural tools—for example, giving fuzzy inputs to a crisp network and extracting fuzzy outputs as well. Recently, Srinivasan (1994) has reported on a forecasting approach using fuzzy inputs to a neural network, in electric load forecasting problems. Expert knowledge represented by fuzzy rules is used for preprocessing input data fed to a neural network. The method effectively deals with trends and special events that occur annually. The fuzzy-neural network was trained on real data from a power system and evaluated for forecasting next-day load profiles based on forecast weather data and other parameters and according to the researchers has demonstrated very good performance.

A fuzzy-neural network approach developed by Cooley, Zhang, and Chen (1994) utilizes a hybrid consisting of a parameter-computing network, a converting layer, and a backpropagation-based one for classification problems with complex feature sets. The approach has been applied to satellite image classification and lithology determination. Lee and Wang (1994) have also proposed a neural network for classification problems with fuzzy inputs.

A fuzzy input is represented as an LR-type fuzzy set, and the network structure is automatically generated with the number of hidden nodes determined by the overlapping degree of training instances. Two sample problems, heart disease and knowledge-based evaluator, have been addressed by the researches to illustrate the working of the model. Sharpe et al. (1994) have also presented a hybrid method using fuzzy logic techniques to adapt a conventional network configuration criteria.

In another interesting hybrid application, fuzzy logic has been used by Hu and Hertz (1994) for controlling the learning processes of neural networks. Since the convergence of multilayer feedforward neural networks using the backpropagation training algorithm may be slow and uncertain due to the iterative nature of the dynamic process of finding the weight matrices with static control parameters, Hu and Hertz use a fuzzy logic controller during the course of training to adjust the learning rate dynamically according to the output error of a neuron and a set of heuristic rules. Comparative tests reported by the investigators have showed that such fuzzy backpropagation algorithms stabilized the training processes of these neural networks and, produced two to three times more converged tests than the conventional backpropagation algorithms. Kuo (1993) has also reported a new learning scheme which integrates the standard backpropagation learning algorithm and fuzzy rules, which are able to dynamically adjust the learning rate, momentum, and steepness of activation function.

A fuzzification layer to a conventional feedforward neural network has been added by Zhang and Morris (1994a, b) for on-line process fault diagnosis. The fuzzification layer converts the increment in each on-line measurement and controller output into three fuzzy sets: "*INCREASE*," "*STEADY*," and "*DECREASE*," with corresponding membership functions. The feedforward neural network then classifies abnormalities, represented by fuzzy increments in on-line measurements and controller outputs, into various categories.

Kwan and Cai (1994) have defined four types of fuzzy neurons similar to those we have seen in Section 12.2, and they have proposed a structure of a four-layer feedforward fuzzy neural network and its associated learning algorithm. The proposed four-layer fuzzy neural network performs well in several pattern recognition problems. In a biotechnology application, a five-layer fuzzy neural network was developed for the control of fed-batch cultivation of recombinant *Escherichia* (Ye et al., 1994).

Karayiannis and Pai (1994) have developed a family of fuzzy algorithms for learning vector quantization and introduced feedforward neural networks inherently capable of fuzzy classification of nonsparse or overlapping pattern classes. On the other hand, a three-layer radial basis function (RBF) network has been developed by Halgamuge et al. (1994) to extract rules and to identify the necessary membership functions of the inputs for a fuzzy classification system.

REFERENCES

Bezdek, J. C., Hybrid Modeling in Pattern Recognition and Control, *Knowledge-Based Systems*, Vol. 8, N. 8, pp. 359–371, 1995.

Carpenter, G., and Grossberg, S., Fuzzy ARTMAP: A Synthesis of Neural Networks and Fuzzy Logic for Supervised Categorization and Nonstationary Prediction, in *Fuzzy Sets, Neural Networks, and Soft Computing*, R. R. Yager and L. A. Zadeh, eds., Van Nostrand Reinhold, New York, 1994.

Cooley, D. H., Zhang, J., Chen, L., Possibility Function Based Fuzzy Neural Networks: Case Study, in *Proceedings of the IEEE International Conference on Systems, Man and Cybernetics*, Vol. 1, IEEE, Piscataway, NJ, 1994, pp. 73–78.

Dubois, D., and Prade, H., *Fuzzy Sets and Systems: Theory and Applications*, Academic Press, Boston, 1980.

Gupta, M. M., Rao, D. H., eds., *Neuro-Control Systems Theory and Applications*, a selected reprint volume, IEEE Press, New York, 1994.

Gupta, M. M., and Knopf, G. K., Fuzzy Neural Network Approach to Control Systems, in *Analysis and Management of Uncertainty: Theory and Applications*, B. M. Ayyub, M. M. Gupta, and L. N. Kanal, eds., pp. 183–197, Elsevier, Amsterdam, 1992.

Gupta, M. M., and Qi, J., Theory of T-Norms and Fuzzy Inference Methods, *Fuzzy Sets and Systems*, Vol. 40, pp. 431–450, 1991.

Gupta, M. M., *Fuzzy Neural Networks: Theory and Applications*, *Proceedings of SPIE*, Vol. 2353, The International Society for Optical Engineering, 1994, pp. 303–325.

Halgamuge, S. K., Poechmueller, W., Pfeffermann, A., Schweikert, P., and Glesner, M., New Method for Generating Fuzzy Classification Systems Using RBF Neurons with Extended RCE Learning, in *Proceedings of the 1994 IEEE International Conference on Neural Networks*, 1994, pp. 1589–1593.

Haykin, S., *Neural Networks: A Comprehensive Foundation* (IEEE Computer Society Press), Macmillan, New York, 1994.

Hirota, K., and Pedrycz, W., Knowledge-Based Networks in Classification Problems, *Fuzzy Sets and Systems*, Vol. 59, No. 3, pp. 271–279, 1993a.

Hirota, K., and Pedrycz, W., Neurocomputations with Fuzzy Flip-Flops, in *Proceedings of the International Joint Conference on Neural Networks*, Vol. 2, pp. 1867–1870, 1993b.

Hu, Q. and Hertz, D. B., Fuzzy Logic Controlled Neural Network Learning, *Information Sciences Applications*, Vol. 2, N. 1, pp. 15–33, 1994.

Ishibuchi, H., Fujioka, R., and Tanaka, H., Neural Networks That Learn from Fuzzy If-Then Rules, *IEEE Transactions on Fuzzy Systems*, Vol. 1, No. 2, pp. 85–97, 1993.

Karayiannis, NB., and Pai, P.-I., Family of Fuzzy Algorithms for Learning Vector Quantization, in *Proceedings of the Artificial Neural Networks in Engineering Conference* (ANNIE '94), St. Louis, MO, November 13–16, 1994.

Kasuba, T., Simplified Fuzzy ARTMAP, *AI Expert*, Vol. 8, N. 11, pp. 18–25, November, 1993.

Keller, J. M., and Tahani, H., Implementation of Conjunctive and Disjunctive Fuzzy Logic Rules with Neural Networks, *International Journal of Approximate Reasoning*, Vol. 6, pp. 221–240, 1992.

Kosko, B., *Neural Networks and Fuzzy Systems*, Prentice-Hall, Englewood Cliffs, NJ, 1992.

Kulkarni, A. D., Coca, P., Giridhar, G. B., and Bhatikar, Y., Neural Network Based Fuzzy Logic Decision System, in *Proceedings of World Congress on Neural Networks*, 1994 INNS Annual Meeting, San Diego, CA, June 5–9, 1994.

Kuo, R. J., Fuzzy Parameter Adaptation for Error Backpropagation Algorithm, in *Proceedings of the International Joint Conference on Neural Networks*, Vol. 3, 1993, pp. 2917–2920.

Kwan, H. K., and Cai, Y., Fuzzy Neural Network and Its Application to Pattern Recognition, *IEEE Transactions on Fuzzy Systems*, Vol. 2, No. 3, pp. 185–193, 1994.

Lee, H-M., Wang, W-T., Architecture of Neural Network for Fuzzy Teaching Inputs, *Proc. of the 5th Intern. Conf. on Tools with A.I., TAI '93*, Publ. by IEEE, Boston, MA, Nov. 8–11, 1993, pp. 285–288.

Likas, A., Blekas, K., and Stafylopatis, A., Application of the Fuzzy min–max Neural Network Classifier to Problems with Continuous and Discrete Attributes, in *Proceedings of the 4th IEEE Workshop on Neural Networks for Signal Processing* (NNSP '94), Ermioni, Greece, September 6–8, 1994.

Lin, J.-N., Song, S.-M., Novel Fuzzy Neural Network for the Control of Complex Systems, in *Proceedings of IEEE International Conference on Neural Networks*, Vol. 3, 1994, pp. 1668–1673.

Maeda, M., and Murakami, S., Steering Control and Speed Control of an Automobile with Fuzzy Logic, *Proc. of the 3rd IFSA Congress*, Seattle, Washington, Aug. 6–11, 1989, pp. 75–78.

NeuralWorks Manual, *Fuzzy ARTMAP Classification Network*, pp. NC-157–NC170.

Pedrycz, W., *Fuzzy Control and Fuzzy Systems*, second (extended) edition, John Wiley & Sons, New York, 1993.

Pedrycz, W., and Rocha, A. F., Fuzzy-Set Based Models of Neurons and Knowledge-Based Networks, *IEEE Transactions on Fuzzy Systems*, Vol. 1, No. 4, pp. 254–266, 1993.

Rueda, A., and Pedrycz, W., Hierarchical Fuzzy-Neural-PD Controller for Robot Manipulators, in *Proceedings of IEEE International Conference on Fuzzy Systems*, Vol. 1, 1994, pp. 673–676.

Saleem, R. M., and Postlethwaite, B. E., Comparison of Neural Networks and Fuzzy Relational Systems in Dynamic Modeling, *Proc. of the International Conference on CONTROL '94*, Vol. 2, No. 389, IEE Conference Publication, 1994, pp. 1448–1452.

Sharpe, R. N., Chow, M. Y., Briggs, S., and Windingland, L., Methodology Using Fuzzy Logic to Optimize Feedforward Artificial Neural Network Configurations, *IEEE Transactions on Systems, Man and Cybernetics*, Vol. 24, No. 5, pp. 760–767, 1994.

Srinivasan, D., Liew, A. C., and Chang, C. S., Forecasting Daily Load Curves Using a Hybrid Fuzzy-Neural Approach, *IEE Proceedings Generation, Transmission and Distribution*, Vol. 141, N. 6, 1994, pp. 561–567.

Sugeno, M., and Nishida, M., Fuzzy Control of a Model Car, *Fuzzy Sets and Systems*, Vol. 16, pp. 103–113, 1985.

Terano, T., Asai, K., and Sugeno, M., *Fuzzy Systems Theory and Its Applications*, Academic Press, Boston, 1992.

Terano, T., Asai, K., and Sugeno, M., *Applied Fuzzy Systems*, Academic Press, Boston, 1994.

Travis, M., and Tsoukalas, L. H., Application of Fuzzy Logic Membership Functions to Neural Network Data Representation, Internal Report, University of Tennessee, Knoxville, TN, 1992. Included as Appendix A in Uhrig et al., Application of Neural Networks, EPRI Report TR-103443-P1-2, January 1994.

Wang, L.-X., *Adaptive Fuzzy Systems and Control*, Prentice-Hall, Englewood Cliffs, NJ, 1994.

Wang, L. X., and Mendel, J. M., Backpropagation Fuzzy Systems as Nonlinear Dynamic System Identifiers, in *IEEE International Conference on Fuzzy Systems*, San Diego, CA, 1992, pp. 1409–1418.

Werbos, P. J., Neurocontrol and Fuzzy Logic: Connections and Designs, *International Journal of Approximate Reasoning*, Vol. 6, pp. 185–219, 1992.

Ye, K., Jin, S., and Shimizu, K., Fuzzy Neural Network for the Control of High Cell Density Cultivation of Recombinant *Escherichia coli*, *Journal of Fermentation and Bioengineering*, Vol. 77, No. 6, pp. 663–673, 1994.

Zhang, J., Morris, A. J., On-Line Process Fault Diagnosis Using Fuzzy Neural Networks, *Intelligent Systems Engineering*, Vol. 3, No. 1, pp. 37–47, 1994a.

Zhang, J., Morris, A. J., Process Fault Diagnosis Using Fuzzy Neural Networks, in *Proceedings of the American Control Conference*, American Automatic Control Council, Vol. 1, 1994, pp. 971–975.

PROBLEMS

1. Explain the assumptions made and describe the forms of the dendritic input, the aggregation operator, and the activation function in the perceptron [Eq. (12.2-4)].

2. Besides summation, how else could the dendritic inputs to a neuron be aggregated? List at least three operators that could be used for aggregation.

3. Assuming a [0, 1] range for the input values to the fuzzy neuron shown in Figure 12.2, show that the output will also be in the [0, 1] range. What happens to the output of the synaptic modifications are made through a max operator or any other S norm? Can the same be said when the synaptic modification is done through a T norm?

4. Explain why and how a bias term may be incorporated in the fuzzy neurons described by Equations (12.5-3) and (12.5-4).

5. Consider a fuzzy neuron having as input the fuzzy set $A = 0.5/1 + 1.0/2 + 0.5/3$ and a weight fuzzy relation given by

$$w_{ij}(x, y) = 0.33/(2,5) + 0.5/(2,6) + 0.5/(2,7) + 0.5/(2,8) + 0.33/(2,9)$$
$$+ 0.33/(3,5) + 0.67/(3,6) + 1.0/(3,7) + 0.67/(3,8) + 0.33/(3,9)$$
$$+ 0.33/(4,5) + 0.5/(4,6) + 0.5/(4,7) + 0.5/(4,8) + 0.33/(4,9)$$

What is the dendritic input to the neuron's soma? State all assumptions clearly.

6. How can excitatory and inhibitory inputs be taken into account in the fuzzy neuron described in Problem 5?

7. Consider a three-layer fuzzy neural network having *AND* neurons in the hidden layer as shown in Figure 12.6a. Show that the network's output is given by Equation (12.6-3). Suppose next that the network has *OR* neurons in the middle layer as shown in Figure 12.6b. What is its output?

8. For the three-layer network of Figure 12.6a using probabilistic sum for S norm and product for the T norm, show that the rate of output change with respect to input weights is given by Equation (E12.1-7).

9. Derive an expression for the rate of output change with respect to input weights in Problem 8 when min and max are used for T norm and S norm, respectively.

10. Develop a fuzzy-neural network representation similar to the one shown in Figure 12.11 for the fuzzy algorithm described in Example 6.3 (rules given in (E6.3-1)).

13

NEURAL METHODS IN FUZZY SYSTEMS

13.1 INTRODUCING THE SYNERGISM

The complementary nature of fuzzy and neural systems forms a basis for their synergistic utilization. In the last chapter we examined ways of fuzzifying neurons and networks for the purpose of making them more expressive. In this chapter we focus on ways of endowing fuzzy systems with neuronal learning capabilities for the purpose of making them more adaptive. Although both fuzzy and neural approaches possess remarkable properties when employed individually, there are great advantages to using them synergistically.

Quite often, a strong point of fuzzy systems turns out to complement nicely a weakness—so to speak—of neural networks and conversely. This may not be so surprising after all, since both neural and fuzzy approaches have had strong biological and cognitive inspirations. In the realm of human intelligence, there is certainly a synergism between the neuronal transduction and processing of sensory signals, on the one hand, and the cognitive, perceptual, and linguistic higher functions of the brain, on the other.

Consider, for example, the processes involved in our awareness of ambient temperature. The sensation of temperature is based on two kinds of *temperature receptors* in the skin—one signaling *warmth*, the other *cold* (Nicholls et al., 1992). Neural fibers from thousands of temperature receptors enter the spinal cord to form synapses with second-order neurons[1]. The second-order axons ascend into the *ventral* and *lateral spinothalamic* tract to end in the *medial* and *ventrobasal thalamus* (in the lower part of the brain) as shown in Figure 13.1. Third-order cells then carry the information to the *cerebral cortex*

[1] Higher-order biological neurons performing communication functions.

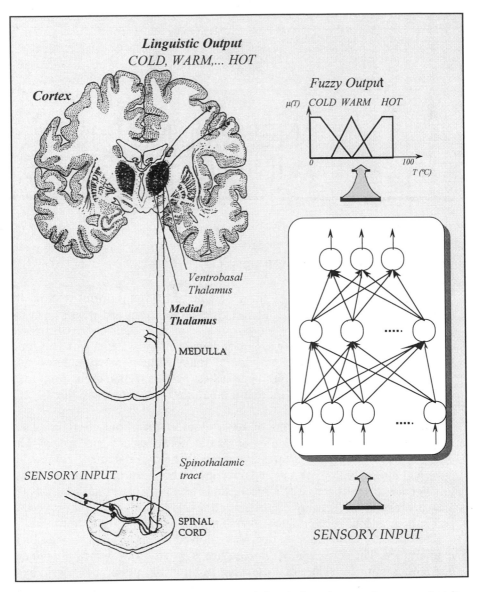

Figure 13.1 Sensory pathways take neuronal signals from temperature receptors in the skin through the spinal cord and the lower brain to the cerebral cortex, where they are ultimately transformed into linguistic categories. Fuzzy-neural hybrids are inspired by such biological-cognitive synergisms.

(the upper part of the brain). In the cerebral cortex, ultimately these temperature sensations get fused and expressed linguistically. Thus, a person may "know" that the temperature in the room is *COLD* or *WARM* or *HOT*, all linguistic (fuzzy) categorizations of the sensation of temperature.[2] This knowledge becomes a basis for human decisions and actions such as, for example, turning off the air-conditioning or a heating system. By analogy, we can train a neural network to cluster and map a set of temperature measurements from the ambience to a set of fuzzy values as shown in Figure 13.1. Of course, compared to the complexities and intricacies of the biological-cognitive system responsible for the sensation of temperature, our fuzzy-neural analog is at best very naive.

In this chapter we bring neural methods into fuzzy systems, both for the purpose of identifying (extracting) rules and membership functions and for adaptation of a fuzzy system (or linguistic description) to a changing physical system and its environment. The approach is known in the literature as *neural-network-driven fuzzy reasoning* (Takagi, 1992). For both *expert knowledge elicitation* and *adaptation,* the underlying strategy is, in essence, to *identify certain parameters of fuzzy systems and use neural networks to induce and/or adjust them.* Generally a fuzzy linguistic description of the kind we examined in Chapters 5 and 6 is computationally identical to a neural net, a fact theoretically proven by Buckley and Hayashi (1993), who demonstrated that neural nets can approximate continuous fuzzy controllers (and conversely) to any degree of accuracy.

Adaptation concerns the maintenance of a fuzzy linguistic description on the face of a changing process. The salient questions here have to do with how to adjust, over time, either the rules or what is involved within the rules, in order to better reflect changes in the actual physical system and its environment. Adaptation relates to the issue of *learning.* An adaptive system (that is, an *adaptive system description*) is one that can learn about the changes in the physical (target) system and modify its internals to improve the correspondence between the physical system and itself and/or its environment.

13.2 FUZZY-NEURAL HYBRIDS

In an abstract manner, a system can be viewed as shown in Figure 13.2*a,* a relation between inputs and outputs (where the relation is not necessarily a function, but a more general relation such as a *many-to-many* mapping). In Figure 13.2*b* and 13.2*c* we have two idealized extremes where either (1) we know exactly how the system should be working but have no example of its

---

[2]It is interesting to note that at a skin temperature of about 33°C (91.4°F) we are usually unaware of any temperature sensation. Raising or lowering skin temperature above this neutral point produces a sensation of *warming* or *cooling.*

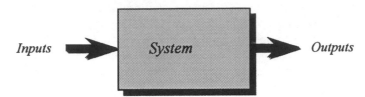

A System is a Relation Between Inputs and Outputs

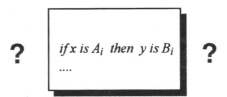

When the System Logic is Known: Use Fuzzy if/then Rules

When Examples of Input/Output are Known: Use Neural Networks

Figure 13.2 Depending on whether the internal relation (a system's logic) or the input–output behavior of a system is known, fuzzy or neural modeling tools may be chosen.

input–output behavior (see Figure 13.2*b*) or (2) we know its input–output behavior but know nothing of the system's internals (i.e., we have a black box) (see Figure 13.2*c*).

In the first case, it is convenient to write fuzzy *if/then* rules, at the appropriate level of precision, to describe (or prescribe) system behavior. In the second case, it is convenient to use the available input–output data to train artificial neural networks to model the internals of the system. Of course, in real-world systems we may have some examples of a system's input–output behavior and some knowledge of what is inside the black (or better yet "gray") box. Hence, we may utilize various hybrids of neural and fuzzy tools to successfully model the system. In the final analysis, however, our choice is made not by a commitment to a particular tool but a desire to adequately model the system at hand in a timely, reliable, and cost-effective manner.

The great array of system conditions that may be encountered calls for a variety of series and parallel combinations of fuzzy and neural systems. Consider the arrangement shown in Figure 13.3 which provides a means of inspecting and testing physically damaged components. Here a neural network is trained to receive three measurements as inputs (electrical and visual data from an automated test station testing electronic components for the purpose of eliminating physical defects (O'inca, 1994)). The input is mapped to two numerical values that serve as input to a fuzzy algorithm. The output of the neural module indicates the degree of a component's *physical damage* (a number between 0 and 1) and the *signal-to-noise ratio* (a number between 0 and 30). These two features are subsequently fed as inputs into a fuzzy system where fuzzy variables map signal-to-noise ratio and physical damage information to the quality of the component. The output of the fuzzy system is an action (decision) to accept or reject the component.

The benefits in using hybrid combinations of neural and fuzzy systems such as the one shown in Figure 13.3 are due to the fact that numerical measurements may actually provide too much detail to be effectively used

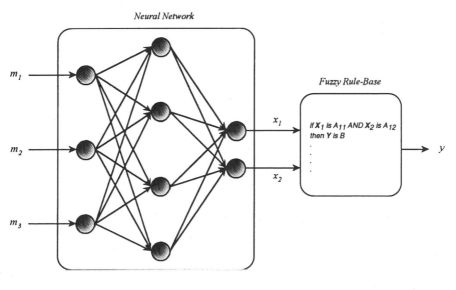

Figure 13.3 A hybrid system involving a neural network *in series* with a fuzzy system where measurements get mapped to features serving as inputs to the fuzzy system.

on-line (in addition to noise and other problems). Hence, neural filtering, smoothing, and mapping of numerical measurements to a feature space (e.g., *physical damage* and *signal-to-noise ratio*) may facilitate quick action by a fuzzy controller. Of course the arrangement can be reversed; that is, fuzzy processing can precede the neural network. Such an approach was taken by Yea and his coworkers in Japan in an interesting project involving odor discrimination where in order to discriminate amongst many kinds of odor species, a system has been developed using multiple gas sensors as sensory input to neural networks (Yea et al., 1994). When the system is presented with a number of inflammable gases, fragrant smells or even offensive odors, it is capable of an almost 100% discrimination of the different odors. The discrimination is performed in two steps: First, classification of the odor group performed by a fuzzy algorithm—that is, determining the groups of inflammable gases, fragrant smells, or offensive odors; second, discrimination of individual odor species in the classified group, performed by neural tools.

13.3   NEURAL NETWORKS FOR DETERMINING MEMBERSHIP FUNCTIONS

An interesting and often advantageous feature of fuzzy systems is that they allow for rather flexible categorization of a domain of interest. For example, when a problem calls for a small number of categories of temperature, we define *SMALL, MEDIUM*, and *LARGE* as the values of the fuzzy variable *temperature*, instead of say 100 categories of natural numbers taking us from 1°C to 100°C. For each and every linguistic value a unique membership function analytically describing the degree of membership to the fuzzy value of each individual crisp element of the universe of discourse is sought. The problem of determining membership functions has occupied a central importance in the history of fuzzy logic with a number of subjectivist, statistical, and (more recently) neural approaches being proposed.

Membership function determination may be viewed as a data clustering and classification problem. Hence, neural clustering and classification algorithms can be brought to bear to solve this problem as illustrated in Figure 13.4. When multidimensional data are clustered, we can extract either one-dimensional membership functions based on a distance metric δ (as shown in the figure) or obtain multidimensional membership functions modeling fuzzy relations (i.e., *if/then* rules). A typical use of neural networks for producing membership functions involved a two-stage process: *clustering* and *fuzzification* (Adeli and Hung, 1995). The first stage is essentially a *classification* stage where a neural network is used to classify or cluster domain data into a certain number of clusters. The second stage is a fuzzification process where fuzzy membership values are assigned (to each training instance) in the set of classified clusters [see also Travis, 1994]. Of course the problem of membership function determination is not totally separate from the problem

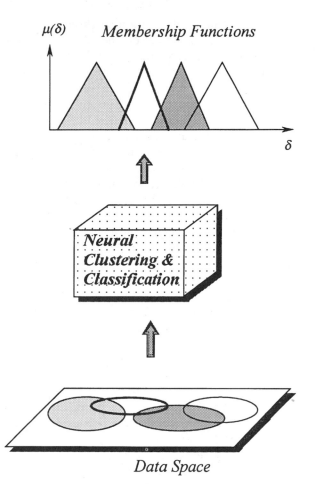

Figure 13.4 Clustering approaches are used to determine membership functions in data-rich applications.

of identifying rules (only the alter one is much more difficult). In practice, both the determination of membership functions and the extraction of rules proceed through some kind of clustering.

The methods used in the categorization of a universe of discourse are typically based on some type of Kohonen or unsupervised learning network. A pattern clustering method based on the Kohonen feature mapping algorithm and the backpropagation multilayer perceptron has been used for membership function determination by Pham and Bayro-Corrochano (1994). The method is applied first to the training data set to divide it into labeled clusters using the Kohonen algorithm and a cluster labeling procedure. The data clusters are then employed to train a three-layer perceptron. The approach is self-organizing by virtue of the Kohonen algorithm, and it produces fuzzy outputs as a consequence of the backpropagation network.

Kuo, Cohen, and Kumara at Penn State have taken a similar yet different approach in developing a novel self-organizing and self-adjusting fuzzy modeling approach with learning capabilities (Kuo et al., 1994). Basically, their approach consists of two stages: a *self-organizing* and a *self-adjusting* stage. In the first stage, the input data are divided into several groups by applying Kohonen's feature maps. Gaussian distribution functions are employed as the standard form of the membership functions. Statistical tools are used to determine the center and width of the membership function for each group. Error backpropagation (see Section 13.5) fine-tunes the parameters involved. Feedforward neural estimation for membership function determination and fuzzy classification have also been investigated by Purushotaman and Karayiannis at the University of Houston (Purushothaman and Karayiannis, 1994), while Higgins and Goodman at MIT developed a different method for learning membership functions and rules from a set of examples (Higgins and Goodman, 1994). Their method is a general function approximation system using a three-step approach: first, learning the membership functions and creating a cell-based rule representation; second, simplifying the cell-based rules using an information-theoretic approach for induction of rules from discrete-valued data; and, finally, constructing a neural network to compute the function value given its independent variables.

A typical use of neural networks for producing membership functions involved a two-stage process. The first stage is essentially a *classification* stage where a neural network is used to classify or cluster domain data into a certain number of clusters. The second stage is a *fuzzification process* where the fuzzy membership values for each training instance in the set of supports, classified clusters, are evaluated. Let us look at the *Adeli–Hung algorithm* (AHA) for determining membership functions (Adeli and Hung, 1994).

Determining Membership Functions Through the Adeli–Hung Algorithm

Suppose that our data consist of N training instances $X_1, X_2, \ldots, X_N$ and we have M patterns in each training instance, $X_i = [x_{i_1}, x_{i_2}, \ldots, x_{i_M}]$. The *mean vector* of these instances may be defined as

$$\overline{X}_N = \frac{1}{N} \sum_{i=1}^{N} X_i \tag{13.3-1}$$

For $N + 1$ training instances the mean vector is found from the mean $\overline{X}_N$

and the instance X_{N+1} as follows:

$$\bar{X}_{N+1} = \frac{1}{N+1} \sum_{i=1}^{N+1} X_i$$

$$= \frac{1}{N+1} \left[\sum_{i=1}^{N} X_i + X_{N+1} \right]$$

$$= \frac{1}{N+1} \left[N\bar{X}_N + X_{N+1} \right]$$

$$= \frac{N}{N+1} \bar{X}_N + \frac{1}{N+1} X_{N+1} \qquad (13.3\text{-}2)$$

Classification by AHA is performed using a topology-and-weight-change, two-layer (flat) neural network where the number of input nodes equals the number of patterns (M) in each training instance and the number of output nodes equals the number of clusters.

The algorithm uses a neural network $NN(M, 1)$ with M inputs and an as yet undetermined number of outputs. The first training instance gets assigned to the first cluster. If the second instance is classified to the first cluster, the output node representing the first cluster becomes active. If the second training instance is classified as a new cluster, an additional output node is added to the network, and so on, until all training instances are classified.

To perform the classification in AHA, a function $diff(X, C)$ is defined, called the *degree of difference*, representing the difference between a training instance X and a cluster C in a $NN(M, P)$ network (P indicates the number of output nodes or equivalently the number of classes). This function maps two given vectors (X and C) to a real number ($diff$). The patterns of each cluster (means of the patterns of the instance in the cluster) are stored in the weights of the network during the classification process. The following procedure for classifying a training instance into an active or new cluster is used in AHA:

Step 1. Calculate the degree of difference, $diff(X, C_i)$, between the training instance, X, and each cluster, C_i. A Euclidean distance is used (in image recognition applications) and the function $diff(X, C_i)$ becomes

$$diff(X, C_i) = \sqrt{\sum_{j=1}^{M} (x_j - c_{ij})^2} \qquad (13.3\text{-}3)$$

Step 2. Find the smallest degree of difference, $diff_{min}(X, C_i)$, and make the cluster with the smallest degree of difference an active cluster:

$$C_{active} = \{C|\min\{diff(X, C_i)\}, i = 1, 2, \ldots, P\} \qquad (13.3\text{-}4)$$

Step 3. Compare the value of $diff_{min}$ with a predefined a threshold value κ. If the value of $diff_{min}$ is greater than the predefined threshold, the training instance is classified as a new cluster (at this point one more output node is turned on).

$$C_{new} = X \qquad \text{if } k < \min\{diff(X, C_i), i = 1, 2, \ldots, P\} \quad (13.3\text{-}5)$$

Suppose the given N training instances have been classified into P clusters. Let us use the symbols C_j to denote the jth cluster and use U to denote the set of all clusters. If the clusters are completely disjoint, each instance in the training set belongs to only one of the classified clusters and a binary matrix Z can be used to record the cluster of each instance. If the instance i belongs to the j cluster we have $z_{ij} = 1$, while if it does not belong we have $z_{ij} = 0$. On the other hand, if the classified clusters are partly overlapping, a given instance in the training set may belong to more than one cluster. Hence the boundaries of the classified clusters are fuzzy rather than crisp. The same binary matrix Z may be used to record the cluster of each instance. The prototype for each cluster is defined as the mean of all instances in that cluster, and the degree of membership of each instance in the cluster is based on how similar this instance is to the prototype one. The similarity can be defined as a function of distance between the instance and the prototype of the cluster. If there are n_p instances in a cluster p, the pattern vector of the ith instance in the cluster p is $X_i^p = [x_{i_1}^p, x_{i_2}^p, \ldots, x_{i_M}^p]$. Then, the vector of the prototype instance (the mean of all instances) in cluster p is defined as

$$C_p = [c_{p_1}, c_{p_2}, \ldots, c_{p_M}] = \frac{1}{n_p} \sum_{i=1}^{n_p} X_i^p \qquad (13.3\text{-}6)$$

where $c_{p_j} = (1/n_p)\sum_{i=1}^{n_p} x_{ij}^p$ and $j = 1, 2, \ldots, M$. Using triangular-shaped membership functions (over the *diff* universe of discourse) the fuzzy membership value of the ith instance in the p cluster is defined as

$$\mu_p(X_i^p) = f\left[D^w(X_i^p, C_p)\right]$$

$$= \begin{cases} 0 & \text{if } D^w(X_i^p, C_p) > \kappa \\ 1 - \dfrac{D^w(X_i^p, C_p)}{\kappa} & \text{if } D^w(X_i^p, C_p) \leq \kappa \end{cases} \qquad (13.3\text{-}7)$$

where a predefined threshold value κ is used as crossover value. The similarity function is defined as the weighted norm $D^w(X_i^p, C_p)$. The weighted norm in the Adeli–Hung algorithm is defined as the Euclidean distance:

$$D^w\left(X_i^p, C_p\right) = \left\| w_p\left(X_i^p, C_p\right)\right\|^w = \sqrt{\sum_{j=1}^{M}\left(x_{ij}^p - c_{pj}\right)^2} \quad (13.3\text{-}8)$$

In image recognition problems, a value of 1 is used for the weight parameters w and w_p. If the Euclidean distance for a given instance is less than the crossover value κ, the instance belongs to the cluster p to a degree given by the membership value.

13.4   NEURAL-NETWORK-DRIVEN FUZZY REASONING

In fuzzy systems employing more than three or four fuzzy variables, it may be practically difficult to formulate fuzzy *if/then* rules, and it would be desirable if they could be extracted automatically out of data from the physical system being modeled. The problem of inducing (extracting) fuzzy rules has been addressed by several researchers and is still undergoing intense investigation (Kosko, 1992; Takagi, 1991; Hayashi et al., 1992; (Keller and Tahani, 1992; (Keller et al., 1994; Khan, 1993; Li and Wu, 1994; Wang, 1994; Wang and Mendel, 1992; Nie, 1994; Jang and Sun, 1995; Werbos, 1992; Yager, 1994; Blanco et al., 1995). In an important paper published in 1991, Matsushita Electric engineers Tagaki and Hayashi presented a comprehensive approach for the induction and tuning of fuzzy rules, known as *neural-network-driven fuzzy reasoning* or the *Takagi–Hayashi (T–H) method.*

Consider the situation shown in Figure 13.5 where we have the data space of two inputs, x_1 and x_2 (e.g., two measurements obtained from sensors), knowledge of the target or desirable output, and a nonlinear partition of this space in three regions. These regions correspond to three fuzzy *if/then* rules. The identified rules R_1, R_2, and R_3 are of the Sugeno variety (see Chapter 6); that is, their consequent is a functional mapping of the antecedent variables, with the mapping actually being performed by specially trained neural networks.[3] The Takagi–Hayashi method consists of three major parts:

Part 1: Partitions the control or decision hypersurface into a number of rules.

Part 2: Identifies a given rule's LHS (antecedent) values (i.e., determines their membership functions).

---

[3]A potential drawback of the T–H method as well as most similar methods is that one has to decide in advance the possible number of rules—for example, three rules in this case.

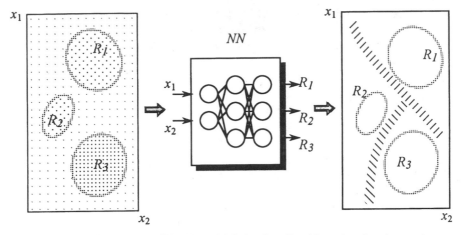

Figure 13.5 Schematic of the Hayashi–Takagi method for extracting fuzzy rules.

Part 3: Identifies a given rule's RHS (consequent) values (the amount of control for each control rule).

Part 1 determines the number of fuzzy inference rules through clustering performed on the data available. Part 2 employs a neural network to derive the membership function for each rule (it therefore identifies the LHS of rules). The T–H method combines all the variables (x_1 and x_2, for example) in the LHS and is based on the theoretical result that an arbitrary continuous function is equivalent to a neural network having at least one hidden layer. Buckley and Hayashi (1993) have shown the computational equivalence between continuous functions, regular neural nets, fuzzy controllers, and discrete fuzzy expert systems and have shown how to build hybrid neural nets numerically identical to a fuzzy controller or a discrete fuzzy expert system. Part 3 of the T–H method determines the RHS parts using neural networks with supervised learning (supervised by the learning data and the control value for each rule as in *Part 2*).

Sugeno-type rules are used (see Chapter 6) where the output is a function of the inputs. Sugeno rules are typically of the form

$$\text{if } x_1 \text{ is } A_1 \ AND \, x_2 \text{ is } A_2 \ldots, \text{ then } y = f(x_1, \ldots, x_n) \quad (13.4\text{-}1)$$

where f is a function of the inputs $x_1, \ldots, x_n$. In the T–H method this function has been replaced by a neural network. For example, an induced rule would be of the form

$$\text{if } (x_1, x_2) \text{ is } A^s, \text{ then } y^s = NN_s(x_1, x_2) \quad (13.4\text{-}2)$$

where $\mathbf{x} = (x_1, x_2)$ is the vector of inputs and $y^s = NN_s(x_1, x_2)$ is a neural

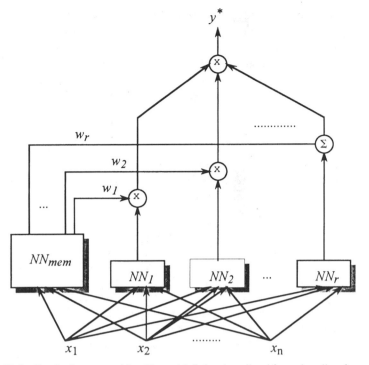

Figure 13.6 Block diagram of the Hayashi–Takagi method for extracting fuzzy rules.

network that determines the output y^s of the sth rule and A^s is the membership function of the antecedent of the sth rule.

A block diagram of the T–H method is shown in Figure 13.6. As may be seen in the figure, several neural networks are used. The neural network labeled NN_{mem} is responsible for generating the membership functions of the antecedents of rules while networks $NN_1, NN_2, \ldots, NN_r$ determine the consequent parts. Networks $NN_1, NN_2, \ldots, NN_r$ provide the RHS function of Sugeno rules shown in equation (13.4-1). In actual applications these are three-layer networks trained by backpropagation. As seen in Figure 13.6, the overall system weighs the output of the RHS networks by the membership values of LHS and computes a final output value. The following eight steps constitute the outline of the procedure used in the Takagi–Hayashi method:

Step 1. We define y_i as the output and define x_j, $j = 1, 2, \ldots, k$, as the input variables. Inputs x_j, $j = 1, 2, \ldots, m$, $m \leq k$, that are related to the observed value of the output are selected by a neural network through a backward elimination method using sum of squared errors as a cost function for the purpose of eliminating input variables attributed to noise. It is important to select only those input variables that have significant correlation to the observed values.

Step 2. The input–output population n is divided into *training data* (TRD) of n_t) and *checking data* (CHD of n_c), where $n = n_t + n_c$.

Step 3. The TRD is partitioned into r groups using a clustering method. Each partition is labeled as R^s, $s = 1, 2, \ldots, r$, and the data within the ith partition is expressed as $(\mathbf{x}_i^s, y_i^s)$, where $i = 1, 2, \ldots, (n_t)^s$ and $(n_t)^s$ are the TRD numbers in each R^s. Partitioning the data space into r partitions implies that the number of inference rules will be taken to be r.

Step 4. The antecedent part of each rule is identified through NN_{mem} (the neural network generating the membership functions, see Figure 13.6). If $\mathbf{x}_i$ are the values for the input layer of NN_{mem}, the weights w_i^s are assigned as the supervised data for the output layer, where

$$w_i^s = \begin{cases} 1, & \mathbf{x}_i \in R^s \\ 0, & \mathbf{x}_i \notin R^s \end{cases} \quad i = 1, \ldots, n_t; i = 1, \ldots, r \quad (13.4\text{-}3)$$

The network NN_{mem} is trained to infer weights w_i^s given an input vector $\mathbf{x}$. NN_{mem} thus becomes capable of computing the degree of membership $\hat{w}_i^s$ of each training data item $\mathbf{x}_i$ to the rule (or partition) R^s. The membership function of the antecedent A^s of the sth rule is defined as the inferred value $\hat{w}_i^s$—that is, the output of NN_{mem}:

$$\mu_{A^s}(\mathbf{x}_i) \equiv \hat{w}_i^s, \quad i = 1, 2, \ldots, n \quad (13.4\text{-}4)$$

Step 5. After identifying the antecedent membership function in step 4, we now identify the *consequent* part of the Sugeno fuzzy *if/then* rules we are looking for. The RHS of each rule is expressed by the input–output relationship. Inputs $x_{i1}^s, \ldots, x_{im}^s$ and outputs y^s, $i = 1, 2, \ldots, (n_t)^s$, from the training data are used as input–output pairs for training the NN_s neural network that models the consequent of the sth rule. Subsequently the checking data $x_{i1}, \ldots, x_{im}$, $i = 1, 2, \ldots, n_c$, are used as input and the sum of squared errors is formed:

$$\Theta_m^s = \sum_{i=1}^{n_c} \left[y_i - u_s(\mathbf{x}_i) \cdot \mu_{A^s}(\mathbf{x}_i) \right]^2 \quad (13.4\text{-}5)$$

where $u_s(\mathbf{x}_i)$ is the calculated output of NN_s, y_i is the target output for the network, and Θ_m^s is sum of squared errors. The sum can also be computed after weighing by $\mu_{A^s}(\mathbf{x}_i)$; that is,

$$\Theta_m^s = \sum_{i=1}^{n_c} \mu_{A^s}(\mathbf{x}_i) \cdot \left[y_i - u_s(\mathbf{x}_i) \cdot \mu_{A^s}(\mathbf{x}_i) \right]^2 \quad (13.4\text{-}6)$$

Takagi and Hayashi use an index to decide the best iteration number during training (to prevent "overlearning" or memorization):

$$I^s = \frac{n_c}{(n_t)^s + n_c} \sum_{i=1}^{(n_t)^s} [y_i - u_s(\mathbf{x}_i)]^2$$

$$+ \frac{(n_t)^s}{(n_t)^s + n_c} \sum_{j=1}^{n_c} [y_j - u_s(\mathbf{x}_j) \cdot \mu_{A^s}(\mathbf{x}_j)]^2 \quad (13.4\text{-}7)$$

If the sth network has overlearned, the error of the TRD becomes small but the error of the CHD becomes large, suggesting that the optimum number of iterations is the one that gives the smallest I^s in equation (13.4-6).

Step 6. In this step a number of variables may be eliminated from the consequent through a backward elimination method. Out of the m input variables of a network inferring the consequent of a rule, one (e.g., x^p) is arbitrarily eliminated, and the neural network for each consequent is trained using the TRD as in step 5. Equation (13.4-8) below gives the squared error Θ_{m-1}^{sp} of the control value of the sth rule in the case where x^p has been eliminated. This sum squared error can be estimated using the *checking data*:

$$\Theta_{m-1}^{sp} = \sum_{i=1}^{n_c} [y_i - u_s(\mathbf{x}_i) \cdot \mu_{A^s}(\mathbf{x}_i)]^2, \quad p = 1, 2, \ldots, m \quad (13.4\text{-}8)$$

After comparing equations (13.4-6) and (13.4-8) and $\Theta_m^s > \Theta_{m-1}^{sp}$, the significance of the eliminated variable x^p can be considered minimal, and x^p can be discarded.

Step 7. The operations in step 6 are carried out for the remaining $m - 1$ input variables until it is no longer true that $\Theta_m^s > \Theta_{m-1}^{sp}$ for any of the remaining input variables. The model that gives the minimum Θ^s value is the best-trained neural network for the sth rule.

Step 8. Equation (13.4-8) below gives the final control value y_i^*:

$$y_i^* = \frac{\sum_{s=1}^r \mu_{A^s}(\mathbf{x}_i) \cdot \{u_s(\mathbf{x}_i)\}_{\text{inf}}}{\sum_{s=1}^r \mu_{A^s}(\mathbf{x}_i)}, \quad i = 1, 2, \ldots, n \quad (13.4\text{-}8)$$

where $[u_s(\mathbf{x}_i)]_{\text{inf}}$ is an inferred value obtained when CHD is substituted in the best *NN* obtained in step 7.

It should be noted that the T–H method allows for nonlinear partitioning of the input space and hence the identification of nonlinear membership functions. Each cluster of input data corresponds to an *if/then* Sugeno-type rule as shown in (13.4-2). Although these rules individually make a good fit for data similar to what they have been trained on, a gradual fitting to multiple rules has to be performed for data near the boundary region (see Figure 13.5). The following example [taken from Hayashi et al. (1992)] serves to illustrate the T–H method.

Example 13.1 COD Density Estimation. Data of *chemical oxygen demand* (COD) density in Japan's Osaka Bay taken over a 10-month interval were used by Takagi and Hayashi (1992) to test their neural-network-driven fuzzy reasoning method. In this application of the T–H method the input–output variables are

y	COD density (ppm)
x_1	Water temperature (°C)
x_2	Transparency (m)
x_3	Dissolved oxygen density (ppm)
x_4	Salinity (%)
x_5	Filtered COD density (ppm)

In accordance with step 2 in the T–H method, the data were divided in training and checking data as shown in Figure 13.7. Thirty-two data points were used for estimation, while 12 data points were used for testing. Performing a backward elimination experiment suggested the use of all input variables for estimation.

For determining the membership functions of the antecedent sets (i.e., NN_{mem}), a four-layer network with five input nodes in the input layer, two

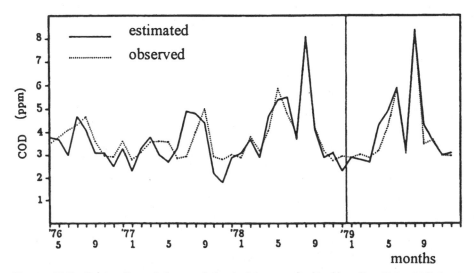

Figure 13.7 Osaka Bay data used for training and checking the Hayashi\Takagi method. (Takaki and Hayashi, 1991).

hidden layers with 12 nodes each, and two output nodes in the output layer was used. Similarly, for determining the consequent part of rules, four-layer networks were used with m input nodes in the input layer ($m = 5, 4, \ldots$), two hidden layers with 12 nodes each, and an output layer with only one node (see Figure 13.6).

Network training took 1500–2000 iterations, and the following rule structure was finally identified:

$$R_1: \quad \text{if } (x_1, x_2, x_3, x_4, x_5) \text{ is } A^1, \text{ then } y^1 = NN_1(x_1, x_2, x_3, x_4, x_5)$$

$$R_2: \quad \text{if } (x_1, x_2, x_3, x_4, x_5) \text{ is } A^2, \text{ then } y^2 = NN_2(x_1, x_2, x_3, x_5)$$

The estimated COD density by the above system was in very good agreement with observed data, and it performed better in comparison with results obtained by other methods. □

13.5 LEARNING AND ADAPTATION IN FUZZY SYSTEMS VIA NEURAL METHODS

In recent years Nomura, Hayashi, and Wakami proposed an approach to fuzzy system adaptation utilizing the gradient-descent error minimization we saw in connection with backpropagation in Chapter 8. A parameterized description of a fuzzy system with symmetric, triangular-shaped membership functions for inputs and crisp outputs was developed, and error minimization through gradient-descent was used (Nomura et al., 1994; Wang, 1994; Jang and Sun, 1995). Ichihashi et al. (1993) used gradient descent with exponential membership functions. Guély and Siarry (1993) have solved the problem more generally—that is, for symmetric as well as nonsymmetric antecedent membership functions and different connectives and consequent forms.

Most fuzzy system adaptation approaches rely on gradient-descent optimization. As is the case in neural learning, an objective function E is sought to be minimized:

$$E = \tfrac{1}{2}(y' - y)^2 \qquad (13.5\text{-}1)$$

where y is the output of the fuzzy system and y' is the reference (target) system output.

Consider the ith zero-order Sugeno rule of a system having n such rules ($i = 1, \ldots, n$):

$$R_i: \quad \text{if } x_1 \text{ is } A_{i1} \text{ AND } \cdots \text{ AND } x_m \text{ is } A_{im} \text{ then } y \text{ is } w_i \quad (13.5\text{-}2)$$

where $A_{i1}, \ldots, A_{im}$ are the fuzzy values of the LHS of the rule and w_i is a constant in the consequent of the ith rule. In a fuzzy system, we combine fuzzy *if/then* rules like the one above to perform a mapping from fuzzy sets of the LHS universe of discourse to constants in the RHS. Let μ_i be the membership function for the fuzzy relation of the ith rule. Then the output of a simplified fuzzy reasoning approach, y, can be obtained through equations

$$\mu_i = \prod_{j=1}^{m} A_{ij}(x_j) \tag{13.5-3}$$

and

$$y = \frac{\sum_{i=1}^{n} \mu_i \cdot w_i}{\sum_{i=1}^{n} \mu_i} \tag{13.5-4}$$

Using input–output data and a gradient-descent algorithm, we can optimize the w_i's by minimizing an objective function E such as the one given by equation (13.5-1). Let us rewrite this squared error function as

$$E = \tfrac{1}{2}(y^{rp} - y^{p})^2 \tag{13.5-5}$$

where y^{rp} is the target for the pth input data $(x_1^p, \ldots, x_m^p)$ and y^p is the calculated output of the system corresponding to the same input data. The learning rule for the real numbers in the RHS of rules (13.5-2) is

$$w_i(t' + 1) = w_i(t') - K \cdot \frac{\partial E}{\partial w_i} \tag{13.5-6}$$

where t' is the number of iteration of learning. Following Nomura et al. (1994) and using the above error function (13.5-5) in (13.5-6), we express the w_i update as

$$w_i(t' + 1) = w_i(t') - K \cdot \frac{\mu_i^p}{\sum_{i=1}^{n} \mu_i^p}(y^p - y^{rp}) \tag{13.5-7}$$

where μ_i^p is the membership value of the ith rule corresponding to the pth input–output example and K is a constant.

Using input–output examples with learning rule (13.5-7) repeatedly, the RHS numbers w_i are updated so as to minimize the error function, ultimately reaching a global minimum since $\partial^2 E / \partial w_i^2 \geq 0$ is obtained for all rules.

As with *neural-network-driven fuzzy reasoning* in the previous section, in this adaptation approach too one has to come up *a priori* with the optimal number of rules, often through a trial-and-error approach. A number of researchers are proposing various genetic approaches to address this issue [see Nomura et al. (1994), Pedrycz (1995), Perneel (1995)][4].

[4]See Chapter 17 of this book.

Using gradient descent, the membership functions of the LHS of rules (15.4-2) may be tuned to better reflect the problem at hand. Consider the symmetric triangular membership function for the jth antecedent of the ith rule shown in Figure 13.8. Such a membership function can be represented by the peak a_{ij} and the support b_{ij}, and therefore the entire rule (13.5-2) can be parameterized through the peaks of antecedent values a_{ij}, their support b_{ij}, and w_i.

Following Guély and Siarry (1993), let us address adaptation for rules with symmetric triangular membership functions in the LHS, product interpretation of *AND*, center of area output calculation, and constant outputs as in rule (13.5-2). As seen in Figure 13.8a, the symmetric triangular membership functions in the LHS are given by

$$\mu_{ij}(x_j) = \begin{cases} 1 - \dfrac{|x_j - a_{ij}|}{5b_{ij}}, & \text{if } |x_j - a_{ij}| \le \dfrac{b_{ij}}{2} \\ 0, & \text{otherwise} \end{cases} \qquad (13.5\text{-}8)$$

As before, we want to adapt the parameters (a_{ij}, b_{ij}, w_i). Let p denote the number of training samples, and y'^p the training sample output. Using gradient descent means that our peak parameters, for example, will be updated in the following manner:

$$a_{ij}(t' + 1) = a_{ij}(t') - \frac{\eta_a}{p} \cdot \frac{\partial E}{\partial a_{ij}}$$

$$= a_{ij}(t') - \frac{\eta_a}{2p} \cdot \sum_{p'=1}^{P} \frac{\partial E_{p'}}{\partial a_{ij}} \qquad (13.5\text{-}9)$$

where η_b is the gradient-descent speed for a_{ij}, the peak parameter, and we use η_b and η_w for the support b_{ij} and the RHS parameter w_i, respectively. Guély and Siarry observed experimentally that learning was sensitive to these parameters.

To simplify (13.5-9) let us use E, y, and y' instead of E_p, y^p, and y'^p for the pth input–output example. Then we have the following derivative that we could use in equation (13.5-9) to obtain the update of the peak parameters:

$$\frac{\partial E}{\partial a_{ij}} = \frac{\partial E}{\partial y} \cdot \frac{\partial y}{\partial \mu_i} \cdot \frac{\partial \mu_i}{\partial \mu_{ij}} \cdot \frac{\partial \mu_{ij}}{\partial a_{ij}} \qquad (13.5\text{-}10)$$

Similarly for the upgrade of the support and RHS parameter, we need to

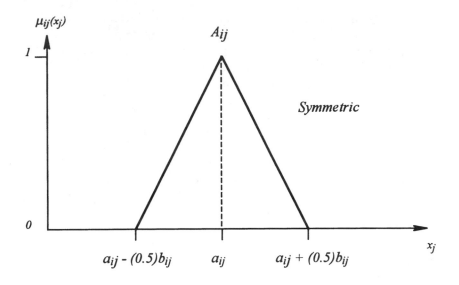

(a)

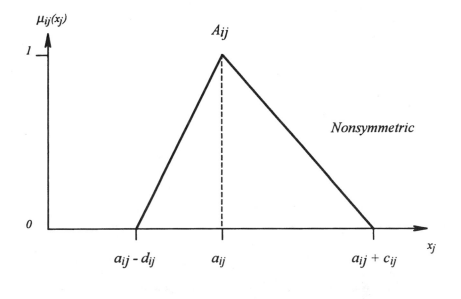

(b)

Figure 13.8 Parameters used to describe the triangular-shaped membership function for the jth antecedent of the ith rule. (a) Symmetric triangular membership functions and (b) nonsymmetric triangular membership functions.

evaluate the following derivatives:

$$\frac{\partial E}{\partial b_{ij}} = \frac{\partial E}{\partial y} \cdot \frac{\partial y}{\partial \mu_i} \cdot \frac{\partial \mu_i}{\partial \mu_{ij}} \cdot \frac{\partial \mu_{ij}}{\partial b_{ij}} \tag{13.5-11}$$

and

$$\frac{\partial E}{\partial w_i} = \frac{\partial E}{\partial y} \cdot \frac{\partial y}{\partial w_i} \tag{13.5-12}$$

Given the symmetric triangular shape of the membership functions [equation (13.5-8)] and the product interpretation of *AND*, the partial derivatives in the above equations are as follows:

$$\frac{\partial E}{\partial y} = y - y' \tag{13.5-13}$$

The partial $\partial y / \partial w_i$ is

$$\frac{\partial y}{\partial w_i} = \frac{\mu_i}{\sum_{i'=1}^{n} \mu_{i'}} \tag{13.5-14}$$

And we also have (treating y' as constant)

$$\frac{\partial y}{\partial \mu_i} = \frac{(w_i - \mu_i)}{\sum_{i'=1}^{n} \mu_{i'}} \tag{13.5-15}$$

$$\frac{\partial \mu_i}{\partial \mu_{ij}} = \frac{\mu_i}{\mu_{ij}(x_j)} \tag{13.5-16}$$

$$\frac{\partial \mu_{ij}}{\partial a_{ij}} = \frac{2 \, sign(x_j - a_{ij})}{b_{ij}} \tag{13.5-17}$$

$$\frac{\partial \mu_{ij}}{\partial b_{ij}} = \frac{1 - \mu_{ij}(x_j)}{b_{ij}} \tag{13.5-18}$$

Equations (13.5-10) to (13.5-18) provide all the terms needed for the learning formula (13.5-9) for the peak, but also for the support and RHS parameters —that is, the entire set of parameters (a_{ij}, b_{ij}, w_i) we use to adapt our fuzzy systems. In general, the system is sensitive to the gradient-descent speeds, and increasing the number of rules makes training more difficult.

Similarly we can train a fuzzy system that uses nonsymmetric antecedent membership functions such as the one shown in Figure 13.8*b* by repeating the above procedure for the set of relevant parameters. Variations of the gradient-descent procedure have been developed and applied when the rules use the min interpretation of *AND* (instead of product) and a polynomial RHS instead of constant [see Guély and Siarry, (1993)]. In general, researchers report considerable advantages in the speed of training adaptive fuzzy systems when compared to regular three-layer neural networks.

13.6 ADAPTIVE NETWORK-BASED FUZZY INFERENCE SYSTEMS

To tackle the problem of parameter identification, Jang and Sun (Jang, 1992; Jang and Sun, 1995; (Jang and Gulley, 1995) have proposed an *adaptive network-based fuzzy inference system* (ANFIS) that identifies a set of parameters through a hybrid learning rule combining the backpropagation gradient-descent and a least-squares method. ANFIS can be built through the fuzzy toolbox available for MATLAB [actually developed by Jang (Jang and Gulley, 1995)]. Applications and properties of ANFIS have been investigated, and a number of methods has been proposed for partitioning the input space and hence address the structure identification problem. Fundamentally, ANFIS is a graphical network representation of Sugeno-type fuzzy systems, endowed with neural learning capabilities. The network is comprised of nodes and with specific functions, or duties, collected in layers with specific functions. To illustrate its representational strength, let us consider two first-order Sugeno rules having outputs which are linear combinations of their inputs:

$$
\begin{aligned}
&\text{if } x \text{ is } A_1 \; AND \, y \text{ is } B_1, \text{ then } f_1 = p_1 x + q_1 y + r_1 \\
&\text{if } x \text{ is } A_2 \; AND \, y \text{ is } B_2, \text{ then } f_2 = p_2 x + q_2 y + r_2
\end{aligned}
\tag{13.6-1}
$$

ANFIS can construct a network realization of rules (13.6-1). Figure 13.9 illustrates the evaluation of these rules (upper part) and the corresponding ANFIS architecture (lower part). The nodes in the same layer of ANFIS are of the same function family and are arranged as follows:

Layer 1. Each node in this layer generates the membership grades of a linguistic label. The ith node for example may perform the following (fuzzification) operation:

$$
O_i^1 = \mu_A(x) = \cfrac{1}{1 + \left[\left(\dfrac{x - c_i}{a_i} \right)^2 \right]^{b_i}}
\tag{13.6-2}
$$

where x is the input to the ith node and A_i is the linguistic value (small, large, etc.) associated with this node. The set of parameters $\{a_i, b_i, c_i\}$ is used to adjust the shape of the membership function.

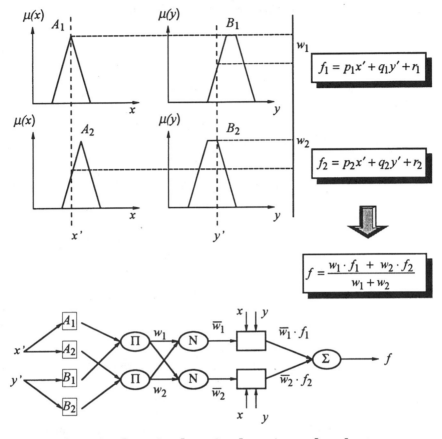

Layer 1   Layer 2   Layer 3   Layer 4   Layer5

Figure 13.9 Evaluation of a network realization of rules (13.6-1) (*top*) and the corresponding ANFIS architecture (*bottom*).

Layer 2. Each node in this layer calculates the firing strength of each rule via multiplication (or min):

$$O_i^2 = w_i = \mu_{A_i}(x) \times \mu_{B_i}(y), \quad i = 1, 2 \qquad (13.6\text{-}3)$$

Layer 3. The ith node of this layer calculates the ratio of the ith rule's firing strength to the sum of all rules' firing strengths:

$$O_i^2 = \overline{w}_i = \frac{w_i}{w_1 + w_2}, \quad i = 1, 2 \qquad (13.6\text{-}4)$$

Layer 4. The *i*th node in this layer has the following function:

$$O_i^2 = \overline{w}_i f_i = \overline{w}_i (p_i x + q_i y + r_i) \tag{13.6-5}$$

where $\overline{w}_i$ is the output of layer 3, and $\{ p_i, q_i, r_i \}$ is the parameter set. Parameters in this layer will be referred to as the *consequent parameters*.

Layer 5. The single node in this layer aggregates the overall output as the summation of all incoming signals:

$$O_1^5 = \text{overall output} = \sum_i \overline{w}_i f_i = \frac{\Sigma_i w_i f_i}{\Sigma_i w_i} \tag{13.6-6}$$

The learning rule of ANFIS is based on gradient descent optimization as with the feedforward neural networks that we have seen in Chapters 8 and 9 [see MATLAB's fuzzy toolbox for more details on ANFIS (Jang and Gulley, 1995)].

REFERENCES

Adeli, H., and Hung S.-L., *Machine Learning—Neural Networks, Genetic Algorithms and Fuzzy Systems*, John Wiley & Sons, New York, 1995.

Blanco, A., Delgado, M., and Ruquena, I., A Learning Procedure to Identify Weighted Rules by Neural Networks, *Fuzzy Sets and Systems*, Vol. 69, pp. 29–36, 1995.

Buckley, J. J., and Hayashi, Y., Hybrid Neural Nets Can Be Fuzzy Controllers and Fuzzy Expert Systems, *Fuzzy Sets and Systems*, Vol. 60, No. 2, pp. 135–142, 1993.

Guély, F., and Siarry, P., Gradient Descent Method for Optimizing Various Fuzzy Rules, in *Proceedings of the Second IEEE International Conference on Fuzzy Systems*, San Francisco, 1993, pp. 1241–1246.

Hayashi, I., Nomura, H., Yamasaki, H., and Wakami, N., Construction of Fuzzy Inference Rules by NDF and NDFL, *International Journal of Approximate Reasoning*, Vol. 6, pp. 241–266, 1992.

Higgins, C. M., Goodman, R. M., Fuzzy Rule-Based Networks for Control, *IEEE Transactions on Fuzzy Systems*, Vol. 2, No. 1, pp. 82–88, 1994.

Ichihashi, H., Miyoshi, T., and Nagasaka, K., Computed Tomography by Neuro-fuzzy Inversion, in *Proceedings of 1993 International Joint Conference on Neural Networks*, Part 1 (of 3), Nagoya, October 25–29, 1993, pp. 709–712.

Jang, J.-S., Self-Learning Fuzzy Controller Based Inference Systems, *IEEE Transactions on Nueral Networks*, Vol. 3, No. 5, pp. 714–723, 1992.

Jang, J.-S., and Gulley, N., *Fuzzy Logic Toolbox for Use with MATLAB*, The Math-Works, Inc., Natick, MA, 1995.

Jang, J.-S., and Sun C.-T., Neuro-fuzzy Modeling and Control, *Proceedings of the IEEE*, Vol. 83, No. 3, March 1995, pp. 378–406.

Keller, J. M., and Tahani, H., Implementation of Conjunctive and Disjunctive Fuzzy Logic Rules with Neural Networks, *International Journal of Approximate Reasoning*, Vol. 6, pp. 221–240, 1992.

Keller, J. M., Hayashi, Y., and Chen, Z., "Additive Hybrid Networks for Fuzzy Logic, *Fuzzy Sets and Systems*, Vol. 66, No. 3, pp. 307–313, 1994.

Khan, E., NeuFuz: An Intelligent Combination of Fuzzy Logic with Neural Nets, in *Proceedings of the International Joint Conference on Neural Networks*, Vol. 3, 1993, pp. 2945–2950.

Kosko, B., *Neural Networks and Fuzzy Systems*, Prentice-Hall, Englewood Cliffs, NJ, 1992.

Kuo, R. J., Cohen, P. H., and Kumara, S. R. T., Neural Network Driven Fuzzy Inference System, in *IEEE International Conference on Neural Networks— Conference Proceedings* 3, IEEE Piscataway, NJ, 1994, pp. 1532–1536.

Li, W., and Wu, Z., Self-Organizing Fuzzy Controller Using Neural Networks, in *Computers in Engineering*, Proceedings of the International Conference and Exhibit, Vol. 2, ASME, New York, 1994, pp. 807–812.

Nicholls, J. G., Martin, A. R., and Wallace, B. G., *From Neuron to Brain: A Cellular and Molecular Approach to the Function of the Nervous System*, third edition, Sinauer Associates, Sunderland, MA, 1992.

Nie, J., Neural Approach to Fuzzy Modeling, in *Proceedings of the American Control Conference*, Vol. 2, American Automatic Control Council, Green Valley, AZ, 1994, pp. 2139–2143.

Nomura, H., Hayashi, I., and Wakami, N., A Self-Tuning Method of Fuzzy Reasoning by Genetic Algorithms, in *Fuzzy Control Systems*, A Kandel and G. Langholz, eds., CRC Press, Boca Raton, FL, 1994, pp. 338–354.

O'inca Design Framework, *User's Manual*, Intelligent Machines, Sunnyvale, CA, 1994, pp. 727–748.

Pham, D. T., Bayro-Corrochano, E. J., Self-Organizing Neural-Network-Based Pattern Clustering Method with Fuzzy Outputs, *Pattern Recognition*, Vol. 27, No. 8, pp. 1103–1110, 1994.

Purushothaman, G., and Karayiannis, N. B., "Feed-Forward Neural Architectures for Membership Estimation and Fuzzy Classification, in *Proceedings of the Artificial Neural Networks in Engineering Conference (ANNIE '94)*, St. Louis, MO, November 13–16, 1994.

Takagi, H., and Hayashi, I., NN-Driven Fuzzy Reasoning, *International Journal of Approximate Reasoning*, Vol. 5, No. 3, pp. 191–212, 1991.

Travis, M., and Tsoukalas, L. H., Application of Fuzzy Logic Membership Functions to Neural Network Data Representation, Internal Report, University of Tennessee, Knoxville, TN, 1992. Included as Appendix A in Uhrig et al., *Application of Neural Networks*, EPRI Report TR-103443-P1-2, January 1994.

Uhrig, R. E., Tsoukalas, L. H., and Ikonomopoulos, A., Application of Neural Networks and Fuzzy Systems to Power Plants, in *Proceedings of the 1994 IEEE International Conference on Neural Networks*, Vol. 2, Part 6 (or 7), Orlando, FL, June 27–29, 1994, pp. 510–512.

Wang, L.-X., *Adaptive Fuzzy Systems and Control*, Prentice-Hall, Englewood Cliffs, NJ, 1994.

Wang, L. X., and Mendel, J. M., Backpropagation Fuzzy Systems as Nonlinear Dynamic System Identifiers, in *IEEE International Conference on Fuzzy Systems*, San Diego, CA, 1992, pp. 1409–1418.

Werbos, P. J., Neurocontrol and Fuzzy Logic: Connections and Designs, *International Journal of Approximate Reasoning*, Vol. 6, pp. 185–219, 1992.

Yager, R. R., Modeling and Formulating Fuzzy Knowledge Bases Using Neural networks," *Neural Networks*, Vol. 7, No. 8, pp. 1273–1283, 1994.

Yea, B., Konishi, R., Osaki, T., and Sugahara, K., Discrimination of Many Kinds of Odor Species Using Fuzzy Reasoning and Neural Networks, *Sensors and Actuators, A: Physical*, Vol. 45, No. 2, pp. 159–165, 1994.

PROBLEMS

1. Show a nontriangular form for the membership function of the ith instance in the p cluster described by Equation 13.3-7 and discuss its potential benefit.

2. Derive Equation (13.4-6) in the Takagi-Hayashi (T-H) method.

3. Explain qualitatively the significance and use of Equation (13.4-7).

4. Explain qualitatively how the number of variables in the consequent of Equation (13.4-1) is controlled in the T-H method.

5. Show that for a fuzzy algorithm comprised of zero-order Sugeno rules the output is given by Equation (13.5-4). Identify the parameters that may be used for training.

6. Using input–output data and gradient descent we can modify (adapt) the parameters of a fuzzy algorithm in a manner analogous to neural learning. Show how the equation for updating parameters such as weights, that is Equation (13.5-7), is obtained.

7. Given a fuzzy algorithm comprised of n first-order Sugeno rules, derive expressions analogous to Equations (13.5-3) and (13.5-4) and identify all parameters that may be used for training.

8. How can the parameters of the fuzzy algorithm of Problem 7 be trained? Describe all assumptions that need to be made and give the learning rule for each parameter.

9. If min instead of product is used in fuzzy algorithms comprised of zero-order Sugeno rules, how would their parameter training be different?

10. Derive expressions for the training parameters involved in fuzzy algorithms that use nonsymmetric triangular membership functions such as shown in Figure 13.8.

14

SELECTED HYBRID NEUROFUZZY APPLICATIONS

14.1 INTRODUCTION

Recent years have seen a rapidly growing number in neural-fuzzy applications and a blossoming bibliography on the subject.[1] Although it is too early for the merits of any particular approach to be comprehensively assessed, it appears that in a number of engineering disciplines the research is maturing and moving toward developmental phases. In this chapter we describe selected hybrid neurofuzzy engineering applications. The task of reporting on a field that is still in a state of flux is difficult and tricky, and unfortunately our selection is incomplete. Nevertheless, we think it may be useful to offer a panoramic view of applications through the neurofuzzy bibliography.

In Part II of this book we have seen that problems associated with obtaining expert knowledge and adapting a system description to changes in itself or its environment can be addressed through neural networks. Since neural descriptions of systems are typically made through example data or some kind of performance function, expert knowledge is not explicitly required. In addition, neural networks are inherently capable of adaptation through the various learning algorithms which were reviewed earlier. It would seem plausible, therefore, to try to overcome the expert knowledge and adaptation problems of fuzzy systems through synergistically exploiting these advantageous features of neural networks.

---

[1] The material presented in this chapter is largely a condensation of research reports in neurofuzzy applications that have appeared in the early 1990's, including material obtained through searches at Purdue University Library's Engineering Index.

Table 14.1   Comparative characteristics of fuzzy and neural systems

Fuzzy Systems	Neural Systems
Linguistic Representation	*Black Box Representation*
Expert Knowledge Required	*Example Data or Performance Function Required*
Some Adaptation	*Adaptation Mechanisms Available*
Fault Tolerant	*Fault Tolerant*
Application-Dependent Computational Cost	*Rather High Computational Cost*
Multiple Descriptions Possible	*Multiple Descriptions Possible*

Neural networks exhibit highly desirable inherent parallelism and fault-tolerant behavior. Of course, they have disadvantages of their own such as, for example, difficulties in inspecting and modifying internal parameters. Whereas fuzzy systems are relatively easy to inspect and modify, neural networks are not as transparent to a user. In addition, there may be situations where adequate data are simply not readily available, which could cause difficulties in training or possibly a high computational cost associated with training. Table 14.1 presents a comparison and a summary of the characteristics of fuzzy and neural systems.

14.2   NEUROFUZZY INTERPOLATION

The notion of interpolation typically refers to a process whereby we estimate the value of a function between values that are already known. More generally, this notion refers to methods for approximating a function with a simpler one, when interpolating values or derivative values are provided, as is the case in spline fitting of Langrange interpolation.

In fuzzy logic, we deal primarily with complex many-to-many mappings rather than the simple many-to-one mappings (or functions). Consider the situation shown in Figure 14.1a, where the empty circles represent known fuzzy rules (see Chapter 5). As seen in the figure, there are regions where such knowledge (i.e., the underlying relations) is missing. We can think of this problem in a manner analogous to crisp interpolation, that is, find a

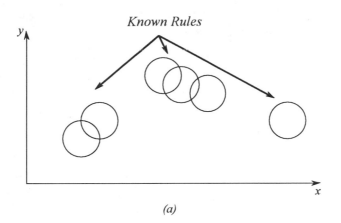

(a)

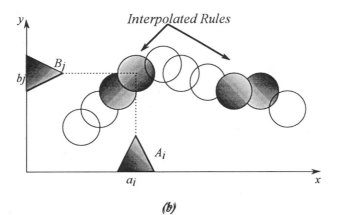

(b)

Figure 14.1 Neurofuzzy interpolation involves the use of neural methods for obtaining the *interpolated rules*.

method through which we can estimate the relation in the missing part. When this is accomplished through neural methods, we have what is known as *neurofuzzy interpolation*. To obtain the rules involves finding the appropriate membership functions as shown schematically in Figure 14.1*b* where the shaded circles represent interpolated rules.

Abe and his colleagues at Hitachi (Abe and Lan, 1993) have developed a method for extracting fuzzy rules directly from numerical input–output data for pattern classification in a manner similar to neural networks and extended it to approximate any arbitrary function. For function approximation, the universe of discourse of an output variable is divided into multiple intervals, and each interval is treated as a class. Then in a manner similar to that used for pattern classification, fuzzy rules are recursively defined by

(a) *activation hyperboxes* which show the existence region of the data for the interval and (b) *inhibition hyperboxes* which inhibit the existence region of data for that interval. Input data are used for each individual interval. The approximation accuracy of the fuzzy system derived by this method has been empirically studied by Abe and Lan (1993) using an operation learning application of a water purification plant and found to be satisfactory. Additionally, it has been reported that the approximation performance of the fuzzy system compares favorably with the function approximation approach based on neural networks.

Blanco and Delgado (1993) have also developed an interpolation method based on a neural network's ability to approximate any function. The methodology involves a neural network learning the information contained in fuzzy rules, as well as expert knowledge found in a set of examples, and directly interpolating from rules through the output of neural networks.

Kosko (1994) has shown that an additive fuzzy system can uniformly approximate any real continuous function on a compact domain to any degree of accuracy. An additive fuzzy system approximates the function by covering its graph with fuzzy patches in the input–output state space and averaging patches that overlap. The fuzzy system computes a conditional expectation $E[Y|X]$ if the fuzzy sets are viewed as random sets. Each fuzzy rule defines a fuzzy patch and utilizes common-sense knowledge with state-space geometry. Neural or statistical clustering systems can approximate the unknown fuzzy patches from training data. Kosko (1994) has reported that these adaptive fuzzy systems approximate a function at two levels. At the local level the neural system approximates and tunes the fuzzy rules. At the global level the rules or patches approximate the function.

14.3   GENERAL NEUROFUZZY METHODOLOGICAL DEVELOPMENTS

Leading a research field from infancy to maturity and technological deployment is a particularly difficult task, and crucial methodological developments obtained through the insight and intuition of experienced researchers make a difference. Let us take a look at some of these pivotal methodological advancements which have contributed to neurofuzzy integration.

Werbos (1993) introduced the concept of *elastic fuzzy logic* as a way of combining neural and fuzzy capabilities. Werbos' methodology uses fuzzy logic as a kind of "translation" technology, to go back and forth between the words of a human expert and the equations of a controller, a classifier, or some other useful system. One can then use neural methods to adapt that system to improve performance. Elastic fuzzy logic translates the words of an expert into an elastic fuzzy logic network, a kind of local neural network which can be plugged directly into a wide range of neural network designs, ranging from pattern classification through the brain-like optimizing control. The words of the expert are used to initialize this network, but neural network methods can then be used to adapt all the weights or parameters. In Werbos' methodology, neural network methods can also be used to prune or grow the network.

Ronald Yager, a prominent researcher in the field of fuzzy systems, has advanced a general framework for developing fuzzy algorithms using neural networks. Yager (1994) interprets the firing level of a neuron as a measure of possibility between two fuzzy sets, the weights of connection and the input, and suggests a way to represent fuzzy production rules in a neural framework. Central to Yager's representation is the notion that the linguistic variables associated with a fuzzy *if/then* rule may be represented as weights in the resulting neural structure. Such a structure allows for learning of the membership functions involved.

Several investigators have proposed neural-network-based fuzzy systems. A leading part of the research and important methodological advancements have come out of the work of Professor Keller and his coworkers at the University of Missouri (Keller et al., 1994; Keller and Tahani, 1992). Over the years they have developed a variety of approaches toward improving the performance of various systems by exploiting the neurofuzzy synergism. They have introduced evidence aggregation networks based on additive fuzzy hybrid operators, for image segmentation, pattern recognition, and general multicriteria decision-making. These networks have excellent properties for decision-making under uncertainty and present advantages in training due to their simple form. Keller's additive hybrid operators are found to be flexible and useful for modeling nodes in a network structure for fuzzy logic inference capable of learning appropriate functional relationships while being rather transparent; that is, after training, individual nodes can be analyzed as a collection of "mini-rules".

Neural networks for the parallel high-speed processing of the rules found in a fuzzy logic controller have been used by Patrikar and Provence (1993) at Southern Methodist University. In the methodology advanced by the researchers, the fuzzy algorithm is replaced by a feedforward neural network with a single hidden layer that is trained using backpropagation and input and output fuzzy values expressed in terms of numerical patterns.

As we have seen in Chapter 13, the automatic categorization of a universe of discourse is typically based on some type of Kohonen network. A pattern clustering method based on the Kohonen feature mapping algorithm and the backpropagation multilayer perceptron has been used for membership function determination by Pham and Bayro-Corrochano (1994). The method is applied first to the training data set to divide it into labeled clusters using the Kohonen algorithm and a simple cluster labeling procedure. The data clusters are then employed to train a three-layer perceptron using the error backpropagation training. Thus, this approach is self-organizing by virtue of the Kohonen algorithm and produces fuzzy outputs as a consequence of the backpropagation network. Results of using the pattern clustering method on standard problems show it to be superior in performance compared to crisp clustering networks such as the Kohonen feature map and the ART-2 network [see also Nie (1994)].

Feedforward neural estimation for membership function determination and fuzzy classification has been proposed by Purushotaman and Karayiannis (1994) at the University of Houston. They have used feedforward neural networks inherently capable of fuzzy classification of overlapping pattern classes such as a feedforward neural network in conjunction with multilevel neurons in two hidden layers called (a) the "quantum neural network" and (b) a "membership estimating network," which is a feedforward network trained with generalized Hebbian learning rules. Professor Karayiannis and his students have offered theoretical and experimental results showing that both architectures are inherently capable of partitioning the feature space in a fuzzy manner.

To a large extent, the successful implementation of neural nets depends on several ancillary techniques for data preprocessing, training, and testing. Some of these techniques were investigated and discussed by Professor El-Sharkawi of the University of Washington [see El-Sharkawi (1994)]. They include genetic algorithms, fuzzy logic theory, query-based learning, and feature extraction. The advantages of the application of these ancillary techniques for neural networks and simulation studies have been performed to assess their role and practicality.

14.4 ENGINEERING APPLICATIONS

Fuzzy and neural approaches have found their way in a variety of engineering applications, including, but not limited to, consumer electronics, various aspects of control, diagnostics, industrial production lines, biotechnology, power generation, chemical processes, power electronics, communications, and software resource management. It is expected that the applications of the fuzzy-neural synergism will increasingly move toward computer applications as well, such as machine learning [see Adeli and Hung (1995)].

It is now rather well established that fuzzy systems aided by neural networks can adequately address the adaptation problems we discussed in Chapters 12 and 13. Ishibuchi and his coworkers have reported on an approach based on empirical research where they examine the ability of trainable fuzzy systems as approximators of nonlinear mappings by computer simulations using real-life data. Fuzzy *if/then* rules of the Sugeno variety (see Chapter 6) are adjusted by a gradient descent method. After examining the capabilities of fuzzy systems by numerical examples, the researchers tested them through an interesting project, involving the development of a six-variable fuzzy relation used in rice tasting. By computer simulations based on a random subsampling technique, they demonstrated that the performance of individual fuzzy systems is comparable to that of neural networks (Ishibuchi et al., 1994).

Researchers at Tohoku University in Japan have used neural networks in conjunction with a fuzzy logic for decision-making (Kozma et al., 1994). They developed a method which can make a distinction between the occurrence of unexperienced events and any inconsistency in the judgments of agents caused by statistical uncertainties in actual data. The method has been applied to the analysis of signals of numerical experiments and also actual measurements in a nuclear reactor.

Several fuzzy-neural methodologies are of special interest to many researchers when integrated with other approaches. Pao (1994) reported on the fusion of three distinct computational intelligence paradigms, neural computing, evolutionary programming, and fuzzy-logic, to support the task of process monitoring and optimization. The resulting computational intelligence has been successfully applied to optimal process planning in electric power utilities that include, but are not limited to, heat rate improvement and NO_x emission minimization.

14.5 DIAGNOSTICS IN COMPLEX SYSTEMS

Hybrid fuzzy-neural systems have been used in several aerospace applications. Raza, Ioannou, and Youssef (1994) reported on the problem of detecting control surface failures of a high-performance aircraft. The detection model is developed using a linear, six-degree-of-freedom dynamic model of an F-18 aircraft. The detection scheme makes use of a residual tracking error between the actual system and the model output in order to detect and identify a particular fault. Two parallel models detect the existence of a surface failure, whereas the isolation and magnitude of any one of the possible failure modes is estimated by a decision algorithm using neural networks and fuzzy logic. Simulation results demonstrate that detection can be achieved without false alarms even in the presence of actuator/sensor dynamics and noise.

In the power industry, neural networks and fuzzy logic systems offer an interesting, challenging, and productive means of addressing many of the problems that occur in the operation of nuclear power plants. Uhrig, Tsoukalas, and Ikonomopoulos (1994) have described how such systems can be used to model nuclear reactor system dynamics and nuclear fission step responses of nuclear plants. They can also help operators in assessing the condition of the plant during abnormal operation or emergencies by analyzing and integrating the process parameters and system interactions (Guo and Uhrig, 1992).

Matsuoka and Blanco (1993) reported on an Electric Power Research Institute (EPRI) survey of recent advances and trends in Japanese power plants. The survey includes case studies of many applications and widespread implementations of advanced technologies such as: (1) a neurofuzzy system for plant monitoring and diagnosis, combined with knowledge-based preventive maintenance systems: (2) fuzzy-logic dynamics schedulers for plant transient operations; (3) fuzzy-expert tuners of dynamic control systems; (4) fuzzy-algorithmic operation guidance systems for major plant equipment; and (5) telepresence with machine vision and robotics; among others. These advanced approaches had to be introduced due to smaller stability margins in the plants, rapid changes toward more efficient thermal cycles and new plant equipment dynamics, coming with stronger nonlinearities and subsystem interconnections.

In the field of nuclear engineering diagnostics is a very important task for the safety of power plants. Moon and his coworkers at the Korean Advanced Institute on Science and Technology have reported on a method for predicting the critical heat flux (CHF)—a quantity with safety significance— based on fuzzy clustering and neural networks [Moon and Chang, (1994)]. The fuzzy clustering classifies the experimental CHF data into a few data clusters (data groups) according to the data characteristics. After classification of the experimental data, the characteristics of the resulting clusters were carefully examined. Using the CHF data in each group, neural networks were trained and successfully predicted the CHF.

14.6 NEUROFUZZY CONTROL SYSTEMS

In another application, Chen and Chen (1994) have investigated the relationship between a piecewise linear fuzzy controller (PLFC), in which the membership functions for fuzzy values and the fuzzy *if/then* rules are all in piecewise linear forms, and a Gaussian potential function network-based controller (GPFNC), in which the network output is a weighted summation of hidden responses from a series of Gaussian potential function units. Systematic procedures were developed for transformations from a PLFC to its GPFNC counterpart, and vice versa. Based on these transformation principles, a series of systematic and feasible steps were developed for the design of an optimized PLFC (PLFC*) using neural network techniques. The optimized GPFNC (GPFNC*) can be implemented directly to actual systems, and the GPFNC* could further be converted into its fuzzy counterpart (PLFC*) if more structural interpretation of the intelligent control strategy is required.

Several self-organizing fuzzy controllers have found their way to field deployment. Li and Wu (1994) developed an interesting a self-organizing fuzzy logic control scheme based on neural networks, which consists of a traditional fuzzy logic controller and a conventional derivative controller.

Neural networks are used to optimize membership functions that are parameterized by the use of the cubic splines in a self-organizing manner.

In another development, Professors Lin and Lee (1994) have proposed a promising approach for constructing a fuzzy system automatically. In their approach a reinforcement neurofuzzy control system with multiple connectionist models with feedforward multilayered networks is used to realize a fuzzy logic controller. One network performs the role of a fuzzy predictor, while the other acts as a fuzzy controller. Using the temporal difference prediction method, the fuzzy predictor can predict the external reinforcement signal and at the same time provide a more informative internal reinforcement signal to the fuzzy controller. During the learning process, both structure learning and parameter learning are performed simultaneously in the two networks using a fuzzy similarity measure, and a reward/penalty signal.

As far as the practical implementation of neurofuzzy control is concerned, there is tremendous variation in the themes and areas of applications. Stylios and Sotomi (1994) have developed a neurofuzzy sewing controller for the next generation of the so-called intelligent sewing machines. The model incorporates discrimination of material characteristics to be stitched and automatic determination of their properties. The fabric–machine interactions at different speeds have been articulated in the form of fuzzy *if/then* and implemented in a neural network to allow for optimization of fuzzy membership functions and, subsequently, self-learning. The controller was successfully applied to an instrumented industrial sewing machine.

Neurofuzzy approaches are expected to play a major role in the development of future fusion reactors. Yamazaki et al. (1994) reported that the world's largest superconducting fusion machine LHD (large helical device), under construction in Japan, will utilize fuzzy logic and neural networks for feedback control of plasma configurations in addition to classical proportional-integral-derivative control. Design studies of the control system and related R & D programs with coil-plasma simulation systems include neurofuzzy control systems.

Foslien and Samad (1993) at Honeywell reported on the general problem of optimizing a fuzzy controller through the use of a neural network model for the process in the optimization procedure. The integration of neural network models with fuzzy control is very appropriate since both techniques are best used when detailed analytical understanding of a process is not available. To illustrate this concept, a fuzzy controller was synthesized for a simple nonlinear process with (1) a feedforward neural network used for modeling the process and (2) an optimization criterion based on setpoint error.

The synergistic utilization of fuzzy and neural systems, often resulting in an entity of its own referred to as *neurofuzzy systems*, is increasingly applicable in many control technologies (Werbos, 1992). As we have seen in Chapter

6, for example, a difficult part in designing an ordinary fuzzy controller is selecting which fuzzy sets are best representing the controlled and controlling variables. Most fuzzy controllers are sensitive to the shapes of the membership functions, and as the number of rules increases, the use of "trial and error" tuning procedures become less and less feasible.

A report in *IEEE Spectrum* magazine (Schwartz and Klir, 1992) described work at Matsushita and Hitachi in Japanese in which a backpropagation neural network learned the needed membership functions from a set of training examples (Hayashi et al., 1992). It is claimed that a tuning task that had previously taken 6 months was accomplished in 1 month. Wakami et al. (1993) at Matsushita Electric reported on recent applications of fuzzy-neural methodologies to home electric appliances. Many appliances produced in Japan have internally encoded expert knowledge for their operation. In order to overcome the problem of extracting the necessary expertise, Matsushita engineers use neural networks in conjunction with fuzzy rules. Applications of their neurofuzzy methods are found in refrigerators, air-conditioning systems, and welding machines. In air-conditioning systems, a thermal sensory system and a fuzzy-image-understanding algorithm are used to identify the number and positions of occupants in a room. Allowing air-conditioning systems to "see" their environment allows them to better and more efficiently produce a comfortable thermal environment.

As far as the industrial merit of neurofuzzy technologies is concerned, Wegmann (1994) at Siemens reported a growing interest in programmable logic controller (PLC) applications and a wide range of possible applications in the area of nonlinear processes, especially those with great parameter fluctuations. Applications in environmental processes, such as sewage and exhaust gas cleaning, appear to be of particular interest. Neural networks alone may be at a disadvantage in operating phases of a process where example data are not readily available, whereas neurofuzzy formulations lend themselves conveniently to such situations allowing the control behavior in such phases to be prescribed by a fuzzy algorithm, with most learning left to neural networks.

In another interesting application, Yen (1994) reported on the design of control algorithms for flexible space structures, possessing nonlinear dynamics which are often time-varying and usually ill-modeled. A hybrid connectionist system was used as a learning controller with reconfiguration capability. Neural networks were used to provide vibration suppression and trajectory maneuvering for precision pointing of flexible structures. Radial basis function networks were employed for capturing spatiotemporal interactions among the structure members. A fuzzy-based fault diagnosis system provided the neural controller with various failure scenarios, and the associative memory incorporated into the adaptive architecture compensated for catastrophic changes of structural parameters by offering a continuous solution space of acceptable controller configurations.

Sharaf and Lie (1994) reported on a novel neurofuzzy hybrid power system stabilizer designed for damping electromechanical modes of oscillation and enhancing power system synchronous stability. The hybrid system comprises a front-end conventional analog power system stabilizer design, an artificial neural network based stabilizer, and a fuzzy logic postprocessor gain scheduler.

In the power electronics field, Professor Bose at the University of Tennessee reported on new applications emerging in the field that exploit fuzzy and neural approaches along with other AI techniques (Bose, 1994).

14.7 NEUROFUZZY CONTROL IN ROBOTICS

In the field of robotics, there is a booming interest in neurofuzzy means for supervisory control, planning, grasping, and guidance, and a variety of applications are found (Kuo, 1993; Kuo et al., 1994). Professor Bourbakis and his colleagues at Binghamton University (Tascillo et al., 1993) have developed a neurofuzzy hand-grasp algorithm for improving the first grasp of a wheelchair robotic arm with two three-joined fingers and a two-jointed thumb. The robotic arm uses pressure and force feedback and a learning mechanism that helps to avoid an extensive search of an optimal grasp each time an object is lifted.

Hanes et al. (1994) reported on an intelligent control architecture for a robotic grasping system capable of acquiring an object into a fully enveloping power grasp. Control of the internal forces of the grasp is provided, along with trajectory control of object position, as the object is picked up. Fuzzy control techniques are used for control of internal forces in the power grasp, and a neural network provides a means of in-process nonlinear friction estimation.

Fatikow and Wohlke (1994) reported on a neurofuzzy architecture for the intelligent control of multifinger robot hands. The control system is based on the combination of a neural network approach for the adaptation of grasp parameters and a fuzzy logic approach for the correction of parameter values given to a conventional controller. A planning component of the system determines initial manipulation parameters, while a neural network performs continual computations of suboptimal grasp forces. On-line learning of fuzzy *if/then* rules is used for parameter adjusting.

A neurofuzzy controller for adaptive tracking in unknown nonlinear dynamic systems and for on-line computation of inverse kinematic transformations of a two-linked robot has been developed by Rao and Gupta (1994). The controller is comprised of a fuzzy algorithm in the feedback configuration and a recurrent neural network in the inverse mode (feedforward) configuration. The controller provides a means for converting a linguistic control strategy into control actions while the neural network provides sensory (low-level) computations and embodies important features such as learning, fault-tolerance, parallelism, and generalization in a manner similar to the one we have seen in Chapter 13.

14.8   PATTERN RECOGNITION AND IMAGE ENHANCEMENT

Neural networks have been extensively used in connection with fuzzy algorithms for edge detection, and in connection with fuzzy means for defining parameters they are producing interesting realizations of neurofuzzy systems. Kim and Cho (1994) reported on an edge relaxation method utilizing fuzzy logic and neural networks where candidates for edge segments are first estimated using a local derivative operator with a window of small size. Fuzzy *if/then* rules, each of which is associated with a neighborhood pattern defined by the spatial relationships among the neighboring edge segments, are used as a computational framework of collecting the evidence for the existence of an edge segment. The fuzzy rules are trained by a specially structured neural network which performs a fuzzy reasoning operation.

Improvements on clustering algorithms are being investigated by many researchers. The extension of neural-net-based crisp clustering algorithms to fuzzy clustering algorithms has been addressed extensively. For a comprehensive review see the excellent compilation of papers in the book *Fuzzy Models for Pattern Recognition*, edited by J. C. Bezdek and A. K. Pal (Bezdek and Pal, 1994). However, many neurofuzzy clustering algorithms developed so far suffer from restrictions in identifying the actual decision boundaries among clusters with overlapping regions. These restrictions are induced by the choice of the similarity measure and the representation of clusters. An integrated adaptive fuzzy clustering algorithm was developed by Kim and Mitra (1994) to generate improved decision boundaries by introducing a new similarity measure and by integrating the advantages of the fuzzy optimization constraint of fuzzy c-means, the control structure of adaptive resonance theory (ART-1), and a fuzzified Kohonen-type learning rule.

Dalton (1994) at Apple Computers reported on a fuzzy-neural approach to image manipulations that allows a user to quantify qualitative aesthetics. Image enhancement and other desired manipulations are thought of as nonlinear transformations from an input space of arbitrary images into an output space of desired aesthetic images. Derivation of imaging manipulations of this type can be viewed as supervised learning problems that can be solved by neural methods. In order to reduce the dimensionality of the transformations involved, descriptors more structured than raw image pixels may be used; hence, imaging transformations between sets of image metrics as opposed to sets of image pixels can be learned by the network (from example images). Alternatively, an adaptive fuzzy algorithm can be used to achieve the underlying functional transformation while providing a link between semantic labeling of qualitative image characteristics and the underlying raw image data.

Kulkarni et al. (1994) have proposed a neural network model for fuzzy logic decisions consisting of six layers; the first three layers map the input variables to membership functions, and the last three layers implement the decision rules. Triangular membership functions are used, and the model learns the decision rules using a supervised gradient descent procedure. The connection strengths between the last three layers encode the decision-rules used in decision-making. Layer 1 is the input layer that receives the input features, while layer 2 represents the linguistic variables (with five values, *VERY LOW, LOW, MEDIUM, HIGH*, and *VERY HIGH*) for each input feature; Hence, layer 2 has five times as many nodes as layer 1. Each node of layer 2 is connected with weights ± 1 to two nodes in layer 3 where the two nodes represent the left and right sides of the triangular membership functions. Each node in layer 4 combines the outputs of the corresponding two nodes in layer 3 so that it now represents the membership values, which is presented to layer 5. Layers 5 and 6 are implementing the inference process. Layers 4, 5, and 6 represent a simple three-layer feedforward network with backpropagation learning. The number of nodes in the output layer is equal to the number of output decisions. During training, only the weights between layers 4, 5, and 6 are adjusted.

The above system has been successfully used to recognize objects in multispectral satellite images based on data obtained from thematic mapper sensors (a multispectral scanner that captures data in seven spectral bands). Five inputs to layer 1 were used, and layers 2, 3, and 4 contained 25, 50, and 25 nodes, respectively, since five linguistic values were used. Layers 5 contained 35 nodes, and layer 6 contained 5 nodes representing output categories. The researchers have reported that results obtained were virtually identical with results from a three-layer conventional neural network classifier and a conventional maximum likelihood classifier. However, the conventional neural network took over 24 hours to train as opposed to about 25 minutes for the fuzzy neural system. Both the conventional and fuzzy neural network systems gave results very rapidly after training. In contrast, the conventional maximum likelihood classifier had to handle each pixel individually and sequentially; as a result, the conventional classifier took excessively long times for classification.

14.9 MEDICAL AND ENVIRONMENTAL IMAGING USING NEUROFUZZY METHODOLOGIES

The extraction of fuzzy values is of particular interest in medical imaging, where a plethora of data-rich situations exist. Computed tomography, magnetic resonance, digital ultrasound, and other forms of computer-assisted radiology provide an unprecedented volume of data that, if interpreted correctly, lead to the visualization of the body's internals and diagnosis of subtle decease processes at a very early stage of development (Ichihashi et al., 1993). Physicians and engineers collaborate in many areas of medicine to develope and use revolutionary computer-assisted techniques for education, visualization, diagnostics, and telesurgery, amongst others.

Brotherton et al. (1994) have developed a neurofuzzy system to automatically classify structures and tissues in echocardiograms. The system performs structure classification as a first step using advanced multiple-feature, hierarchical, fuzzy neural network fusion approach. It learns to classify tissue types by examination of image training data. Classification assigns each image pixel a fuzzy membership measure for each structure or tissue type. Final hard classification, if required, is delayed until the system's output stage. This allows important information to be retained throughout the system. The first layer in the hierarchy of networks determines gross spatial relationships and texture classes, while the second layer fuses the spatial and textual net outputs to make final classifications.

In a related medical imaging problem, T. Chen, W-C. Lin, and C-T. Chen (Chen et al., 1994) at Argonne National Laboratory have developed a fuzzy neural network based approach to 3-D heart motion understanding using expert cardiologist knowledge to specify different classes of motion and obtain classification rules. The objective is to find the decisions for all possible classes of motion in the form of possibilities. Experiments on real data have been conducted to corroborate the neurofuzzy approach.

On the environmental side of neurofuzzy applications, Barbosa et al. (1994) in Brazil reported on a neural system for deforestation monitoring through automatic interpretation of satellite images of the Amazon region. Their approach is based on a combination of image segmentation and classification techniques, the latter employing a neural network architecture that works on a fuzzy model of classification. It appears that such an approach has a range of advantages over more traditional, pixel-based approaches employing statistical techniques, ranging from the possibility of treating transition and interference phenomena in the images to the ease with which complex information related to a region's geometry, texture, and contextual setting can be used.

14.10 TRANSPORTATION CONTROL

In the field of transportation engineering, efforts to manage freeway congestion have been seriously impeded by the inability to promptly and reliably detect the presence of traffic incidents. Traditional incident-detection algorithms distinguish between congested and uncongested operations by comparing measured traffic-stream parameters with predefined threshold values. Given the range of possible operating conditions in a certain traffic stream, selecting a single threshold value may be a difficult and uncertain decision. A system called *fuzzy logic incident patrol system* was developed by Hsiao, Lin, and Cassidy (1994) to solve many of the problems inherent in traditional incident-detection algorithms. The *fuzzy logic incident patrol system* is a hybrid neurofuzzy system constructed from training examples to find the optimal input–output membership functions. Threshold values, implicitly obtained by *if/then* rules and membership functions, are treated as dependent variables, which change according to prevailing traffic-stream parameters measured by detectors.

14.11   ADAPTIVE FUZZY SYSTEMS

Neural-network-based adaptive fuzzy systems have been used in the field of seismic evaluation. Chu and Mendel (1994) have developed a method for solving the so-called "first break picking" problem in seismic signal processing, one that requires much human effort and is difficult to automate. The goal has been to reduce the manual effort in the picking process and accurately perform the picking. A backpropagation fuzzy logic system has been used for first break picking by employing derived seismic attributes as features. Experimental results reported by Chu and Mendel have indicated that this neurofuzzy system achieves about the same picking accuracy as a feedforward neural network that is also trained using a backpropagation algorithm; however, it is trained in a much shorter time because there is a systematic way in which initial parameters can be chosen, as opposed to the random way in which the weights of the neural network are chosen.

Mitra and Pal (1994) have proposed a self-organizing artificial neural network, based on Kohonen's model of self-organization, which is capable of handling fuzzy inputs and of providing fuzzy classification. Unlike conventional neural net models, this algorithm incorporates fuzzy set-theoretic concepts at various stages. The input vector consists of membership values for linguistic properties along with some contextual class membership information which is used during self-organization to permit efficient modeling of fuzzy (ambiguous) patterns. A new definition of gain factor for weight updating has been proposed by the researchers. Incorporation of the concept of fuzzy partitioning allows natural self-organization of the input data, especially when they have ill-defined boundaries. The output of unknown test patterns is generated in terms of class membership values. Incorporation of fuzziness in input and output is seen to provide better performance than a Kohonen model.

Fei-Yue Wang and D. D. Chen (Wang and Chen, 1994) have investigated general principles involved in the design of adaptive fuzzy controllers via neural networks and proposed a method that implements a rule-based fuzzy control system via a neural network consisting of two subnetworks: one for pattern recognition and the other for fuzzy reasoning and control synthesis. The neural network is arranged in such a way that the structure and operations of the original fuzzy control system can be fully retrieved from its network implementation. Equipped with the learning capability of neural networks, this implementation provides a mechanism for refining the existing rules and generate new rules for fuzzy control (Wang, 1994)

14.12 INSPECTION USING NEUROFUZZY METHODS

In the area of fault diagnosis, Goode and Chow (1994) presented a novel hybrid fuzzyneural fault detector that will use the learning capabilities of the neural network to detect if a motor has an incipient fault. Once the neurofuzzy fault detector is trained, heuristic knowledge about the motor and the fault detection process can also be extracted. With better understanding of heuristics through the use of fuzzy rules and fuzzy membership functions, a better understanding of the fault detection process of the system is obtained.

Moganti, Dagli, and Ercal (1994) developed a fuzzy-neural method for automatically inspecting printed circuit boards for defects. The process involves a two-level classification of the board image subpatterns into either standard nondefective patterns or defective patterns. The patterns that are identified as being defective in the first level are thoroughly checked for defects in the second level, and the patterns that are nondefective are checked for dimensional verification for the classes that a board has been identified and assigned to the correct class.

14.13 NEUROFUZZY METHODS IN FINANCIAL ENGINEERING

Another application where the use of neural network technology is introducing fuzzy concepts is in the financial community. It is well known that neural networks have been used for several years in the selection of investments because of their ability to identify patterns of behavior that are not readily available. Much of this work has been proprietary for the obvious reason that the users want to take advantage of their insight into the market gained through the use of neural network technology.

In the past year, some financial work has incorporated neurofuzzy technology. Hobbs and Bourbakis (1995) have described a neurofuzzy simulator used for stock investing that identifies patterns associated with whether a stock is underpriced or overpriced. Since stock prices are determined by what a buyer will pay, most stocks tend to be underpriced or overpriced at one time or another. Eventually, the price corrects itself, but there is an opportunity for an investor who can recognize these conditions to make money by buying an underpriced stock or selling an overpriced stock and perhaps buying it back later. Buying and selling options is another way of making money on the use of this information.

The fuzzy neural network used by Hobbs and Bourbakis is a modification of S. Y. Kung's fuzzy-based neural network (Kung, 1993). The inputs are 13 market indexes provided by the various financial services and institutions that reflect the average stock price. The program has consistently averaged over 20% annual return.

14.14   COMMERCIAL NEUROFUZZY SYSTEM SOFTWARE

Several software products are currently available to help with neurofuzzy problems. Four of these systems will be briefly described on the basis of information provided to the authors by the commercial organization involved. They are listed alphabetically by their commonly accepted name.

ANFIS. Jang has described ANFIS, an acronym for *adaptive neuro-fuzzy inference system.* It has an architecture that is equivalent to a two-input first-order Sugeno fuzzy model with nine rules, where each input is assumed to have three associated membership functions (Jang and Sun, 1995). Its two-dimensional input space is partitioned into nine overlapping fuzzy regions, each of which is governed by fuzzy *if-then* rules, where the premise part of a rule defines a fuzzy region, and the consequent part specifies the output within this region. ANFIS can achieve a highly nonlinear mapping. It consists of fuzzy rules which are local mappings instead of global ones. It can also be used as a neurofuzzy controller. ANFIS is implemented in the Fuzzy Systems Toolbox of MATLAB, a commercial software package produced by MathWorks, Inc.

CUBICALC. CUBICALC is a "fuzzy shell" that has great flexibility in building various kinds of fuzzy systems for decisions, inference, and control. Its primary basis for being listed here is that it has a library of neural network subroutines that can be utilized in a way that makes it possible to construct neurofuzzy systems (Watkins, 1993).

NEUFUZ. Khan (1993) reported a novel method of combining neural nets with fuzzy logic. The combined technology, NeuFuz, generates membership functions as well as fuzzy *if/then* rules by learning the system behavior using input–output data. The generated rules and membership functions are then processed using new fuzzy logic algorithms for defuzzification, rule evaluation, and antecedent processing which are developed based on neural network architecture and learning. These fuzzy logic algorithms replace conventional heuristic fuzzy logic algorithms and enable full mapping of neural net to fuzzy logic. Full mapping provides an important key feature of generating fuzzy rules and membership functions to meet a prespecified accuracy level. Simulation results have shown the approach to significantly improve performance and reliability while reducing design time and computational cost.

O'INCA. Intelligent Machines, Inc. has produced O'INCA (1994), an integrated platform for the development of fuzzy logic, neural networks, and neurofuzzy systems (O'Inca, 1993). It allows for user-defined modules in the same framework as the other systems. It combines graphical user interface, design validation, simulation and debugging, C code generation, and design documentation. In the fuzzy logic module, intermediate results after fuzzification, rulebase evaluation (inference), and defuzzification can be examined. "Fired" (active) rules can be isolated, and the results of the rule antecedent and consequent parts, rule weights, and the effects can also be examined. In the neural network module, output and bias values of all neurons, as well as all link weights, can be examined. Fixed weights and biases can be modified during simulation.

REFERENCES

Abe, S., and Lan, M. S., Function Approximator Using Fuzzy Rules Extracted Directly from Numerical Data, in *Proceedings of the International Joint Conference on Neural Networks*, Vol. 2, IEEE, Piscataway, NJ, 1993, pp. 1887–1892.

Adeli, H., and Hung S. L., *Machine Learning-Neural Networks, Genetic Algorithms and Fuzzy Systems*, John Wiley & Sons, New York, 1995.

Barbosa, V. C., Machado, R. J., dos Liporace, F., Neural System for Deforestation Monitoring on Landsat Images of the Amazon Region, *International Journal of Approximate Reasoning*, Vol. 11, No. 4, 1994, pp. 321–359.

Bezdek, J. C., and Pal, S. K. (eds.), *Fuzzy Models for Pattern Recognition*, IEEE Press, New York, 1992.

Blanco, A., Delgado, M., and Requena, I., A Learning Procedure to Identify Weighted Rules by Neural Networks, *Fuzzy Sets and Systems*, Vol. 69, 1995, pp. 29–36.

Blanco, A. and Delgado, M., Direct Fuzzy Inference Procedure by Neural Networks, *Fuzzy Sets and Systems*, Vol. 58, No. 2, pp. 133–141, 1993.

Bose, B. K., Expert System, Fuzzy Logic, and Neural Network Applications in Power Electronics and Motion Control, *Proceedings of the IEEE*, Vol. 82, No. 8 Aug pp. 1303–1323, 1994.

Brotherton, T., Pollard, T., Simpson, P., and DeMaria, A., Echocardiogram Structure and Tissue Classification Using Hierarchical Fuzzy Neural Networks, *Proceedings —ICASSP, IEEE International Conference on Acoustics, Speech and Signal Processing*, Vol. 2, IEEE, Piscataway, NJ, 1994, 94CH3387-8. pp. 573–576.

Chen, C.-L, and Chen, W.-C., Fuzzy Controller Design by Using Neural Network Techniques, *IEEE Transactions on Fuzzy Systems*, Vol. 2, No. 3, pp. 235–244, 1994.

Chen, T., Lin, W.-C., and Chen, C.-T., Fuzzy Neural Networks for 3-D Heart Motion Understanding, in *Proceedings 12th International Conference on Pattern Recognition*, Vol. 2, Jerusalem, Oct. 9–13, 1994, pp. 510–512.

Chu, C.-K. P., and Mendel, P., and Jerry, M., First Break Refraction Event Picking Using Fuzzy Logic Systems, *IEEE Transactions on Fuzzy Systems*, Vol. 2, No. 4, pp. 255–266, 1994. Conference (ANNIE '94); St. Louis, MO, November 13–16, 1994.

Dalton, J. C., Adaptive Learning Systems and Qualitative Manipulation of Digital Imagery, in *Proceedings of SPIE—The International Society for Optical Engineering* Vol. 2179 Society of Photo-Optical Instrumentation Engineers, Bellingham, WA, 1994, pp. 440–451.

El-Sharkawi, M. A., and Huang, S. J., Ancillary Techniques for Neural Network Applications, *Proceedings of the IEEE International Conference on Neural Networks*, Vol. 6, Orlando, FL., June 27–29, 1994, pp. 3724–3729.

Foslien, W. and Samad, T., Fuzzy Controller Synthesis with Neural Network Process Models, in *Proceedings of the 1993 IEEE International Symposium Intelligence Control*, 1993, pp. 370–375.

Fatikov, S., and Wohlke, G., Neuro-Fuzzy Control Approach for Intelligent Micro-robots, *Proceedings of the IEEE International Conf. of Systems, Man and Cybernetics*, Vol. 4, La Touquet, France, October 17–20, 1993, pp. 441–446.

Goode, P, V., and Chow, M.-Y., Hybrid Fuzzy/Neural System Used to Extract Heuristic Knowledge from a Fault Detection Problem, in *Proceedings of the 3rd IEEE Conference on Fuzzy Systems*, Part 3 (of 3), Orlando, FL, June 26–29, 1994.

Guo, Z., and Uhrig R. E., Using Modular Neural Networks to Monitor Accident Conditions in Nuclear Power Plants, in *Proceedings of the SPIE Technical Symposium on Intelligent Information Systems*, Application of Artificial Neural Networks III, Orlando, FL, April 20–24, 1992.

Hanes, M. D., Ahalt, S. C., and Orin, D. E., Intelligent Control of Object Acquisition for Power Grasp, *IEEE International Symposium on Intelligent Control—Proceedings 1994*, pp. 303–308, 1994.

Hayashi, I., Nomura, H., Yamasaki, H. and Wakami, N., Construction of Fuzzy Inference Rules by NDF and NDFL, *International Journal of Approximate Reasoning*, Vol. 6, pp. 241–266, 1992.

Hobbs A., and Bourbakis, N. G., A NeuroFuzzy Arbitrage Simulator for Stock Investing, in *Proceedings of the IEEE/IAFE 1995 Computational Intelligence for Financial Engineering*, New York, April 9–11, 1995.

Hsiao, C.-H.; Lin, C.-T., and Cassidy, M., Application of Fuzzy Logic and Neural Networks to Automatically Detect Freeway Traffic Incidents, *Journal of Transportation Engineering* Vol. 120, No. 5, pp. 753–772, 1994.

Ichihashi, H., Miyoshi, T., and Nagasaka, K., Computed Tomography by Neurofuzzy Inversion, in *Proceedings of 1993 International Joint Conference on Neural Networks*, Part 1 (of 3), Nagoya, Japan, Oct. 25–29 1993, pp. 709–712.

Intelligence Machines, *NeuFuz, Neural Fuzzy for Intelligence Systems*, National Semiconductor leaflet #633100, 1993.

Ishibuchi, H., Nozaki, Ken., Tanaka, H., Hosaka, Y., and Matsuda, M., Empirical Study on Learning in Fuzzy Systems by Rice Taste Analysis, *Fuzzy Sets and Systems*, Vol. 64, No. 2, pp. 129–144, 1994.

Jang, J-S. R., and Sun, C.-T., Neuro-fuzzy Modeling and Control, in *Proceedings of the IEEE*, 1995, pp. 378–406.

Keller, J. M., and Tahani, H., Implementation of Conjunctive and Disjunctive Fuzzy Logic Rules with Neural Networks, *International Journal of Approximate Reasoning*, Vol. 6, pp. 221–240, 1992.

Keller, J. M., Hayashi, Y., and Chen, Z., Additive Hybrid Networks for Fuzzy Logic, *Fuzzy Sets and Systems*, Vol. 66, No. 3, pp. 307–313, 1994.

Khan, E., NeuFuz: An Intelligent Combination of Fuzzy Logic with Neural Nets, in *Proceedings of the International Joint Conference on Neural Networks*, Vol. 3, IEEE, pp. 2945–2950, 1993.

Kim, J. S., and Cho, H. S., Fuzzy Logic and Neural Network Approach to Boundary Detection for Noisy Imagery, *Fuzzy Sets and Systems*, Vol. 65, No. 2–3, pp. 141–159, 1994.

Kim, Y. S., Mitra, S., Adaptive Integrated Fuzzy Clustering Model for Pattern Recognition, *Fuzzy Sets and Systems*, Vol. 65, No. 2–3, pp. 297–310, 1994.

Kosko, B., Fuzzy Systems as Universal Approximators, *IEEE Transactions on Computers*, Vol. 43, No. 11, pp. 1329–1333, 1994.

Kozma, R., Sato, S., Sakuma, M., and Kitamura, M., Detecting Unexperienced Events Via Analysis of Error Propagation in a Neurofuzzy Signal Processing System, *Proceedings of the Artificial Neural Networks in Engineering Conference* (ANNIE '94), St. Louis, MO, Nov 13–16, 1994.

Kulkarni, A. D., Coca, P., Giridhar, G. B., and Bhatikar, Y., Neural Network Based Fuzzy Logic Decision System, in *Proceedings of World Congress on Neural Networks*, 1994 INNS Annual Meeting, San Diego, CA, June 5–9, 1994.

Kung, S. Y., *Digital Neural Networks*, Prentice-Hall, Englewood Cliffs, NJ, 1993.

Kuo, R. J., Intelligent Robotic Die Polishing System Through Fuzzy Neural Networks and Multisensor Fusion, in *Proceedings of the International Joint Conference on Neural Networks*, Vol. 3, 1993, pp. 2925–2928.

Kuo, R. J., Cohen, P. H., and Kumara, S. R. T., Neural Network Driven Fuzzy Inference System, *IEEE International Conference on Neural Networks—Conference Proceedings* 3, IEEE, Piscataway, NJ, 1994, 94CH3429-8, pp. 1532–1536.

Li, W., Wu, Z., Self-Organizing Fuzzy Controller Using Neural Networks, Computers in Engineering, in *Proceedings of the International Conference and Exhibit*, Vol. 2, ASME, New York, pp. 807–812.

Lin, C.-T., and Lee, C. S. G., Reinforcement Structure/Parameter Learning for Neural-Network-Based Fuzzy Logic Control Systems, *IEEE Transactions on Fuzzy Systems*, Vol. 2, No. 1, pp. 46–63, 1994.

Lin, Y., and Cunningham, G. A., A New Approach to Fuzzy-Neural System Modeling, *IEEE Transactions on Fuzzy Systems*, Vol. 3, No. 2, pp. 190–198, 1995.

Matsuoka, K., Blanco, M. A., Neural-Fuzzy Systems for Real-Time Control of Large-Scale Utility Power Plants, in *Intelligent Control Systems American Society of Mechanical Engineers, Dynamic Systems and Control Division*, Publication DSC Vol. 48 1993, ASME, New York, 1993, pp. 75–85.

Mitra, S., Pal, S. K., Self-Organizing Neural Network as a Fuzzy Classifier, *IEEE Transactions on Systems, Man and Cybernetics*, Vol. 24, No. 3, pp. 385–399, 1994.

Moganti, M., Dagli, C., and Ercal, F., PCB Inspection Using Competitive Learning and Fuzzy Associative Memories, *Proceedings of Artificial Neural Networks in Engineering (ANNIE '94)*, St. Louis, MO, Nov. 13–16. 1994, pp. 421–426.

Moon, S. K., Chang, S. H., Classification and Prediction of the Critical Heat Flux Using Fuzzy Theory and Artificial Neural Networks, *Nuclear Engineering and Design*, Vol. 150, No. 1, pp. 151–161, 1994.

Nie, J., Neural Approach to Fuzzy Modeling, in *Proceedings of the American Control Conference*, Vol. 2, American Automatic Control Council, Green Valley, AZ, 1994, 94CH3390-2, pp. 2139–2143.

O'INCA Design Framework, *User's Manual*, Intelligent Machines, Sunnyvale, CA, 1994, pp. 727–748.

Pao, Y.-H., Process Monitoring and Optimization for Power Systems Applications, in *Proceedings of the 1994 IEEE International Conference on Neural Networks*, Part 6 (of 7), Orlando, FL, June 27–29, 1994.

Patrikar, A., Provence, J., Control of Dynamic Systems Using Fuzzy Logic and Neural Networks, *International Journal of Intelligent Systems*, Vol. 8, No. 6, 1993, pp. 727–748.

Pham, D. T., Bayro-Corrochano, E. J., Self-Organizing Neural-Network-Based Pattern Clustering Method with Fuzzy Outputs, *Pattern Recognition*, Vol. 27, No. 8, pp. 1103–1110, 1994.

Purushothaman, G., Karayiannis, N. B, Feed-Forward Neural Architectures for Membership Estimation and Fuzzy Classification, in *Proceedings of the Artificial Neural Networks in Engineering Conference (ANNIE '94)*, St. Louis, MO, November 13–16, 1994.

Rao, D. H., Gupta, M. M., Neuro-fuzzy Controller for Control and Robotics Applications, *Engineering Applications of Artificial Intelligence*, Vol. 7, No. 5 1994, pp. 479–491.

Raza, H., Ioannou, P., and Youssef, H. M., Surface Failure Detection for an F/A-18 Aircraft Using Neural Networks and Fuzzy Logic, in *IEEE International Conference on Neural Networks—Conference Proceedings*, Vol. 5, IEEE, Piscataway, NJ, 1994, 94CH3429-8, pp. 3363–3368.

Schwartz, D. G., and Klir, G. J., Fuzzy Logic Flowers in Japan, *IEEE Spectrum*, July 1992.

Sharaf, A. M., and Lie, T. T., Neuro-fuzzy Hybrid Power System Stabilizer, *Electric Power Systems Research*, Vol. 30, No. 1, pp. 17–23, 1994.

Stylios, G. and Sotomi, O. J., Neuro-fuzzy Control System for Intelligent Sewing Machines, in *IEE Conference Publication*, No. 395, IEE, Stevenage, England, 1994, pp. 241–246.

Tascillo, A., Crisman, J., and Bourbakis, N., Intelligent Control of a Robotic Hand with Neural Nets and Fuzzy Sets, in *Proceedings IEEE Conference on Intelligent Control*, Chicago, August 1993.

Tsoukalas, L. H., Ikonomopoulos, A., and Uhrig, R. E., Virtual Measurements Using Neural Networks and Fuzzy Logic, in *Proceedings of American Power Conference*, Vol. 54-II, pp. 1437–1442, 54th Annual Meeting, Chicago, April 13–15, 1992.

Uhrig, R. E., Tsoukalas, L. H., and Ikonomopoulos, A., Application of Neural Networks and Fuzzy Systems to Power Plants, in *Proceedings of the 1994 IEEE International Conference on Neural Networks*, Vol. 2, Part 6 (of 7), Orlando, FL, June 27–29, 1994, IEEE, Piscataway, NJ, 1994, 94CH3440 5, pp. 510 512.

Wakami, N., Araki, S., and Nomura, H., Recent Applications of Fuzzy Logic to Home Appliances, Plenary Session, Emerging Technologies, and Factory Automation *IECON Proceedings* (Industrial Electronics Conference), Vol. 1, IEEE, Computer Society Press, Los Alamitos, CA, 1993, 93CH3234-2, pp. 155–160.

Wang, F.-Y., Chen, D. D., in *Proceedings of the IEEE International Conference on Systems, Man and Cybernetics*, Vol. 2, IEEE, Piscataway, NJ, 1994, 94CH3571-5, pp. 1803–1808.

Wang, L.-X., *Adaptive Fuzzy Systems and Control*, Prentice Hall, Englewood Cliffs, NJ, 1994.

Watkins, F., "Tutorial, Neural fuzzy Systems" notes, *World Congress on Neural Networks, 1993 International Neural Network Society Annual Meeting*, Portland OR, July 1993.

Wegmann, H., Fuzzy Control and Neural Networks Industrial Applications in the World of PLCs, in *Proceedings of the IEEE Conference on Control Applications*, Vol. 2, IEEE, Piscataway, NJ, 1994, 94CH3420-7, pp. 1245–1249.

Werbos, P. J., Neurocontrol and Fuzzy Logic: Connections and Designs, *International Journal of Approximate Reasoning*, Vol. 6, pp. 185–219, 1992.

Werbos, P. J. Elastic Fuzzy Logic: A Better Way to Combine Neural and Fuzzy Capabilities, in *Proceedings of the IJCNN*, Vol. II, Portland, OR, pp. 623–626, 1993.

Yager, R. R., Modeling and Formulating Fuzzy Knowledge Bases Using Neural Networks, *Neural Networks*, Vol. 7, No. 8, pp. 1273–1283, 1994.

Yamazaki, K., Kaneko, H., Yamaguchi, S., Watanabe, K. Y., Taniguchi, Y., and Motojima, O., Design of the Central Control System for the Large Helical Device (LHD), in *Nuclear Instruments and Methods in Physics Research*, Section A: Accelerators, Spectrometers, Detectors and Associated Equipment, Vol. 352, No. 1–2, December 15, 1994, pp. 43–46.

Yen, G. G., Hybrid Learning Control in Flexible Space Structures with Reconfiguration Capability, in *IEEE International Symposium on Intelligent Control—Proceedings 1994*, IEEE, Piscataway, NJ, 1994, 94CH3453-8, pp. 321–326.

15

DYNAMIC HYBRID NEUROFUZZY SYSTEMS

15.1 INTRODUCTION

In the decade of the 1990s, the authors and their graduate students[1] have carried out research activities that utilize hybrid systems involving various combinations of neural network, fuzzy systems, genetic algorithms, and expert systems. These hybrid systems were utilized to acquire and process data from engineering systems, ranging from nuclear and fossil power plants to steel rolling mills, as well as their various components (check valves, compressors, rolling element bearings, control systems, etc.) The purpose of this chapter is not to describe the result of this work, but rather to convey the essence of the methodologies developed and how they were used advantageously in comparison with more conventional technologies. Hence, only those details necessary to illustrate the use of hybrid artificial intelligence techniques are presented. Although the results of the work are not given, they can be obtained from the references cited.

Various aspects of the analysis of data from three specific sets of experiments are involved in the work described here. The first is a set of vibration spectral data provided by Électricité de France on accelerated testing of bearings with faults deliberately introduced in some bearings to induce early

[1] Special acknowledgment must be given to Dr. Andreas Ikonomopoulos, on leave from the Demokritos Nuclear Research Laboratory, Athens, Greece; Dr. Anna Loskiewicz-Buczak, now with AlliedSignal Corporation, Morristown, NJ, and Dr. Israel E. Alguindigue of the University of Tennessee, Chattanooga, who worked with the authors on many of the activities described in this chapter while pursuing their doctoral degrees at the University of Tennessee, Knoxville, TN.

failure.[2] The diagnosis of faults in roller bearings is based on the relative magnitude of the peaks occurring in spectra at characteristic frequencies (and their harmonics) associated with the bearing geometry and the basic rotating frequency. A comparison between the results using crisp magnitudes of the amplitudes and two types of fuzzy representation of these amplitudes is presented. The goal was to detect which bearings had faults and to identify the magnitude and location (inner race, outer race, or ball) of the faults from the vibration spectra measured by accelerometers mounted on the frame supporting the bearing (Loskiewicz-Buczak, 1993).

The second is a set of vibration spectral data[3] taken from "laminar flow" table rolling machines in a steel sheet manufacturing mill. Data were taken from sensors at nine locations on each of 163 table rolling machines, 49 of which had one or more faults identified. In this example, a composite diagnosis of single and multiple faults in the machines is obtained based on the fusion of nine tentative diagnoses (which were usually all not the same) indicated by nine neural networks processing data from nine sensors placed on the individual machine. The fusion of these decisions is performed by a fuzzy logic connective called the generalized mean (a generalized type of fuzzy variable). The goal of the study was to fuse the data together using neural networks and fuzzy systems methodologies to identify the faults (Loskiewicz-Buczak and Uhrig, 1994).

The third is a set of data taken from the High Flux Isotopes Reactor (HFIR) at Oak Ridge National Laboratory[4] during startup from source level to full power. The goal here is to demonstrate the interrelationship between the various output variables at the HFIR and to infer the value of a variable that cannot be measured directly (Tsoukalas, 1993; Ikonomopoulos et al., 1994).

The final section deals with various aspects of neurofuzzy control, including some discussion of neurofuzzy approaches to anticipatory control. Fuzzy logic and neural networks are complementary technologies, and both are well suited for controlling nonlinear and time-varying system. This discussion reviews the benefits of integrating these two methodologies in an advantageous way. Anticipatory control is one of the areas that can benefit from a neurofuzzy approach. The ability to predict the future faster than real time

[2] The authors are indebted to Électricité de France for providing the data used here from tests carried out at their laboratory facilities near Paris. This work was performed as part of a contract with Électricité de France carried out by one of the authors and his graduate students at the University of Tennessee.
[3] The authors are indebted to the U.S. Steel Corporation, Gary, IN, for permission to utilize vibration spectra data from rolling mills in their plants and to Technology for Energy Corporation, Knoxville, TN, for gathering these data and making them available to the authors and their graduate students.
[4] The authors want to acknowledge the efforts made by many individuals at Oak Ridge National Laboratory to make this project possible.

enables us to take steps to correct a deteriorating situation (Tsoukalas et al., 1994a,b).

15.2 FUZZY-NEURAL DIAGNOSIS FOR VIBRATION MONITORING

Vibration monitoring of engineering systems involves the collection of vibration data from system components and detailed analysis to detect features which reflect the operational state of the machinery. The analysis leads to the identification of potential failures and their causes and makes it possible to perform timely preventive maintenance. A hybrid neurofuzzy system for vibration monitoring of rolling element bearings is discussed. The system takes advantage of the learning and generalization abilities of neural networks. The ambiguity that accompanies fault diagnosis is handled by means of fuzzy membership functions. The combination of neural networks and fuzzy logic contributes to the high speed and flexibility of the system.

For many machines, the vibration frequency spectrum has a characteristic shape when the machine is operating properly, and it has other features for different faults that may appear. Recognition of faults can be accomplished in many cases by detecting specific features in the frequency spectrum which are known to be related to particular faults. All vibration monitoring techniques are based fundamentally on the recording and quantification of small vibration impulses (Zwingelstein and Hamon, 1990). Often, spectral features associated with specific defects are generated at frequencies that can be calculated from formulae derived from bearing geometry.

However, the task of recognition is complicated by a series of factors, such as noise, presence of multiple faults, severity of the fault and speed changes. Fault recognition is also complicated by the fact that fundamental frequency components often disappear at advanced stages of the defect, while harmonic components remain. Furthermore, when performing vibration monitoring of rolling element bearings, the emphasis is more on the content of the spectrum than on its amplitude (Hewlett Packard, 1983; Berry, 1990) (Jackson 1979). Amplitudes of bearing characteristic frequencies often begin to decrease as conditions worsen. Therefore, more importance should be attributed to the fact that a multiple number of fault frequencies are appearing in the spectrum than to the exact amplitude. This fact led to incorporation of fuzzy logic into the classification system. The soft boundaries in fuzzy logic environments, obtained by membership functions, are of special interest because their use results in flexible, more human-like classifications. On the other hand, neural networks provide a viable technique for the analysis of vibration data because of their inherent ability to operate on noisy, incomplete, or sparse data and to model processes from actual system parameters. Some previous University of Tennessee work (Loskiewicz-Buczak and Uhrig, 1992, 1993a, b; Alguindigue et al., 1993) deals with the problem of vibration monitoring by neural network technology alone. By combining neural networks and fuzzy logic, we are able to take advantage of the strengths of both approaches (Loskiewicz-Buczak, 1993a, b).

Vibration Signatures

To perform spectral monitoring of components in an operating engineering system, signatures are collected from plant machinery and analyzed to detect features which reflect the operational state of the machinery. The data consist of vibration measurements collected from SKF ball bearings of type 6206 during an aging simulation process. The rolling element under test is mounted on a horizontal shaft and loaded radially by means of a jack, imposing a vertical force on the bearing. These severe conditions generate scaling faults on the component. Data are collected using an accelerometer placed in the radial direction to the loading zone of the bearing. From these measurements, spectra are generated using fast Fourier technique transform (FFT) techniques (see Figure 15.1). Spectra are averaged over 16 samples with a Hanning window, and each contains 397 points in the range 5 Hz to 1 kHz.

Methodology

For this project, first the characteristic frequencies for a flaw in the inner and outer races and in one of the balls can be calculated in terms of its rotational

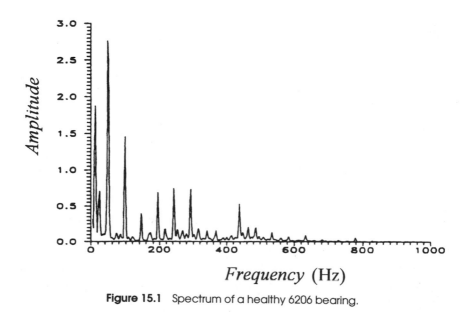

Figure 15.1 Spectrum of a healthy 6206 bearing.

speed from its dimensions (Berry, 1990). These critical frequencies for the type 6206 bearing manufactured by SKF were calculated. Then the location of the peaks at these characteristic frequencies as well as at their harmonics was investigated. The exact value of the amplitude of each peak is not important for the classification process. Instead, we need to know if there is a peak at a given frequency and whether it is small or big. Therefore a transformation of peaks' amplitudes by fuzzy membership functions was performed. At the beginning three triangular membership functions—"none," "small," and "big" (see Figure 15.2)—were used. Then four membership functions were tried using "non," "small," "medium," and "big" (See Figure 15.3). These values of membership functions are the input to the Kohonen self-organizing map with categorization. The output of the network is the fault (or faults) present. The effect of different number of membership functions on the final classification was investigated.

For the analysis of spectral signatures, neural networks may be used both as classifying and clustering systems. To perform classification it is necessary to attach to each signature a label which describes the operational state of the machine at the time of collecting the signature. The input to the network is a spectrum, or some features from it, and the output is the class label. The network is trained to identify an arbitrary pattern as a member of a state among a set of possible states. Clustering involves the grouping of patterns according to their internal similarity and is performed in an unsupervised mode. The aim of clustering is to distribute the set of patterns into states such that the patterns in each state have similar statistical and geometrical properties.

For this project a two-dimensional self-organizing map (SOM) neural network was used (Kohonen, 1990; NeuralWare, 1991). In order for the SOM

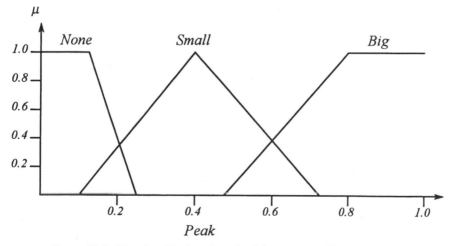

Figure 15.2 Membership functions for ``none,'' ``small,'' and ``big.''

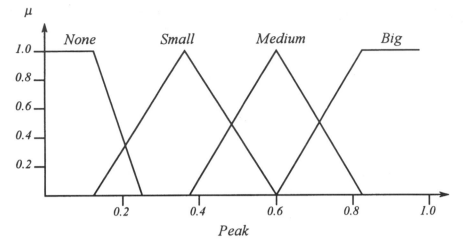

Figure 15.3   Membership functions for "*none*," "*small*," "*medium*," and "*big*."

network to solve categorization problems, an output layer is added to the network. During training the SOM is given a sufficient number of iterations in which to stabilize with the learning rate of the weights going to the output layer set to zero. Hence, the network begins training in an unsupervised mode and then uses supervised training for the output layer.

Results

For this project the NeuralWorks™ version of SOM with a categorization network (NeuralWare, 1991) was used. Three different kinds of inputs to the networks were compared. As the first type of input, amplitudes transformed by three membership functions (see Figure 15.2) were used. The second type of input consisted of amplitudes transformed by four membership functions. (see Figure 15.3.). The last type of input were "raw" amplitudes (without the membership function transformation). In each case, 21 signatures (29.6% of the whole set) were used for training, and the entire set (71 signatures) was used for recall. The output layer of the network was the same in all the situations: six output nodes, each one corresponding to one fault. An activation of 0.5 in an output node indicated the presence of the corresponding fault. Activation in more than one node corresponded to a multiple fault, while no activation was perceived as no fault.

For the transformation of amplitudes by three membership functions the network has 48 inputs ($16 * 3$). This corresponds to three membership values for each of 16 amplitudes at frequencies related to faults. Different sizes of the Kohonen map were tried: 4×5, 5×5, 6×5, 6×6, 6×7. The best results were obtained with the 6×5 Kohonen layer. In this case, only one misclassification occurred (98.5% accuracy). One of the signatures used in training was classified as nonfaulty, whereas it was an instance of a generalized scaling fault of all the components. Had only the signatures not used in training been classified, the accuracy would have been 100%.

When using four membership functions for the transformation process, the input layer had 64 ($16*4$) nodes. The best results were obtained with the Kohonen layer 6×5, resulting in misclassification of one signature. This signature, instead of being classified as a localized fault on the outer race, was classified as nonfaulty. The results for three and four membership function transformations give the same accuracy. However, for four membership functions many of the nonfaulty signatures have one of the outputs with value about 0.3, whereas when three membership functions are used the outputs for nonfaulty signatures are at most 0.05. This means that the transformation by three membership functions gives more robust results than the transformation by four functions.

When using the raw amplitudes as the input to the neural network, the input layer has 16 nodes. The best results were obtained also for the 6×5 Kohonen layer. However, even in this case, 12 of the 71 signatures were misclassified, which gives only 83.1% classification accuracy. The transformations of both fuzzy membership functions results in great improvements of the final classification over a neural network that uses raw amplitudes as inputs.

Fuzzy c-Means Clustering Algorithm

The fuzzy c-means algorithm has been used for the clustering and classification of vibration signatures in the frequency domain (Alguindigue et al., 1992). The fuzzy c-means algorithm is a variant of the fuzzy clusteing algorithms pioneered by Bezdeck since the late 1970s. It attempts to cluster measurement vectors by searching for local minima of the generalized within group sum of squared errors functions (WGSSE). It was proposed by Trivedi and Bezdeck (1986) and is given by

$$J_m(\mathbf{U}, \mathbf{v}) = \sum_{k=1}^{n} \sum_{i=1}^{c} (u_{ik})^m |x_k - v_i|_A^2, \qquad 1 < m < \infty \qquad (15.2\text{-}1)$$

where c is the number of clusters, n is the number of vectors, x_k is a kth measurement vector, v_i is the ith centroid vector, m is the fuzzy coefficient, $|\cdot|_A$ is an inner product norm, $|Q|_A^2 = Q^T A Q$, and A is a $d \times d$ positive definite matrix where d is the dimension of the pattern vectors.

When $m = 1$ the objective function J_m in (15.2-1) is the classical WGSSE function, and the algorithm reduces to the crisp k-means clustering algorithm. For $m > 1$ under the assumption that $x_k \neq v_i$, $(\mathbf{U}, \mathbf{v})$ may be a local minimum of J_m only if

$$u_{ik} = \frac{1}{\displaystyle\sum_{j=1}^{c} \left(\frac{\|x_k - v_i\|_A}{\|x_k - v_j\|_A} \right)^{2/(m-1)}} \qquad \forall\, i, k \qquad (15.2\text{-}2)$$

and

$$v_j = \frac{\displaystyle\sum_{k=1}^{n} (u_{ik})^m x_k}{\displaystyle\sum_{k=1}^{n} (u_{ik})^m}, \qquad \forall\, i \qquad (15.2\text{-}3)$$

The fuzzy c-means algorithm consists of the following steps (Trivedi and Bezdeck, 1986):

1. Fix the number of clusters c. Select the inner product norm. Fix the fuzzy coefficient m. Set $p = 1$ and initialize $U^{(0)}$.
2. Calculate fuzzy cluster centers $\{v^{(p)}\}$ using $U^{(p-1)}$ and the condition specified in equation (15.2-3).
3. Update $U^{(p)}$ using $v^{(p)}$ and the condition specified in equation (15.2-2).
4. If $\|U^{(p)} - U^{(p-1)}\|_A < \varepsilon$ then terminate; else set $p = p + 1$ and go to step 2.

15.3 DECISION FUSION BY FUZZY SET OPERATIONS

Fusion of information from multiple sources for object recognition and classification is an increasingly important area of research and application. Information fusion is employed in robotics, computer vision, managerial decision-making systems, and many engineering systems. Fusion of information is often made more difficult by problems of uncertainty characterized by vagueness, inexactness, and ill-definedness. This is the reason to employ fuzzy set theory in information fusion systems.

Vibration Signatures

Data used for this project consist of vibration signatures from 163 identical "laminar flow" table rolls in a steel sheet manufacturing mill. Data were collected with sensors attached to the plant machinery at the same nine locations on each machine. Spectra acquired from the nine sensors are correlated but not identical due to different vibration levels throughout the machine and to the fact that the faults which are particular to a bearing located near one sensor are not necessarily recorded by the other more distant sensors. The 150-point spectrum of each sensor output is generated using FFT techniques, and the coefficients are stored in a database. The data set contains signatures from 49 machines for which the types of faults had been identified. For some machines, one to three sensor readings were missing. The data sets reflected faulty operating conditions such as misalignment (M), looseness (L), wear (W), outboard bearing damage (O), lubrication (C), and their combinations (double and triple faults.) Data from machines operating properly were not included in the data set used here.

The first step is classification of signatures coming from each sensor separately using recirculation neural networks to reduce dimensionality, and backpropagation or probabilistic neural networks for classification of faults. This classification process has been adequately described in the literature (Alguindigue et al., 1993; Loskiewicz-Buczak and Uhrig, 1993a, b, 1992) and will not be repeated here. The second step is information fusion from these nine classifications, performed by means of fuzzy set operations. Information fusion is used whenever several sensors are employed in a system, in order to reduce uncertainty and resolve the ambiguity often present in classifications from several sensors. In this approach, a confidence factor of the fused decision is determined and the data are fused only from the sensors which cause the confidence factor to grow.

Information Fusion by Means of Fuzzy Logic

Among the approaches for information fusion that have been proposed in the literature are probability theory, Dempster–Shafer theory, neural networks theory and fuzzy set theory. Fuzzy set theory provides several advantages due to the fact that there are numerous ways of combining fuzzy sets in addition to the union (e.g., the "max" operator) and intersection (e.g., the "min" operator) used in traditional theories. Numerous fuzzy set connectives can be used for the purpose of aggregation (Krishnapuram and Lee, 1992; Zimmermann, 1987). The requisites of the decision-making process and the character as well as relative importance of criteria determine the particular connective to be chosen. The requisite may be that all the criteria be satisfied for which an intersection connective should be used, or any one of the criteria be satisfied for which a union connective should be used. When the criteria are mutually compensatory, a mean operator is the most appropriate. Usually in decision-making based on several criteria, a certain amount of compensation is desirable. Zimmermann (1987) showed that human decisions and evaluations almost always show some degree of compensation and that the "generalized mean" used here very closely matches the human-decision making process. In almost all categorization problems the final classification that the system should give is the one that humans would give. This is the reason to use the aggregation connectives that match the best the human decision-making process for fusion of evidence.

For this project the "generalized mean" operator was chosen for the fusion process. It was proposed first by Djumovic (1974) and later Dyckhoff and Pedrycz (1984) and defined by

$$g(x_1, x_2, \ldots, x_n; p, w_1, w_2, \ldots, w_n) = \left(\sum_{i=1}^{n} w_i x_i^p \right)^{1/p} \qquad (15.3\text{-}1)$$

where p is the degree of fuzziness, and the w_i's can be thought as the relative importance factors for the different criteria where

$$w_1 + w_2 + \cdots + w_n = 1 \qquad (15.3\text{-}2)$$

The behavior of the generalized mean with p is shown in Figure 15.4 where the amplitude has been scaled between 0.1 and 0.9. The attractive properties of the generalized mean are as follows:

- $\min(a, b) \leq \text{mean}(a, b) \leq \max(a, b)$;
- mean increases with an increase in p; by varying the value of p between $-\infty$ and $+\infty$, one can obtain all values between min (intersection) and max (union) respectively.

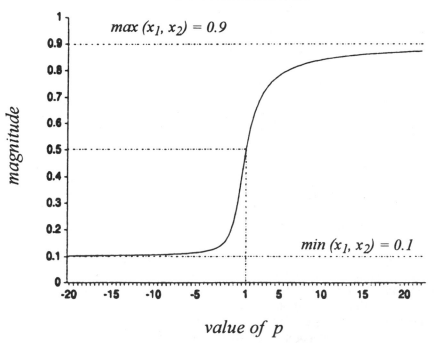

Figure 15.4 The generalized mean operator.

Therefore in the extreme cases the generalized mean operator can be used as union or intersection. Also, it can be shown that $p = -1$ gives the harmonic mean, $p = 0$ gives the geometric mean, and $p = 1$ gives the arithmetic mean. The rate of compensation for the generalized mean can be controlled by changing p. When using larger values of p, the partition becomes more fuzzy.

The definition of the confidence factor can affect the fusion results significantly. For classification problems, the confidence factor (CF) is defined as:

$$CF = \frac{1}{\text{average error}} \quad (15.3\text{-}3)$$

$$\text{average error} = w_1 \cdot \text{error}^1 + w_2 \cdot \text{error}^2 \quad (15.3\text{-}4)$$

$$\text{error}^k = \sqrt{\sum_{i=1}^{n} \sum_{j=1}^{c} \left(\text{mean}_{ij} - \text{pattern}_{ij}^k\right)^2} \quad (15.3\text{-}5)$$

where n is the number of patterns, c indicates the number of classes of faults, and $k = 1, 2$ is the sensor number. The average error [equation (15.3-4)] is a measure of how different the aggregated decision is from the earlier decisions that were the input to the aggregation process. This error will become small when enough decisions are aggregated, because the decision from the next sensor will tend to be redundant with the decisions already fused.

The weights used in the calculation of the average error are the weights used for the fusion process. These weights should describe the relative confidence that we have in those sensor measurements. If there is no reason to bias the decision, all the weights should be the same. For this project, there is no information on the precision of the sensor readings, and therefore the weights for the fusion process for each sensor are the same when fusing information from the first two sensors. At later fusion steps involving n decisions, when one decision is already an aggregated decision from $n - 1$ sensors and the other is a decision from only one sensor, the weights are calculated as $(n - 1)/n$ and $1/n$, respectively. This ensures that the decision from each sensor is given the same importance.

"Active" Information Fusion Scheme

For the fusion process, one can choose a larger p value for fusion of information from complementary sensors and choose a smaller p value for fusion of information from redundant sensors. In this case, information on the degree of complementarity/redundancy of the sensors is required. However, if no such information is available (but the number of sensors is large), it is reasonable to presume that at the beginning of the fusion process [i.e., when fusing information from a small number of sensors (1, 2, 3...)], this information is complementary; as the number of sensors increases, the information should become more and more redundant. Therefore using large p (union-like operation) at the beginning of the fusion process, and decreasing p as the number of sensor increases, seems to be the most appropriate method. In the project this method was used.

Our goal is the best classification possible. We want the fusion process to be "active," meaning that the next step of the fusion process is determined by the results of the previous one. The fusion scheme is the following:

1. Fuse the decisions from two sensors.
2. Evaluate the confidence factor of the fused decision.
3. Fuse the decision from the next sensor.
4. If the confidence factor has decreased undo step 3 (do not fuse data from this sensor.)
5. If there are data from more sensors, repeat steps 3–5; otherwise, this is the aggregated decision.

As the number of sensors involved increased, the value of p in equation (15.3-1) changed from a large value for complementary sensors to a smaller value when the addition sensor information was considered redundant. Subsequent work involved the use of genetic algorithms to optimize the sequence in which the sensor decisions are fused. The concern here is that the choice of a bad sensor decision for the first fusion step could lead to the rejection of good sensor decisions in later fusion steps. Each advancement in methodology improved the resultant identification of the faults.

The final decision has to be obtained from the aggregated decision by some defuzzification method. The method chosen was α-cuts. After fixing the value of α, an α-cut is performed on the aggregated decision. For each of the five faults (M, L, W, O, C) there is a corresponding α-level set (M_α, L_α, W_α, O_α, C_α). Each of these sets includes all the patterns that are manifesting a given fault. If a pattern belongs only to one α-level set, it means that the final decision is that it is exhibiting only this fault (single fault pattern). If a given pattern belongs to more than one α-level set, it means that the final decision is that it is a multiple fault pattern, manifesting the faults to which α-level sets the pattern belongs.

15.4 HYBRID NEUROFUZZY METHODOLOGY FOR VIRTUAL MEASUREMENTS

A method of generating fuzzy numbers representing the values of system-specific variables such as performance has been developed (Ikonomopoulos et al., 1992, 1994). It constitutes essentially the fuzzification (symbolization) of measurements (and predictions), and thus we refer to it as *virtual measurement*. It should be remembered, however, that virtual measurements are simply predictions involving fuzzy numbers where the notion of a measuring device has been extended to incorporate significant modeling capability at the level of the instrument.

In virtual measurements, neural networks are used to perform a mapping

$$f\colon M \to E \qquad\qquad (15.4\text{-}1)$$

where the domain M is the hyperspace of accessible variables such as temperatures and pressures in an engineering system, and the output range E is a set of fuzzy numbers that constitute our predictions of fuzzy values referred to as *virtual measurement values* (VMVs). (VTMs are the fuzzy analogs of the units of measure, e.g., volts, pounds, degrees, etc.) As discussed in Chapter 4, a fuzzy number is a normal and convex fuzzy set on the real numbers which models the value of a fuzzy variable at any given time, uniquely represented by a membership function. The fuzzy numbers used here had a trapezoidal shape. Trapezoidal membership functions are uniquely described by a set of four numbers—for example, a given number $C = \{o_1, o_2, o_3, o_4\}$, where $O \le o_1, o_2, o_3, o_4 \le 1$ and $\{o_1, o_2, o_3, o_4\}$ (from left to right) represents the universe of discourse components of the four corners of the trapezoid (from left to right). Such representations offer considerable advantage to computing speed.

The methodology for predicting fuzzy numbers used here has been described elsewhere (Ikonomopoulos et al., 1994), and its main points may be summarized in the following steps:

1. Decide how many fuzzy values are necessary to adequately cover the range of the fuzzy variable to be predicted.
2. Determine the number and the type of physically measurable variables that will be the basis (i.e., the input) of the virtual instrument.
3. Train one neural network per VMV, for example, a program trained on five VMVs will require five trained neural networks as shown in Figure 15.5.
4. Design an appropriate logic using the *index of dissemblance* to select which membership function will be the predicted value of the instrument at any given time.

The networks $N_1, N_2, \ldots, N_n$ comprising the virtual instrument are trained (in a process analogous to "calibration") with time series as input vectors, and vectors $\{o_1, o_2, o_3, o_4\}$ representing fuzzy numbers are trained as outputs. Each network learns to map a constellation of input patterns to a particular linguistic label. The situation is illustrated in Figure 15.5, where five inputs to each of the n networks are used; hence this virtual instrument is calibrated

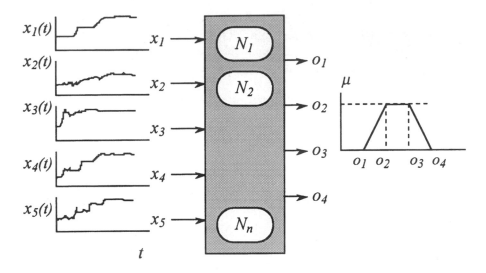

Figure 15.5 Each neural network in a virtual instrument maps a time-series input vector onto a vector $\{o_2, o_2, o_3, o_4\}$ representing a trapezoidal fuzzy number.

with n fuzzy numbers. After training, all networks $N_1, N_2, \ldots, N_n$ receive on-line time signals as inputs and produce a set of membership functions as outputs. Generally the outputs will be somewhat different from the membership functions the networks were trained for (the prototypes); moreover, one or at most two (if we allow overlap of membership functions) will represent correct values while the rest need to be ignored. It is thus important to identify the correct output. Since we consider each network's output to be a fuzzy number, we use a *dissemblance index* (Kaufmann and Gupta, 1991) to estimate the output membership functions that are closest to the set of prototype membership functions on which we trained the networks and select one as the predicted fuzzy value. The dissemblance index, $\delta(A, B)$, of two fuzzy numbers A, B gives the distance between the fuzzy numbers. When $\delta(A, B) = 0$ we can infer that the fuzzy numbers are identical. When $\delta(A, B) = 1$ we infer that the fuzzy numbers are totally different.

Using physically observable quantities to predict fuzzy values offers some unique advantages. A set of complicated time series is mapped to the universe of discourse of human linguistics through a neural network which acts as an interpreter of vital information supplied from the system. The information encoded in a time series is in the form of rate of increase/decrease and maximum/minimum values attained over a period of time. The network is trained to represent this kind of "hidden" information in the form of membership functions which can be used for fuzzy inferencing as shown by Sugeno and Yasukawa (1993). The membership function provides sufficient information to predict the value of a fuzzy variable in the near future.

Furthermore, a network trained to recognize a specific complicated time pattern (i.e.,the time series has "crisp" values) will lose much of its ability to deal with noisy input signals since it will tend, for distorted inputs, to produce averaged forms of the desired output, missing therefore vital pieces of information.

As an example of the prediction method, consider the following experiment. Actual data obtained during a start-up of the high flux isotope reactor (HFIR) was used in order to test the methodology for predicting fuzzy values. HFIR is a three-loop pressurized water research reactor operated at the Oak Ridge National Laboratory. A flow control valve on the secondary side of the system is used as the main mechanism for control (there is also a "trim flow control valve" for finer flow adjustments, as well as control rods) as shown in Figure 15.6. Although the signal sent to the motor of the valve is known, the actual position of the secondary flow control valve is not known and is rather hard to predict. The disk position is something that the operators of the plant "learn" how to estimate intuitively on the basis of experience. However, valve aging and varying plant operating conditions as well as operator experience are major factors for substantial variations in the estimate of valve position.

Five parameters in the form of time series were chosen as the basis for predicting the secondary flow control valve position: *neutron flux, primary flow pressure variation (DP), core inlet temperature, core outlet temperature*, and *secondary flow*. All but the last one of the above-mentioned time series contain average values of the corresponding parameters of the three-loop system. Figure 15.7 shows the secondary flow signal normalized in the range between zero and one. These five parameters were selected in order to provide sufficient description of conditions in both the primary and secondary sides of HFIR during start-up. The time series of these five parameters are used to train five neural networks (i.e., $n = 5$, but it can be any number dependent upon the virtual measurement values (VMV). Each one

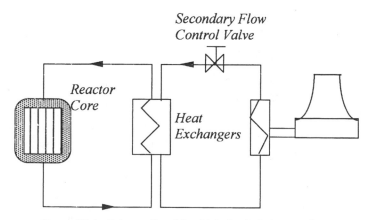

Figure 15.6 Schematic of the high-flux isotope reactor.

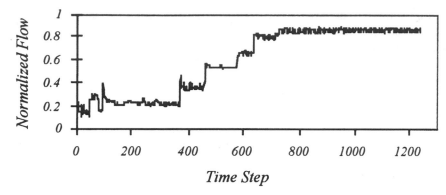

Figure 15.7 Secondary flow signal during start-up.

of them has five neurons at the input layer and four neurons at the output layer and 10 neurons in the hidden layer of each network. In each network N in Figure 15.5, there are five input neurons each receiving a time series from the five physically measurable variables; four output neurons represent the four corners of a trapezoidal membership function. The output is a membership function uniquely labeling a fuzzy value of the fuzzy variable describing the position of the secondary flow control valve, referred to as *valve_position*. The data for each of the time series used for network training is scaled to the interval 0.1 to 0.9 and sampled every 16 seconds, with a total of 1240 samples available.

Designing a virtual instrument to predict *valve_position* requires first the partition of its membership of discourse with the appropriate number of VMVs. In this example, five values—*CLOSED, PARTIALLY_CLOSED, MEDIUM and PARTIALLY_OPEN* and *OPEN*—were chosen. (The choice of five VMVs had nothing to do with the fact that there are five input variables.) Each value is represented by a membership function, namely, μ_{CLOSED}, $\mu_{PARTIALLY\_CLOSED}$, μ_{MEDIUM}, $\mu_{PARTIALLY\_OPEN}$, and μ_{OPEN}. These five membership functions describe the position of the valve at every instant during the start-up period. The universe of discourse on which these membership functions are defined is the interval [0, 1]. Thus, μ_{OPEN} associates each point in the universe of discourse with the fuzzy value *OPEN* at this point.

The membership functions representing the output of the predictive instrument in this particular study have trapezoidal shape or the degenerated (triangular) form of it, which is very useful for computations in the fuzzy control area. The membership function for *CLOSED*, namely μ_{CLOSED} is defined by a trapezoid with peak coordinates {(0.02, 0), (0.05, 1), (0.10, 1), (0.2, 0)}. Similarly, *PARTIALLY_CLOSED* is represented by the trapezoid with coordinates {(0.15, 0), (0.2, 0), (0.30, 1), (0.4, 0)}, *MEDIUM* by {(0.35, 0), (0.4, 1), (0.50, 1), (0.6, 0)}, *PARTIALLY_OPEN* by {(0.5, 0), (0.6, 1), (0.7, 1),

(0.75, 0)}, and *OPEN* by {(0.7, 0), (0.82, 1), (0.85, 1), (0.90, 0)}. It is evident from the above geometrical schemes that there is an overlap between the membership functions used. The reason for the overlap is the fuzziness in the definition of the different states of valve position.

Figure 15.8 shows the prediction of the instrument during a start-up of the reactor (1240 time steps). The valve is initially *CLOSED* as seen by the membership function in the origin of the 3-D graph. It goes through the "medium" range rather quickly in the vicinity of 400–500 time steps, and finally it becomes fully open after the 800th time step. Note that this confirms rather well the trend shown in Figure 15.7 where the secondary flow reaches its maximum value after about the 800th time step.

To test the ability of each network to predict the valve position by calculating the right membership function at any particular time step, different levels of noise were introduced in the input signals. Initially up to 10% noise was introduced to all five input signals, and the set of networks was tested with the "noisy" vectors. The appropriate networks fired at the corresponding time steps, calculating the coordinates of the peaks of the corresponding membership functions with 98% accuracy. Henceforth there was an excellent prediction of the position of the disk valve during the whole time interval under consideration. In addition, 20% noise was introduced to

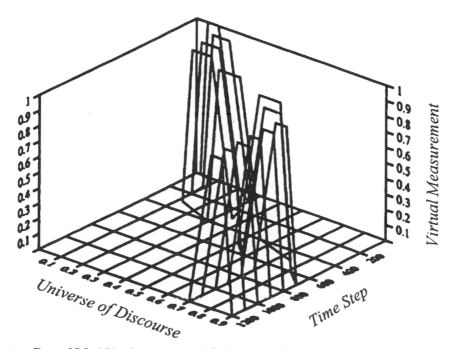

Figure 15.8 Virtual measurement for time steps 0–1240 in 200-step intervals.

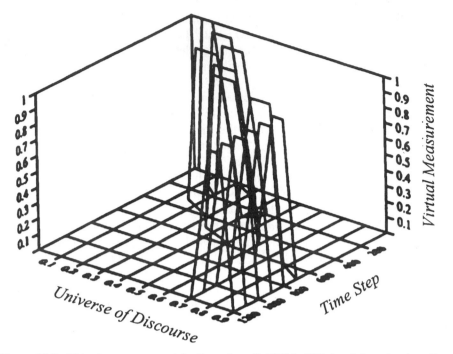

Figure 15.9 Virtual measurement for time steps 0–1240 in 200-step intervals when the secondary flow input signal has been substituted with 100% noise.

all five input signals, and the networks were tested again. The response of the system was indistinguishable form the previous case.

Even when the most closely related input signal was replaced with random noise, the predictive instrument still predicted the valve position rather accurately. Figure 15.9 shows the output of the instrument when the secondary flow signal has been replaced with random noise. Comparison with Figure 15.8 shows that the virtual instrument still indicates the valve position rather well. A series of statistical tests were conducted to confirm that the output of the instrument is actually within random error of the previous case. This represents a significant tolerance to informational hazards to which the instrument was exposed. Even with about 20% of its input information lost, it still rather accurately measured the valve position. Similar results were obtained by replacing the other input signals one by one with random noise.

15.5 NEUROFUZZY APPROACHES TO ANTICIPATORY CONTROL

Anticipatory systems are systems where change of state is based on information pertaining to present as well as future states. Cellular organisms, industrial processes, and global markets provide many examples of behavior where global output is the result of anticipated, as well as current, states. In the global economy, for example, the anticipation of an oil shortage or of a significant default of foreign loans can have profound effects upon the course of the economy, whether or not the anticipated events come to pass (Holland, 1988). Participants in the economy build up models of the rest of the economy and use them to make predictions. The models are more *prescriptive* (prescribing what should be done in a given situation) than *descriptive* (describing the options of a given situation) and involve strategies appropriately formulated in terms of *lookahead*, or anticipation of market conditions. In an industrial process, the prescriptions are typically standard operating procedures (SOPs), dictating actions to be taken under specific conditions. The accumulated experience of various decision-makers at all levels of the process provides increasingly refined SOPs and progressively more sophisticated interactions amongst them and computer tools designed to assist the operators. As another example, consider a car driven on a busy highway. The driver and the car taken together are a simple, everyday example of an anticipatory system. An automobile driver makes decisions on the basis of predicting what may be happening in the future, not simply reacting to what happens at the present. Driving requires one to be aware of future system inputs by observing the curvature and grade of the road ahead, road conditions, and the behavior of other drivers. Perceptual information received at the present may be thought of as input to internal predictive models. Such a system, however, is very difficult to model using conventional approaches. In part, the difficulty relates to the fact that conventional predictive models are unduly constrained by excessive precision. Generally, in situations like the driver–car system, it is important for a decision-maker (the driver) to use a *parsimonious description* of the overall situation—that is, a model with the appropriate level of precision. Predictions about the future are not very precise, and, of course, they may be wrong. Yet, their efficacy does not rest on *precision* as much as on the more general issue of *accuracy* and their successful utilization. High levels of precision may not only be unnecessary for problems utilizing predicted values, they may very well be counterproductive. An overprecise driver may actually be a dangerous driver.

Although anticipatory systems have been studied by a number of researchers in the context of mathematical biology (Rosen, 1985), it should be noted that automata theory (Trachtenbrot and Barzdin, 1973), preview control (Tomizuka and Whitney, 1975), and their epistemological roots may be traced back to Aristotle's views on causality. It is only recently that the advent of modern computing technologies makes it possible to employ them for complex system regulation and management (Berkan et al., 1991; Tsoukalas et al., 1990, 1994b). In Japan, the automatic train operator (ATO) used in Sendai's subway system, as well as some tunnel ventilation systems and elevator control systems employ anticipatory control strategies (Yasunobu, 1985); and researchers at Tohoku University and Mitsubishi Research Institute have studied an innovative anticipatory guidance and control system for computer-assisted operation of nuclear power plants (Washio, 1993).

Probabilistic Predictions

In preview control, future information was considered as probabilistic in kind, and the control problem was seen as a problem of time delay (Tomizuka and Whitney, 1975). The situation is illustrated in Figure 15.10a where a discrete control problem that lasts n time steps, presently at time i, is

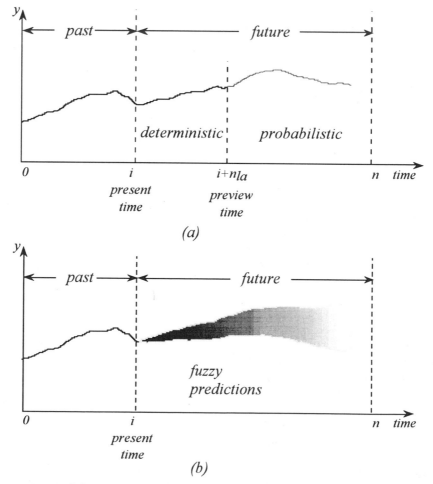

Figure 15.10 (*a*) The future in finite preview problems is modeled deterministically and stochastically. (*b*) The future in anticipatory systems is modeled fuzzily.

considered. Tomizuka postulated that up to certain time n_{1a} beyond i, predictions can be made and utilized by the controller at i. Thus, the future is divided into *deterministic* and *probabilistic* parts as seen in Figure 15.10a. The controller is assumed to make use of preview information with respect to a command signal (desired trajectory) from the present time i up to n_{1a} time units into deterministic future. The quantity n_{1a} is the *preview time* (or *length of anticipation*) and is usually shorter than n, the problem duration, often by one or two time steps. To make the solution applicable to a broader class of problems, measurements of time delay, observation noise, and driving noise were included in formulating the problem. The solution showed how to utilize the local future information obtained by finite preview (n_{1a}) in order to minimize an optimality criterion evaluated over the problem duration n. It was found that preview dramatically improved the performance of a system relative to nonpreview optimal performance, and a heuristic criterion about the preview time, n_{1a}, was suggested, that is, $n_{1a} \approx 3 \times$ (longest closed-loop plant time constant).

Fuzzy Predictions

The point of departure for our formulation is that future information is essentially *fuzzy* in nature, that is, predicted values are not imbued with stochastic or probabilistic type of uncertainty. Whatever can be said about the future does not come from measurements but instead from models; hence, such predictions are fuzzy numbers—that is, linguistic categorizations of information pertaining to the future of the system. Generally, fuzziness is a property of language, whereas randomness is a property of observation; and since there is no physical measurement pertaining to the future, the mathematics of fuzzy sets may be more appropriate for anticipatory systems. Consider, for example, the process depicted in Figure 15.10b. At any time i we have available information from the present as well as information from the output of some predictive model. According to our formulation, this is a fuzzy prediction. Therefore, the mathematical tools for utilizing it at time i ought to be fuzzy as well. The time Δt into the future, the anticipatory time step, depends on the nature of the problem and the predictive model used and, generally, need not be one or two time steps as is often the case in preview control. As is suggested in Figure 15.10b, the fuzziness of a prediction is postulated to depend on time in the future in the sense that for greater time we get fuzzier predictions.

Issues of Formalism in Anticipatory Systems

A system that makes decisions in the present on the basis of what may be happening in the future is thus envisioned to be different in two important respects: in the *language* used to formulate models of its behavior and in the method of *measurement* used to access future states. The first we call the *issue of formalism* and we address it in this section, while the latter we examined in the previous section through the concept of virtual measurement.

Consider the typical systems formulation in modern control theory. A system is described by a set of difference (differential) equations of the form

$$x(t + 1) = Ax(t) + Bu(t) + w(t), \qquad x(t_0) = x_0$$
$$y(t) = Cx(t) + v(t)$$

(15.5-1)

where $\{u(t)\}$ is an $r \times 1$ input sequence. $\{y(t)\}$ is an $m \times 1$ output sequence, $\{x(t)\}$ is an $n \times 1$ state sequence, A, B, and C are appropriate transition matrices, x_0 some initial state, and, $w(t)$ and $v(t)$ are noise terms.

A system is called *anticipatory* if $x(t + 1)$ and $y(t)$ are not uniquely determined by $x(t)$ and $u(t)$ alone, but use information pertaining to some future state $x(t + \Delta t)$ and/or input $u(t + \Delta t)$.

Looking at equations (15.5-1) we observe that it is rather difficult to include future information in these equations except by containing it within the noise terms as in the case of nondeterministic systems. In such a case, one obtains sets of values $x(t + 1)$ and $y(t)$ with each pair $[x(t), u(t)]$. Suppose that the values of x and y are subsets of some larger sets X and Y. If we denote these subsets of X and Y by X^{t+1} and Y^t, we obtain mappings of the form

$$X^{t+1} = F[x(t), u(t)]$$
$$Y^t = G[x(t), u(t)]$$

(15.5-2)

Of course, one could fuzzify this system by assuming that these X^{t+1} and Y^t are fuzzy subsets on X and Y, respectively, and obtain a fuzzy system determined by conditional membership functions

$$\mu[x(t + 1)|x(t), u(t)]$$
$$\mu[y(t)|x(t), u(t)]$$

(15.5-3)

Subsequently the compositional rule of inference may be used to calculate the fuzzy response of the fuzzy system to any fuzzy input. The problem, however, of involving future information in the formulation of equations (15.5-1) still remains. Generally, if we do so, the mappings in equations (15.5-2) cease to be *many-to-one* mappings (i.e., *functions*) but instead become more general *many-to-many* mappings such as we now have in *fuzzy relations*.

Consider again equations (15.5-1). Another approach is to consider the equal signs " $=$ " as the assignment operators " $:=$ ", that is

$$x(t + 1) := Ax(t) + Bu(t) + w(t), \qquad x(t_0) = x_0$$
$$y(t) := Cx(t) + v(t) \tag{15.5-4}$$

where the assignment operator " $:=$ " is an *if/then* rule, which assigns the right-hand side (RHS) of equation (15.5-4) to the left-hand side (LHS) upon update. Now we are in the realm of logical implications and we can easily include terms such as $x(t + \Delta t)$, and $u(t + \Delta t)$ in our *if/then* rules. The calculus of fuzzy *if/then* rules is rather well known and provides an interesting alternative and enhancement of formulations such as equation (15.5-1), particularly for the purpose of qualitative and complex system modeling. Thus, an anticipatory system can be described by a collection of fuzzy *if/then* rules

$$R^N = \{R^1, R^2, \ldots, R^n\} \tag{15.5-5}$$

Each rule is a *situation/action* pair, denoted as $s \rightarrow a$, where both *present* and *anticipated situations* are considered in the LHS and *current action* is considered in the RHS. The rules of equation (15.5-5) may be rewritten as

$$R^N = \{s^1 \rightarrow a^1, s^2 \rightarrow a^2, \ldots, s^n \rightarrow a^n\}$$
$$= \overset{n}{\underset{j=1}{\phi_\alpha}}(s^j \rightarrow a^j) \tag{15.5-6}$$

where ϕ_α is an appropriate implication operator (Terano et al., 1992). In many cases we can further partition the set of rules in equation (15.5-6) into rule bases (RB), with each rule base being responsible for one action; that is,

$$R^N = \overset{r}{\underset{j=1}{\bigcup}} [RB^p] \tag{15.5-7}$$

Rule bases (15.5-7) can be made to reflect temporal partitions that is, we can have rules that describe the state of the system at t, that is,

$$s(t) \rightarrow a(t) \tag{15.5-8}$$

and we can also have rules that describe the possible state of the system at some time later, that is,

$$s(t + \Delta t) \rightarrow a(t) \tag{15.5-9}$$

Thus an anticipatory fuzzy algorithm can infer the current action $a(t)$ on the basis of the present state $s(t)$ as well as anticipated ones $s(t + \Delta t)$.

Generally, the rules of (15.5-5) describe relations of a more general type than that of functions, i.e., *many-to-many* mappings (see Chapter 5). Such mappings have the linguistic form of fuzzy *if/then* rules—for example,

$$\text{if} \quad x \text{ is } A \quad \text{then } y \text{ is } B \tag{15.5-10}$$

where x is a fuzzy variable whose arguments are fuzzy sets denoted as A, and y is a fuzzy variable whose arguments are the fuzzy sets B. Similar rules pertaining to future states are of the form

$$\text{if} \quad x \text{ will be } A, \quad \text{then } y \text{ is } B \tag{15.5-11}$$

where x is thought of as a situation variable and y is the corresponding action variable. Evaluation of formulations using rules such as (15.5-10) and (15.5-11) can be done through *generalized modus ponens* as we have seen in Chapter 5.

Anticipatory control strategies may be based on global fuzzy variables such as *performance* where a decision at each time t is taken in order to maximize current as well as anticipated performance pertaining to $t + \Delta t$. *Performance* in this case is a fuzzy variable (with an appropriate set of fuzzy values) that summarizes information about the system, thereby allowing the system to make decisions about its change of state. The observation/prediction of such variables can be addressed by the methodology presented in Section 15.4.

Alternatively we may use fuzzy *if/then* rules to generate a decision from (15.5-7) and call a predictive routine to anticipate the effect of the proposed decision on the system output (Yasunobu and Miyamoto, 1985). Additional rules may be called if the current decision will result in system behavior which is unacceptable. Consider, for example, the following rule:

If the current decision (u_c) *will cause the difference between the current and anticipated states to be big, then*

$$u = u_c(1 - \beta \cdot bigt) \tag{15.5-12}$$

where β is a user-chosen parameter between 0 and 1 and *bigt* is the fulfillment function for the anticipated difference states. The parameter β may also be chosen by employing a predictive neural network (McCullough, 1993).

REFERENCES

Alguidingue, I. E., Loskiewicz-Buczak, A., and Uhrig, R. E., Clustering and Classification Techniques for the Analysis of Vibration Signatures, *Applications of Artificial Neural Networks III*, Proceedings of the SPIE, Orlando, FL, April 20–24, 1992.

Alguindigue, I. E., Loskiewicz-Buczak, A., and Uhrig, R. E. Monitoring and Diagnosis of Rolling Element Bearings using Artificial Neural Networks, *IEEE Transactions on Industrial Electronics*: *Special Issue on Applications of Intelligent Systems to Industrial Electronics*, April 1993.

Berkan, R. C., Upadhyaya, B. R., Tsoukalas, L. H., Kisner, R. A., and Bywater, R. L, Advanced Automation Concepts for Large-Scale Systems, *IEEE Control Systems*, Vol. 11, No. 6, pp. 4–12, 1991.

Berry, J., *Tracking of Rolling Element Bearing Failure Stages Using Vibration Signature Analysis*, Technical Associates of Charlotte, Inc., 1990.

Broch, J. T., *Mechanical Vibrations and Shock Measurements*, Bruel & Kjaer, Naerum, Denmark, 1984.

Djumovic, D. Weighted Conjunctive and Disjunctive Means and Their Application in System Evaluation, *Publikacije Elektrotehnickog Faculteta Beograd, Serija Matematika i Fizika*, No. 483, 1974.

Dyckhoff, H., and Pedrycz, W., Generalized means as Model of Compensative Connectives, *Fuzzy Sets and Systems*, Vol. 14, 1984.

Hewlett Packard, *Effective Machinery Maintenance Using Vibration Analysis*, Application Note 243-1, Dynamic Signal Analyzer Applications, 1983.

Holland, J. H., The Global Economy as an Adaptive Process, in *The Economy as an Evolving Complex System*, P. W. Anderson, K. J. Arrow, D. Pines, Eds. A Proceedings Volume in the Santa Fe Institute in the Sciences of Complexity, Addison-Wesley, Reading, MA, 1988.

Ikonomopoulos, A., Tsoukalas, L. H., and Uhrig, R. E., Integration of Neural Networks with Fuzzy Reasoning, for Measuring Operational Parameters in a Nuclear Reactor, *Nuclear Technology*, Vol. 104, pp. 1–12, 1994.

Ikonomopoulos, A., Uhrig, R. E., and Tsoukalas, L. H., A Methodology for Performing Virtual Measurement in a Nuclear Reactor System, *Transactions of American Nuclear Society 1992 Winter Meeting*, Chicago, IL, November 15-20, 1992, pp. 106–109.

Jackson, C., *The Practical Vibration Primer*, Gulf Publishing Company, Houston, TX, 1979.

Kaufmann, A., and Gupta, M. M., *Introduction to Fuzzy Arithmetic*, Van Nostrand Reinhold, New York, 1991.

Kohonen, T., The Self-Organizing Map, *Proceedings of the IEEE*, Vol. 78, No. 9, 1990.

Krishnapuram, R., Lee, J., Fuzzy-Connective-Based Hierarchical Aggregation Networks for Decision Making, *Fuzzy Sets and Systems*, Vol. 46, pp. 11–27, 1992.

Loskiewicz-Buczak, A, and Uhrig, R. E., Probabilistic Neural Network for Vibration Data Analysis, in *Intelligent Engineering Systems Through Artificial Neural Networks*, Vol. 2, ASME Press, New York, 1992.

Loskiewicz-Buczak, A., and Uhrig, R. E., Aggregation of Evidence by Fuzzy Set Operations for Vibration Monitoring, *Proceedings of the Third International Conference on Industrial Fuzzy Control & Intelligent Systems*, R. Langari, J. Yen, and J. Painter, eds., Houston, TX, December 1–3, 1993a.

Loskiewicz-Buczak, A., and Uhrig, R. E., Neural Network—Fuzzy Logic Diagnosis Systems for Vibration Monitoring, *Proceedings of ANNIE-93, Artificial Neural Networks in Engineering Conference*, St. Louis, MO, November 14–17, 1993b.

Loskiewicz-Buczak, A., and Uhrig, R. E., Decision Fusion by Fuzzy Set Operation, in *Proceedings of IEEE World Congress on Computational Intelligence, Fuzzy Systems Conference,* Orlando, F. L., June 26–30, 1994, pp. 1412–1417.

Loskiewicz-Buczak, A., Alguindigue, I., and Uhrig, R. E., Monitoring and Diagnosis of Rolling Element Bearings Using Artificial Neural Networks, in *IEEE Transactions on Industrial Electronics: Special Issue on Applications of Intelligent Systems to Industrial Electronics,* April 1993b.

Loskiewicz-Buczak, A., Alguindigue, I. E., and Uhrig, R. E., Vibration Analysis in Nuclear Power Plants Using Neural Networks, in *Proceedings of Second International Conference on Nuclear Engineering,* San Francisco, CA, March 21–24, 1993a.

McCullough, C. L., Anticipatory Neuro-Fuzzy Control: A Powerful New Method for Real World Control, in *Proceedings of IEEE International Workshop on Neuro Fuzzy Control,* Muroran, Japan, March 22–23, 1993, pp. 267–272.

NeuralWare, Neural Computing-NeuralWorks Professional II/PLUS and Neural-Works Explorer, *User Manual,* NeuralWare, Inc., Pittsburgh, 1991.

Rosen, R., *Anticipatory Systems,* Pergamon Press, New York, 1985.

Sugeno, M., and Yasukawa, T., A Fuzzy-Logic-Based Approach to Qualitative Modeling, *IEEE Transactions on Fuzzy Systems,* Vol. 1, No. 1, pp. 7–31, 1993.

Terano, T., Asai, K., and Sugeno, M., *Fuzzy Systems Theory and Its Applications,* Academic Press, Boston, 1992.

Tomizuka, M., and Whitney, D. E., Optimal Finite Preview Problems (Why and How Is Future Information Important), *Journal of Dynamic Systems, Measurement, and Control,* pp. 319–325, December 1975.

Trachtenbrot B. A., and Barzdin Y. M., *Finite Automata Behavior and Synthesis,* North-Holland, Amsterdam, 1973.

Trivedi, M. M., and Bezdeck, J. C., Low-Level Segmentation of Aerial Images with Fuzzy Clustering, *IEEE Transactions on Systems, Man and Cybernetics,* Vol. SMC-16, No 4., 1986.

Tsoukalas, L. H., Ikonomopoulos, A., and Uhrig, R. E., Generalized Measurements in Neuro-Fuzzy Systems 1993, in *Proceedings of the International Workshop on Neural Fuzzy Control,* Muroran, Japan, March 22–23, 1993.

Tsoukalas, L. H., Berkan, R. C., and Ikonomopoulos, A., A Methodology for Uncertainty Management in Knowledge-Based Systems Employed Within the Anticipatory Paradigm, in *Proceedings of the Third International Conference of Information Processing and Management of Uncertainty in Knowledge-Based Systems,* IPMU, Paris, July 2–6, 1990, pp. 77–80.

Tsoukalas, L. H., and Ikonomopoulos, A., Uncertainty Modeling in Anticipatory Systems, in *Analysis and Management of Uncertainty,* B. M. Ayyub, M. M. Gupta and L. N. Kanal eds., Machine Intelligence and Pattern Recognition Series, Elsevier/North Holland, Amsterdam, 1992, pp. 79–91.

Tsoukalas, L. H., Ikonomopoulos, A., and Uhrig, R. E., Fuzzy Neural Control, in *Artificial Neural Networks for Intelligent Manufacturing,* C. H. Dagli, ed., pp 413–434, Chapman & Hall, London, 1994.

Tsoukalas, L. H., Ikonomopoulos, A., and Uhrig, R. E., Neuro-Fuzzy Approaches to Anticipatory Control, in *Artificial Intelligence in Industrial Decision Making, Control and Automation*, S. G. Tzafestas and H. B. Verbruggen eds., Kluwer Academic Publishers, The Netherlands, 1995, pp. 405–419.

Washio, T., and Kitamura, M., General Framework for Advance of Computer-Assisted Operation of Nuclear Plants—Anticipatory Guidance and Control for Plant Operation, *Proceedings of Japan Atomic Energy Society*, Kobe, Japan, October 1993 (in Japanese).

Yasunobu, S., and Miyamoto, S., Automatic Train Operation by Predictive Fuzzy Control, in *Industrial Applications of Fuzzy Control*, M. Sugeno, ed., pp. 1–18, North Holland, Amsterdam, 1985.

Zadeh L. A., Fuzzy Sets, *Information and Control*, Vol. 8, 1965.

Zimmermann, H. J., *Fuzzy Sets, Decision Making, and Expert Systems*, Kluwer Academic Publishers, Boston, 1987.

Zwingelstein G., and Hamon, L., EDF Studies on the Condition Monitoring of Rolling Element Bearings, in *Proceedings of the 4th Incipient Failure Detection Conference*, Philadelphia, October 1990.

IV

OTHER ARTIFICIAL INTELLIGENCE SYSTEMS

16

EXPERT SYSTEMS
IN NEUROFUZZY SYSTEMS

16.1 INTRODUCTION

Artificial intelligence is a branch of computer science that attempts to emulate certain mental processes of humans by using computer models. In expert systems, perhaps the first field of artificial intelligence to be commercially recognized in its own right, one of the primary objectives is to mimic human expertise and judgment using a computer program by applying knowledge of specific areas of expertise to solve finite, well-defined problems. These computer programs contain human expertise (called *heuristic knowledge*) obtained either directly from human experts or indirectly from books, publications, codes, standards, or databases, as well as general and specialized knowledge that pertains to specific situations. Expert systems have the ability to reason using formal logic, to seek information from a variety of sources including databases and the user, and to interact with conventional programs to carry out a variety of tasks including sophisticated computation.

The principal use of expert systems in neurofuzzy systems is to ensure that the unique capabilities of neural networks and fuzzy logic systems are implemented in the proper way including the fuzzy rules of fuzzy algorithms, sometimes called "fuzzy associate memory" (FAM) matrices (see Figure 6.2). While it may be possible to use ordinary software programs to do this, the ability of expert systems to adapt to and deal with unforeseen situations is very important when the outputs may not be precise or the model may be based on less-than-perfect data.

16.2   CHARACTERISTICS OF EXPERT SYSTEMS

A number of characteristics of expert systems are unique and generally advantageous [see, for example, Van Horn (1986) and Feigenbaum et al., (1988)]:

1. Experts need not be present for a consultation; expert systems may be delivered to remote locations where expertise may not be otherwise available.
2. Expert systems do not suffer from some of the shortcomings of human beings (e.g., they do not get tired or careless as the work load increases) but, when properly used, continue to provide dependable and consistent results.
3. The techniques inherent in the technology of expert systems minimize the recollection of information by requesting only relevant data from the user or appropriate databases (i.e., data encountered in the reasoning path).
4. Expert knowledge is saved and readily available because the expert system can become a repository for undocumented knowledge that might otherwise be lost (e.g., through retirement).
5. The development of expert systems forces documentation of consistent decision-making policies. The clear definition of these policies makes the overall decision-making process transparent and the implementation of policy changes instant and simultaneous at all sites.

On the other hand, expert systems have disadvantages that affect their use:

1. They usually deal only with static situations.
2. They must be kept up to date as conditions change.
3. They often cannot be used in novel or unique situations.
4. Results are very dependent on the adequacy of the knowledge incorporated into the expert system.
5. Perhaps most important, they do not benefit from experience except through updating of the knowledge base (based on human experience).
6. Expert systems are unable to solve problems outside their domain of expertise. In many cases they are unable to detect the limitations of their domain (Swartout and Smoliar, 1987; Ricker, 1986).

The domain of an expert system refers to the scope of the knowledge contained within the knowledge base. If the expert system operates outside its domain, it is possible that it may generate incorrect results by utilizing nonapplicable, irrelevant knowledge while searching for a solution. The inability of expert systems to recognize the limitations of their knowledge has been identified as a very serious shortcoming.

Expert systems can, under certain circumstances, deal with imprecise or "fuzzy" information, missing information, and even a certain amount of conflicting information through the use of "certainty factors" or Bayesian probabilities (Kaplan et al., 1987). Certainty factors represent a measure of belief of the user that a piece of evidence is true. These are not probabilities but rather simply a subjective judgment on the degree of truth or validity of an assertion. Some of the information used in development and application of an expert system may not be absolutely certain, and the use of certainty factors allows this subjective evaluation to be incorporated into the expert system. The final results in these cases may be the "most probable" solution or the "best" solution, but there is no absolute guarantee that the solution is the "correct" solution. Recent work incorporating "fuzzy logic" and "reasoning under uncertainty" into expert systems has greatly improved the performance of expert systems when dealing with complex systems.

A comparison of human and artificial expertise will help convey the strengths and weaknesses of expert systems. Human expertise is perishable and difficult to transfer, whereas artificial expertise is permanent and easy to transfer. Human expertise is not always consistent, whereas artificial expertise is consistent. (If you give an expert system the same problem on two occasions you will get the same answer unless stochastic processes are involved; this is not necessarily true of a human expert.) On the other hand, human expertise is creative and has a broad focus, whereas artificial expertise is uninspired and usually has a very narrow focus. Above all, human expertise is adaptive and demonstrates common sense, characteristics usually lacking in expert systems because the knowledge is entirely technical or objective in nature. For instance, artificial expertise does not know that the objects cannot occupy the same space unless it is told.

16.3 COMPONENTS OF AN EXPERT SYSTEM

The principal components of an expert system are the *inference engine*, the *knowledge base*, and the *interface* between the expert system and humans (users, knowledge engineers, and experts). The inference engine is a computer program that gathers the information needed from the knowledge base, associated databases, or the user, guides the search process in accordance with a preselected strategy, uses rules of logic to draw inferences or conclusions for the processes involved, and presents these inferences or conclusions (where warranted) with explanations or bases.

The knowledge base consists of information stored in retrievable form in the computer, usually in the form of rules or frames. The correctness and completeness of the information within the knowledge base is the key to obtaining correct results or solutions using expert systems. Knowledge bases may contain models of systems which produce real-time results or certain learning systems (such as neural networks) that provide new knowledge.

The interface between the human and the expert system must translate user input into the computer language, and it must present conclusions and explanations to the user in clear written or graphical form. It should also include an editor to assist in adding to or changing the knowledge base.

One of the major breakthroughs in development of expert systems came in the mid-1970s with the expert system MYCIN, a diagnostic system for infectious diseases of the blood. The MYCIN architecture completely separated the knowledge base from the inference engine, which permitted modification of the knowledge base without any influence on the inference engine. Hence, it was possible to start with a simple expert system and incrementally add features and complexity as needed. Such a separation is common today even in conventional software, but it was a significant advancement in the mid-1970s.

The knowledge base of an expert system contains the expertise (facts and heuristics) collected from experts, books, publications, and other sources and encoded into rules, frames, or other computer representations of knowledge. This information describes a methodology for solving the problem as a human expert would solve it. Collecting adequate knowledge from experts and translating it into computer code (a process called "knowledge acquisition") has proven to be a very difficult task. All too often, experts really do not understand the processes by which they reason or solve problems. In other cases, experts are reluctant to give up their expert knowledge because they perceive that the availability of an expert system with their expertise may lessen their value to their employer or clients. Because an expert system is only as good as its knowledge base, proper collection and representation of knowledge is critical for the successful implementation and operation of expert systems.

Some expert systems contain a degree of self-awareness or self-knowledge that allows them to reason about their own operation and to display inference chains and traces of the rationale behind their results (Waterman, 1986). These abilities (the explanation facilities) have been recognized as one of the most valuable features of expert systems. The user can take advantage of explanation facilities to request a complete trace for a consultation, request an explanation on how a particular goal or subgoal was inferred, or request an explanation of why a particular piece of information is needed. These facilities can be used to obtain information on the status of a system. Explanation-generating facilities are also of great use in debugging expert systems and may play a key role in verification and validation of expert systems.

The performance of mature expert systems has shown that the reliability of an expert system in a given subject area asymptotically approaches the reliability of the expert as the knowledge base approaches the expert's knowledge in that area. In some cases the reliability of an expert system exceeds the reliability of the expert, not because the expert system is "smarter" than the expert, but rather because the expert system does not forget anything contained in the knowledge base and is capable of rapidly carrying out analytic and mathematical operations.

An expert system "shell" is a computer program used to develop an expert system. Early shells were expert systems from which the domain-specific knowledge bases had been removed and the mechanism for creating a new knowledge base of the user's choice had been made "user friendly." Often a shell also has provisions for changing the reasoning processes of the inference engine to adapt to the specific problem. The first shell was EMYCIN (essential MYCIN), in which the knowledge base on infectious diseases of the blood was removed from MYCIN and knowledge bases on cancer treatment and pulmonary diseases were used to create two new expert systems (ONCONIN and PUFF, respectively) to assist doctors in these fields. The pioneering efforts of Stanford University on EMYCIN paved the way for virtually all modern expert system shells. Indeed, only in the last few years have expert system shells begun to deviate significantly from the overall structure developed for MYCIN.

Expert system shells today differ significantly from each other and offer the user a wide variety of capabilities. Some have sacrificed size of knowledge base to improve ease of updating the knowledge base, and vice versa. Certain expert systems (e.g., 1ST CLASS and VP EXPERT) have the ability to derive the knowledge base from a series of examples by induction. Such ability to extract information from databases and experimental results are one of the strengths of artificial neural networks. Hence, the use of a hybrid consisting of an artificial neural network in the knowledge base of an expert system is feasible and often advantageous. Recently, an expert system shell was introduced with HYPERTEXT as part of the knowledge base. Selection of an expert system to fit a specific need is almost a research project in itself and has, in fact, been the topic for an expert system.

16.4 KNOWLEDGE REPRESENTATION AND INFERENCE

There are a variety of approaches to encode human expertise in expert systems, the most common one being *if/then* rules. Semantic networks, frames, and logical expressions are alternative paradigms of knowledge representation, although the majority of industrial expert systems use the rule-based paradigm. [For a discussion of the subject see (Gonzalez and Dankel, 1993).)

The three basic constituents of a rule-based expert system are: *rule base*, *working memory* and *rule interpreter*. The rule base is often partitioned into groups of rules, called *rule clusters*. Each rule cluster encodes the knowledge required to perform a certain *task* or a fraction of a task, usually referred to as a *subtask*. There may also be rules for internal control purposes—for example, to signal which rule cluster to select as holding potentially relevant knowledge at a given time. Collectively, these rules are referred to as the *control structure* of an expert system. Another class of rules, called *demons*, may be present; they are designed to function outside the control structure of the program for the purpose of enhancing its ability to respond quickly to the occurrence of an event requiring some immediate action. Demons address inefficiency issues that may arise from excessive control over the rule base (Cooper and Wogrin, 1988).

Working memory is a database holding input data, inferred hypotheses, and internal information about the program. In an on-line expert system with monitoring functions, for example, the state of working memory at any given time reflects changes occurring in the process being monitored as well as internal changes due to the reasoning process of the program itself.

The mechanism through which rules are selected to be fired is called the *rule interpreter*. It is based on a pattern matching algorithm whose main purpose is to associate at any given time the state of the system (input data, inferred hypotheses, etc.) with applicable rules from the rule base.

The inference engine of an expert system is in charge of manipulating the data presented to the system and arriving at a conclusion. In expert system technology the two most widely used reasoning techniques are forward chaining (forward reasoning) and backward chaining (backward reasoning). In forward chaining the system reasons forward from a set of known facts and tries to infer the conclusions or goals. Design of a complex system is a forward-chaining application where the expert system starts with the known requirements, investigates the very large array of possible arrangements, and makes a recommendation based on criteria specified by the user.

In backward chaining the system works backward from tentative conclusions or goals and attempts to find supporting evidence to verify their correctness. Solving a crime is a backward chaining application where the expert system identifies the possible suspects, looks for evidence indicating the guilt and innocence of each suspect, and makes a recommendation regarding which suspect is the most likely criminal. In many cases, backward-chaining systems are more efficient than true forward-chaining systems because they tend to reduce the search space and arrive at a conclusion more quickly.

Many advanced expert systems use a combination of both forward and backward chaining. Different search strategies, such as "depth first" or "breadth first", may be incorporated into either backward or forward chaining.

Data enter an expert system either through a user interface or from other programs such as databases, data acquisition systems, simulation packages, and so on, and form the initial facts (or assertions or evidence) available to the rules. From the input data, conclusions are drawn in a process called *inferencing*. The two basic inferencing strategies, *forward chaining* and *backward chaining*, are also referred to as *modus ponens* and *modus tollens*, respectively. These are crisp versions of GMP and GMT introduced in Chapter 5 for fuzzy systems.

In *modus ponens*, if we have the following rule

$$\text{if} \quad A \text{ is true}, \quad \text{then } B \text{ is true}$$

Hence if it is known that "*A* is true," then we can infer that "*B* is true." Most expert systems use this powerful inferencing strategy. In *modus tollens*, if we know that the rule is true and we also know that "*B* is false," then we can infer that "*A* is false." We often simply write *A* instead of "*A* is true" and NOT *A* instead of "*A* is false." The requirement for an exact match between input data and what is stated in a rule is relaxed in fuzzy expert systems where fuzzified versions of the basic inferencing strategies have been developed.

16.5 UNCERTAINTY MANAGEMENT

An important issue in expert systems is uncertainty management. Representing or combining uncertain data and drawing reliable inferences from it has been extensively investigated over the past two decades, and several theories of uncertainty have provided tools for solving uncertainty problems (Kruse et al., 1991). Probability theory and certainty factors have been frequently used; but also possibility (fuzzy) theory, Dempster–Shafer belief measures, Cohen's theory of endorsements, and subjective Bayesian methods are uncertainty management paradigms that have found important applications.

The oldest approach to uncertainty management has been the probabilistic approach which essentially ascribes probabilities to facts and rules and uses Bayes' rule and a rather large amount of statistical data to construct the various probabilities in the knowledge base (Kruse et al., 1991). A drawback of the probabilistic approach is its difficulty in distinguishing between *absence of belief* and *doubt* or to represent how *ignorance* is related to the lack of knowledge.

Certainty Factors

In order to overcome the difficulties of Bayesian probabilities (e.g., requiring a large volume of data or distinguishing between *absence of belief* and *doubt*), certainty factors may be used. In the certainty factor (*CF*) formalism, knowledge is expressed as a set of rules having the form

$$\text{if} \quad E, \quad \text{then } H \quad \text{with } CF(H|E)$$

where *E* is the *evidence*—that is, one or more facts known to support the *hypothesis H*, and $CF(H|E)$ is the certainty factor for the rule, a measure of belief in H, given that E has been observed. The value of CF ranges from -1 to $+1$. When $CF = -1$ the hypothesis H is totally denied, while at $CF = +1$, the hypothesis H is totally confirmed.

Certainty factors are obtained from *measures of belief*, $MB(H, E)$, and *measures of disbelief*, $MD(H, E)$, both taking values between 0 and 1. A measure of belief $MB(H, E)$ represents the degree to which the belief in hypothesis H is supported by observing evidence E, and it is computed by

$$MB(H, E) = \begin{cases} 1 & \text{if } p(H) = 1 \\ [p(H|E) - (H)]/[1 - p(H)] & \text{else} \end{cases} \quad (16.5\text{-}1)$$

A *measure of disbelief* $MD(H, E)$, on the other hand, represents the degree to which the disbelief in hypothesis H is supported by evidence E. It is computed by

$$MD(H, E) = \begin{cases} 1 & \text{if } p(H) = 1 \\ [p(H) - p(H|E)]/[1 - p(H)] & \text{else} \end{cases} \quad (16.5\text{-}2)$$

The certainty factor CF is defined in terms of $MB(H, E)$ and measure of disbelief $MD(H, E)$

$$CF = [MB(H, E) - MD(H, E)]/\{1 - \text{MIN}[MB(H, E), MD(H, E)]\} \quad (16.5\text{-}3)$$

During the execution of a knowledge base, multiple rules are typically capable of deriving the same hypothesis or conclusion, resulting in modification of the CF's involved. Consider, for example, a case where two different evidences E_1 and E_2 lead to the same hypothesis H. In such cases, certainty factors of the same or opposite signs can be combined directly by the following formulas (Gonzalez, 1993; Kruse et al., 1991):

Case 1. When both $CF(H|E_1)$ *AND* $CF(H|E_2) > 0$

$$CF(H|E_1, E_2) = CF(H|E_1) + CF(H|E_2) - CF(H|E_1) * CF(H|E_2)$$

Case 2. When $-1 < CF(H|E_1) * CF(H|E_2) \leq 0$

$$CF(H|E_1, E_2)$$
$$= [CF(H|E_1) + CF(H|E_2)]/\{1 - \text{MIN}[|CF(H|E_1)|, |CF(H|E_2)|]\}$$

Case 3. When $CF(H|E_1) * CF(H|E_2) = -1$

$$CF(H|E_1, E_2) = undefined$$

Case 4. When both $CF(H|E_1)$ *AND* $CF(H|E_2) < 0$

$$CF(H|E_1, E_2) = CF(H|E_1) + CF(H|E_2) + CF(H|E_1) * CF(H|E_2)$$

It has been assumed in the above equations that we have absolute confidence in the evidence of premises used to derive various values. In expert systems, often (but not always) a hypothesis from a rule is used as evidence for another rule and hence we should not actually have absolute confidence in the evidence, and the certainty factor approach does not materially contribute to the final results. An additional drawback of certainty factors seems to be the complexity of maintaining them. When, for example, new knowledge is added or deleted from the knowledge base, the certainty factors of existing knowledge change as well, making the maintenance of the system rather complicated. For these reasons and others, use of fuzzy set theory in the form of reasoning under uncertainty is more commonly encountered today.

16.6 STATE OF THE ART OF EXPERT SYSTEMS

The impact of expert systems technology has been felt in many areas of science, education, and industry. In the past decade a great many applications have been initiated, and many are now operational or in the prototype stage. (Uhrig, 1988; Hertz, 1988). The extent of the potential application of this technology is not yet known, because expert systems in the future may be used in completely new settings to solve quite different problems. However, the introduction of fuzzy rules has greatly enhanced the usefulness of expert systems.

It is very difficult to gain a true picture of just how widespread the use of expert systems has become. In many cases, organizations are using expert systems internally. Even the fact that they are used, let alone the details of the expert systems, are treated as proprietary for the simple reason that the company or organization wants to gain competitive advantage. By one analyst's estimate, about half of the companies listed in the Fortune 500 are developing expert systems (Coates, 1988). One automobile manufacturer is reportedly insisting that manufacturers supply diagnostic expert systems with the equipment they provide.

Expert systems may change the manner in which many organizations operate, and they could change the workplace in general. In large organizations such as government, big corporations, and associations, one expert predicts that 60–90% of all jobs are candidates for augmentation, displacement, or replacement by expert systems (Coates, 1988). Coates further predicts that by about the turn of the century the capabilities of expert systems will have grown to such a degree that their impact will be felt throughout most occupations and workplaces.

16.7   USE OF EXPERT SYSTEMS

Generally, but not always, problems that are amenable to a numerical solution should be solved using conventional computer programs. However, there are many situations in which expert systems offer unique advantages over conventional programs. Most applications of expert systems today can be classified into the following six categories: (1) monitoring systems, (2) control systems, (3) configuring systems, (4) planning systems, (5) scheduling systems, and (6) diagnostic systems.

Monitoring Systems

Monitoring systems are dedicated to data collection and analysis over a period of time. The collected values are compared against expected performance, and if discrepancies are identified the expert system generates recommendations and/or notifies the operator.

Control Systems

Control systems are monitoring systems in which action (e.g., opening a valve, adjusting a bias, turning on a heater, etc.) is taken as a result of the discrepancy identified by the monitoring system.

Configuring Systems

Configuring systems address problems in which a finite set of components is to be arranged in one of many possible patterns. The classical example in this category is XCON, an expert system used by a large computer manufacturer to configure its equipment in accordance with its own rules and the user specifications.

Scheduling and Planning Systems

Scheduling and planning expert systems coordinate the capabilities or components within an organization to optimize production and/or increase efficiency. The difference between planning and scheduling systems is that the components for a task are not always known in planning systems.

Diagnostic Systems

Diagnostic systems observe and analyze data and map the analysis results to a set of problems. Once the problems have been identified, the expert system usually recommends a solution based on facts in its knowledge base and on the other information it can acquire. Expert systems have been used to solve many different problems in a variety of fields. Some of these areas are listed in Table 16.1, which is intended to give a brief overview of the breadth of applications that has developed. One area in which there has been extensive efforts to utilize expert systems is the nuclear power field, many of which could affect safety and safety-related systems. The scope of these applications has been documented by Bernard and Washio (1989).

Table 16.1 Applications of expert systems

FIELD	*USE*
Design and engineering	Collecting and storing knowledge of best designers speeding the design process
Computer applications	Configuring equipment to user specifications Diagnosing problems with computer equipment
Manufacturing	Managing human and machine resources Facilitating factory automation
Finance	Decision support tools Providing tax and other business advice Processing loan and mortgage applications Analyzing financial risk
Science and medicine	Providing medical advice in hospitals Providing diagnostic assistance to medical personnel Patient monitoring
Geological applications	Advising regarding mineral deposit and oil locations Advising drillers regarding stuck bits
Training	Interface for computer-aided instruction Assisting in computer-based training

16.8   EXPERT SYSTEMS USED WITH NEURAL NETWORKS AND FUZZY SYSTEMS

Neural Network in the Knowledge Base of an Expert System

Neural networks, in spite of the extraordinary usefulness, have relatively limited capabilities. They are trained using available data, tested, and put into use. All they can do is recall an output when presented with an input consistent with the training data. They cannot reason, seek data from available databases to assist their operation, or provide an explanation of their outputs. They need a structured environment in which to operate, which can be provided in some cases by conventional software programming. However, recent experience indicates that usefulness of a neural network can be enhanced significantly if an expert system is used to provide this operating environment. Indeed, an expert system can retrain a neural network to adapt this hybrid system to new situations, or it can intermittently update the training of the neural network to adapt to changing situations. Some recent work indicates that expert systems can be used to provide explanations for why a neural network gives the output it does.

Perhaps the most direct combination of these two artificial intelligence technologies is the use of a neural network in the knowledge base of an expert system. This gives the expert system the ability to learn from data presented to it. The training may be on-line or performed during an initialization period. Multiple and/or modular neural networks may be incorporated into the knowledge base, and neural network outputs may be combined within the knowledge base. Control of the neural network is carried out by the inference engine in the same way that it seeks additional information from a database or initiates a logic reasoning step.

Fuzzy Rules in the Knowledge Base

One of the most popular methods of storing information in the knowledge base is through the use of *if/then* rules. Both the antecedent and the consequent or action of the rules may have multiple statements connected by conjunctions such as *AND* and/or *OR*. For simple systems, the rules can be relatively simple and straightforward. If the individual components of a system are independent and follow a "logic tree" structure, the rules proceed in a monotonic manner; that is, the inferencing process always proceeds forward. However, if the components are interconnected, the logic trees interact with the result that the rules become very long (more qualifying conditions connected by conjunctions), more complex, and more numerous. Hence, it is increasingly harder to prevent rules from conflicting with each other. Indeed, it has been the experience of many investigators that when the number of rules gets beyond about 200, it is virtually impossible to write a meaningful rule that does not conflict with previously written rules. This paralysis of the knowledge base for complex systems caused interest in expert systems to decline in middle to late 1980s. With the advent of fuzzy rules, based on fuzzy set technology, expert systems are again being introduced in high-technology systems. For instance, an autonomous navigation system using sensor signals to navigate between moving objects was almost abandoned when 450 rules did not provide a satisfactory system. However, the replacement of the navigation system's 450 rules with 15 fuzzy rules provided a system with outstanding performance (Pin, 1992). Comparable results in the reduction in size of expert system knowledge bases by the introduction of fuzzy rules have been reported by many investigators (Terano et al., 1994).

It is this use of fuzzy rules in an expert system, a combination that is often called "fuzzy expert systems," that has reawakened interest in expert systems. In a traditional expert system, the number of rules necessary to unambiguously define a situation tended to grow in an exponential-like fashion as the complexity of the system increased. For complex problems that were amenable to monotonic reasoning (i.e., the reasoning proceeded forward directly toward a goal with a reversal), a large number of rules simply meant a slow and cumbersome process. For complex problems that involved searching many paths with reversals and many-to-many mappings, use of a rule-based knowledge base was simply not feasible. The ability of fuzzy rules to drastically reduce the number of rules has been the secret of success in using expert systems in most complex situations.

Even in fuzzy expert systems, a major effort must be made to minimize the number of rules without deteriorating the operation. Consider the case where there are three inputs and one output that utilize five membership functions each. This could lead to 5^4 (625) fuzzy rules. Fortunately, in most situations, all the rules do not contribute equally to the solution. Many methods are available to reduce the number of rules involved. One way would be to use a genetic algorithm optimization. Usually, however, a statistic-based processor can analyze the situation and give the contribution of each rule to the solution. Then the user can set the threshold for including rules at a level consistent with the specifications for precision and speed.

16.9 POTENTIAL IMPLEMENTATION ISSUES FOR EXPERT SYSTEMS

Potential problems in implementing expert systems in complex engineering systems can be projected from past experience with the introduction of new and innovative systems.

General Implementation Issues

1. Most complex engineering systems, as presently built and operated, are considered by the operators to be safe enough. With the possible exception of the severe accidents (i.e., Chernobyl, Bopough. etc.),

expert systems are not perceived to be needed to provide additional safety functions.

2. Introduction and use of an expert system must not introduce a new operational or safety problem. A thorough analysis of what could go wrong and what effect it could have on the plant and its safety system would be essential before implementation. The ultimate criterion in judging any new system is whether its failure can, in any way, lead to a challenge of existing plant protection systems.

Implementation Issues That Need To Be Addressed

A number of issues regarding the implementation of expert systems in complex engineering systems need to be addressed. These include, but are not limited to, the following:

1. *Quantitative and Objective Performance Guidelines for Expert Systems.* The primary concern about the introduction of any new system into a complex engineering system appear to be the impact it can have on the safety system when something goes wrong. The ultimate question in judging any new system must be "Can the failure of the expert system lead to a challenge of the existing safety systems?" Above all, replacement of an existing system with an expert system must not introduce new unresolved issues (i.e., new unreviewed safety hazards).

Introduction of a new system must not lead to confusion of operators or other plant personnel. New tools may be needed to evaluate and measure the performance of expert systems and the impact of these systems on human performance. Objective criteria that are quantitative in nature are needed.

2. *Validation and Verification* ($V\&V$). In conventional software programming, verification and validation have well-established meanings; verification is a determination that software has been developed in a formally correct manner in accordance with a specified software engineering methodology; validation means demonstrating that the completed program performs the functions in the requirements specification and is usable for the intended purposes. However, expert systems go beyond the procedures of conventional software engineering, and a modularized, top-down, hierarchically decomposed design that makes conventional V & V possible may not be achievable. Expert systems, especially those operating under uncertainty or with incomplete data, may have so many states as to make exhaustive testing unfeasible. Hence, new approaches to V & V are needed for expert systems.

The inference engine may be considered simply as another digital computer program, and its V & V can be dealt with in the same way as with other digital computer programs (e.g., IEEE-5.3.2.1). The real problem is the adequacy of the knowledge base—that is, the qualifications of the expert whose expertise is incorporated into the knowledge base, the method used for acquisition of this expertise, and the method used to represent this expertise in the knowledge base. Except for relatively simple expert systems, exhaustive testing of the expert system or the knowledge base to cover all likely situations may not be adequate or feasible.

Generally, as a matter of policy, V & V should always be carried out by a group completely independent of the group that developed the expert system. Because V & V in expert systems is so intimately related to the design, true independence may extremely difficult to achieve. To the extent possible, the independence of the group that does V & V should be ensured by quality assurance procedures and organization policy.

3. *Human Factors.* A primary human factors concern is that the expert system should present information to the user in a way that is comprehensible and understandable. Information must mesh well with the perspectives used by the human, and the way in which the information is displayed should correspond to the user's mental model of the plant. The user should be able to understand the expert system's behavior.

Another concern is user reaction to the expert system. Will they like the system and accept it? Will they be comfortable with an expert system and use it when needed? Will they believe that the system will work and that it is useful? Above all, will they trust and have confidence in the information presented by the expert system? On the other hand, the user could become too dependent upon the guidance of an expert system and ignore other indications that might not agree with the conclusion of an expert system.

The function allocation and division of responsibility between the expert system and the human is another important issue. Humans should be assigned only those functions that they are most capable of performing and that utilize their abilities. Expert systems should relieve some of the physical and cognitive workload on users to avoid overload of the operators. The system should make human jobs more efficient. The expert system should be integrated with the other hardware, software, and tools in the user's work environment. Clearly, users should be involved in the design and analysis of the expert system and its interface with users.

REFERENCES

Bernard J., and Washio, T., *Expert Systems in the Nuclear Power Reactors*, ANS Publishing, La Grange Park, IL, 1989.

Coates, J., Artificial Intelligence: Observations on Applications and Control, *Computer Security Journal*, Vol. 5, No. 1, 1988.

Cooper, T. A., and Wogrin, N., *Rule-based Programming with OPS5*, Morgan Kaufmann, San Mateo, CA, 1988.

Feigenbaum, E. P., McCorduck, P., and Nii, H. P., *The Rise of the Expert Company*, Times Books, New York, 1988.

Gonzalez, A. J., and Dankel, D. D., *The Engineering of Knowledge-Based Systems*, Prentice-Hall, Englewood Cliffs, NJ, 1993.

Hertz, D. B., Boeing Has High Hopes for AI, *AI Week*, University of Miami, Coral Gables, FL, July 1, 1988.

Kaplan, S., Frank, M. S., Bley, D. C., and Lindsay, D. G., Outline of COPILOT, Expert System for Reactor Operational Assistance Using a Bayesian Diagnostic Module, *Proceedings of the International Post SMiRT-9 Seminar on Accident Sequence Modeling: Human Actions, System Response Intelligent Decision Support*, Munich, August 24–25, 1987.

Kruse, R., Schwecke, E., and Heinsohn, J., *Uncertainty and Vagueness in Knowledge Based Systems*, Springer-Verlag, Berlin, 1991.

Pin, F., Private communication, Oak Ridge, TN, 1992.

Ricker, M., An Evaluation of Expert System Development Tools, *Expert Systems*, Vol. 3, No. 3, 1986.

Sackett J. I., ed., *Proceedings of the ANS Topical Meeting on Artificial Intelligence and Other Innovative Computer Applications in the Nuclear Industry*, Snowbird, UT, August 31–September 2, 1987.

Swartout, W. , and Smoliar, S., On Making Expert Systems More Like Experts, *Expert Systems*, Vol. 4, No. 3, 1987.

Terano, T., Asai, K., and Sugeno, M., *Applied Fuzzy Systems*, Academic Press, Boston, 1994.

Uhrig, R. E., Applications of Artificial Intelligence in Nuclear Power Plants, *POWER Magazine*, McGraw-Hill Energy Systems, June 1988.

Van Horn, M., *Understanding Expert Systems*, Bantam Books, New York, 1986.

Waterman, D., How do Expert Systems Differ from Conventional Programs, *Expert Systems* Vol. 3, No. 1, 1986.

PROBLEMS

1. Discuss how expert systems can be verified and validated in the sense that software undergo verification and validation. Consider the different requirements for the inference engine and the knowledge base.

2. It is well known that the use of fuzzy rules has revived the use of expert systems. Explain what you believe to be responsible for this resurgence of interest of expert systems. Is the reason the same for all types of expert systems? If not, why?

3. Discuss the legal ramifications of using expert systems. If an expert system (or a neural network) fails in service and causes damages, who is responsible? The company selling the expert system? the user? The person who wrote the software for the expert system? The expert who supplied the information to the expert system? All of the above?

4. Discuss the relative advantages and disadvantages of having a deterministic model of a system in the knowledge base compared to having a neural network model in the knowledge base.

17

GENETIC ALGORITHMS

17.1 INTRODUCTION

Genetic algorithms as a field of study was initiated and developed in the early 1970s by John Holland (Holland, 1975, 1992) and his students, but its applications to real-world practical problems was almost two decades in developing. In one way or another, the primary purpose of using genetic algorithms is optimization. The specific nature of the problem or system to which optimization is being applied will determine the approach, the type of genetic algorithms used, and especially the evaluation or fitness function. There is no guarantee that a genetic algorithm will give an optimal solution or arrangement, only that the solution will be near-optimal in the light of the specific fitness function used in the evaluation of the many possible solutions generated.

In this chapter, the terms *chromosomes* and *genes* may appear to be used synonymously, but they are not. The meaning of these terms as used here is the same as that used by Goldberg (1989). Chromosomes are composed of genes which define the characteristics of the chromosomes and may take on several values called *alleles*. The position of a *gene* (its *locus*) is identified separately from the gene's function. Hence we can have a particular gene with a locus of position 12 whose allele value is brown. Generally, the *strings* of artificial genetic systems are analogous to chromosomes in biological systems and are often called chromosomes.

17.2 BASIC CONCEPTS OF GENETIC ALGORITHMS

Genetic algorithms mimic some of the processes of natural evolution. In doing so, some of the inherent features of evolution are utilized in fields far beyond genetics. However, there is no necessity that genetic algorithms as we use them mimic in detail the behavior of the evolutionary process. Indeed, users are free to utilize those features that are useful and discard aspects that seem unimportant in their applications. Since normal evolution processes are quite slow, biased reproduction, based on an aggressive "survival of the fittest" philosophy, is used to speed up the evaluation process.

The mechanisms that induces evolution are not well understood, but the features of evolution have been investigated thoroughly. First, evolution takes place in chromosomes, the genetic units that encode the features and structure of living creatures. The specific descriptive features of a living creature is determined by the chromosomes of the previous generation, and evolution influences only these chromosomes, not the living creature from which they came. Since evolution is limited to chromosomes, living creatures do not evolve during their lifetime; their features, which are presumably set at the time of conception, are different from the previous generation only because the chromosomes of their parents changed through evolution.

Evolution, Natural Selection, and the Gene Pool

Natural selection is a process by which nature causes those chromosomes that encode better characteristics (by some criteria) to reproduce more often than those that encode poorer characteristics. Natural selection is the process that causes genetic algorithms to produce near-optimal solutions when the selected chromosome is decoded. This process involves creation of many chromosomes by *reproduction, mating (crossover), mutation,* and the survival of the chromosomes with the better characteristics. Successive generations of chromosomes improve in quality, provided that the criteria used for survival is appropriate. This process is often referred to as *Darwinian natural selection* or the *survival of the fittest*. In nature, this process of evolution occurs over many years, even hundreds or thousands of years. In the computer, the representations of chromosomes can undergo literally thousands of generations of change in a few seconds.

Historically, the characteristics of the chromosomes in genetic algorithms have been represented by 0s and 1s. All of Holland's work used this representation, and we will use it here. However, chromosomes can be represented by real numbers, permutations of elements, a list of rules, or other symbols. Binary representation is still the most common representation because the behavior of bit strings are more familiar and better understood.

Evolution takes place through the process of reproduction, which involves mutations and recombination after mixing of the parent's chromosomes to produce a creature that may have entirely different characteristics from the

String A   **10001101100|010101110**   **10001101100|011100011**

String B   **01100011101|011100011**   **01100011101|010101110**

      Original Pair of Strings   Pair of strings after crossover
                                            at one location

(a) A Pair of Strings with Crossover at One Location.

String A   **10001101100|01010|1110**   **10001101100|01110|1110**

String B   **01100011101|01110|0011**   **01100011101|01010|0011**

      Original Pair of Strings   Pair of strings after crossover
                                            at two locations

(b) A Pair of Strings with Crossover at Two Locations.

                           *   *

String A   **10001101100|010101110**   **10001101000|011100011**

String B   **01100011101|011100011**   **01100011101|010101110**

      Original Pair of Strings   Pair of strings after crossover
                                            at one location and
                                          mutation at *

(c) A Pair of Strings with Crossover at one Location and a Mutation at *.

Figure 17.1 Demonstration of crossover at one and two locations and *Mutation.*

previous generation. The chromosomes of the two parents are mixed by a process called "crossover," in which two new chromosomes are produced, each having some of the characteristics of the two parents. The two parent chromosomes split at some point, and one part of one parent chromosome is exchanged for the corresponding part of the other parent chromosome. The location of the crossover point at which the parents' chromosomes divide is apparently a uniform random process. If one of the new chromosomes obtains 75% of its characteristics from one parent and 25% from the other, the second new chromosome gets 25% form the first parent and 75% from the other. This process is illustrated schematically in Figure 17.1, where the chromosome is illustrated as a string of 0s and 1s.

Both of the resultant chromosomes go into the gene pool (where all the alleles reside), where they either replace poorer-quality chromosomes or are discarded. Once a chromosome is discarded, its unique features (good or bad) are lost forever. It is only through the processes of mutation (described below) that such a chromosome might possibly be recreated.

Mutation is a process by which a single component of a chromosome is changed randomly. It occurs in only a very small fraction (typically less than a fraction of a percent) of the chromosomes. It represents an abrupt change in the nature of the chromosome and influences all subsequent generations of chromosomes containing this mutated component. Of course, if this mutation results in a poorer-quality chromosome, it will be discarded and lost from the gene pool.

Each population has a gene pool consisting of a large number of chromosomes generated by the process of natural selection. The choice of which two chromosomes are mated and subject to the crossover process is somewhat random, but the fitter chromosomes are more likely to be selected first. New chromosomes are constantly being reproduced by the mating process described above, and those with better characteristics are retained while those with poorer characteristics are discarded. Generally, but not always, the mixing of chromosomes with quite different characteristics produce better chromosomes. As the process proceeds, the average quality of the gene pool improves because the poorer-quality chromosomes are discarded.

Objective Function–Fitness Function

The function on which an optimization algorithm operates—that is, seeking its maximum or minimum—is called the *objective function*. In neural networks, the objective function to be minimized is the mean square error over the entire training set. In genetic algorithms, the "fitness" is the quantity that determines the quality of a chromosome, from which a determination can be made as to whether it is better or worse than other chromosomes in the gene pool. The fitness is evaluated by a "fitness function" that must be established for each specific problem. This fitness function is chosen so that its maximum value is the desired value of the quantity to be optimized. Its importance cannot be overemphasized, because it is the only connection between the genetic algorithm and the problem in the real world. A fitness function must reward the desired behavior; otherwise the genetic algorithm may solve the wrong problem. Fitness functions should be informative and have regularities. However, they need not be low-dimensional, continuous, differentiable, or unimodal.

17.3   BINARY AND REAL-VALUE REPRESENTATIONS OF CHROMOSOMES

Weight representation in the chromosome has used both binary and real-value encodings, with binary being the more prevalent method. Binary coding of the weights can be implemented using either an ordinary binary representation of real-value weights or a corresponding Gray-scale binary encoding.

Table 17.1   Comparison of hamming distances in binary and gray coding

Decimal Number	Binary Coding	Hamming Distance	Gray Coding	Hamming Distance
0	0000		0000	
1	0001	1	0001	1
2	0010	2	0011	1
3	0011	1	0010	1
4	0100	3	0110	1
5	0101	1	0111	1
6	0110	2	0101	1
7	0111	1	0100	1
8	1000	4	1100	1
9	1001	1	1101	1
10	1010	2	1111	1
11	1011	1	1110	1
12	1100	3	1010	1
13	1101	1	1011	1
14	1110	2	1001	1
15	1111	1	1000	1
16	10000	5	11000	1

Binary and Gray-Scale[1] Representations

Gray scaling has the characteristic that the Hamming distance (the number of binary digits that change between successive decimal numbers) is always 1 compared to binary coding where the hamming distance is 4 in a four-bit representation (i.e., all bits change) as a decimal number changes from 7 to 8 (see Table 17.1). Such so-called Hamming cliffs can make genetic algorithms less stable. The reason is that change of a single digit due to mutation will usually cause a smaller change if the Hamming distance is small. In the binary code, half the changes have a Hamming value of 2 or more, whereas all changes in the gray scale have a Hamming distance of 1. Empirical studies (Caruana and Schaffer, 1988) on algorithms indicate that gray coding improves the process for some functions and performs no worse than binary coding in all cases.

There are several gray codings for any number, but the most commonly used Gray coding is the binary-reflected gray code. One simple scheme for

---

[1] Gray encoding or gray scale refers to the use of a binary code developed by F. Gray (1953), U. S. Patent #2-632-058 issued March 17, 1953.

generating such a gray code sequence is "start with all bits set equal to zero and then successively flip the rightmost bit that produces a new string." Table 17.1 compares binary coding with Gray coding for decimal numbers from 0 to 16. (At 16, the binary and gray codes must go to 5 digits since all possible combinations of 4 digits have been exhausted.) Interestingly, the 4-digit combinations used for representation of numbers from 0 to 15 are exactly the same for the binary and gray encodings, except that a specific combination of 4 digits represent different decimal numbers in the two codes. For instance, 0111 represents 7 in the binary code and 5 in the gray code.

Real-Valued Representations of Chromosomes

In real-valued encodings, the network weights are encoded as lists of real-valued weights. Crossover occurs across whole weights instead of occurring across bit strings representing weights. In mutations, incremental changes (plus or minus) are introduced into the real values. Davis (1991) discusses the advantages and disadvantages of real-valued encodings and argues that such encodings can yield superior results. Perhaps the main disadvantages of real-valued encodings are that robust parameters are not known and that specialized genetic algorithms may need to be tailored for each problem. Both of these disadvantages should be lessened as we gain more experience with genetic algorithms using real-valued representations of chromosomes.

17.4 IMPLEMENTATION OF GENETIC ALGORITHM OPTIMIZATION

Living creatures are probably the most complex systems in the universe. Hence, if evolution that involves reproduction with crossover, mutation, and natural selection can result in improvement of the species, it seems reasonable that the process will work to optimize other complex systems. Indeed, this has been the case, and the range of applications where genetic algorithms can optimize a process or system is limited only by the ingenuity of the user. Applications have now reached the point where many users are no longer versed in the details of how genetic algorithms operate; rather, they are concerned only with how the powerful capability of genetic algorithms, as presented in commercial software, can be utilized to optimize their particular problem or system.

Genetic algorithms do not rely on any analytical properties of the function to be optimized (such as the existence of a derivative). They are well suited to a wide class of problems, including optimization over parameter sets as well as global optimization of functions. However, before the genetic algorithm process can be carried out, two steps are necessary: (1) encode the variable to be optimized into a string of binary bits (or other appropriate representations) and (2) create an appropriate fitness function.

Bit-String Representation

It is necessary to structure bit strings to represent practical problems before undertaking a search for optimal conditions. Individual bit strings are organized to form an initial population of chromosomes. They can be generated randomly, but its is advantageous if the initial population of chromosomes can be somewhat related to the nature of the system being optimized. Genetic algorithms guide the string population to propagate from generation to generation to improve the survival probability of the entire population.

There are two approaches to determining which chromosomes to delete after a cycle of reproduction, crossover, and mutation. These are (a) the *generational approach*, where the entire population is replaced after each cycle, and (b) the *steady state approach*, where the members of both the old and the new gene pools with the highest fitness factor are retained. The generational approach tends to speed convergence, perhaps at the expense of diversity in the gene pool. The steady-state approach tends to produce somewhat better performance by retaining the best-performing bit strings, but the best solution may be missed because new genes that are the precursors of high-performing genes may be eliminated prematurely.

The convergence criteria for stopping the genetic algorithm is somewhat arbitrary. Generally, genetic algorithms converge rapidly, typically in a few hundred cycles or less. Stability in the value of the average fitness function from one generation to the next is generally the most appropriate criterion.

A related issue is the reproduction process where there are several options for selecting which genes should be reproduced. The two most common methods are *proportional selection* and *rank-based selection*. In proportional selection (discussed below) the number of times the gene can be reproduced is proportional to its fitness function. This technique, which was used by Holland, involves selecting the top performers and allowing multiple reproductions of the best performers. A sampling algorithm is usually used to allocate the number of reproductions to the various genes. The proportional method sometimes tends to give undue emphasis to superior performing chromosomes whose fitness functions may be 10 times the average fitness function. If such a super chromosome is reproduced 10 times in a pool of 50 genes, it would clearly distort the gene pool. In the rank-based selection process, each gene is typically reproduced only once, although there are variations of this algorithm that allow multiple reproduction of a single gene. Rank-based selection tends to converge slowly with less premature convergence and better diversity of the gene pool.

Reproduction

In the implementation of genetic algorithms, the reproduction process consists in the copying of individual strings according to the priority established by their objective function or fitness function f. (We will use the latter term in this chapter.) Copying strings according to their fitness function values means that candidates with higher fitness values have a greater probability of contributing one or more offsprings in the next generation. This is the "proportional" selection method discussed above. The selection probability for an individual string i (the ith string in the population) may be defined as

$$p_i = \frac{f_i}{\Sigma f_i} \tag{17.4-1}$$

where f_i is the fitness value of the ith individual in the population N.

The mating pool of the next generation is selected according to the probability p_i. Once an individual has been selected for reproduction, it is then entered in the mating pool, for further genetic operation action. If there is no overlapping between populations (i.e., the population size remains constant when a new generation replaces the old or parent generation), the expected number of reproductions of ith individual string is

$$n_i = N \cdot p_i = N \cdot \frac{f_i}{\Sigma f_i} = \frac{f_i}{\Sigma \left(\dfrac{f_i}{N} \right)} = \frac{f_i}{\bar{f}} \tag{17.4-2}$$

where $\bar{f}$ is the average fitness of the population. This agrees with our earlier thesis that the best chains are more likely to be reproduced.

17.5   FITNESS FUNCTIONS

The fitness function of a genetic algorithm can be designed to perform different search tasks of optimization. The value of fitness function is the quantity to guide the reproduction process in the genetic algorithms for creating the next generation. Usually, the fitness function is designed in a way that its values are all positive, and the higher the value of the fitness function, the better the performance of the individual bit string in the population. A higher value of the fitness function also means that the individual bit string gets the better chance to be selected for production of the next generation. These guidelines indicate that the fitness function is a function of the number of inputs selected (the fewer the number, the greater the fitness) and the network training error (the smaller the value, the greater the fitness).

To illustrate the role of the fitness function and the general process involved in using genetic algorithms for optimization, two examples are provided. In the first, Example 17.1, a quadratic function $y(x) = 1 - x/10 + x^2/200$ is to be optimized for its maximum or minimum value. In Example 17.2, an actual problem is used to illustrate the selection of the fitness function.

Example 17.1   Simple Hand-Calculated Example of the Genetic Algorithm Process. Let us assume that we want to optimize (i.e., find the minimum or maximum value of a function)

$$y(x) = 1 - \frac{1}{10}x + \frac{1}{200}x^2 \qquad (E17.1\text{-}1)$$

using a genetic algorithm process. Of course, we can readily determine that this function has a minimal value of 0.5 at $x = 10$ by other means. Please understand that this problem has been contrived to demonstrate the methodology of genetic algorithms. Because of the small size and small number of binary strings, the behavior of the process is not representative of that in real world genetic algorithms.

We can create the initial population by flipping a coin to select our mating pool, which in this simple example consists of five 5-bit random binary sequences. These are listed in Table 17.2 as "String x," and their binary values converted to the base 10 are listed as "Value x." The function $y_i(x)$ is then evaluated using the above formula. Clearly, $y(x)$ is related to the fitness function since it represents the quantity we want to optimize. In this case, the optimal value is a minimum, and the genetic algorithm process gives the maximum value of the fitness function. Hence, it seems reasonable to let the fitness function be the reciprocal of $y(x)$; that is, $f_i(x) = 1/y_i(x)$. Generally, however, we do not have a formula of the quantity to be optimized, and the fitness function has to be selected on the basis of data available and the nature of the problem involved (see Example 17.2). Note that the last two columns in Table 17.2 are the selection probability for an individual string and the expected number of reproductions of the ith individual string as given by equations (17.4-1) and (17.4-2) respectively.

Table 17.2   Hand calculations for a genetic algorithm

	String x	Value x	$y_i(x)$	$f_i(x)$	$f_i/\sum f_i$	$f_i/\bar{f}$
v_1	10111	23	1.345	0.743	0.101	0.504
v_2	01100	12	0.520	1.923	0.261	1.304
v_3	10100	20	1.000	1.000	0.136	0.677
v_4	00110	6	0.580	1.732	0.235	1.173
v_5	01001	9	0.505	1.980	0.267	1.342
Sum				7.378	1.000	5.000
Avg				1.476	0.200	1.000
Max				1.980	0.267	1.342

Now, let us carry out *reproduction, crossover,* and *mutation* operations on string x. This is shown in Table 17.3. The strings are randomly mated with other strings at the indicated crossover points to produce new strings. Furthermore, mutation takes place in the middle digit of value v_1 (indicated by an asterisk in Table 17.3), where a 1 changes to a 0 after crossover has taken place.

This mutation changes the fitness function of v_1 from a 1.000 to a 1.471, a 47.1% increase. Table 17.3 shows the results of reproduction, crossover, and mutation on the group of five chromosome strings. Crossover between each mated pair of genes in the pool produces two new chromosomes which are given in the column headed "new population." These new genes are evaluated to give their fitness functions listed under the column labeled $f_i(x)$. The double asterisk indicates the five genes with the highest fitness functions that are to be reproduced in the next generation if rank-based selection is used. Of these five genes, three have much higher values of $f_i(x)$. Hence, if proportional selection is used, the two genes with lower values of $f_i(x)$ would probably be replaced with duplicates of the two genes with the highest values of $f_i(x)$.

It is interesting to note that the average fitness function of the original five strings is 1.476 (see Table 17.2) compared to an average fitness function of 1.459 (see Table 17.3), a 1.1% decrease for the 10 new chromosomes produced by crossover and mutation. However, the fitness function of the five new chromosomes selected for reproduction is 1.821, an increase of 23% over the original five chromosomes. This increase is unusually high for one cycle due to the small length of the strings, which tends to increase the impact of even a change in a single digit.

An examination of the new set of strings indicates that new v_5 could be a "super" string that could dominate future cycles of reproduction. This is not desirable, particularly in the early part of the optimization process, because it could lead to a premature selection of an optimum which was not a true optimum. The concern here is prematurily limiting the gene pool which could cause the process to select a local minimum or maximum rather than a global value. There are a number of techniques available to avoid this problem which the reader can find in literature that specializes in genetic algorithms (Holland, 1975, 1992: Goldberg, 1989; Davis, 1991).

The generation, crossover, and mutation processes continue until there is no significant change in the average value of the fitness functions. At that point, it is necessary to decode the string to identify the optimal value, a minimum in the function in this case. Let us use the largest value of fitness function (2.000) to represent the optimal case. Because of the reciprocal relationship, the optimal value of $y(x)$ is 0.500. If we substitute this value into the equation for y, we get

$$y(x) = 1 - \frac{1}{10}x + \frac{1}{200}x^2 = 0.5 \qquad \text{(E17.1-2)}$$

Table 17.3 Reproduction, crossover, and mutation on Strings v_i

	MATING POOL (REPRODUCTION)	MATE RANDOMLY SELECTED	CROSSOVER SITE RANDOMLY SELECTED	NEW POPULATION	VALUES			
					x_i	$y_i(x)$	$f(x)$	$f_i(x)$ **
v_1	101|11	2	3	10000 *	16	0.680	1.471	
				01111	15	0.625	1.600**	1.600
v_2	01|100	5	2	01001	9	0.505	1.980**	1.980
				01100	12	0.520	1.923**	1.923
v_3	1010|0	1	4	10101	21	1.105	0.905	
				10110	22	1.220	0.820	
v_4	0|0110	3	1	00100	4	0.680	1.471	
				10110	22	1.220	0.820	
v_5	010|01	4	3	01010	10	0.500	2.000**	2.000
				00101	5	0.625	1.600**	1.600
					Sum		14.589	9.103
					Avg		1.459	1.821
					Max		2.000	2.000

* MUTATION

**GENES SELECTED FOR NEW GENE POOL

INFLUENCE OF MUTATION UPON CHARACTERISTICS OF v_1

		x_i	$y_i(x)$	$f(x)$
BEFORE MUTATION	10100	20	1.000	1.000
AFTER MUTATION	10000	16	0.680	1.471

549

We can then solve the quadratic equation (E17.1-2) for the value of $x = 10$ as the location of the optimal value, which is then calculated to be 0.5.

Clearly, this example was contrived to give good results with only a single cycle. The short length of the bit string and the small size of the gene pool accentuate the effect of the processes involved. In a practical problem, the typical gene pool may have 50 to 200 chromosomes and go through hundreds of cycles. However, pools of over 50,000 chromosomes and tens of thousands of cycles have been used. A large population gives more diversity and better final solutions, but longer computational times are involved. Pools of less than 30 chromosomes are subject to premature convergence because stochastic effects tend to dominate the behavior of the genetic algorithm. □

Example 17.2 Evolution of a Fitness Function.[2] In this example, a large neural network with 25 inputs (which are instantaneous values of 25 different measured parameters) and 8 outputs (representing 7 different transients and a no transient state) is used to diagnose transients in a nuclear power plant. Every half-second, the 25 measured values are applied to the neural network whose output indicates almost instantaneously which transient is occurring or that there is no transient. The neural network is trained on transients generated in a full scope, high fidelity nuclear power plant simulator. A complex recurrent backpropagation neural network was needed to model the plant dynamic behavior because of the complex interrelations between the variables.

A sensitivity analysis as described in Section 8.5 was used to determine the most important inputs (typically 4 to 6 inputs) needed for the detection of each specific transient. This allowed the use of "modular" neural networks,[3] small backpropagation networks without recurrent connections with only a few inputs and a single output for each transient. Subsequent tests indicated that the modular neural networks were equally as effective in detecting transients as the large master neural network with 25 inputs and 8 outputs. The problem was that the master neural network had to be created before the sensitivity analysis could be used to determine the most important inputs for the "modular" neural networks.

Genetic algorithms were selected as an alternate method of determining the most important (optimal) variables for the modular networks without having to create and train the master neural network. The fitness function needs to be defined to guide the search for the best combination of inputs for the individual modular networks. The fitness function may have different forms for different optimal search tasks. It needs to be defined to guide the

---

[2] This example was developed by Zhichao Guo in his Ph.D. dissertation in Nuclear Engineering entitled "Nuclear Power Plant Diagnostics and Thermal Performance Studies using Neural Networks and Genetic Algorithms," University of Tennessee Library, Knoxville, TN, 1992.

[3] The term "modular" as used here does not refer to modular neural networks as described in Section 8.8.

search for the best combination of inputs for different modular networks. The criterion chosen was that the fitness function should be designed to (a) select fewer variables while making the training error smaller for a fixed number of training cycles and (b) penalize the selection which chooses more than necessary inputs and/or makes training error larger. This can be expressed in equation form as

$$\text{fitness} = \frac{c_1 \cdot (\text{total number of inputs})}{\text{number of inputs selected}} + \frac{c_2}{\text{network training error}} \quad \text{(E17.2-1)}$$

where c_1 and c_2 are constants based on the conditions of the problem. Tests carried out using the fitness function of equation (E17.2-1) failed because the resultant neural network training error was too large to be accepted. The fitness function was the sum of two terms which was dominated by the first term associated with a small number of inputs without regard for the influence of the training error.

This experience led to the proposal of a second trial fitness function of the form

$$\text{fitness} = (1 - e^{-(x-1)^{1.5}}) \cdot e^{-0.01 y^{1.5}} \quad \text{(E17.2-2)}$$

where

$$x = \frac{\text{total number of inputs}}{\text{total number of inputs selected}}$$

and

$$y = \text{network training error } (\%)$$

Equation (E17.2-2) is the product of the two exponential terms. The first term reflects the rewarding and penalizing for the number of inputs. Its value is exponentially decreasing with increasing number of inputs. The second term reflects the rewarding for a small training error and penalizing of large errors. The numerical coefficients were selected after experimentation involving the expected minimum number of inputs and expected smallest value of error. Tests indicated that this fitness function gave improved results but still had difficulties. For instance, reduced training error continuously improved the fitness function until an error of about 4% was reached, after which further reductions did not change the fitness function. Furthermore, strings which exhibited large fitness values in the early stages tended to dominate the next generation gene pool. In the case of the fitness function of equation (E17.2-2), one specific string provided 14 of the 20 strings in the pool.

The experience with the second fitness function led to the proposal of a third fitness function of the form

$$\text{fitness} = \left(1 - e^{-(x-1)^{0.15(x+1)}}\right) \cdot e^{-0.01 y^{0.7(x+1)^{1/3}}} \qquad \text{(E17.2-3)}$$

where, x and y are defined as in the previous fitness function. This fitness function was found to provide the appropriate influence of the number of inputs and the training error without the undue influence of early strings having large fitness functions.

Subsequent comparison of the input variables selected for the "modular" neural networks by sensitivity analysis and by genetic algorithms showed that the two most important inputs for each of the seven transients and the normal states were almost always the same. Beyond the second most important variable, there were a number of inconsistencies. However, tests showed that the training errors after a prescribed number of cycles for the networks selected by the two methods were substantially the same and that the networks performed equally well.

It is seen that this fitness function was arrived at by a series of "trials and errors." Clearly, experience in working with fitness functions gives insight into the form of the fitness function. However, if specific information that is useful in forming the fitness function is available, it should be used. □

17.6 APPLICATION OF GENETIC ALGORITHMS TO NEURAL NETWORKS

A number of researchers have tried to connect the genetic algorithms with neural networks in recent years. Whitely and co-workers (Whitely and Bogart, 1989, 1990; Whitely and Starkwerther, 1990) used genetic algorithms to guide a backpropagation based neural network in finding the necessary connections instead of full connections in the GENITOR II software in order to enhance the speed of training. They also used this software to optimize small networks with the result that the resultant networks learned much faster and much more consistently than fully connected networks. Koza (1990) used genetic algorithms to guide search for the time-optimal "bang-bang" control strategy for the cart-centering problem, a version of the broom balancing problem, with the additional constraint that the cart be located at a specific location during operation, by genetically breeding populations on control strategy. Maricic and Nikolov (1990) used the neural network designer GENNET to find the most appropriate network architecture for solving a given problem. In GENNET, a genetic algorithm block is responsible for generating population architectures, which are used to create a set of stand-alone backpropagation networks. Interactions between genetic algorithms and neural networks generate the best network architecture. Garis

(1990) used a genetic algorithm to train modular neural networks by finding the proper weights for the full connections. Genetic algorithms have been used to guide the design of neural control circuits, which combine the neural modules to form functional hierarchies. Muselli and Ridella (1990) proposed a combination method of genetic algorithms and simulated annealing to generate and choose the set of points in the network connection weight space to speed up reliability and convergence.

Combining Neural Networks and Genetic Algorithms[4]

Genetic algorithms typically encode the parameters of artificial neural networks as a string or list of the network's properties. The algorithm requires that there be a large population of these lists or strings (chromosomes) representing many possible parameter sets for the given network. The utilization of genetic algorithms for optimization lends itself easily to parallel computers. Advantages of using parallel techniques include the ability to search the entire weight space versus localized search in the weight space via a gradient descent technique. This global aspect of the search helps avoid local minima which can occur with other gradient descent techniques. Combined genetic algorithm–neural network technology (sometimes called GANN) have the ability to locate the neighborhood of an optimal solution quicker than backpropagation methods due to its global search strategy, but once in the neighborhood of the optimal solution, the GANN algorithm tends to converge to the optimal solution slower than backpropagation methods. This is because the final convergence of the genetic algorithm from the optimal neighborhood to the optimal solution is controlled mainly by the mutation operators. Drawbacks of the GANN technology are the large amount of memory required to maintain a viable population of chromosomes for a given network and some question as to whether this technique scales to larger network sizes.

The most common implementations of neural networks and genetic algorithms use direct encoding strategies that directly encode network parameters, such as weight values and network connectivity to optimize the weights and/or architecture of a given artificial neural network. When optimization is confined to the weights of a given neural network (i.e., the structure is fixed), the network weights are encoded as genetic strings (chromosomes) or lists of parameters. A large population of these strings, where each string represents an instance of a network's parameters, is then combined using the genetic operators of crossover and mutation to form the next generation of chromosomes based on their fitness function. These fitness functions are often taken as the inverse of the network error (yielding a large number for good weights) scaled by the sum total error of the population.

[4] Part of this section was taken from a class report prepared by James R. Cain, a graduate student at the University of Tennessee in 1995–1996.

Using genetic algorithms to optimize the architecture of an artificial neural network is carried out in a similar fashion with the network connectivity of the neurons being encoded into the chromosomes. Because of the large and initially diverse population, a larger area of the weight space is much more likely to be searched compared to more traditional gradient descent techniques.

A combination of neural network and genetic algorithm training methods (called the Lamarkian learning method) involves periods of genetic optimization in between periods of backpropagation training. This method provides a powerful method for combining gradient descent techniques (like backpropagation) with evolutionary optimization techniques encompassed in the genetic algorithms.

An alternative approach is the Baldwin learning method which utilizes the backpropagation algorithm to adjust the fitness value for chromosomes. Hence, chromosomes that show the ability to learn through the backpropagation algorithm are considered to be more fit and therefore more likely to be selected to pass their genetic material onto subsequent generations.

Most common coding strategies employ connection-based systems which allow for weight and connectivity optimization of a predefined architecture. Issues which must be addressed before the start of training include population size, binary versus real valued weight representation, how many bits to use if binary chromosomes are used, type of crossover used, the prevalence of mutation, whether to use rank-based roulette wheel or proportional selection, and the criterion for stopping the process.

17.7   FUZZY GENETIC MODELING

As discussed earlier, fuzzy systems are made up of fuzzy sets, defined by their membership functions and fuzzy rules that determine the action of the fuzzy systems. Fuzzy systems can model general nonlinear mappings in a manner similar to feedforward neural networks since it is a well-defined function mapping of real-valued inputs to real-valued outputs. Kosko (1992) has shown that fuzzy systems, like feedforward neural networks, are universal approximators in that they are capable of approximating general nonlinear functions to any desired degree of accuracy. All that is needed for practical application is a means of adjusting the system parameters so that the system output matches the training data. Genetic algorithms can provide such a means. Furthermore, fuzzy systems are effectively transparent in that everything that happens is clearly apparent. Each fuzzy associate memory (FAM) matrix entry is just a fuzzy rule that is easy to understand. This is very different from neural networks where the weight matrix, the most visible parameter, is virtually uninterpretable. Furthermore, a fuzzy system has the capability to analyze the distribution of training data versus the distribution of test data. If these are radically different, then one knows in advance that the results of the mapping will not be satisfactory.

Fuzzy rules can be concisely represented with one or more FAM matrices. In some cases, the FAM matrix can be established on the basis of a person's knowledge of the system. If such information is not available, then genetic algorithms can be used to establish the FAM matrix.

Optimizing a FAM Matrix Using Genetic Algorithms

Earlier the FAM matrix in fuzzy systems was discussed as an alternative to a neural network to relate or model complex inputs and outputs when they were represented by fuzzy sets in a fuzzy variable. This arrangement was particularly attractive when there were two inputs, and the overall behavior was intuitively obvious or the relationship could be derived from simple experiments. When this was not the case, the FAM matrix has to be trained from data available in ways similar to the training of neural networks. This section discusses the use of genetic algorithms to optimize the training of the FAM matrix.

A fuzzy system has a number of parameters that define fuzzy sets that are candidates for optimization. While optimization of several variables simultaneously is possible, it is much simpler and more practical to optimize only one variable at a time. This is usually possible if the general nature of most variables are known or at least bounded.

A fuzzy system has several parameters that can be optimized using genetic algorithms. Included are fuzzy sets used for input and output variables, the membership functions that define fuzzy sets, the structure and entries in the FAM matrix, and, in some cases, the weight assigned to each rule. Welsted (1994) presents such an example where the FAM matrix entries are optimized because they have the most influence in determining system output. In that example, adaptation is accomplished through the minimization of an error function. The approach used by Welsted is to convert the matrix entries into a long binary string. Since each matrix entry is a string of 1s and 0s, the linking together end to end of these entries creates a very long binary vector. This is the chromosome used in the optimization in genetic algorithms.

Welsted's example problem (interest rate modeling) is structured to use a single FAM matrix that deals with all five inputs simultaneously. However, with five input variables, the FAM matrix is a five-dimensional hypercube. With three fuzzy sets per variable (negative, zero, and positive represented by a "left shoulder," a trapezoid, and a "right shoulder," respectively), the FAM matrix has 3^5, or 243, entries. This is the "curse of dimensionality" referred to by Kosko (1992). The number of fuzzy sets per input determines how finely we look at a problem and how much data we have to have for training. Each FAM matrix is an output fuzzy set represented by a three-bit representation; hence there are eight (2^3) output fuzzy sets.

If only one item in each matrix entry is activated, we need 8×243 or 1944 items of information just for training. This problem is equally serious in neural networks where increasing the number of inputs increases the number of data sets needed for training to cover the dynamic range over which the variables may change. Inadequate data in either fuzzy systems or neural networks will result in regions of the state space not being covered.

The training (adapting) and testing phases of this process are similar to the training phase of neural networks. The system is "initialized" by setting initial input values and corresponding initial output values. The quantity to be minimized is the error accumulated over the training set between the fuzzy system output and the desired output. Since the fitness function is to maximized, it is defined as a constant minus this accumulated error. Training is accomplished by running the genetic algorithm operating on the fuzzy system fitness function. The outputs of the genetic algorithm training process are the coordinates of the defining values of the fuzzy sets. Every time a new optimal value is attained, the fuzzy system is saved to file.

Since the genetic algorithm is used only to minimize the error in the training process, it is not used after training is complete. Use of the FAM matrix to relate inputs and outputs proceeds in a normal manner.

17.8   USE OF GENETIC ALGORITHMS IN THE DESIGN OF NEURAL NETWORKS

At the present time, there is no generally accepted theory or methodology for the design of neural networks, and the process used is generally a trial-and-error approach based on the experience of the designer. The complexity of neural network design arises from the high-dimensional, heterogeneous space that must be explored by the system. The primary features that are of concern in the design of neural networks are the structure of the network, the inputs to the networks, and the specification of the learning algorithm parameters. All of these quantities are problem-specific. While there are guidelines based on experience that can be very helpful in the design, some mathematical-based procedure would be very helpful. Optimization of the design based on the use of genetic algorithms offers such a methodology.

Use of Genetic Algorithms in Selecting Neural Network Structure

The Honeywell Technology Center (Harp and Samad, 1994) has developed an approach for designing and utilizing genetic algorithms for optimizing neural networks for use in modeling of complex systems. Their experience shows that the simultaneous optimization of network inputs, structure, and learning parameters is crucial for accurate modeling.

Usually, all of a network's parameters are encoded as genes in a chromosome in the form of a string of bits. Genetic algorithms procedures are then used to manipulate these chromosomes to produce improved parameters represented by the bit strings. Honeywell personnel (Harp and Samad, 1994) extended the concept of chromosomes and bit strings to a tree-structured entity with three generic families of genes: bytes, sequences, and structures. Byte genes represent scalar-valued parameters (e.g., learning rates). Sequence genes are ordered collections of other genes in which all the genes within a sequence are of a given specified type. Structure genes are fixed length ordered collections of genes of given types, with the type being determined by position. In an analogy to trees, the leaves are byte genes while the branches are formed by sequence and structure genes. An individual genetic tree is a sequence of structure genes representing areas and related connectivity that correspond loosely to a layer of a neural network but are more broadly applicable. Each area structure gene parameterizes the area in terms of an address, its number of neurons, the connecting weights, the learning parameters, and so on.

Harp et al. (1989, 1990) used a "blueprint" scheme to manipulate genetic algorithm representations of how sets of neurons are connected. In this work, network characteristics are represented by a blueprint, defined as a data structure that encodes various characteristics of the network including structural properties, input selection, and learning algorithm parameter values. A blueprint is instantiated into an actual network, and the neural network is trained using a learning algorithm and the learning parameters specified in the blueprint. The trained neural network is then evaluated using testing data, including an evaluation of its robustness by disabling some neural units or perturbing the learned weight values. Then its fitness is computed. The fitness estimate can be an arbitrarily complex function, such as the weighted linear sum of relevant criteria such as the number of nodes and weights in the network, accuracy, learning speed, efficiency, average and maximum number of outgoing weights from a node, and the various test scores.

After the evaluation, the next generation of the network is formulated in the blueprint. This process is mediated by a number of genetic operators (crossover, mutation, etc.) in which two blueprints are spliced together to produce a child blueprint. In effect, the genetic operators are being applied to these blueprints on a macroscale whereas genetic algorithms apply these genetic operators to bit strings on a microscale. The advantage of using the overall approach described here compared to "manual optimization" is that it allows the developer of the neural networks to explore large amounts of design space that would otherwise be left unexplored.

REFERENCES

Caruana, R. A., and Schaffer, J. D., Representation and Hidden Bias: Gray vs. Binary Coding for Genetic algorithms, in *Proceedings of the Fifth International Congress on Machine Learning*, J. Laird, ed., Morgan Kaufmann, San Mateo, CA, 1988.

Davis, L., ed., *Handbook of Genetic Algorithms*, Van Nostrand Reinhold, New York, NY, 1991.

Garis, H. de, Modular Neural Evolution for Darwin Machine, *Proceedings of IJCNN-90-Wash.-DC*, Washington, DC, January 1990.

Goldberg, D. E., *Genetic Algorithms in Search, Optimization, and Machine Learning*, Addison-Wesley, Reading, MA, 1989.

Gray, F., Pulse Code Communication, U. S. Patent #2-632-058, March 17, 1953.

Gruau, F., and Whitley , D., Adding Learning to the Cellular Development of Neural networks: Evolution and the Baldwin Effect, in *Evolutionary Computation*, Vol. 3, No. 1, MIT Press, Cambridge, MA, 1993, pp. 213–233.

Guo, Z., Nuclear Power Plant Diagnostics and Thermal Performance Studies Using Neural Networks and Genetic Algorithms, University of Tennessee Library, Knoxville, TN, 1992.

Harp, S., Samad, T., Genetic Optimization of Neural Network Architectures for Electric Utility Applications, Final Report, Electric Power Research Institute, Research Project No. 8016-04, Palo Alto, CA March 1994.

Harp, S., Samad, T., and Guha, A., Designing Application Specific Neural Networks Using the Genetic Algorithm, in *Advances in Neural Information Processing Systems 2*, D. S. Touretzky, ed., Morgan Kaufmann, San Mateo, CA, 1990.

Harp, S., Samad, T., and Guha, A., Towards the Genetic Synthesis of Neural Networks, in *Proceedings of the Third International Conference on Genetic Algorithms*, Morgan Kaufmann, San Mateo, CA, 1989.

Holland J. H., *Adaptation in Natural and Artificial Systems*, University of Michigan Press, Ann Arbor, MI, 1975.

Holland J. H., Genetic Algorithms, *Scientific America*, Vol. 267, pp. 66–72, 1992.

Kosko, B., *Neural Networks and Fuzzy Systems*, Prentice-Hall, Englewood Cliffs, NJ, 1992.

Koza, J. R., and Keane, M. A., Cart Centering and Broom Balancing by Genetically Breeding Populations of Control Strategy Programs, in *Proceedings of IJCNN-90-Wash.-DC*, Washington, DC, January 1990.

Moricic, B., and Nikolov, Z., GENNET—System for Computer Aided Neural Network Design Using Genetic Algorithms, *Proceedings of IJCNN-90-Wash.-DC*, Washington, DC, January 1990.

Muselli, M., and Ridella, S., Supervised Learning Using a Genetic Algorithm, *Proceedings of INNC-90*, Paris, France, July 9–13, 1990.

Welsted, S. T., *Neural Network and Fuzzy Logic Applications in C/C++*, John Wiley & Sons, New York, 1994.

Whitely, D., and Bogart, C., Optimizing Neural Networks using Faster, More Accurate Genetic Search, *Proceedings of the Third International Conference on Genetic Algorithms*, Morgan Kaufmann, San Mateo, CA, 1989.

Whitely, D.,and Bogart, C., The Evolution of Connectivity: Pruning Neural Networks using Genetic Algorithms, *Proceedings of IJCNN-90-Wash.-DC*, Washington, DC, January 1990.

Whitely, D., and Starkwerther, Optimizing Small Neural Networks Using a Distributed Genetic Algorithm, *Proceedings of IJCNN-90-Wash.-DC*, Washington, DC, January 1990.

PROBLEMS

1. Discuss the relative merits of the "proportional selection" and the "rank-based selection" of the surviving chromosomes. Which is easier to implement?

2. The "fitness function" is by far the most important quantity in the use of genetic algorithms. One method of selecting this function was illustrated in Section 17.5. Discuss other means of determining the most appropriate "fitness function." Can you envision a method of optimizing the selection of the "fitness function" used in an optimization process?

3. Consider the data presented in Table 10.1 for welding tests. If you wanted to use genetic algorithms to optimize the travel speed and arc current for a given configuration (thickness, bead width, and bead penetration) of a weld, how would you go about it? What kind of "fitness function" would you develop? How would you go about ensuring that the "fitness function" was appropriate?

4. Carry the reproduction, crossover, and mutation processes on for another cycle using the information provided in Table 17.3. Is there significant improvement (or deterioration) of the 'fitness" of the chromosomes? If so, why? If not, why?

5. An alternate fitness function for Example 17.1 is

$$f_i(x) = 1 - \frac{y_i(x)}{2.71}$$

The denominator 2.71 is $y_i(x)$ when $x = 31$, the largest possible value of x for a 5-bit binary string. Evaluate $f_i(x)$ and $f_i(x)^{**}$ in Table 17.3. Are the new genes selected the same? If so, why? If not, why not?

18

EPILOGUE

18.1 INTRODUCTION

In the late 1960s, the American Society for Engineering Education (ASEE) issued a special report entitled "Goals of Engineering Education" in which the authors looked into their crystal balls to the year 2000 and predicted the types of projects on which engineers would be working during the next one-third century. The point of the study was that the engineering students in school at the time of the study (1967–1968) would still be active in the engineering profession at the turn of the century, and it was the engineering educators' responsibility to see to it that educational experiences at engineering colleges constitute proper preparation to meet the challenges that would arise in the rest of the century. Among the projects they predicted were:

Large-scale ocean farming

Fabrication of synthetic protein

Controlled thermonuclear power (fusion energy)

Regional weather control

Correction of hereditary defects by molecular (genetic) engineering

Automated high-IQ machines (expert systems and artificial intelligence in general)

Universal language through automated communications

Mining and manufacturing on the moon

Directed energy (microwave and laser) beams

Commercial global ballistic transports (the "Tokyo Express")

With less than half a decade to go, many of these predictions for the year 2000 are well on their way toward reality, while others, though feasible, are not being given serious consideration today. Perhaps artificial intelligence in the more general form of "soft computing" has had as much, if not more, impact than any of the other technologies listed above. Yet, we have only seen the tip of the iceberg as far as its influence on the future.

As far out as some of the items listed by the ASEE seemed in 1967, most of them are accepted as legitimate areas for engineering involvement today. However, some of the things appearing on the horizon today virtually defy our imaginations. For example, it has recently been reported that a team of scientists at the Max Planck Institute has opened a two-way communication link between a silicon chip and a biological neuron, effectively establishing a signaling channel that works in both directions. (ACM, 1995). The chip stimulates a leech's nerve cell through induced charges, and while capable of communication, no electrical current flows between the neuron and the chip (an essential requirement for any prosthetic limb controlled by the brain through a living nervous system).

18.2 IS ARTIFICIAL INTELLIGENCE REALLY INTELLIGENT?

In her book entitled *In Our Own Image*; *Building an Artificial Person*, Maureen Caudill (Caudill, 1994) examines the current state of technology and the accelerating trend in developments of robots, computer vision, understanding speech, sensing, diagnostics, and so on, and concludes that the construction of an "artificial person" is closer than most of us believe. Clearly, the first such "artificial person" would not be a Commander Data of Star Trek fame, but the essential processes to sustain "artificial life" are perceived by Caudill to be feasible in the twenty-first century. While we take no position on this well-documented but controversial thesis, we will point out that most of the advances in the technologies listed under soft computing have their origin in biological processes of humans, especially physiological processes and psychological behavior of the brain.

Many scientists and engineers are somewhat skeptical about artificial intelligence, especially in the light of the "excessive claims" of some of the pioneers in expert systems and neural networks. Fuzzy systems avoided this pitfall only because it was quietly developed in Japan without much fanfare until successful systems were being demonstrated. The skepticism comes from deep and privately held gut feelings that computers can never "be like" or "live like" humans. They are perceived as "just machines," and hence by definition they cannot be intelligent. We often tend to use anthropomorphic terms like "intelligence" to describe their workings in the age-old tradition of projecting something of ourselves and nature to the artifacts we made. It may well be that computers are no more intelligent than locomotives are iron horses. Artificial intelligence may simply be an inspiring metaphor for pursuing the enhancement of human intelligence. Indeed, concepts that have sprung out of artificial intelligence are revolutionizing the workplace, the office, the school, the marketplace, and the laboratory. In the manufacturing sector, soft computing is changing not only the design and analysis, but also the physical manufacturing with added flexibility and benefits in scheduling, production, maintenance and managment. The most often heard terms in manufacturing these days are *agile manufacturing* and the *virtual company*, concepts that are considered to map directly to advancements in *soft computing* and its implementations.

The fundamental characteristics of emerging new products and systems may be quite different from these we deal with today. Agile or flexible systems will, of necessity, be more *proactive*. Whether they are intelligent agents or subjective objects (like today's machines that respond only to present conditions), is less important than the fact that they get the job done for you. Already a trend is underway in Japan toward predictive and anticipatory systems using predictive fuzzy control. Examples include control systems in many Japanese elevators, the Sendai metroliner, and the *Fugen* nuclear power plant.

In the future, technology that is user-friendly to people and capable of self-adjustment to custom fit individual needs will become important. In order to bring this flexibility, we see neurofuzzy technologies becoming part of the man-machine interface. They truly hold considerable promise to harmonize and enhance the relation between humans and machines and make it possible to incorporate actions, judgments, and thoughts that are near-human into a wide variety of devices and systems. Neurofuzzy technologies that respect the users' subjective desires, backgrounds, and idiosyncracies are expected to find their way into a great variety of products, making it possible for machines to say things like "Is this what you are trying to say?" or "Is this what you really want?" In industry, this may be particularly suitable for addressing bad structure problems for which computerized control has until now been difficult; it may indeed make automatic operation equivalent to operation by skilled operators.

18.3 THE ROLE OF NEUROFUZZY TECHNOLOGY

The principal topic in this book that is not commonly covered in other books or in university courses in the science and engineering fields is the *neurofuzzy* methodologies of Chapters 12 through 15. In Chapter 12, artificial neurons that utilize fuzzy operations (e.g., max, min, etc.) in place of multiplication and addition, as well as neural networks that also utilize fuzzy processes, are described. Applications such as "Fuzzy ARTMAP" and "fuzzy clustering" are already widely known in the artificial intelligence community and beginning to be utilized in engineering research and development.

In Chapter 13, we introduced neural methods into fuzzy systems. The overriding issue in fuzzy systems, whether they be used for expert systems, decision-making, or control, is defining the linguistic *if/then* relationships that constitute the algorithm on which the process is based. Neural networks and/or neural processes with their ability to extract information from examples (learning) can play an essential role in providing a better basis for fuzzy algorithms. Defining membership functions for fuzzy variables by using neural network is a very valuable process.

Chapter 14 is the result of a computerized literature search of scientific and technical journals for titles of articles that include both the words *fuzzy* and *neural*. Out of about 700 such publications, we chose about 50 examples from 12 fields where neurofuzzy systems were used advantageously. The purpose of this chapter was to illustrate the wide range of applications of neurofuzzy systems.

In Chapter 15, examples of research carried by graduate students working under the authors have demonstrated the advantage of utilizing fuzzy systems, neural networks, and genetic algorithms as semi-integrated processes. More complete integration of these methodologies will bring additional benefits when we learn to control the integrated fuzzy neurons and networks and the neurofuzzy systems in a straightforward manner. Indeed, the main reason for using these various methodologies in a semi-integrated (and usually sequentially) manner is to keep the processes under control.

We cannot overemphasize this last point. Neurofuzzy systems are at the state of development as neural networks before the rediscovery of backpropagation in 1968. We are proceeding on a trial-and-error basis with little guidance as to which is the best way to apply neurofuzzy concepts. The potential payoff for using neurofuzzy systems properly can be enormous. Indeed, the "fuzzy neuron" or the "neurofuzzy system" are the modern analogs of the perceptron and the adaline processing units, and they are at about the same stage of development today as the perceptron and adaline in 1960. What is now needed is the creation or discovery of an integrated training/control process.

18.4 LAST THOUGHTS

Tom Peters, coauthor of *In Search of Excellence* (Peters and Watermann, 1982), wrote the Foreword of *The Rise of the Expert Company*, an exposition on the benefits of expert systems in industry by Feigenbaum, McCorduck, and Nii (1988). The closing two paragraphs stated the following:

> I came to this book and to the task of writing this foreword interested, even fascinated, by the topic about which I am largely naive. I leave the process of digesting the manuscript and writing the foreword mesmerized. The emerging world, brilliantly and pragmatically described in *The Rise of the Expert Company*,

is not the world we now know. The consequences are exciting and a bit frightening—and clearly monumental.

I conclude that any senior manager in any business of almost any size who isn't at least learning about AI and sticking a tentative toe or two into AI's waters is simply out of step, dangerously so.

In the eight years since that foreword was written, neural networks and fuzzy systems have achieved equally important status as expert systems in 1988. Neurofuzzy technology is the next big step because of the synergistic benefits of the merging these two important technologies. It is our hope that we have taken that first step—that is, put that first tentative toe into neurofuzzy technology's waters, lest we too get out of step, dangerously so.

REFERENCES

ACM, *Communications of the ACM*, Newstrack, Vol. 38, No. 10, pp. 11–12. 1995.

American Society for Engineering Education, *Goals of Engineering Education*, Washington, DC, 1967.

Caudill, M., *In Our Own Image*: *Building an Artificial Person*, Oxford Press, Cambridge, MA, 1992.

Feigenbaum, E., McCorduck, P., and Nii, H. P., *The Rise of the Expert Company*, Times Books, New York, 1988.

Peters, T. H., and Watermann, R. H., Jr., *In Search of Excellence*, Harper and Row, New York, NY, 1982.

APPENDIX

T NORMS AND S NORMS

Fuzzy set operations such as the *complement* (*NOT*), *union* (*OR*), and *intersection* (*AND*) as well as operations associated with the evaluation of linguistic descriptions and neurofuzzy models can be parameterized through *T* norms and *S* norms. The concepts of *T* norm and *S* norm were originally developed by mathematicians in connection with probability theory, but they have found considerable application in fuzzy-neural approaches. The operators can be thought of as the extension of fuzzy operations. For example, the *T* norm (*T* stands for triangular) can be thought of as the extension of *AND*. Following Terano et al. (1994) we present here semi-formally the basic ideas involved.

A.1 T NORMS

A *T* norm is a two-place function $T: [0,1] \times [0,1] \to [0,1]$ satisfying the following four axioms:

$$x\,T\,1 = x, \quad x\,T\,0 = 0 \qquad \forall x \in [0,1] \qquad \text{(A-1)}$$

$$x_1\,T\,x_2 = x_2\,T\,x_1 \qquad \forall x_1, x_2 \in [0,1] \qquad \text{(A-2)}$$

$$x_1\,T(x_2\,T\,x_3) = (x_1\,T\,x_2)T\,x_3 \qquad \forall x_1, x_2, x_3 \in [0,1] \quad \text{(A-3)}$$

$$\text{if } x_1 \le x_2, \text{ then } x_1\,T\,x_3 \le x_2\,T\,x_3 \qquad \forall x_1, x_2, x_3 \in [0,1] \quad \text{(A-4)}$$

Equation (A-1) is a boundary condition refering to crisp *AND*, while equations (A-2) and (A-3) are *commutative* and *associative* laws; Equation (A-4)

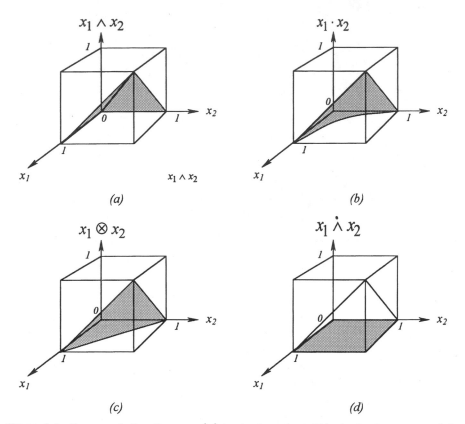

Figure A.1 Representative *T* norms: (*a*) Logical product, (*b*) algebraic product, (*c*) bounded product, and (*d*) drastic product.

requires the preservation of order, and it guarantees that the order of evaluation cannot be reversed at the third evaluation.

The *logical product* produced by a min operation is a representative *T*-norm operation.

$$x_1 \, T x_2 = x_1 \wedge x_2 \tag{A-5}$$

It corresponds to the intersection of fuzzy sets. It is easy to verify that equation (A-5) above satisfies equations (A-1)–(A-4). If (A-5) is expressed graphically, we obtain Figure A1. It is easy to see in Figure A.1*a* that the four corners of the $x_1 \wedge x_2$(shaded) square are the value of the crisp *AND* operation.

What other kinds of operations can the *T* norm produce? The most important ones for neurofuzzy applications are the *algebraic product* $x_1 \cdot x_2$, the bounded product $x_1 \otimes x_2$, and the drastic product $x_1 \overset{\cdot}{\wedge} x_2$ defined as follows

$$x_1 \cdot x_2 = x_1 x_2 \tag{A-6}$$

$$x_1 \otimes x_2 = (x_1 + x_2 - 1) \vee 0 \tag{A-7}$$

$$x_1 \stackrel{.}{\wedge} x_2 = \begin{cases} x_2 & x_1 = 1 \\ x_1 & x_2 = 1 \\ 0 & \text{otherwise} \end{cases} \tag{A-8}$$

The graphs of these are given in Figure A.1b–d. Using these graphs it is easy to show that

$$0 \leq x_1 \stackrel{.}{\wedge} x_2 \leq x_1 \otimes x_2 \leq x_1 \cdot x_2 \leq x_1 \wedge x_2 \tag{A-9}$$

The T norm can produce an infinite number of other operations which can be placed in order between the *drastic product* and the *logical product* as seen in Figure A.1.

A.2 S NORMS

The S norm, which is the extension of OR, is also called the T conorm.

An S-norm is a two-place function $S: [0,1] \times [0,1] \to [0,1]$ satisfying the following four axioms:

$$x \, S \, 1 = 1, \quad x \, S \, 0 = x \qquad\qquad \forall x \in [0,1] \tag{A-10}$$

$$x_1 \, S \, x_2 = x_2 \, S \, x_1 \qquad\qquad \forall x_1, x_2 \in [0,1] \tag{A-11}$$

$$x_1 \, S(x_2 \, S \, x_3) = (x_1 \, S \, x_2) \, S \, x_3 \qquad \forall x_1, x_2, x_3 \in [0,1] \tag{A-12}$$

$$\text{if } x_1 \leq x_2, \text{ then } x_1 \, S \, x_3 \leq x_2 \, S \, x_3 \qquad \forall x_1, x_2, x_3 \in [0,1] \tag{A-13}$$

A representative S norm is the *logical sum* produced by a *max* operation.

$$x_1 \, S \, x_2 = x_1 \vee x_2 \tag{A-14}$$

Others include the algebraic sum $x_1 \stackrel{.}{+} x_2$, the bounded sum $x_1 \oplus x_2$, and the drastic sum $x_1 \stackrel{.}{\vee} x_2$:

$$x_1 \stackrel{.}{+} x_2 = x_1 + x_2 - x_1 x_2 \tag{A-15}$$

$$x_1 \oplus x_2 = (x_1 + x_2) \wedge 1 \tag{A-16}$$

$$x_1 \stackrel{.}{\vee} x_2 = \begin{cases} x_2 & x_1 = 0 \\ x_1 & x_2 = 0 \\ 1 & \text{otherwise} \end{cases} \tag{A-17}$$

The properties of these are given in Figure A.2 *a–d*. As is obvious from these figures, we have

$$x_1 \vee x_2 \le x_1 \dotplus x_2 \le x_1 \oplus x_2 \le x_1 \mathbin{\dot\vee} x_2 \qquad \text{(A-18)}$$

It is easy to see by inspecting Figure A.2 and equations (A-18) that the order is the reverse of the *T* norm.

The *T* norm shown in Figure A.1, and *S* norm shown in Figure A.2 must meet the boundary conditions shown in Figure A.3. It is easy to show that the smallest *S* norm is the logical sum and the largest *S* norm is the drastic sum, as also seen in Figure A.3.

Various fuzzy *negations*, *T* norms and *S* norms have been proposed, but it is convenient to employ the ones that meet the following conditions:

$$\neg(x_1 \, S \, x_2) = \neg x_1 \, T \, \neg x_2 \qquad \text{(A-19)}$$
$$\neg(x_1 \, T \, x_2) = \neg x_1 \, S \, \neg x_2 \qquad \text{(A-20)}$$

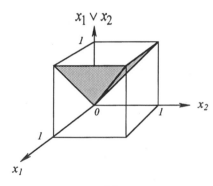

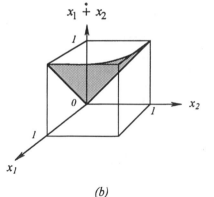

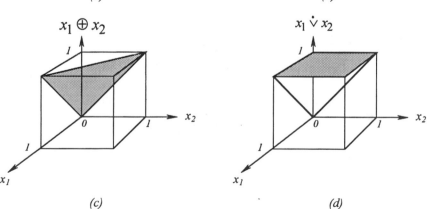

Figure A.2   Representative *S* norms: (*a*) Logical sum, (*b*) algebraic sum, (*c*) bounded sum, and (*d*) drastic sum.

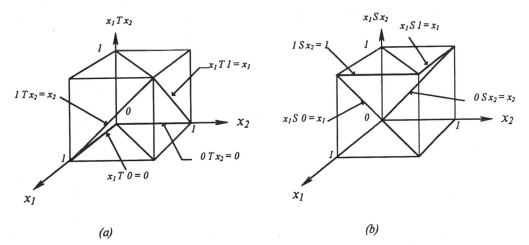

Figure A.3 Boundary conditions for (a) general T norm and (b) general S norm.

These correspond to de Morgan's laws for crisp operations and are called *fuzzy de Morgan's laws*. When these equations above arise, the T norm and S norm are dual with respect to fuzzy negation, and it can be shown that the logical product and logical sum, algebraic product and sum, bounded product and sum, and drastic product and sum all show duality of the variance from 1 in the fuzzy negation. In practical applications, the logical pair is standard, and the algebraic and bounded pairs are used occasionally. The drastic pair has the property of being discontinuous and is important in terms of the lower bound for T norms and the upper bound for S norms, but it is not often used in practical applications.

The logical pair of T norms and S norms is most often used due to its explicit physical meaning and the fact that the

$$([0,1], \leq ,1 - , \wedge , \vee) \tag{A-21}$$

system gives complete pseudo-Boolean algebra (and thus good mathematical characteristics). Only complements do not arise:

$$x \vee (1 - x) \geq 0, \qquad x \vee \leq 1 \tag{A-22}$$

and all the other properties found in crisp logic arise just the same. With crisp logic, the equality of the above equation arises. The complements are called the *laws of contradiction and exclusion*. The law of contradiction—that is, that no property and its negation can exist at the same time—and exclusion—that is, that both the property and its negation exist with no ambiguous intermediates—are properties unique to crisp logic and not found in fuzzy logic.

A.3 *T* NORMS AND FUZZY IMPLICATIONS

Let us look at the mathematical basis for modeling the fuzzy relations involved in fuzzy *if/then* rules, that is, expression of the form

$$\text{if } X \text{ is } A, \text{ then } Y \text{ is } B \tag{A-23}$$

with various implications. If the elements in the total space are fixed and we confine our discussion to evaluations within $[0, 1]$, fuzzy implications are two-variable functions or two-item relations of $[0, 1]$:

$$\phi: [0, 1] \times [0, 1] \to [0, 1] \tag{A-24}$$

The implication "if X is A, the n Y is B" is described by $(X$ is $(NOT\ A))$ $OR(Y$ is $B)$ in crisp cases. Therefore if we replace *NOT* with fuzzy negation and *OR* with max, the most standard fuzzy logic operations, we get the implication operator

$$\phi = \left(1 - \mu_A(x)\right) \vee \mu_B(y) \tag{A-25}$$

which we have called *Zadeh's implication operator* in Chapter 5.

   In order to look into the meaning of the equation (A-25), let us graph the *T* norm and the *S* norm as shown in Figure A.4. The four crisp points (shaded areas in the figure) are preserved, and the figure is composed of two triangular planes. However, given just the coordinate axis and the four crisp points, imagine Figure A.4*b* for the same graph of two triangular planes if

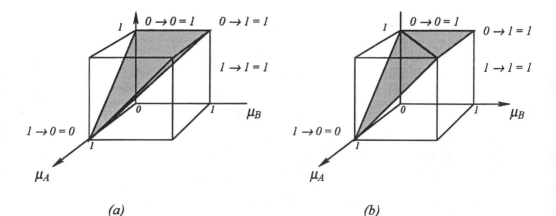

$$\mu_A \to \mu_B = (1 - \mu_A) \vee \mu_B \qquad\qquad \mu_A \to \mu_B = (1 - \mu_A + \mu_B) \wedge 1$$

(a)   *(b)*

Figure A.4 Two fuzzy implications (*a*) Zadeh and (*b*) Lukasiewicz.

asked to interpolate. Expressing Figure A.4*b* as an equation, we get

$$\phi = (1 - \mu_A(x) + \mu_B(y)) \wedge 1 \qquad \text{(A-26)}$$

This is the limited sum operation for fuzzy negation of the variance from 1 of *x* and *y*, and it is an equation known as the *Lukasiewicz implication in multiple logic*.

REFERENCE

Terano, T., Asai, K., and Sugeno, M., *Applied Fuzzy Systems*, Academic Press, Boston, 1994.

INDEX

Aarts, 306
Abe, 473
absolute magnitude of data, in neural
  networks, 395
absolute values, in neural networks, 393
absorption, 28
accretive associative memories, 297
accurate analysis, 7
action, as fuzzy variable, 153
action side of rules, 106
activation function, 194, 411
activation functions in fuzzy neurons, 416
activation hyperboxes, 474
active information fusion, 503
Adaline, 4, 213, 217, 220, 333
adaptation, 447
adaptation, fuzzy neural networks, 423
adaptive control, 353
adaptive control systems, 371
adaptive critic reinforcement learning control
  system, 361
adaptive critics, 353
adaptive critics method of neural control, 360
adaptive fuzzy systems, 485
adaptive inverse model control, 372
adaptive linear combiner, 341
adaptive model control, 372
adaptive network-based fuzzy inference
  systems, 466–468
adaptive neural network, 345
adaptive processors and neural networks, 345
adaptive resonance theory neural networks,
  328–331
adaptive signal processing, 341–345
adaptive system description, 447
adaptive techniques for fuzzy controller, 162
adding fuzzy numbers, 77, 84–90

addition of discrete fuzzy numbers, example
  of, 85–87
addition of fuzzy numbers, extension principle,
  87–90
Adeli, 450, 476
Adeli–Hung algorithm 452–455
adjusting coefficient in sigmoidal term, 251
aggregation of neuronal inputs, 410
aggregation operation, 412
aggregation operations in fuzzy neurons, 416
agile manufacturing, 563
algebraic product, 417
algebraic sum, 417
algorithmic relation, 110, 127, 137
Alguidingue, 276, 493, 499
alleles, 539
alpha cuts, 34–37
α-cut of a fuzzy relation, 61
α-cuts, and fuzzy numbers, 77
Amari, 229
Aminzadeh, 4
analog computer, 4
analytic approaches to modeling systems, 105
analytical forms, 106–107
AND fuzzy neurons, 413, 418–421
ANFIS architecture, 467
ANFIS, 487
antecedent propositions, 106
anticipatory systems, 1, 510–516
antireflexive fuzzy relation, 58
antireflexive relation, 57
antisymmetric relation, 57
applications of adaptive neural networks, 345
a priori probability, 319
arctangent, activation function, 232–233
area-cum-point notion of fuzzy rules, 108–109
arithmetic implication, 122

575

arithmetic implication and *ELSE*, 137
ART, *see* adaptive resonance theory neural
    networks
ART neural network, general operations, 329
ART-1, properties of, 330–331
artificial intelligence, 1, 191
artificial neural networks, 196–203
artificial person, 562
Assilian, S, 145
associative memories, 296–306
associative neural memory, 212
associative neural networks, 8, 211, 257–258
associative property, 28
associativity, 19
asymmetric relation, 57
autoassociative neural net, systemwide
    monitoring, 262
autoassociative neural network, 197
autoassociative neural network for filtering,
    260
autocorrelation function, 334
automatic gain control, 233
axiomatic, interpretation of probability, 37
axon, 410

backpropagation, 3, 8, 229
backpropagation algorithm, variations of, 252
backpropagation and related training
    paradigms, 8
backpropagation, factors that influence,
    248–255
backpropagation through time, 353, 360
backpropagation training, 229, 238, 401
backward-chaining, and fuzzy inference, 112
Bailey, 385
BAM, 299
BAM, operation of, 301
bandwidth of P-shaped functions, 117
Barbosa, 484
Barto, 374
Bayesian probability, 320
Bayro-Corrochano, 451, 475
bearing failure in large motor pump, 379
bearings fault detection, 494
belief measures, 39
Berkan, xviii, 511
Bernard, 176
Bezdeck, 435, 482, 499
bias, neurons, 248
bidirectional associative memory, 299–302
binary relations, 50, 52
binary representation of chromosomes, 542,
    545
binary representation of data, in neural
    networks, 391

biological basis of neural networks, 192–193
biotechnology, fuzzy-neural applications, 439
biunique relation, 58
black-box identification, 353
Blanco, 548, 455, 474, 477
Boltzmann machine, 306
Boolean implication, 122
Boolean implication and *ELSE*, 137
Boolean logic, 4
bottleneck layer, 257
bottom-up weights, 274
bounded product, 417
bounded product implication, 122
bounded product implication and *ELSE*, 137
bounded sum, 417
Bourbakis, 481, 486
Brotherton, 484
Buckley, 447
Buckley and Hayashi, 456
Butler, 213

C++, 4
Cai, 413, 439
Cain, 549
calculation of weights, hidden neurons, 242
calculation of weights, output neurons, 240
capacity and efficiency of crossbar network,
    302–303
Carpenter, 328, 431,432
Cartesian plane, 32
Cartesian product, 50, 53
Cartesian product, and fuzzy numbers, 77
Cartesian product, continuous, 53
Cartesian product, discrete, 53
Cartesian product, fuzzy, 55
cascade-correlation neural nets, 280
cascade-correlation training process, 280
Cassidy, 485
categories, 14
Cauchy machine, 306
Caudill, xviii, 213, 326, 402, 562
cellular automata, 1
center of area (*COA*) defuzzification, 164
center of sums (*COS*), defuzzification, 165
centroid defuzzification, *COA*, 164
certainty factors, 529
Chang, 478
change in action, as fuzzy variable, 153
change in error, as fuzzy variable, 153
changes in values, in neural networks, 393
chaotic systems, 1
characteristic function, 14, 41
check-valve monitoring, 378
chemical oxygen demand density, 460
Chen, 438

Chen and Chen, 478
Chen, Lin, and Chen, 484
choice of neural network type, 386
choice of output, 387
chromosomes, 539
clarity through fuzzy description, 151
classical conditioning, 290
classical sets, 14
classical set theory, 13
clipping of the consequent, 131
closed-loop adaptive operation, 346
closed-loop control system, 354
closed-loop operation, 345
clustering, 450
clustering data, 389
COA defuzzification, 164
Cohen, 452
Cohen–Grossberg learning, 290–296
coherence function, 339
collection of intervals, and fuzzy arithmetic, 83
commutativity, 28
competitive learning, 204, 306
competitive neural networks, 8, 306–315
complementary nature of fuzzy-neural systems, 409, 445
complementation, 14
complement coder, 433
complement decoder, 433
complement of a fuzzy set, 20
compositional rule of inference, 126
composition, examples of, 69, 71
composition of fuzzy relations, 49, 65–74
composition operations, 61
compound fuzzy values, 117
compressed representation of the input, 258
compression of information, 257
compression ratio, 206
CON, see concentration
concentration, CON, 21, 24
configurations of adaptive neural networks, 347
connected relation, 58
connectionist systems, 193
connectives, AND, OR, 106
connectives for compound values, 117–120
consequent propositions, 106
consistency principle, 38
content addressable memories, 297
continuous-valued data, in neural networks, 391
contrast intensification, 25
control hypersurface, 170, 171, 173
controller autotuning, 371
controller gain, in PID, 147
controller, practical example of, 146
controlling parameter, in process control, 147

control valve, 150
conventional approaches to modeling systems, 105
convergence criteria, in genetic algorithms, 545
convex fuzzy set, 77, 79
convexity, 79
Cooley, 438
corrected reading from failed sensors, 263
COS defuzzification, 165
counterpropagation networks, 315
counterpropagation networks, characteristics of, 319
coupling coefficients, 376
crisp restriction, 41
Crooks, 390
crossbar associative network, see Hopfield network
crossbars, disadvantages of, 303
crossbar structure of networks, 297
cross-correlation, 336
cross-correlation functions, 2, 340
crossover, 541
cross-spectral density, 2, 336
cross-spectral density measurements, 338
CUBICALC, 487
cutoff frequency, 262
Cybenko, 267
cylindrical extension, 60–64

Dagli, 486
Dalton, 482
DARPA, 213
Dartmouth Conference, 4, 214
Dartmouth Summer Research Conference, 213
Darwinian natural selection, 540
data acquisition, 391
databases, errors in, 389
data compression, 206
data compression-expansion via neural nets, example, 210
data-driven inference, 112
data kinds, in neural networks, 399
data normalization, in neural networks, 395
data representation, in neural nets, 391
data scaling, in neural networks, 396
data selection for training and testing, 399
data sources for neural networks, 389–391
Dean, 214
decision fusion, 494, 500
defuzzification, and adaptive approaches, 166
defuzzification, effects on fuzzy controller quality, 181–184
defuzzifying the output of a controller, 152
defuzzyfication methods, 163–166
degree of fulfillment (DOF), 131, 157

degree of fulfillment, and fuzzy controller quality, 179
degree of fulfillment, examples in control 168, 169, 174
degrees of freedom, 207
Delgado, 474
delta-bar-delta networks, 254
delta-bar-delta networks, extended, 255
delta rule, 204, 218
delta vector, 219, 237
demapping layers, 259
De Morgan's laws, 28
De Morgan's laws and T norms, 417
Dempster-Shafer Theory of Evidence, 39
dendrites, 192, 410
dendritic inputs, 411
dendritic inputs, and neuronal fuzzification, 412
denominational fuzzyfier, 26
derivative gain constant, in PID, 147
derivative term, in PID, 147-148
designing fuzzy controllers, issues of, 176
design of the neural network, 386-389
desired response and error, 344
diagnosis, fuzzy, 52
diagnosis, fuzzy-neural, 495
difference in data requirements in supervised and unsupervised learning, 400
differential Hebbian learning, 290
DIL, see dilation
dilation, DIL, 21, 24
directed graph, 51, 56
discrete universe of discourse, 16
dissemblance index, 506
distortion correction via neural net, 211
distributed information, 394
distributed memory, 211
distributed neural memory, 212
distributive property, 28
distributivity, 19
divisibility relation, example of, 50
division of fuzzy numbers, 77, 99-101
division of fuzzy numbers, example of, 99-101
division of x-y plane, 224
DOF (degree of fulfillment), 131
Dong, 368
double negation law, 28
drastic product, 417
drastic product implication, 123
drastic product implication and ELSE, 137
drastic sum, 417
Driankov, 145
driver reinforcement learning, 296
Dubois, 176, 417
Dubois and Prade, 15, 17, 30, 42

dynamic hybrid neurofuzzy systems, 8, 493-516
dynamic neural nets and control systems, 8, 332-382
dynamic variable, in process control, 147

EBR-2, 262, 264
Efstathiou, 152
elastic fuzzy logic, 474
electrocardiography, 2
ELSE, interpretation under various implications, 136-137
El-Sharkawi, 476
empty fuzzy set, 19
encoding data, in neural networks, 394
energy surface representation, 297, 305
environmental imaging, 483-484
epistemic descriptions, 43
epistemic variables, see fuzzy variables
epoch, 209
equality of fuzzy sets, 20
equalization and deconvolution, 351
equal-percentage valve, 150
equivalence relation, 58
Ercal, 486
error, as fuzzy variable, 153
error function, using different ones, 254
error, in process control, 147
error rate, 38, 39
errors in databases, 389
error squared, 234
error squared, minimum, 236
Eryurek, 262
evaluating control rules, schematically, 158, 169, 174
evaluating functions, analogy to fuzzy inference, 126-127
evaluating fuzzy relations through composition, 67
evolution, 540
evolutionary processes, 540
excitatory, 412
excitatory effects in fuzzy neurons, 415
expert knowledge elicitation, 447
expert systems, 1, 523-537
expert systems, characteristics, 524
expert systems, components, 525
expert systems, implementation issues, 535
expert systems in fuzzy-neural systems, 8
expert systems, state of the art, 531
expert systems, use, 532
expert systems, with neural nets and fuzzy rules, 534-535
exponential activation function, 322

exponential fuzzifier, 26
extension principle, 30–34
extension principle, addition of fuzzy numbers, 84, 88
extension principle, and fuzzy numbers, 77
external inputs to a neuron, 410

factors influencing fuzzy controller quality, 179
Fahlman, 253, 281
failure and error possibilities, 39
failure rate, 38, 39
fairly continuous data, 392
Farber, 267
fast backpropagation, 253
Fatikow, 481
fault-tolerance of neural nets, 211, 213
feature extraction layers, 259
features of artificial neural networks, 211
feedback competition representation, 298
feedback connections, 197, 201
feedback, in neural networks, 200
feedforward network, 204
Feigenbaum, 524, 564
Filev, 145, 152
filtering application, 262
financial engineering, neurofuzzy methods, 486
finite impulse response, *FIR*, 345
first-order Sugeno controller, 162
first projection, 63
fitness function, 542
fitness value, 546
Folger, 55
Fontana, xviii
formalism in anticipatory systems, 513
forward-chaining inference, 112
Foslien, 479
foundations of fuzzy approaches, 7
Fourier transform, 2, 334
frequency domain, 2
frequency response methods, 148
frequentistic interpretation, 37, 43
Fujii, xviii
fully connected network, 197
Funahashi, 267
functional links, 279
fundamentals of neural networks, 8
fusion reactors, and neurofuzzy systems, 479
fuzzification, 27
fuzzification and clustering, 450
fuzzification of aggregation, 414
fuzzification of dendritic inputs, 414
fuzzification of neuronal outputs, 415
fuzzification of neurons, 409
fuzzification of synapses, 414
fuzzifying the input of a controller, 152

fuzziness, 1, 13
fuzziness, schools of thought, 39
fuzzy algorithms, 49, 106, 110, 136–141
fuzzy algorithms, and composition, 67
fuzzy ARTMAP, 431–435
fuzzy categorization of a universe of discourse, 114
fuzzy chips, 112
fuzzy *c*-means clustering algorithm, 499
fuzzy control, 8, 145–185
fuzzy control of backpropagation, 437
fuzzy controller emulating *PID* modes of control, 160–162
fuzzy controller, example of level control, 166–169
fuzzy controller, two-input example, 169–176
fuzzy event, 43, 44
fuzzy expert systems, 535
fuzzyfication, 25–27
fuzzyfier function, 25
fuzzy genetic modeling, 554
fuzzy *if/then* rules 49
fuzzy implication operators , 121–125
fuzzy implication operators, table of ϕ, 124
fuzzy inference and composition 125–136
fuzzy inference, 111
fuzzy inference, schematic representation, 158, 169, 174
fuzzy inputs and outputs to neural networks, 435
fuzzy kernel, 27
fuzzy linguistic controllers, 151–163
fuzzy linguistic descriptions, 105–141,
fuzzy mean, 44
fuzzy measure, 39
fuzzy methods in neural networks, 8, 409–443
fuzzy microprocessors, 112
fuzzy multiplication, 95
fuzzy-neural hybrid data presentation, 434
fuzzy-neural hybrids, 447
fuzzy neural networks, 412–413
fuzzy neural, survey of engineering applications, 437–439
fuzzy neuron, 412
fuzzy neuron, and *AND*- gate, 413
fuzzy neuron, and *OR*-gate, 412
fuzzy numbers, 8, 77
fuzzy output as indication of controller quality, 177
fuzzy partial ordering, 59
fuzzy probabilities, 38–54
fuzzy process control system schematic, 152
fuzzy propositions, 106
fuzzy relation, 7, 49, 52, 50, 67
fuzzy relation, representation of, 55

fuzzy relations, notation for, 53
fuzzy relations, operations, 60
fuzzy relations, properties of, 58
fuzzy restriction, 41
fuzzy set, 13, 15
fuzzy set theory, 13
fuzzy similarity relation, 59
fuzzy supervisor for *PID* controller, 162
fuzzy systems, 1, 105
fuzzy values, 106, 112

Gaussian activation function, 325
gene pool, 540
generalized fuzzy neuron and networks,
    414–416
generalized mean operator, 502
generalized modus ponens (*GMP*), 111, 112,
    125
generalized modus ponens (*GMP*), in control,
    156
generalized modus tollens (*GMT*), 111, 112,
    125
generalized regression neural network, 326
generational approach, genetic algorithms, 545
genes, 539
genetic algorithm optimization, 544
genetic algorithms, 1, 8, 539–557
genetic algorithms application to neural
    networks, 548
genetic algorithms in neural network design,
    556–557
genetic computing, xiii
Girosi, 238
goal-driven inference, 112
Goals of Engineering Education, (ASEE), 561
Gödelian implication and *ELSE*, 137
Gödelian implication operator, 123
Goldberg, 539
Goode and Chow, 486
Goodman, 452
Gougen implication, 123
Gougen implication and *ELSE*, 137
graded learning, 203
grades of membership in fuzzy relations, 56
gradient descent, for adapting membership
    functions, 463
granularity of a description, 112
Gray-scale representation, 543
GRNN, see generalized regression neural
    network
Grossberg, 289, 328, 431, 432
Grossberg learning in outstars, 294
Grossberg outstar learning, 295
Guély, 461, 463
Guo, 255, 362, 477

Gupta, 409, 417
Gyftopoulos, 340

Halberstam, 5
Halgamuge, 439
Hamming distance, 543
Hanes, 481
Harp, 557
Harris, 145
Harston, 213
Hashem, 255
Hayashi, xviii, 447, 455, 461
Haykin, 7, 437
heat rate of a power plant, monitoring of,
    361–363
Hebb, 204, 289
Hebbian learning, 203, 215, 289–290, 294
Hebb's law, 289
Hecht-Nielsen, 213, 303, 306, 315
hedging, as semantic constraint, 117
Hertz, 438
heteroassociative neural network, 197
heuristic knowledge, 523
hidden layers, 274
Higgins, 452
high flux isotope reactor (HFIR), 507
Hines, J. W., xviii, 8, 264, 270
Hinton, 229, 250, 275
Hirota, 145, 409, 438,
Hobbs, 486
Hoff, 217, 314
Hojny, xviii
Holland, 539
Hooker, 340
Hopfield, 303
Hopfield layer, 304
Hopfield learning rule, 304
Hopfield networks, 303'
Hornik, 267
Houstis, xviii
Hsiao, 485
Hu, 438
human factors, expert systems, 537
Hung, 450
Hush, 325
hybrid AI systems, 1
hybrid neurofuzzy applications, 1, 8, 471–487
hybrid systems, 493
hyperbolic tangent, activation function,
    232–233

Ichihashi, 461
idempotency, 28
Identity fuzzy relation, 54

iff (if and only if), 14
if/then (crisp) rule approximation of functions, 107
if/then (fuzzy) rule approximation of functions, 108
if/then rules, 3
if/then rules, and composition, 67
Ikonomopoulos, 375, 477, 493
Illingsworth, 213
image analysis, and fuzzification, 27
image enhancement, 482–483
implication operators, 121–125
implication operators and controller quality, 177–184
implication relation of fuzzy rule, 109
implication relations, 121, 120–125, 127
imprecision, 3, 39
impulse response function, 334
incomplete data sets, 390
indexed center of gravity method of defuzzification, 165
industrial merit of neurofuzzy technologies, 480
inference engine, 525
inferencing procedures, and composition, 67
inferential measurements, 266
inferring the value of a variable, neurofuzzy approach, 494
infimum, 18
infinitesimal weight adjustments, 250
influence of membership function shape in control, 181
influence of noise on measurements, 337
informal linguistic descriptions, 105
information fusion, through fuzzy logic, 501
inhibition hyperboxes, 474
inhibitory, 412
inhibitory effects in fuzzy neurons, 415
inner product, 202
input interface, of fuzzy controller, 152
input layer, 196
input-output relationships, 343
input transformations, 398
input variables in fuzzy controllers, 153
input vector, in neural nets, 198
inspection using neurofuzzy methods, 486
instars, 291
instars learning law, 294
instrumentation system, 337
integral term, in *PID*, 148
integration of fuzzy and neural systems, 1
intelligence, 4
intelligent management, 5
intelligent management of large complex systems, 5

interface, expert systems, 525
interference canceling, 351
intermediate layer of neurons, *see* hidden layer
interpolative associative memories, 297
interpretation of *ELSE* in fuzzy algorithms, 137
intersection, 14
intersection, of fuzzy relations, 61
intersection of fuzzy sets, 20
interval arithmetic, and fuzzy numbers, 82–102
interval representation, of fuzzy numbers, 83
inverse control, 353, 360
inverse fuzzy relation, 57
inverse modeling, 349, 369
inverse modeling systems, 350
inverse relation, 54
Ioannou, 477
irreflexive, fuzzy relation, 58
Ishibuchi, 476
Ishikawa, xviii
iterative procedure for improving plant performance, 363

Jacobs, 254
Jamshidi, 4
Jang, 455, 461
Jang and Sun, 487
Jianqin, 176

Kandel, A., 55, 81
Karayiannis, 439, 452, 476
Karonis, 145
Kartalopoulos, 7
Kasuba, 433
Kaufmann, 15, 53
Keller, 455, 475
Khan, 455, 487
Khrisnapuran, 475
Kim, 482
King, 145, 152, 176
Kiszka, 176
Kitamura, xviii
Klir, 55, 480
Klir and Folger, 17, 67, 58, 82
k-means algorithm, 326
knowledge acquisition, 526
knowledge base, 525
knowledge representation-inference, expert systems, 527
Kohonen, 306
Kohonen feature mapping, 451
Kohonen layer, 306
Kohonen learning rule, 308
Kohonen networks, 306

Kohonen neurons, 307
Kohonen self-organizing systems, 306–315
Kohonen training law, 314
Kolmogorov's theorem, 238
Korn, 306
Korst, 306
Kosko, 15, 299, 455, 474, 554
Kozma, xviii, 477
Kramer, 257, 259
Krishnapuram and Lee, 501
Kulkarni, 483
Kumara, 452
Kung, 487
Kuo, 438, 439, 452, 481
Kwan, 413, 439

Labierre, 281
Lan, 473
Lapedes, 267
Lapp, 5
Larkin, 152
Larsen product implication, 122
Larsen product implication and *ELSE*, 137
Larsen product implication and *GMP* example,
  132–126
Larsen product implication, in control, 156
lateral connections between neurons, 197
lateral inhibition, 306
law of contradiction, 29
law of the excluded middle, 29
Lawrence, 392
learning, 203
learning algorithm, for perceptrons, 216
learning and adaptation in fuzzy systems via
  neural networks, 461
learning by example, 211
learning constants, 252
learning, fuzzy neural networks, 423
learning machine by Minsky and Dean, 214
learning vector quantization, 314, 439
Lee, 121, 438, 475
left-hand side fuzzy variables in controllers,
  152–153
left unique relation, 58
level fuzzy sets, 36
level sets, 34
Li and Wu, 455, 478
Lin, 438, 485
Lin and Lee, 479
linear associator, 201–203
linear control theory, 344
linearly separable variables, 220
linear systems theory, 333
linear versus nonlinear systems, 345
linguistic description of a function, 109
linguistic description of fuzzy control, 145, 177

linguistic descriptions, 105–141
linguistic descriptions and analytical forms, 8,
  105–141
linguistic forms, 106–107
linguistic hedges, 117
linguistic modifiers, 114
linguistic modifiers and fuzzy neurons, 417
linguistic modifiers and fuzzy set operations,
  120
linguistic terms, 3
linguistic values, 112
linguistic variable, 106, 113
LISP, 4
Lo, 268, 270
local minima, dealing with, 251, 402
logical product, 417
logical sum, 417
logistic function, 194, 230
logistic function and its derivative, 231
Loskiewicz-Buczak, 375, 493
low-pass filter, autoassociative neural net, 262
LVQ, see learning vector quantization

machine learning, 112
Madaline, 220
Mamdani, 145, 152, 176
Mamdani min implication, 122
Mamdani min implication and *ELSE*, 137
Mamdani min implication and *GMP*, example,
  127–132
Mamdani min implication, in control, 156
Mamdani rules, in fuzzy control, 162
manipulated variable, in control, 147
many-input-many-output (MIMO) systems, 163
many-input-single-output controllers, 163
many-to-many mappings, 7, 50, 110–111, 447,
  472
many-to-many relations, 7
many-to-one mapping, 7, 31, 50, 107, 110–111,
  147, 472
mapping, 50
mapping layers, 259
mapping the alphabet to five-bit code, 206–209
Maren, 213
Maricic, 548
Masters, 257, 268, 269
mating (crossover), 540
MATLAB, 267, 466
MATLAB Supplement, 8
matrix notation in neural nets, 198, 297
Matsuoka, 477
max, as operator, 18
max-average composition, 67, 69, 73, 74
max fuzzy neuron, 413
max-* composition, 67, 68, 126–127

maximum, 31
maximum of fuzzy numbers, 101–102
max (*OR*) fuzzy neuron, 413
max–min composition, 67, 125–126, 127–141
max–min composition, example of, 69
max–min composition, in control, 156–161
max–min transitive relation, 59
max-product composition, 67–68, 126–127
max-product transitivity, 59
McAvoy, 368
McCarthy, 4, 213
McClelland, 275
McCorduck, 564
McCulloch–Pitts model of a neuron, 4, 213, 215
McCullough, 516
mean of maxima (*MOM*) defuzzification, 165
meaning of function, 13
medical imaging, 483–484
membership function, 15
membership matrix, 54
memorization in neural nets, 257
memory capacity of Hopfield nets, 305
Mendel, 437, 455, 485
metarules, 163
Miller, 213, 368
Minai, 255
min (*AND*) fuzzy neuron, 413
min, as operator, 18
min fuzzy neuron, 411
minimization of least squares errors, 3
minimum of fuzzy numbers, 101–102
Minsky, 4, 213, 214, 215, 229,
Minsky and Papert's *Perceptrons*, 216
MITI, 5
Mitra, 482
Miyamoto, 145, 516
Mizumoto, 121, 125
modeling, 1, 2, 347
model reference adaptive control, 373
models of dynamic systems, 366
modes of control, in PID, 148–149
modular neural networks, 270–273
Moganti, 486
MOM defuzzification, 165
momentum, 249
monitoring, 262
monitoring heat rate of a power plant, 361–363
monitoring network training process, 403
Moody, 325
Moon, 478
MORE OR LESS, 21
mounted devices, 112
Mouselli, 549
m-tuples, 31

multidimensional universe of discourse, and fuzzy numbers, 79
multi i/o system, example of neural control, 356–358
multilayer fuzzy neural networks, 421
multiple-input adaptive linear combiner, 341
multiple parallel slabs, 388
multiplication of fuzzy numbers, 77, 95
multiplication of fuzzy number with crisp number, 95
multiplication of two fuzzy numbers, example of, 96
multiplying a fuzzy set by crisp number, 21
multivariate fuzzy algorithms, 139–141
multivariate fuzzy implications, 139
mutation, 540, 542
MYCIN, 526

Nabeshima, xviii
Naredra, 359, 368
n-ary fuzzy relation, 53
n-ary relations, 50, 52
natural selection, 540
nearest-neighbor heuristic, 326
necessity measures, 39
Nelson, 213
Neo–Hebbian learning, 289
network paralysis, 250
NEUFUZ, 487
neural adaptive control, 359
neural and fuzzy systems in series, 449
neural control, 333–382
neural control systems, implementation of, 368
neural methods in fuzzy systems, 8, 445–468
neural network control, 353–363
neural network driven fuzzy reasoning, 447
neural networks, 1, 191
neural network training, alternative approach, 266–270
NeuralWare, 254
neurocomputing, xiii
neurodes, 193
neurofuzzy control in robotics, 481
neurofuzzy interpolation, 472
neurofuzzy systems, 7, 564, 565
neurofuzzy technology, role of, 563
neuron, 193
neuron activation function, 194
neuronal input signals, 193
neuron, biological, 192
Nguyen 368
Nie, 475
Nii, 564
Nikolov, 548
nodes, *see* neurons
noise analysis, 375

noise analysis, applications of neural networks, 374–380
noise input, 338
Nomura, 461, 462
nondistributed information, 394
nonlinear characteristics of devices, 151
nonlinearly separable variables, 221
nonlinear principle components, 259
nonlinear relationships, 3
nonlinear systems, 3
nonparametric identification, 363
normal fuzzy set, 19, 77, 79
normalization, 152
normalization of data, in neural networks, 395
n-tuples, 52
null relation, 54
number of layers, in neural design, 387
number of neurons per layer, 388

O'INCA, 449, 488
objective function, in genetic algorithms, 542
observational conditioning, 290
Occam's Razor, 3
odor discrimination, 450
operational conditioning, 290
ordered pair, in relations, 50, 55
ordering of variables, 393
OR fuzzy neuron, 413, 418–421
Ostergaard, 145
output interface, of fuzzy controller, 152
output (total) of fuzzy controller, 157
output variables in fuzzy controllers, 153–154
outstar learning network, 295
outstars, 291
overtraining a neural network, 209

Pai, 439
Pal, 48
Panas, xviii,
Pao, 213, 319, 477
Pap, 213
Papanikolopoulos, 162
Papert, 215, 229
paraboloid of revolution, 236
parallel features of fuzzy control, 152
parameter identification, 176
parametric identification, 364
parametrizing, fuzzy numbers, 81
Parker, 229, 250
parsimonious description, 3, 108
parsimony, 3
Parthasarathy, 359, 368
partial ordering, 58
Patrikar, 475
pattern recognition, 112, 482–483

pattern recognition, and fuzzification, 27
pattern recognition capabilities of neural nets, 211, 213
Pavlov's experiments, 290
Pavlovian learning, 290
Pedrycz, 145, 409, 419, 438, 462
Pedrycz and Dyckhoff, 502
perceptron, 4, 214, 303, 411
Perceptrons, 215, 217
performance, as fuzzy variable, 153
performance guidelines, expert systems, 536
Perneel, 462
Peters, 564
Pham, 451, 475
physical system, 334
pictorial data, 390
PID control, 146–151, 353, 355, 360, 372,
PID level control, example of, 148
PID tuning, 148
piecewise linear fuzzy controller, 478
Pin, 112
plantwide monitoring system, 264
plaus*ibility measures*, 39
PNN, see probabilistic neural networks
Poggio, 238
point approximation of functions, 107
possibility distribution, 39, 40, 41, 42, 43, 44
possibility distribution function, 41
possibility measures 43,44
possibility theory, 38–45
Postman, 5
power of a fuzzy set, 21
Powers, xviii
power spectral density, 334
practical aspects of neural networks, 8
Prade, 417
prediction, 352
preview control 512
primary set of values, 114
primary values of fuzzy variables, 115
Principia, 7
proactive, 563
probabilistic methods, 38
probabilistic neural nets,
   advantages-disadvantages, 323
probabilistic neural networks, 319
probability measures, 39
probability of fuzzy event, 43
probability/possibility consistency principle 43
process control, 112
process control system block diagram, 152
process control system, practical example of, 146
process relation, 147

process, practical example of, 146
processing elements, *see* neurons
product of fuzzy sets, 20
product space, *see* Cartesian product
projection, 60–64
properties of fuzzy sets, 28–30
proportional-integral-derivative control,
  146–151, 344
proportional selection, 545
proportional term, in *PID*, 148
Provence, 475
PRUF, 42
pseudorandom binary maximum length shift
  register sequence, 340
pseudorandom binary variables, 339
Π-shaped functions in fuzzy control, 181
Π-shaped fuzzy values, 116–117
Purushotaman, 452

Qi, 417
qualitative computations, 114
quality of fuzzy control algorithm, 176–177
quality of measurement, 339
quantitative computations, 114
quantization, 152
quantizer, in Adalines, 218
quickprop training, 253
quotient of fuzzy numbers, 99–101

Radecki, 36
radial basis function network, 325–326, 439
Ragheb, xviii, 42
random noise analysis, 2
rangeability of a valve, 151
rank-based selection, 545
Rao, 409
Rao and Gupta, 481
Raza, 477
RBFN, *see* radial basis function network
real-valued representation of chromosomes,
  542, 544
recall, 203
recirculation neural networks, 274–278
recirculation neural networks, application of,
  276
recurrent neural networks, 281–285
reduced-dimensional representation, 259
redundancy in input data, neural networks,
  399
reference value, *see* setpoint
reflexive relation, 57
regression to solve for the weight matrix, 268
regulatory control, 355
reinforcement learning, 374
relations, importance of order, 50

relations, properties of, 57
relations, representation of, 52
reliability engineering, 112
reliable intelligence, 7
reproduction, 540, 545
reset constant, in *PID*, 147
resolution principle, 37–38
resolution principle, and fuzzy numbers, 77, 83
resolution principle, and fuzzy relations, 61
resonance, in ARTMAP, 434
Ridella, 549
right-hand side fuzzy variables in controllers,
  152–154
right unique relation, 58
robotics, 481
robust autoassociative neural net, 264
Rocha, 419, 438
Rochester, 213
role of hidden layer in training, 403
Rosen, 511
Rosenberg, 250
Rosenblatt, 4, 214, 215
Rosenblatt's Perceptron, 214
Ruano, 371
Rueda, 419, 438
rule extraction, identification, 447
rules and inference in fuzzy control, 154–163
Rumelhart, 229, 250

Saleem, 435
Samad, 253, 368, 479, 557
sampling time interval, 349
Sanchez, 77
scaling, 152
scaling, in control, 147
scaling of data, in neural networks, 395
scaling of the consequent by *DOF*, 135–136
Schwartz, 145, 480
second-order backpropagation, 250
second projection, 63
seismology, 2
Sejnowski, 250
selection of neural networks, 385
self-organizing map, 310
self-organizing map, and membership
  functions, 497
semantics of compound fuzzy values, 119
sensitivity analysis, 2, 255, 362
sensitivity coefficients, experimental
  evaluation, 256
sensors, failed or deteriorated, 263
sequential order, 390
setpoint, 147
Shannon, 213
shift register systems, 340

Shimazaki, xviii,
Shinobu, xviii
Shinohara, xviii, 39
Siarry, 461, 463
sigmoidal activation function, 194, 195, 232
signal-to-noise-ratio, 220, 449
signum function 411
simple open-loop control system, 354
simplified fuzzy ARTMAP, 433
Simpson, 213
simulated annealing, 252, 305–306
single-input adaptive transverse filter, 342
singleton, 15, 55
singular value decomposition, 269
situation side of rules, 106
size of neural networks, 386
small numbers, fuzzy set of, 16
smoothing, 152
smoothing, in control, 147
smoothing parameter, probabilistic neural
    nets, 323, 324
S norms, 409, 417
Sofge, 213, 368
soft computing, xiii, 3, 563
software, neurofuzzy, 487
SOM, see self-organizing map
soma, 192, 409
Song, 438
Sotomi, 477
Specht, 319, 326
squashing function 194
Srinivasan, 438
S-shaped functions, in fuzzy control, 181
S-shaped fuzzy values, 116–117
stability, 250
stability, of fuzzy controllers, 176
standard sequence implication, 123
standard sequence implication and *ELSE*, 137
Starkwerther, 548
statistical regression, 2
steady state approach, genetic algorithms, 545
stochastic learning automata, 374
stochastic neural networks, 306
strings of artificial genetic systems, 539
structure identification, 176
structure of probabilistic neural nets, 320
Stylios, 477
subjectivistic, interpretation of probability, 37
subtracting fuzzy numbers as intervals,
    example, 91
subtracting fuzzy numbers with continuous
    membership functions, example, 91
subtraction of fuzzy numbers, 77, 90–95
Sugeno, 39
Sugeno rules, 456, 459, 466
Sugeno rules, in fuzzy control, 162

sum of errors, as fuzzy variable, 153
Sun, 455, 461
super chromosome, 545
supervised control, 353, 359–360
supervised learning 203, 204–211
supremum, 18
survival of the fittest, 540
Sutton, 213
Suzudo, xviii
Suzuki, xviii
Swiniarski, 368, 371
syllogism, 127
symbolic computations, 114
symmetric, fuzzy relation, 58
symmetric relation, 57
synapse, 192, 410
synaptic junction, *see* synapse
synaptic strength of biological neurons and
    learning, 192
synaptic weight, 193
synergistic interactions, 1
synergistic utilization of fuzzy-neural systems,
    445
system behavior and fuzzy-neural choices,
    447–448
system description, 2
system identification, 347, 363
system response function, 334

tabular representation of fuzzy numbers,
    85–92
Tahani, 455, 475
Takagi, 447, 455
Takagi–Hayashi method, 455–461
Tascillo, 481
Terano et al, 17, 55, 145, 417,
term set, 114
tertiary relations, 50, 52
T-H method, *see* Takagi-Hayashi method
Theoharis, xviii
Thomson, 385
threshold, in neurons, 194
time delay, 349
time domain, 2
time variations in data, 390
time-series prediction, 380–382
T norms, 409, 417, 567–572
T norms and fuzzy implications, 571
T norms and S norms, 567–572
T norms in fuzzy control, 159,
Tomizuka, 512
top-dawn weights, 274
total linear ordering, 58
total projection, 63
training algorithms, 229
training neural networks, 401–404

training of a Kohonen network, 310–311
training the perceptron, 216
transfer function in neurons, *see* activation function
transfer function, systems, 334
transitive relation, 58
transitivity, 59
transportation control, 484
trapezoidal fuzzy values, 116
trapezoidal membership function, 116
Travis, 435, 450
triangular fuzzy values, 116
triangular membership function, 37
Trivedi, 499
TSK rules in fuzzy control, 162
Tsoukalas, 435, 511
Tsukamoto fuzzy model, 163
Tsund, 368
tuning, in *PID* controllers, 148
tuning tasks, 480
two-sensor technique, neural nets in noise analysis, 375
Tzafestas, 162

Uhrig, 255, 264, 333, 340, 362, 375
uncertainty, 4, 39
uncertainty management, in expert systems, 529
unidirectional counterpropagation network, 315
union, 14
union, of fuzzy relations, 61
union of fuzzy sets, 21
universal approximator, 267
universal set, 14
universe, 14
universe of discourse, 14
universe relation, 54
universes of discourse, higher dimensional, 52
unsupervised learning, 203
Upadhyaya, 263
updating weights, example of backpropagation, 246
Usuii, xviii

validation and verification, expert systems, 536
valuation set, 14
valve status classification using Kohonen networks, 311
Van Horn, 524
variable, notion of, 41
vector notation in neural nets, 198
Venn diagrams, 14
VERY, 21
vibration analysis, 2
vibration monitoring, 495

vibration signatures, 495, 500
vigilance, 330
virtual company, 563
virtual measurement values, 505
virtual measurements, using neurofuzzy method, 504
visible layers, 274
von Neuman, 4

Wakami, 461, 480
Wang, 145, 437, 438, 455
Wang and Chen, 485
Washio, xviii, 512
Wasserman, 213, 326
Watanabe, xviii
Watkins, 487
Wegman, 480
weight assignment problem, 229
weight matrix, 198
weight matrix representation of BAM, 299
weights, 193
weights, cumulative update of, 253
weights in Adaline, 219
Welsted, 555
Werbos, 213, 239, 353, 360, 368, 435, 455, 474, 479
Werbos, and backpropagation, 229
White, 213, 368
Whitely, 548
white noise, 337, 340
Widrow, 4, 217, 333, 341
Widrow–Hoff learning rule, 203, 213, 214, 234
Williams, 229, 250, 255
Wohlke, 481
Wrest, 264

XOR problem, and perceptrons, 217

Yager, 145, 152, 455, 475
Yamakawa, 145
Yamazaki, 479
Yasunobu, 145, 511
Ye, 439
Yea, 450
Yen, 480
Youssef 461, 477

Zadeh, xv, 13, 15, 30, 42–44, 59, 114
Zadeh diagrams, 17
Zadeh max–min implication, 122
Zadeh max–min implication and *ELSE*, 137
Zadeh, possibility theory, 38
zero-order Sugeno controller, 162
Zhang, 438
Ziegler–Nichols method, 148
Zimmerman, 15, 81, 501
Z scores, 398